FUNDAMENTALS OF NETWORK ANALYSIS

PRENTICE-HALL INTERNATIONAL SERIES
IN INDUSTRIAL AND SYSTEMS ENGINEERING

W. J. Fabrycky and J. H. Mize, Editors

AMOS AND SARCHET *Management for Engineers*
BEIGHTLER, PHILLIPS, AND WILDE *Foundations of Optimization, 2nd ed.*
BLANCHARD *Logistics Engineering and Management, 2nd ed.*
BLANCHARD AND FABRYCKY *Systems Engineering and Analysis*
BROWN *Systems Analysis and Design for Safety*
BUSSEY *The Economic Analysis of Industrial Projects*
FABRYCKY, GHARE, AND TORGERSEN *Industrial Operations Research*
FRANCIS AND WHITE *Facility Layout and Location: An Analytical Approach*
GOTTFRIED AND WEISMAN *Introduction to Optimization Theory*
HAMMER *Product Safety Management and Engineering*
KIRKPATRICK *Introductory Statistics and Probability for Engineering, Science, and Technology*
MIZE, WHITE, AND BROOKS *Operations Planning and Control*
MUNDEL *Motion and Time Study: Improving Productivity, 5th ed.*
OSTWALD *Cost Estimating for Engineering and Management*
PHILLIPS AND GARCIA-DIAZ *Fundamentals of Network Analysis*
SIVAZLIAN AND STANFEL *Analysis of Systems in Operations Research*
SIVAZLIAN AND STANFEL *Optimization Techniques in Operations Research*
THUESEN, FABRYCKY, AND THUESEN *Engineering Economy, 5th ed.*
TURNER, MIZE, AND CASE *Introduction to Industrial and Systems Engineering*
WHITEHOUSE *Systems Analysis and Design Using Network Techniques*

FUNDAMENTALS OF NETWORK ANALYSIS

DON T. PHILLIPS

Department of Industrial Engineering
Texas Transportation Institute
Texas A&M University

ALBERTO GARCIA-DIAZ

Department of Industrial Engineering
Texas A&M University
College Station, Texas 77843

Prentice-Hall, Inc., Englewood Cliffs, N.J. 07632

Library of Congress Cataloging in Publication Data

Phillips, Don T
Fundamentals of network analysis.

Includes bibliographies and index.
1. Network analysis (Planning)
I. Garcia-Diaz, Alberto, joint author.
II. Title.
T57.85.P53 658.4'032 80-19187
ISBN 0-13-341552-X

Editorial/production supervision by Lori Opre
Manufacturing buyer: Anthony Caruso

Printed in the United States of America

10 9 8 7 6 5 4 3 2 1

PRENTICE-HALL INTERNATIONAL, INC., *London*
PRENTICE-HALL OF AUSTRALIA PTY. LIMITED, *Sydney*
PRENTICE-HALL OF CANADA, LTD., *Toronto*
PRENTICE-HALL OF INDIA PRIVATE LIMITED, *New Delhi*
PRENTICE-HALL OF JAPAN, INC., *Tokyo*
PRENTICE-HALL OF SOUTHEAST ASIA PTE. LTD., *Singapore*
WHITEHALL BOOKS LIMITED, *Wellington, New Zealand*

To *Donnie* and *Ronnie*—
they never understood
why Daddy worked so hard,
and to *Candy* and *Kathy*—
who also suffered without complaint.

CONTENTS

Preface xiii

1 Introduction 2

1.1 Definitions, Notation, and Symbolism 5
1.2 Matrix Representation of Networks 11
1.3 Conservation of Flow 14
1.4 Contribution of the Maximum-Flow Minimum-Cut Theorem 15

2 Deterministic Network Flows 22

2.1 Applications of Network-Flow Models 24
2.1.1 Equipment Replacement 24
2.1.2 Project Planning 25
2.1.3 Scheduling Tanker Voyages 26
2.1.4 The Maximum-Flow and Minimum-Cost Flow Problems 27
2.1.5 The Transportation Model 29
2.1.6 A Caterer Problem 30
2.1.7 Employment Scheduling 31
2.1.8 A Fleet Scheduling Problem 32
2.1.9 A Production Planning Model 35
2.1.10 Conclusions 37

2.2	Linear Programming and Its Relationship to Network Flows	39
2.3	The Shortest-Route Problem—Dijkstra's Algorithm	46
	2.3.1 An Iterative Procedure	47
	2.3.2 Example of Dijkstra's Algorithm	47
	2.3.3 Shortest-Route Problem: Buying a New Car	49
	2.3.4 Computer Code for Dijkstra's Algorithm	51
	2.3.5 A Problem in Oil Transport Technology	52
2.4	Multiterminal Shortest-Chain Route Problems	53
	2.4.1 Example of the Multiterminal Shortest-Chain Problem	56
	2.4.2 Multiterminal Shortest-Route Problem: Design of a Mail Distribution System	61
	2.4.3 Computer Code for the Multiterminal Shortest-Route Algorithm	64
	2.4.4 Routing of Rail Cars	65
2.5	Shortest-Path Models with Fixed Charges	68
2.6	The *K*-Shortest-Path Problem	72
	2.6.1 The Double-Sweep Method	73
	2.6.2 Sample Problem for the Double-Sweep Method	77
	2.6.3 Computer Code for the Double-Sweep Algorithm	82
	2.6.4 Computational Results	84
	2.6.5 Computer Example for a Four-Shortest Path Problem	86
2.7	Analysis and Complexity of Shortest-Path Algorithms	88
	2.7.1 Computational Complexity of Dijkstra's Method	89
	2.7.2 Computational Complexity of Floyd's Algorithm	89
	2.7.3 Computational Complexity of the Double-Sweep Method	90
2.8	Minimal Spanning Tree Problems	91
	2.8.1 Minimal Spanning Trees—Computational Procedure	91
	2.8.2 Example of a Greedy Solution	92
	2.8.3 Computer Code for the Minimal Spanning Tree Algorithm	94
	2.8.4 Allocation of Highway Maintenance Funds	95
	2.8.5 Applications of Minimal Spanning Trees	96
2.9	The Traveling Salesman Problem	97
	2.9.1 Construction of a Lower Bound	98
	2.9.2 Branching	101
	2.9.3 Computational Procedure	101
	2.9.4 Final Remarks	106
	2.9.5 Computer Code for the Traveling Salesman Algorithm	108
	2.9.6 Scheduling International Travel	109
2.10	The Transportation Problem	111
	2.10.1 Mathematical Model	112
	2.10.2 The Transportation Simplex Algorithm	115

2.10.3 A Network Interpretation for the Transportation Simplex Algorithm 126
2.10.4 A Production-Distribution Model 128
2.11 The Transshipment Problem 130
2.12 The Assignment Problem 132
2.12.1 Mathematical Model 132
2.12.2 The Hungarian Algorithm 133
2.12.3 Example of the Hungarian Algorithm 135
2.12.4 Remarks 139
2.12.5 A Product-Plant Allocation Problem 139
2.13 The Assignment Problem and the Traveling Salesman Problem 141
2.14 The Maximum-Flow Problem 144
2.14.1 A Labeling Procedure for the Maximum-Flow Algorithm 146
2.14.2 Example of the Labeling Algorithm 147
2.14.3 The Maximum Flow/Minimum Cut Theorem 149
2.14.4 Design of a Centralized Sewage Waste Treatment Plant 151
2.14.5 Computer Code for the Maximum-Flow Algorithm 155
2.14.6 A Problem in Grain Shipment and Storage 156
2.15 The Multiterminal Maximal-Flow Problem 158
2.15.1 The Gomory–Hu Algorithm 159
2.15.2 Justification of the Algorithm 160
2.15.3 Example of a Multiterminal Maximal-Flow Problem 162
2.16 The Multiterminal Maximum-Capacity Chain Problem 165
2.16.1 Optimal Movement of Bulk Freight 167
2.16.2 Computer Code for the Multiterminal Maximum-Capacity Route Algorithm 172
2.16.3 Transporting the Space Shuttle 173
2.17 Node and Arc Failures in Networks 179

3 The Out-of-Kilter Algorithm: Generalized Analysis of Capacitated, Deterministic Networks 204

Part I: Network-Flow Optimization with the Out-of-Kilter Algorithm—Theoretical Concepts 205

3.1 Basic Terminology 206
3.2 Basic Theory 208
3.3 Fundamental Theorems 212

3.4 The Out-of-Kilter Algorithm for Solution of Minimal-Cost Circulation Problems 213
3.5 The Labeling Procedure 214
3.6 A Graphical Interpretation of the Out-of-Kilter Algorithm 219
3.6.1 Horizontal Displacements 220
3.6.2 Vertical Displacements 222
3.7 Algorithmic Steps 223
3.8 A Numerical Example 224
3.8.1 A Numerical Example for the Out-of-Kilter Algorithm 224
3.9 Summary and Conclusions 229

Part II: Network Flow Optimization with the Out-of-Kilter Algorithm—Modeling Concepts 230

3.10 Problem Solution Using the Out-of-Kilter Algorithm 230
3.11 The Transportation Problem 231
3.11.1 Example of the Transportation Formulation 233
3.12 The Assignment Problem 234
3.12.1 Example of the Assignment Formulation 234
3.13 Maximum Flow Through a Capacitated Network 235
3.14 The Shortest-Path Problem 235
3.15 The Shortest-Path Tree Problem 236
3.16 The Transshipment Problem 237
3.17 Nonlinear Costs 238
3.18 A Production-Distribution Problem 239
3.19 Summary and Conclusions 246

Part III: Application of the Out-of-Kilter Algorithm 247

3.20 The "Bottleneck" Assignment Problem 247
3.21 Scheduling Workers to Time-Dependent Tasks 250
3.22 A Wholesale Storage and Marketing Problem 252
3.23 Computer Code for the Out-of-Kilter Algorithm 254
3.24 Crude Oil Production and Distribution 255

4 Project Management Procedures 268

Part I: Project Management with CPM and PERT 269

4.1 Origin and Use of PERT 270
4.2 Origin and Use of CPM 271
4.3 Problem Characteristics 271
4.4 Network Construction 273
4.4.1 A Manufacturing Problem 275
4.5 Earliest Possible Times for Each Event 276

4.6 Latest Allowable Times for Each Event 277
4.7 Slack Times and the Critical Path 278
4.8 Tabular Calculation of the Early Start and Latest Finish Times 279
4.9 Four Float Measures for Critical Path Scheduling 281
4.9.1 The Computational Procedure 281
4.9.2 Float Calculations 283
4.9.3 Free Float 285
4.9.4 Independent Float 285
4.9.5 Safety Float 286
4.10 An Activity-on-Node Formulation 287
4.10.1 Network Construction 287
4.10.2 Computational Procedure 288
4.11 Program Evaluation and Review Technique (PERT) 291
4.11.1 A PERT Example 293
4.11.2 Probabilities of Completing the Project 294

Part II: Resource Allocation in Project Networks 295

4.12 Time vs. Cost: Dollar Allocations 296
4.12.1 A Flow-Network Algorithm for Time/Cost Trade-offs in CPM Projects 300
4.12.2 Applications of Time/Cost Procedures 314
4.13 Resource Loading 316
4.14 Resource Leveling 318
4.15 Fixed Resource Limits 320
4.16 Constrained Resources 322
4.16.1 Heuristic Approaches 322
4.16.2 Optimal Approaches 324

Part III: A Comparison of Commercially Available CPM/PERT Computer Programs 327

5 Advanced Topics 342

Part I: Generalized Networks—Networks with Gains and Losses 343

5.1 Application of Generalized Networks 344
5.2 GNP as a Linear Programming Problem 345
5.3 Network Characteristics 346
5.4 Case I: Generalized Networks with No Flow-Generating or Flow-Absorbing Cycles 347
5.4.1 Example Formulation 348
5.5 Case II: Generalized Networks with Flow-Generating and/or Flow-Absorbing Cycles 350
5.5.1 Phase I: Initial Flow 350
5.5.2 Phase II: The Marginal Network 350
5.5.3 Phase III: The Flow-Augmentation Process 351

5.6 Phase I: The Minimum-Cost Flow-Augmenting Chain 351
5.7 Phase II: Constructing the Marginal Network 352
5.8 Phase III: Executing the Flow-Augmentation Process 353
5.9 A Numerical Example with Gains and Losses 355
5.10 Summary and Conclusions 362

Part II: Stochastic Networks—Graphical Evaluation and Review Technique (GERT) 363

5.11 Network Representation 363
 5.11.1 Input Functions 364
 5.11.2 Output Functions 364
5.12 GERT Basic Procedures 365
 5.12.1 Arcs in Series 367
 5.12.2 Branches in Parallel 368
 5.12.3 Self-Loops 368
5.13 Basic Concepts of Flowgraphs 370
5.14 Definitions 371
5.15 Mason's Rule for Closed Flowgraphs 373
5.16 Mean and Variance Calculation 376
5.17 GERT Applications 376
 5.17.1 Production of a High-Risk Item 376
 5.17.2 Material Processing 377
 5.17.3 Determination of Probabilistic Time Standards for Tasks Performed under Uncertainty 380

Part III: Multicommodity Network Flows 386

5.18 Linear Programming Formulations 388
5.19 A Special Class of Integer Multicommodity Networks 390
 5.19.1 Shipment of Automotive Transmissions 392
5.20 Approximate Solution of Multicommodity Transportation Problems through Aggregation 393
 5.20.1 A Fruit Distribution Problem 397
5.21 Error Bounds for Aggregation 398
5.22 Maximal Flows in Multicommodity Networks 400
5.23 Multicommodity Flows in Undirected Networks 403
 5.23.1 A Two-Commodity Flow Problem 407
5.24 Maximal Flows and Funnel Nodes 409
5.25 Applications of Multicommodity Networks 410
 5.25.1 Tanker Scheduling 411
 5.25.2 Urban Transportation Planning 412
 5.25.3 Computer-Communication Models 413
5.26 Notes and Remarks 414

Appendix The Network Optimization Computer Code (NETOPT) 425

Index 467

PREFACE

This book is a comprehensive treatment of the theories, algorithms, and computational nature of deterministic network flows. The book was developed from 10 years of classroom lecture notes compiled by ourselves, numerous technical articles, and contributions from leading experts in the field. The fundamental contribution of this textbook is a simple, practical approach to the design and implementation of network-flow algorithms. At the risk of criticism from fellow colleagues, we have sometimes sacrificed theoretical development and mathematical proofs for explanation by way of computational presentations. When appropriate, each technique is first presented as a mathematical programming problem, and then followed by an algorithmic procedure that guarantees problem resolution through faster and more efficient computational techniques as compared to direct use of mathematical models. Every algorithm is used to solve one or more real-world examples. A practical feature of the book is the documentation and listing of a FORTRAN IV network optimization computer program that can be used either for teaching or research purposes, on small- to medium-sized problems.

This book is intended to be an introductory treatment of network flows, and should prove useful as a textbook in a first couse in network analysis at either the undergraduate or graduate level. The treatment is self contained, and no particular degree of mathematical sophistication is necessary to understand the material developed in this book. Limited exposure to linear programming notation and FORTRAN programming would enhance the learning experience, but are not necessary to understand the algorithms as presented.

It has been our pleasure during the last 10 years to note a major explosion in the interest and application of network-flow algorithms. There does not seem to be an educational area that does not embrace network analysis; its applications are continuously surfacing in business administration, all fields of engineering, transportation analysis, project planning and control, sequencing and scheduling, and other areas too numerous to mention. The basic appeal is the great flexibility of network representation and the visual/graphical interpretation of network models. In addition, network solution procedures have recently been developed that are significantly more efficient than conventional linear programming.

The major barrier to widespread acceptance and use of network analysis is undoubtedly technical communication. With only a few exceptions, prior publications in the field and major developments have been presented through a mathematical programming or graph theoretic base. These results have usually surfaced in technical articles or sophisticated reports. This book attempts to avoid some of the theoretical and mathematical sophistication and present the fundamental aspects of network-flow analysis in a nontechnical fashion.

The book is divided into five basic chapters and one appendix. Chapter 1 presents the notation and symbolism used throughout the text, and establishes definitions that pertain to developments in latter chapters. Chapter 2 is a comprehensive treatment of deterministic network flows, and begins with a presentation of several examples illustrating network formulations of practical problems. Chapter 2 utilizes a wide range of examples and introduces FORTRAN IV computer programs to aid practical solution of larger problems. Chapter 3 presents a unified and comprehensive treatment of the elegant out-of-kilter algorithm, and contains a detailed development of the theoretical and computational aspects of this powerful technique. Numerous applications are suggested in this chapter, and several example problems are used to illustrate modeling procedures. Chapter 4 is a complete treatment of project management and control procedures based upon PERT and CPM. Computational procedures, resource balancing/leveling, and computerized procedures are thoroughly discussed in this chapter. Chapter 5 is a treatment of more advanced topics, including networks with gains and losses, GERT procedures for special forms of stochastic networks, and multicommodity network flows. The Appendix provides a complete listing and operating instructions for the network optimization program.

We have attempted to draw upon the knowledge and expertise of many individuals in this field. Indeed, several have contributed directly to the material contained in this text. Portions of Chapter 2, particularly many of the applied examples, were contributed by Dr. G. E. Bennington. Portions of Chapter 3, dealing with the out-of-kilter algorithm, were taken from original lecture notes by Dr. Paul A. Jensen, and several examples were contributed by Dr. R. E. D. Woolsey and Dr. Hunter Swanson. Dr. Warren Thomas contributed most of the material on the computational aspects of PERT/CPM in Part I of Chapter 4. The entire section dealing with resource control in Chapter 4 was contributed by Dr. Edward Davis, and the computer software survey in Chapter 4 was taken from an article by Dr. Larry A. Smith and Mr. Peter Mahler. Material in Chapter 5 dealing with the theory of generalized networks was adopted from original research by Drs. Gora Bhaumik and Paul Jensen, with several examples cited from works of Dr. Darwin Klingman and Dr. Fred Glover. Theories and computational aspects of GERT are attributed to Dr. A. Alan B. Pritsker. Finally, the material on multicommodity network flows was contributed totally by Dr. James Evans.

In addition to direct individual contributions in Chapters 3 and 4, this book obviously depends upon and draws from technical material developed by many of our fellow colleagues, too numerous to acknowledge. They will recognize their individual contributions. We are grateful to have benefited from their work. In conclusion, we would like to specifically acknowledge the individual contributions of Dr. Paul A. Jensen and Dr. R. E. D. Woolsey. They will understand and accept our special recognition for their unique support in writing this book. The one individual who deserves grateful allocades for his comprehensive criticisms and contributions to this book is Dr. James Evans. Dr. Evans reviewed our first draft and supplied many important suggestions, for which we are very grateful. We would also be remiss if we didn't thank Mrs. Jan Bertch and Mrs. Candy Phillips for typing preliminary drafts of this manuscript, and suffering through dozens of angry outbursts as the material was composed. Last but not least, we are indebted to the American Institute of Industrial Engineers for permission to reproduce printed material.

DON T. PHILLIPS, PH.D., P.E.
ALBERTO GARCIA-DIAZ, PH.D.

College Station, Texas

FUNDAMENTALS OF NETWORK ANALYSIS

INTRODUCTION

"Cheshire-Puss", she began, rather timidly,
"Would you tell me, please, which way I
ought to go from here"?
"That depends a great deal on where you
want to get to", said the cat.
"I don't much care where—," said Alice.
"Then it doesn't matter which way you go,"
said the cat.
"—so long as I get somewhere," Alice added.

From *Alice in Wonderland*
Lewis Carroll

As the Cheshire Cat so astutely observed in Wonderland, progress can often be made if one wanders around long enough. However, there are often better ways to search for an optimal solution than the aimless wandering of Alice. Network modeling techniques often provide the framework and computational structure to greatly improve many traditional approaches to systems analysis. The purpose of this chapter is to present the necessary machinery to both understand and apply the fundamental algorithms pre-

1

sented in this text. Of course, the application of network analysis techniques often requires not only "where you want to get to" but also "which way you ought to go." Hopefully, this chapter will aid in defining both strategies.

A modern society can be viewed, in part, as a system of networks for transportation, communication, and distribution of energy, goods, and services. The complex structure and cost of these subsystems demand that existing facilities be efficiently used and that new facilities be rationally designed. Network analysis techniques can be of great value in the design, improvement, and rationalization of complex large-scale systems.

Network-flow models and solution techniques provide a rich and powerful framework from which many engineering problems can be formulated and solved. The visual and logical structure of network-flow analysis often provides a fresh and natural approach from which further engineering analysis can proceed. Once only a small segment in the field of operations research, network analysis techniques have recently emerged as a viable and computationally tractable approach to solving significant problems faced by modern engineering analysts.

The origins of network analysis are old and diverse. Network analysis relies heavily on graph theory, a branch of mathematics that evolved with Leonhard Euler's formulation and solution of the famous Konigsberg bridge problem in 1736 [7]. More than a century later, James Clerk Maxwell and

Gustav Robert Kirchhoff discovered certain basic principles of network analysis in the course of their studies of electric circuits. Since then, network analysis has become an important tool in the investigation of electrical systems. Early in the twentieth century, telephone engineers in Europe and the United States devised network methods to determine the best capacity of telephone trunk lines and switching centers in order to guarantee specified levels of customer serviceability. In the 1940s, during the period of World War II, the development of operations research yielded a number of techniques for the mathematical study of large-scale systems. Pioneering work in modern network analysis was conducted by Hitchcock [11] in 1941 and Koopmans [15] in 1947. Since then, network analysis has been a very active and productive research area with well over 1000 published papers. The emphasis of research in the 1950s and early 1960s was on formulation of new models and development of new algorithms. Later, emphasis shifted to the extension, computer implementation, and analysis of previously developed models and algorithms. Survey papers have been written by Fulkerson [10], Elmaghraby [6], Bradley [2], and Magnanti and Golden [16], among others. As the field of network analysis has developed over the years, the need has arisen for updated books providing different orientations at various levels of discussion. Books with extensive treatment of networks have been contributed by Ford and Fulkerson [8], Charnes and Cooper [4], Dantzig [5], Busacker and Saaty [3], Hu [12], Frank and Frisch [9], Whitehouse [21], and, more recently, Jensen and Barnes [14], Bazarra and Jarvis [1], and Minieka [17].

Network models and analysis are widely used in operations research for diverse applications, such as the analysis and design of large-scale irrigation systems, computer networks, cable television networks, transportation systems, and ground and satellite communication networks. Efficient network methodologies have been implemented to solve industrial problems, such as the warehousing and distribution of goods, project scheduling, equipment replacement, cost control, traffic studies, queueing analysis, assembly-line balancing, inventory control, and manpower allocation, to name a few.

Pritsker [20] provides some insight as to the recent surge in the application of network analysis techniques:

> Networks and network analyses are playing an increasingly important role in the description and improvement of operational systems primarily because of the ease with which systems can be modeled in network form. This growth in the use of networks can be attributed to:
>
> 1. The ability to model complex systems by compounding simple systems.
> 2. The mechanistic procedure for obtaining system figure-of-merits from networks.

3. The need for a communication mechanism to discuss the operational system in terms of its significant features.
4. A means for specifying the data requirements for analysis of the system.
5. A starting point for analysis and scheduling of the operational system.

Item 5 was the original reason for network construction and use. The advantages that accrued outside of the analysis procedure soon justified the network approach. Considerable work is motivated by the need for extending present analysis procedures to keep pace with applications of networks.

Network analysis is not a discipline confined to only one branch of academia or industry. Indeed, the real strength of the network approach lies in the fact that it can be successfully applied to almost any problem when the modeler has enough knowledge and insight to construct the proper network representation. The advantages of using network models can be stated as follows:

1. Network models accurately represent many real-world systems.

2. Network models seem to be more readily accepted by nonanalysts than perhaps any other type of models used in operations research. This phenomenon appears to stem from the notion that "a picture is worth a thousand words." Managers seem to accept a network diagram more easily than they do abstract symbols. Additionally, since network models are often related to physical problems, they can be easily explained to people with little quantitative background.

3. Network algorithms facilitate extremely efficient solutions to some large-scale models.

4. Network algorithms can often solve problems with significantly more variables and constraints than can be solved by other optimization techniques. This phenomenon is due to the fact that a network approach often allows the exploitation of particular structures in a model.

1.1 Definitions, Notation, and Symbolism

A *network* consists of a set of *nodes* and a set of *arcs* connecting the nodes. The nodes are also referred to as vertices or points. The arcs are also called edges, links, lines, or branches. A network can be represented by the notation $G = (\mathbf{N}, \mathbf{A})$, where $\mathbf{N}$ is the set of nodes and $\mathbf{A}$ the set of arcs of the network G.

The nodes of a network can represent highway intersections, power stations, telephone exchanges, railroad yards, airline terminals, water reservoirs, computers, well-defined occurrences in time, or simply project milestones. In general, a node can represent a point where some kind of flow is originated, relayed, or terminated. For this reason, a node can be viewed as a branching point in a typical network.

The arcs of a network can represent roads, power lines, telephone lines, airline routes, water mains, or generalized channels through which entities flow. In some instances, the arcs have no physical meaning but serve to direct flow in a logical sequence, or to maintain a specified precedence relationship.

In an effort to preserve the individuality of the nodes and keep track of them in the process of analysis, they are ordinarily assigned numbers. Although it sometimes helps the mental process to number the nodes in an orderly fashion, this obviously has no significance whatsoever and, in general, the numbering is assumed to be arbitrary.[1] An arc between node i and node j is said to have an orientation from i to j if 1 unit of flow sent from node i to node j along the arc would increase the current value of the flow on the arc, but would decrease it if sent in the opposite direction. In a typical network, the arcs can be classified as undirected, directed, or bidirected. An arc is *undirected* if it does not have a specified orientation, *directed* if it has exactly one orientation, and *bidirected* if it has two orientations.

To represent networks graphically, we use circles to indicate nodes, lines to indicate arcs, and arrowheads to indicate arc orientation. We also use the notation (i, j) to refer to the directed arc leading from node i to node j. Figure 1.1(a) shows an undirected arc, Figure 1.1(b) a directed arc, and Figure 1.1(c) a bidirected arc.

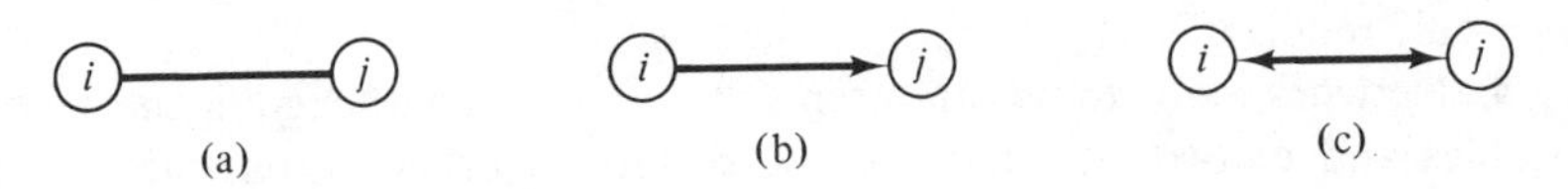

Figure 1.1. Node-arc representations: (a) undirected arc; (b) directed arc; (c) bidirected arc.

For most practical purposes there is no necessity to distinguish between undirected and bidirected arcs. Further, it is evident that a bidirected arc between two arbitrary nodes i and j can be represented by two directed arcs (i, j) and (j, i), as shown in Figure 1.2.

As previously mentioned, the arcs of a network can be considered as channels through which a generalized commodity can flow. The net amount

[1] A notable exception to this rule occurs in CPM/PERT project analysis. As explained in Chapter 4, it is advantageous to sequentially number the nodes in increasing order.

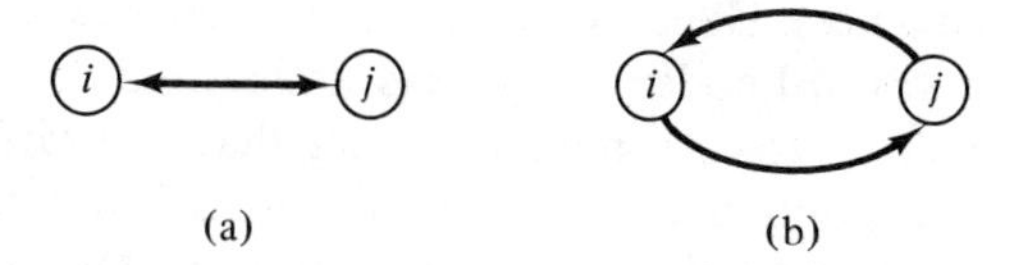

Figure 1.2. Representation of a bidirected arc by two directed arcs with opposite orientations: (a) bidirected arc; (b) equivalent representation.

of commodity at a given node is equal to the difference between the amount of flow leaving the node and the amount of flow arriving at the node. If the net amount of commodity is strictly positive, the node is called a source node. If it is strictly negative, the node is called a sink or terminal node. A network may contain more than one sink or source node. For convenience, we will frequently refer to a *source node* as any node that only originates flow, and a *sink node* as any node that only terminates flow. For example, in the network of Figure 1.3, nodes 1 and 2 are source nodes and nodes 6 and 7 are sink nodes. In the development of some algorithms it is assumed that the network contains a single source node and a single terminal node. Such a network can be constructed by creating a *super* source node and a *super* sink node, and then connecting the multiple sources to the super source node and the multiple sinks to the super sink node by dummy arcs.

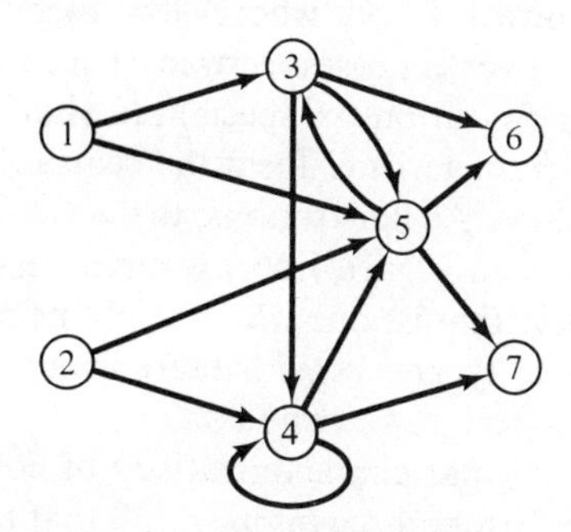

Figure 1.3. Illustrative network.

In representing real-world problems in terms of network flows, it is often desirable to categorize network segments in terms of flow orientation. An important concept in network analysis is that of connecting two nodes i and j with a sequence of arcs, $\mathbf{E} = \{a_1, a_2, \ldots, a_k\}$. A *chain* from node i to node j is a sequence of arcs and nodes such that the terminal node of each arc coincides with the initial node of the succeeding arc (excluding the first and last node in the sequence). Hence, each arc in the sequence is oriented "away from node i" and "toward node j." A *path* is similar to a chain except one or more arcs are oriented "toward node i" rather than "toward node j." This means that in a path, unlike a chain, arcs are not all oriented in the same direction. Using these definitions, a chain cannot contain a bidirected arc but might contain an undirected arc. Normally, chains and paths are only defined relative to flow through directed arcs, and in this book we will usually follow this convention. In the network of Figure 1.3, the sequence of arcs (1, 3), (3, 5), (5, 3), and (3, 6) forms a chain, whereas the sequence of arcs (2, 4), (3, 4), and (3, 6) forms a path.

A network is said to be undirected if no arc has a specified orientation. In an undirected network the terms "chain" and "path" are used interchangeably to refer to a sequence of arcs connecting two arbitrary nodes.

A *cycle* is a finite path that begins and terminates at the same node. In Figure 1.3 the sequence of arcs (4, 7), (4, 5), and (5, 7) forms a cycle. A special class of cycle is a circuit. A *circuit* is defined as a finite chain with the first and last nodes being coincident. Categorically, cycles are closed paths and circuits are closed chains. A *self-loop* is simply a single node and arc, and is therefore both a circuit and a cycle. In the network of Figure 1.3, the sequence of arcs (3, 4), (4, 5), and (5, 3) forms a circuit. A degenerate form of a cycle, i.e., a self-loop, is sometimes called a one-stage cycle because it contains only one arc. In the network of Figure 1.3, arc (4, 4) is a loop. A network is said to be *circuitless* if it does not contain circuits. In other words, a network with only directed arcs whose flow forms no circuits is called *circuitless*. Circuitless networks possess certain desirable characteristics which are exploited in some applications of specialized shortest-route and longest-route problems. Such characteristics form the central logic in the derivation of CPM/PERT algorithms. As illustrations, the network in Figure 1.4(a) has circuits, but the network in Figure 1.4(b) is circuitless. Figure 1.4(c) represents an acyclic network, that is, one which contains no cycles.

A network is said to be *connected* if there is at least one path or chain joining any pair of arbitrary distinct nodes.

Trees represent particular characterizations of connected undirected or directed networks. For further discussions, recall that the set of all nodes and arcs is given by $G = (\mathbf{N}, \mathbf{A})$, and define any subset of G as $\bar{G} = (\bar{\mathbf{N}}, \bar{\mathbf{A}})$. A tree is defined to be a connected subset $\bar{G}$ containing no cycles. Hence, between any two nodes in a tree there is a unique path. For a network containing n nodes, any two of the following three conditions will serve to define a subgraph of k nodes, $k \leq n$, as a tree.

1. The subgraph is connected.
2. The subgraph has no cycles.
3. The number of arcs in the subgraph is $k - 1$.

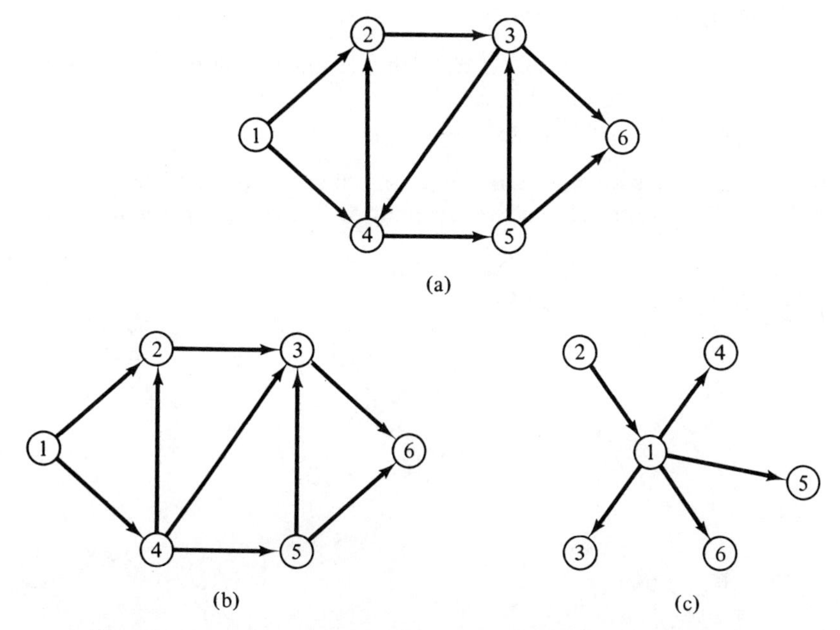

Figure 1.4. Cyclic and acyclic networks: (a) network with a circuit; (b) network without a circuit; (c) acyclic network.

A *spanning tree* is a tree that contains every node in the network. Hence, if a network contains n nodes, a tree with n nodes and $n - 1$ arcs is a spanning tree.

If we associate with each arc of the network a number such as distance, cost, or any other additive parameter to indicate the natural limitations or capabilities of the arcs, then we can develop the notion of a *minimum* (or a maximum) *spanning tree*. The *weight* of a tree is defined as the sum of the arc parameters of the arcs in the tree. A minimum (maximum) spanning tree is defined as the tree with minimum (maximum) weight.

Trees and spanning trees are defined for both directed and undirected networks. However, if a network contains only directed arcs, then the concept of an arborescence is needed. An *arborescence* is easy to visualize; it is a tree that consists of only directed arcs. However, an arborescence always possesses a root. The *root* of an arborescence is a single node that only originates directed arcs. An arborescence cannot have more than one root. Hence, an arborescence of k nodes will have $k - 1$ nodes that terminate only one arc, and a single node that only originates arcs.

A tree that is also an arborescence will be called an *arborescent tree*, and a spanning tree that is also an arborescence will be called an *arborescent spanning tree*.

Formally, an arborescence with root $x_1 \in \mathbf{N}$ is defined as follows:

1. Every node different from x_1 is the terminal node of a single arc.
2. x_1 is not the terminal node of any arc.
3. The network contains no circuits.

Note that there is a single chain connecting node x_1 to every other node.

Figure 1.5 illustrates the fundamental definitions given for undirected

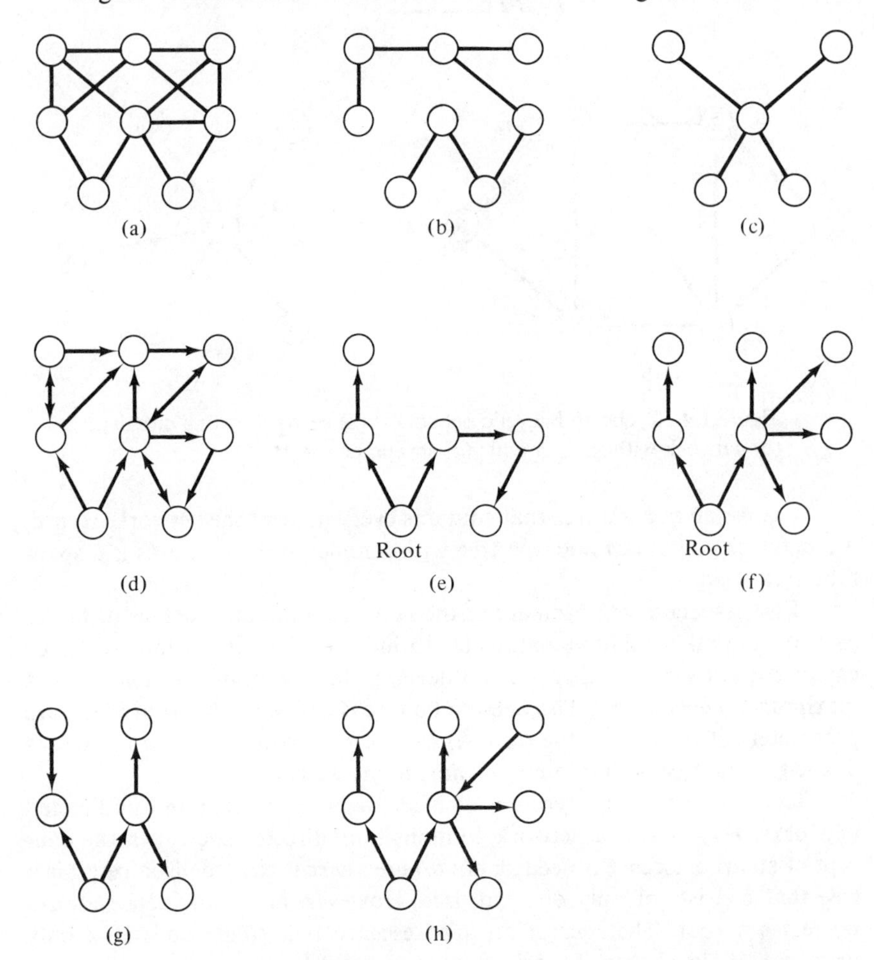

Figure 1.5. Trees and arborescences: (a) undirected network; (b) spanning tree for network given in (a); (c) tree for the network given in (a); (d) directed network; (e) arborescent tree for network given in (d); (f) arborescent spanning tree for network given in (d); (g) tree for the network given in (d); (h) spanning tree for the network given in (d).

networks, directed networks, trees, and arborescences. Figure 1.5(a)–(c) shows an undirected network, a spanning tree for such a network, and a (nonspanning) tree for the same network. Figure 1.5(d)–(f) shows a directed network, an arborescence for that network, and an arborescent spanning tree, respectively. Figure 1.5(g) and (h) shows a tree for a directed network and a spanning tree, respectively.

1.2 Matrix Representation of Networks

It is evident that there are different network configurations which could represent the same elements and relationships of a given system. In fact, for larger networks the equivalence of two network forms can be extremely difficult to recognize. In graph theory, two networks which have a one-to-one correspondence with respect to their arcs and nodes are said to be *isomorphic*.

If two graphs are isomorphic, then any equation defined for one will also be defined for the other. Some common network representation forms often make computations easier or reveal problem characteristics that facilitate the elimination of certain calculations or further analysis. Graphic or matrix representations are useful from both a computational and a pedagogical standpoint. The quantitative characteristics of the arcs, as well as the interconnections between the nodes of a network, can be represented by means of a distance matrix or cost matrix. The interconnections of the nodes can be separately represented by an adjacency matrix or by a node–arc incidence matrix.

To simplify our discussion on matrix representation of a network, we consider a network $G = (\mathbf{N}, \mathbf{A})$, where the nodes in the set $\mathbf{N}$ are numbered $1, 2, \ldots, n$, and where each arc (i, j) in the set $\mathbf{A}$ is assigned a quantitative parameter c_{ij}, usually referred to as the "generalized cost" or "length" of the arc. The *cost matrix* of the network is defined as the array $\mathbf{C} = [c_{ij}]$, for all $(i, j) \in \mathbf{A}$.

The elements of matrix $\mathbf{C}$ represent the natural limitations or capabilities of the system, such as distance, transportation costs, capacities of water mains, peak loads for power lines, reliabilities of system components, and so on. If G is an undirected network, then $c_{ij} = c_{ji}$ for each $(i, j) \in \mathbf{A}$. In this case, $\mathbf{C}$ is a symmetric matrix.

The adjacency matrix of a directed network G is defined as an array $\mathbf{X} = [x_{ij}]$, where

$$x_{ij} = \begin{cases} 1, & \text{if arc } (i, j) \in \mathbf{A} \text{ is directed from node } i \in \mathbf{N} \text{ to node } j \in \mathbf{N} \\ 0, & \text{otherwise} \end{cases}$$

If network G is undirected, each arc $(i, j) \in \mathbf{A}$ can be replaced by two directed arcs with opposite orientations. In this case, the adjacency matrix **X** would be symmetric.

The *node–arc incidence matrix* of a directed network G is defined as an array $\mathbf{Z} = [z_{ik}]$, where

$$z_{ik} = \begin{cases} +1, & \text{if node } i \in \mathbf{N} \text{ is the starting node of arc } a_k \in \mathbf{A} \\ -1, & \text{if node } i \in \mathbf{N} \text{ is the ending node of arc } a_k \in \mathbf{A} \\ 0, & \text{otherwise} \end{cases}$$

If network G is undirected, z_{ik} can alternatively be defined as

$$z_{ik} = \begin{cases} 1, & \text{if node } i \in \mathbf{N} \text{ is connected to arc } a_k \in \mathbf{A} \\ 0, & \text{otherwise} \end{cases}$$

Certain important characteristics can be deduced for both undirected and directed networks from their matrix representations. As an illustration, let us first consider the undirected network given in Figure 1.6. For the network $G = (\mathbf{N}, \mathbf{A})$ given in Figure 1.6, the sets **N** and **A** are defined as follows:

$$\mathbf{N} = \{i\} = \{1, 2, 3, 4\}$$
$$\mathbf{A} = \{a_k\} = \{a_1, a_2, a_3, a_4, a_5, a_6\}$$

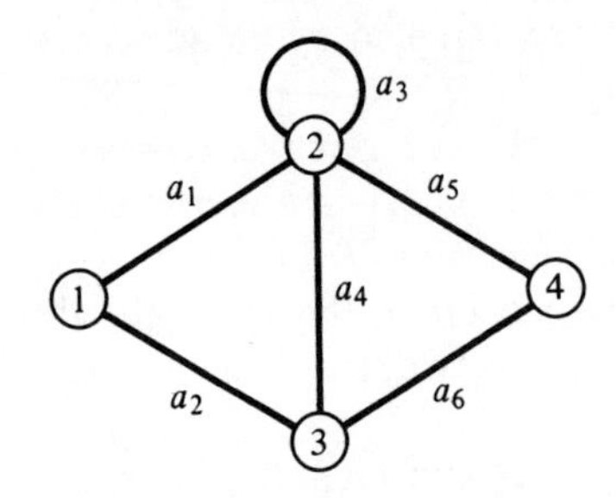

Figure 1.6. Undirected network.

The adjacency matrix **X** and the node–arc incidence matrix **Z** in this example are given by

$$\mathbf{X} = [x_{ij}] = \begin{array}{c} \\ 1 \\ 2 \\ 3 \\ 4 \end{array}\begin{array}{c} \begin{array}{cccc} 1 & 2 & 3 & 4 \end{array} \\ \begin{bmatrix} 0 & 1 & 1 & 0 \\ 1 & 1 & 1 & 1 \\ 1 & 1 & 0 & 1 \\ 0 & 1 & 1 & 0 \end{bmatrix} \end{array}$$

and

$$\mathbf{Z} = [z_{ij}] = \begin{array}{c} \\ 1 \\ 2 \\ 3 \\ 4 \end{array} \begin{array}{c} \begin{array}{cccccc} a_1 & a_2 & a_3 & a_4 & a_5 & a_6 \end{array} \\ \begin{bmatrix} 1 & 1 & 0 & 0 & 0 & 0 \\ 1 & 0 & 1 & 1 & 1 & 0 \\ 0 & 1 & 0 & 1 & 0 & 1 \\ 0 & 0 & 0 & 0 & 1 & 1 \end{bmatrix} \end{array}$$

The following deductions can be made from the adjacency matrix **X**:

1. If an unconnected node is added to the network, a row of zeros and a column of zeros would be added to the matrix.
2. The matrix is symmetric.
3. Any nonzero element on the main diagonal represents a loop.

As an exercise, the reader is invited to draw similar deductions from the node–arc incidence matrix **Z**.

Now let us consider the matrix representation of a directed network, such as the one given in Figure 1.7. For this network the set of nodes **N** and the set of arcs **A** are defined by

$$\mathbf{N} = \{i\} = \{1, 2, 3, 4, 5, 6\}$$
$$\mathbf{A} = \{a_k\} = \{a_1, a_2, a_3, a_4, a_5, a_6, a_7, a_8, a_9\}$$

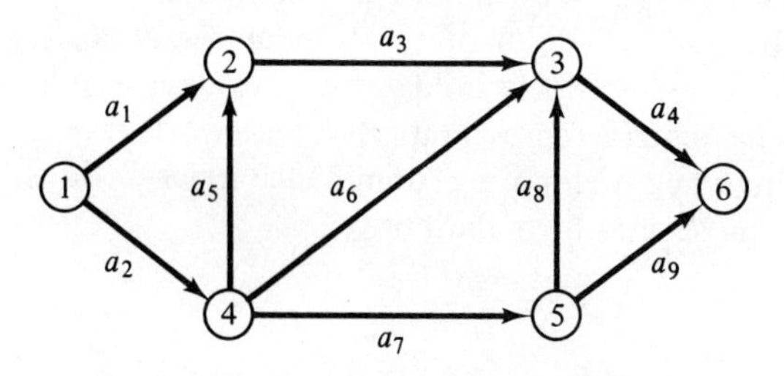

Figure 1.7. Directed network.

The adjacency matrix **X** and the node–arc incidence matrix **Z** are in this example given by

$$\mathbf{X} = [x_{ij}] = \begin{array}{c} \\ 1 \\ 2 \\ 3 \\ 4 \\ 5 \\ 6 \end{array} \begin{array}{c} \begin{array}{cccccc} 1 & 2 & 3 & 4 & 5 & 6 \end{array} \\ \begin{bmatrix} 0 & 1 & 0 & 1 & 0 & 0 \\ 0 & 0 & 1 & 0 & 0 & 0 \\ 0 & 0 & 0 & 0 & 0 & 1 \\ 0 & 1 & 1 & 0 & 1 & 0 \\ 0 & 0 & 1 & 0 & 0 & 1 \\ 0 & 0 & 0 & 0 & 0 & 0 \end{bmatrix} \end{array}$$

and

$$\mathbf{Z} = [z_{ij}] = \begin{array}{c} \\ 1 \\ 2 \\ 3 \\ 4 \\ 5 \\ 6 \end{array} \begin{array}{c} \begin{array}{ccccccccc} a_1 & a_2 & a_3 & a_4 & a_5 & a_6 & a_7 & a_8 & a_9 \end{array} \\ \begin{bmatrix} +1 & +1 & 0 & 0 & 0 & 0 & 0 & 0 & 0 \\ -1 & 0 & +1 & 0 & -1 & 0 & 0 & 0 & 0 \\ 0 & 0 & -1 & +1 & 0 & -1 & 0 & -1 & 0 \\ 0 & -1 & 0 & 0 & +1 & +1 & +1 & 0 & 0 \\ 0 & 0 & 0 & 0 & 0 & 0 & -1 & +1 & +1 \\ 0 & 0 & 0 & -1 & 0 & 0 & 0 & 0 & -1 \end{bmatrix} \end{array}$$

The following facts can be deduced from the structure of the adjacency matrix **X** of the directed network under consideration:

1. Any column containing all zero elements represents a source node.
2. Any row containing all zero elements represents a sink node.
3. If the main diagonal contains all zero elements, there are no loops in the network. Conversely, a nonzero element on the main diagonal indicates a loop.
4. The matrix is no longer symmetric.

As an exercise, the reader is invited to draw similar deductions from the node–arc incidence matrix **Z** of the directed network.

It should be noted that all of the foregoing characteristics for both undirected and directed networks hold when c_{ij} values are considered instead of $x_{ij} = 0, 1$. The matrix representations of network models often prove extremely useful when computing dominance relationships or verifying equivalence among several formulations.

1.3 Conservation of Flow

In considering the performance of a network it is often necessary to calculate the optimal value of a function of the flow between the source node s and the terminal node t. This calculation is usually referred to as a *one-commodity* flow problem because the flows on the arcs of the network correspond to a single entity, such as electric power, water, information, air traffic, and so on.

Let β_i be the set of nodes connected to node i by arcs oriented toward node i, and let α_i be the set of nodes connected to node i by arcs oriented away from node i. An integer-valued function f_{ij} defined on **A** is said to be a *flow* for a directed network $G = (\mathbf{N}, \mathbf{A})$ if

$$f_{ij} \geq 0, \qquad \text{for all } (i, j) \in \mathbf{A} \tag{1-1}$$

$$\sum_{j \in \alpha_i} f_{ij} - \sum_{j \in \beta_i} f_{ji} = 0, \qquad \text{for } i \in \mathbf{N} \quad i \neq s, i \neq t \tag{1-2}$$

$$f_{ij} \leq c_{ij}, \qquad \text{for all } (i, j) \in \mathbf{A} \tag{1-3}$$

In the present discussion f_{ij} can be interpreted as a volume of commodity flowing along the arc (i, j) from node i to node j, and not exceeding the arc capacity c_{ij}. If j is neither the source node nor the terminal node, the amount flowing into node j must be equal to the amount flowing out of that node. This condition is known as the *flow-conservation* property. Constraints (1-2) correspond to the flow-conservation condition for network G. The conservation-of-flow property will always hold for the single-commodity network-flow problems discussed in this book. The flow along an arc may be altered, as in the treatment of networks with gains and losses, discussed in Section 5.2.

1.4 Contribution of the Maximum-Flow Minimum-Cut Theorem

Let V be the value of the flow that can be shipped from the source node s to the terminal node t through the arcs of a network $G = (\mathbf{N}, \mathbf{A})$. Under the assumption that the arcs in the set $\mathbf{A}$ have a finite capacity, the maximal value of the flow V is limited by the capacities of the arcs of the network.

The maximum flow is determined by a fundamental structural property of the network called a *cut*. A cut can be defined as a set of arcs whose removal would disconnect a set of nodes from all the remaining nodes. As an illustration, in the network shown in Figure 1.8, a cut consisting of arcs (2, 4) and (3, 4) disconnects node 4 from the group of nodes 1, 2, 3. When studying the maximum-flow problem, we are only interested in cuts that would disconnect the source from the terminal.

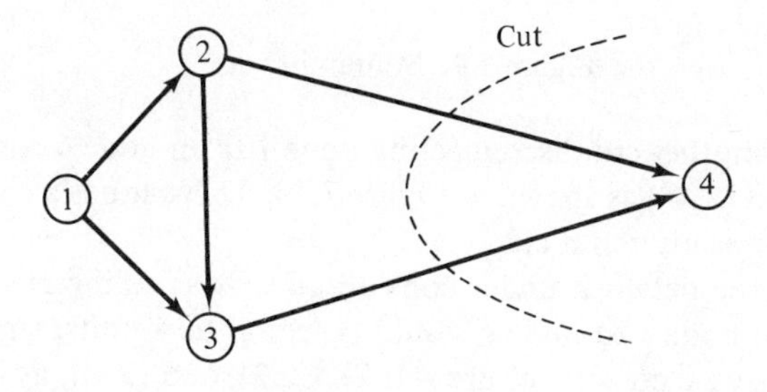

Figure 1.8. Cut set.

The capacity of a cut, or *cut value*, is defined as the sum of the flow capacities of the arcs (in the specified direction) of the cut. In the example network of Figure 1.8, the capacity of the cut shown is equal to $c_{24} + c_{34}$.

The principle behind the algorithms to find the maximum flow in a network is quite simple. Let **X** be a subset of **N** such that $s \notin \mathbf{X}$ and $t \in \mathbf{X}$. The set $\mathbf{A}_X$ of arcs incident to and directed into **X** is, by definition, a cut of the network G. If $c(\mathbf{A}_X)$ represents the capacity of this cut, then the maximal value of flow which can be shipped from node s to node t must satisfy the condition that

$$V \leq c(\mathbf{A}_x) \tag{1-4}$$

Thus, if for a flow V and a cut $\mathbf{A}_V$ we have that $V = c(\mathbf{A}_V)$, we may be sure that the flow is maximum.

One of the most important results in the theory of network flows, the *maximum flow/minimum cut theorem*, proved by Ford and Fulkerson [8] can now be stated: For any network with single source and sink, the maximum feasible flow from source to sink equals the minimum-cut value among the cuts of the network. A complete proof of this theorem will be given in Chapter 2. As an illustration, it can be verified that the maximal flow that can be shipped from node s to node t in the network of Figure 1.9 is equal to 3. In this figure, the numbers shown on the arcs are the corresponding flow capacities in the specified directions. The minimum cut consists of arcs (1, 2) and (3, 4), and has a capacity equal to 3 units.

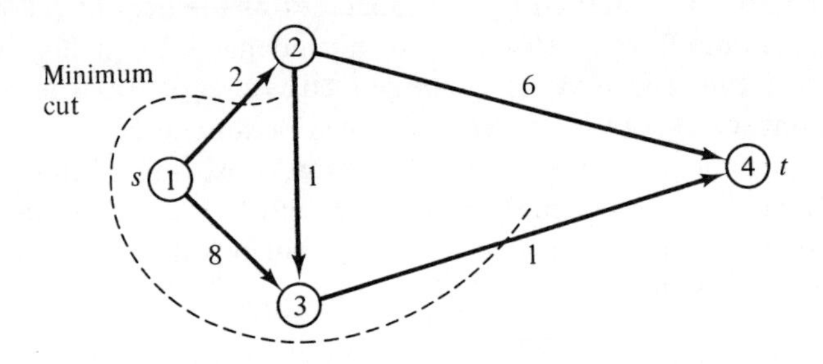

Figure 1.9. Minimum cut.

Note that another cut disconnecting node 1 from node 4 consists of arcs (1, 3), (2, 3), and (2, 4), as shown in Figure 1.10. This cut has a capacity equal to 15 and is not a minimum cut.

Finally, if the network under consideration was undirected, the maximum flow from node s to node t would be equal to 4 units, since the minimum cut in this case consists of arcs (1, 2), (2, 3), and (3, 4), as illustrated in Figure 1.11.

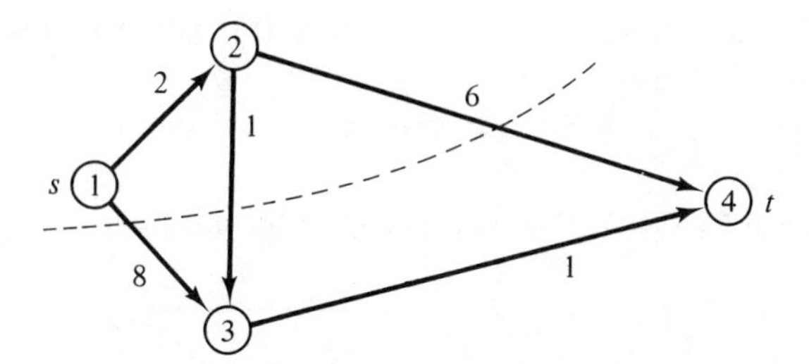

Figure 1.10. Non-minimum cut.

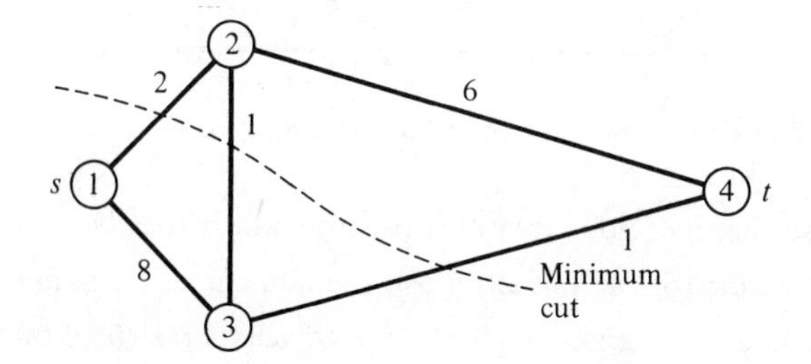

Figure 1.11. Minimum cut for an undirected network.

EXERCISES

1. List four reasons for the preference of network analysis techniques over other operations research techniques.

2. Define the following terms:
- (a) Arc
- (b) Node
- (c) Directed arc
- (d) Bidirected arc
- (e) Undirected arc
- (f) Source node
- (g) Sink node
- (h) Super source
- (i) Super sink

3. Define a chain and a path. What is the difference between the two?

4. Define a cycle and a circuit. What is the difference between the two?

5. Define a cyclic and an acyclic network.

6. If one was trying to minimize network-flow costs, why would a cyclic network cause problems in a minimum-cost algorithm?

7. Define a tree and a spanning tree: (a) in words; (b) mathematically; (c) using graph-theoretic definitions.

8. Define an arborescence. What is the difference between this and a spanning tree?

9. Using the network shown, draw (a) a tree; (b) a spanning tree.

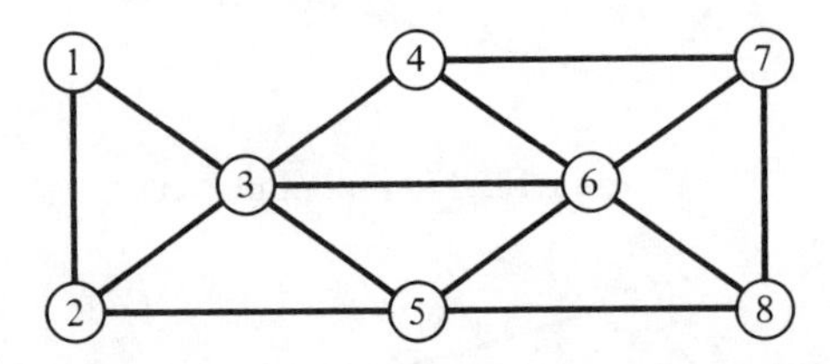

10. Deduce as many facts as you can from the node–arc incidence matrix of the network given in Figure 1.7.

11. Define conservation of flow: (a) in words; (b) mathematically.

12. What is (a) a cut; (b) cut value; (c) a maximum cut; (d) a minimum cut?

13. Find the node–arc incidence matrix and the adjacency matrix of the undirected network shown.

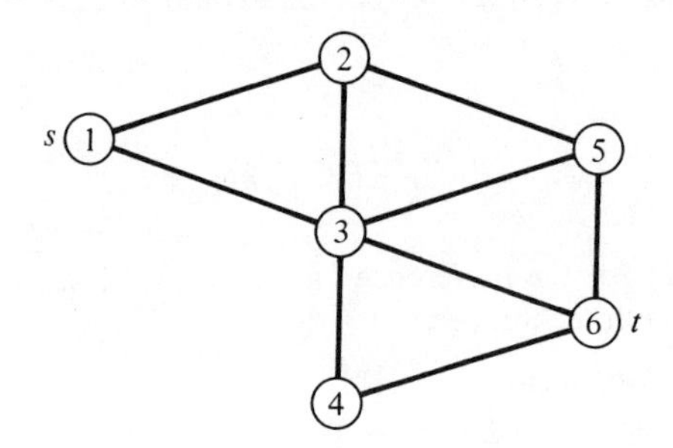

14. Find the node–arc incidence matrix and the adjacency matrix of the directed network shown.

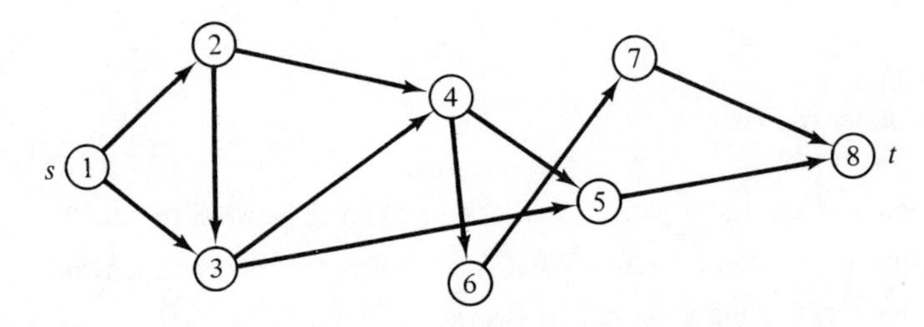

15. For the network given in Exercise 13, find a cycle, a circuit, a path from s to t, and a chain from s to t. Repeat this for the network given in Exercise 14.

16. Consider an undirected network $G = (\mathbf{N}, \mathbf{A})$ with $\mathbf{N} = \{1, 2, \ldots, 9\}$, $\mathbf{A} = \{a_1, a_2, \ldots, a_{12}\}$, and node–arc incidence matrix as shown.

	a_1	a_2	a_3	a_4	a_5	a_6	a_7	a_8	a_9	a_{10}	a_{11}	a_{12}
1	1	1	0	0	0	0	0	0	0	0	0	0
2	0	1	0	0	1	0	0	0	0	0	0	0
3	1	0	1	1	0	0	0	0	0	0	0	0
4	0	0	0	1	0	1	1	0	0	0	1	0
5	0	0	0	0	1	0	1	0	0	0	1	0
6	0	0	0	0	0	0	0	0	0	1	0	1
7	0	0	0	0	0	1	0	0	1	1	0	0
8	0	0	1	0	0	0	0	1	0	0	0	0
9	0	0	0	0	0	0	0	1	1	0	0	1

(a) Draw the network.
(b) Find the adjacency matrix.

17. Consider a directed network $G = (\mathbf{N}, \mathbf{A})$ with $\mathbf{N} = \{1, 2, 3, 4, 5\}$ and $\mathbf{A} = \{(1, 2), (1, 3), (3, 1), (2, 3), (2, 4), (2, 5), (3, 5), (4, 5)\}$. (a) Draw the network. (b) Find the adjacency matrix. (c) Find the node–arc incidence matrix.

18. In the network shown, what is the minimal value allowed for parameter a if node 4 is wanted in the shortest cycleless path from node 1 to node 7? Under what condition on a can the length of the shortest path be reduced as much as desired?

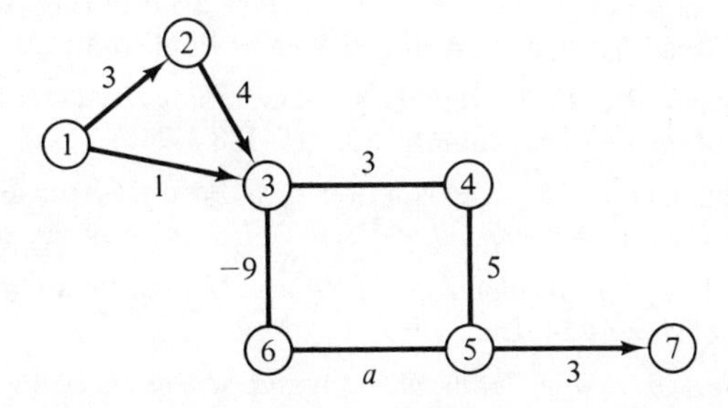

19. What is the value of the maximum flow allowed from node s to node t in the network shown? Identify the minimum cut. Each number represents the capacity of the corresponding arc.

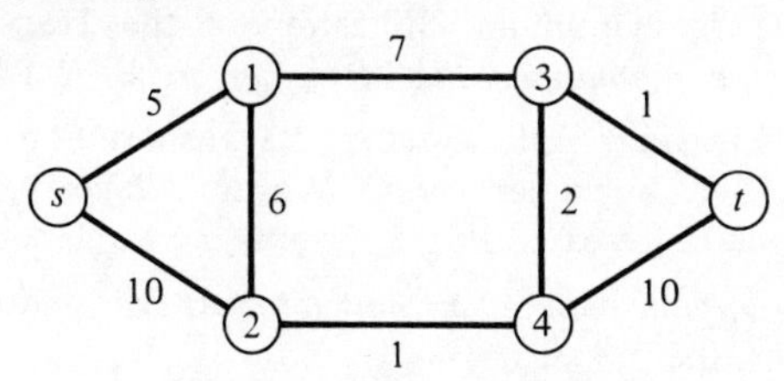

20. Consider the network given in Exercise 19. Write the equations corresponding to the flow-conservation constraints for nodes s, 3, and t.

REFERENCES

[1] BAZARRA, M., AND J. J. JARVIS, *Linear Programming and Network Flows.* New York: John Wiley & Sons, Inc., 1978.

[2] BRADLEY, G. H., "A Survey of Deterministic Networks," *AIIE Transactions, 7*, 222–234 (1975).

[3] BUSACKER, R. G., AND T. SAATY, *Finite Graphs and Networks.* New York: McGraw-Hill Book Company, 1965.

[4] CHARNES, A., AND W. W. COOPER, *Management Models and Industrial Applications of Linear Programming*, Vols. 1 and 2. New York: John Wiley & Sons, Inc., 1961.

[5] DANTZIG, G. L., *Linear Programming and Extensions.* Princeton, N.J.: Princeton University Press, 1963.

[6] ELMAGHRABY, S., *Some Network Models in Management Science.* New York: Springer-Verlag, Inc., 1970.

[7] EULER, L., "The Konigsberg Bridges," *Scientific American, 189*, 66–70 (1953).

[8] FORD, L. R., AND D. R. FULKERSON, *Flows in Networks.* Princeton, N.J.: Princeton University Press, 1962.

[9] FRANK, H., AND I. T. FRISCH, *Communication, Transmission, and Transportation Networks.* Reading, Mass.: Addison-Wesley Publishing Co., Inc., 1971.

[10] FULKERSON, D. R., "Flow Networks and Combinatorial Operations Research," *American Mathematical Monthly, 73*, 115–138 (1966).

[11] HITCHCOCK, F. L., "The Distribution of a Product from Several Sources to Numerous Localities," *Journal of Mathematics and Physics, 20*, 224–230 (1941).

[12] HU, T. C., *Integer Programming and Network Flows.* Reading, Mass.: Addison-Wesley Publishing Co., Inc., 1969.

[13] IRI, M., *Network Flow, Transportation and Scheduling.* New York: Academic Press, Inc., 1969.

[14] JENSEN, P. A., AND W. BARNES, *Network Flow Programming.* New York: John Wiley & Sons, Inc., 1980.

[15] KOOPMANS, T. C., "Optimum Utilization of the Transportation System," *Proceedings of the International Statistical Conference*, Washington, D.C., 1947.

[16] MAGNANTI, T. L., AND B. L. GOLDEN, "Transportation Planning: Network Models and Their Implementation," Working Paper 77-008, University of Maryland, General Research Board, Faculty Research Award, 1977.

[17] MINIEKA, E., *Optimization Algorithms for Networks and Graphs.* New York: Marcel Dekker, Inc., 1978.

[18] PHILLIPS, D. T., A. RAVINDRAN, AND J. J. SOLBERG, *Operations Research: Principles and Practice*. New York: John Wiley & Sons, Inc., 1977.

[19] POTTS, R. B., AND R. M. OLIVER, *Flows in Transportation Networks*. New York: Academic Press, Inc., 1972.

[20] PRITSKER, A. A. B., AND W. W. HAPP, "GERT: Part I—Fundamentals," *Journal of Industrial Engineering*, *17* (5), 267 (1966).

[21] WHITEHOUSE, G. W., *Systems Analysis and Design Using Network Techniques*. Englewood Cliffs, N.J.: Prentice-Hall, Inc., 1973.

DETERMINISTIC NETWORK FLOWS

Peter Rabbit made a face at Jimmy Skunk.
"I don't like being preached to."
"I'm not preaching; I'm just telling you
what you ought to know without being told,"
replied Jimmy Skunk.

From *The Adventures of Peter Cottontail*
T. W. Burgess

The material presented in this chapter is concerned with the study of certain deterministic network-flow models of frequent use in the representation and solution of significant problems in both business and engineering. For each problem an efficient algorithm is discussed and a complete FORTRAN program is provided to implement the solution methodology on a digital computer. Our discussion will include the following algorithms:

1. Dijkstra's algorithm for solving the shortest-route problem [11].
2. A multiterminal shortest-chain/route algorithm [31].
3. A shortest-route algorithm for a class of fixed-charge problems [41].
4. A double-sweep algorithm to solve the K-shortest-route problem [47].

2

5. A minimal-spanning-tree algorithm [31].
6. A branch-and-bound algorithm for solving the traveling salesman problem [40].
7. A maximum-flow algorithm [15].
8. A multiterminal maximum-flow algorithm [31].
9. A multiterminal maximum-capacity chain algorithm [32].
10. Algorithms for disconnecting networks through node and arc failures [18].

In addition, several other specialized algorithms will be discussed which extend the basic structure of these algorithms.

The approach used to present the material throughout this chapter is one of indoctrination by example. Each topic is introduced, mathematically defined, and then dealt with through an algorithmic framework. One or more examples are then formulated and solved by hand to illustrate the fundamental concepts and computational procedures inherent in the algorithmic process. Finally, an example is formulated and solved using a specialized computer code (FORTRAN, ANSI). Each program contains a section with documentation and operating instructions. In many instances, extensions to the basic problem are explored and illustrated through model formulation or solution. The FORTRAN programs are listed in the Appendix.

2.1 Applications of Network-Flow Models[1]

In order to acquaint the reader with network modeling and provide a framework through which the network algorithms can be effectively used, this section will discuss and model several problems employing network flow methodologies.

2.1.1 Equipment Replacement

In the fields of engineering design and analysis, there are many applied problems that can be formulated as *shortest-route-problem* models. The shortest-route problem can be simply stated in the following manner. Given a collection of nodes and arcs, with a set of positive arc parameters c_{ij}, measuring the cost of shipping 1 unit of flow along the arc (i, j), find the path from source node s to sink node t that minimizes the cost of shipping 1 unit of flow from the source to the sink. As an illustration, we consider a simple model for equipment that must be replaced periodically. For convenience, it is assumed that after a machine breakdown it is decided to replace the machine rather than attempt to repair it. Further, it is assumed that the current product line will be obsolete in 3 years and that this type of machine will not be used after that time. The salvage value for differing ages and the operating and maintenance expenses for each period are also assumed known. Consider the beginning of each period of the planning horizon to be represented by a node; under this representation, a given period of useful service life assigned to the machine can be modeled by an arc. With this information it is possible to compute the cost c_{ij} of purchasing a new machine at the beginning of period i and selling it at the end of period $j - 1$ (or equivalently, the beginning of period j). This cost is then defined as the "length" of (i, j).

The set of all replacement plans for $n - 1$ years can be represented by the chains in the network in Figure 2.1 ($n = 4$). Each node represents the beginning of a period, with node 4 representing the end of the planning

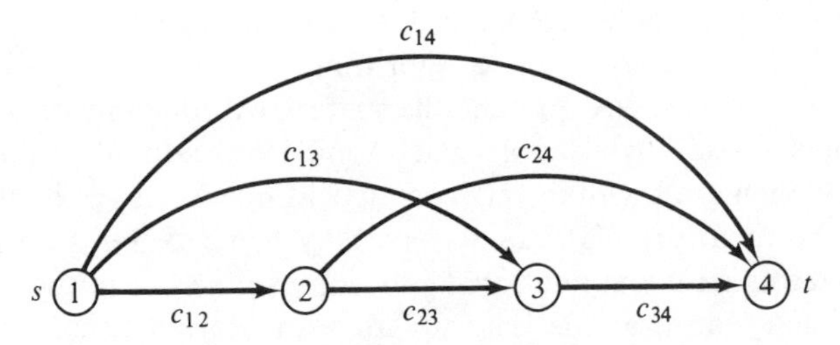

Figure 2.1. Equipment replacement network.

---

[1]Portions of this section were contributed by G. E. Bennington, and are reproduced with the permission of the American Institute of Industrial Engineers [4].

horizon. Each directed chain from node 1 to node 4 represents a replacement plan for the 3-year planning horizon. The chain (1, 3), (3, 4) represents buying a machine at the beginning of period 1, keeping the machine 2 years and selling it at the end of period 2. A new machine is then purchased at the beginning of period 3 and sold at the end of period 3. The total cost of this plan is $c_{13} + c_{34}$. The shortest route in this network from node 1 to node 4 will yield a replacement plan with a minimum cost for the three periods considered.

Several observations are possible at this point. A temporal or dynamic problem has been modeled by representing the time periods as nodes in a network. The initial condition of being forced to buy a machine at the beginning of period 1 is incorporated in the construction of the network. In a like manner, the final condition of selling the machine at the end of the planning horizon requires that all chains terminate at node 4. Models of this type are called *finite-planning-horizon models* and are generally sensitive to the number of periods included in the planning horizon. Shortest-route problems are considered in Sections 2.3 and 2.4.

2.1.2 Project Planning

An important class of network problems centers around the planning and scheduling of large projects, such as missile countdowns, research and development projects, and construction management projects. For purposes of this model, a project is envisioned to be a set of tasks and precedence relationships among the tasks. Each task has a known duration d_i. In addition, the precedence relationships require that a set of tasks must be completed before some other job can be started. For example, suppose that there are five tasks and that 1 precedes 3; 1 and 2 precede 4; and 1, 2, 3, and 4 precede 5. This can be represented by the project network in Figure 2.2. A dummy node is introduced to represent the start of the project, and another is used to represent the project completion. The other nodes represent the beginning of each of the five tasks. An arc (i, j) with length d_i is included in the network if task i must immediately precede task j. Since 1 precedes 3, and 3 precedes 5, the arc (1, 5) is implied and is not included. The longest chain from the start node to any other node represents the earliest start time for the task

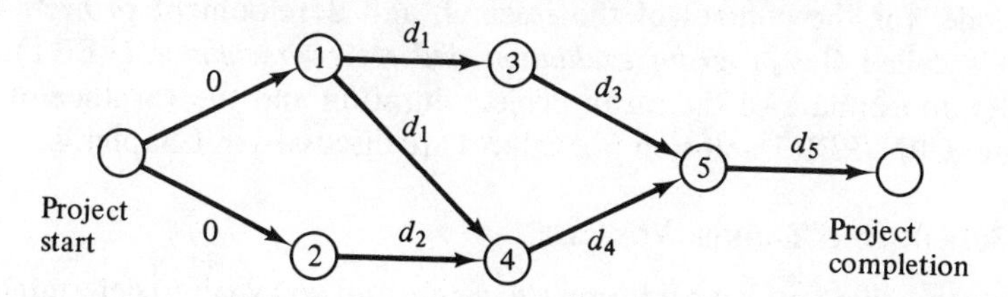

Figure 2.2. Activity-on-node representation of a project network.

beginning at the given node. Therefore, the longest chain from the start node to the completion node represents the minimum project duration.

When the network has circuits with positive lengths, the problem of finding the longest acyclic chain is by no means simple. Fortunately, the project network under consideration is circuitless, and the determination of its longest chain is not difficult.

The previous representation is called an *activity-on-node model*, since the beginning of each task is uniquely identified with a node. An alternative network representation, called an *activity-on-arc model*, represents each task by an arc. This representation is illustrated in Figure 2.3 for the example of Figure 2.2. The number adjacent to each arc is a label used for task identification.

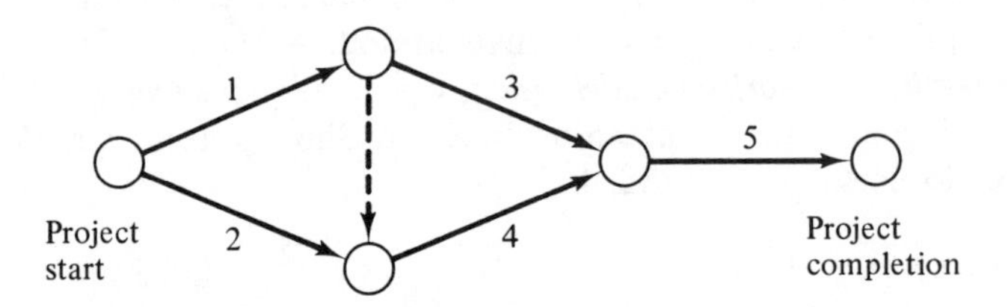

Figure 2.3. Activity-on-arc representation of a project network.

The dashed arc represents a dummy task of zero duration and is included to ensure that task 1 precedes task 4. The nodes in this network can be interpreted as "events," such as the completion of the basement in a building project. In general, an event can be interpreted as a well-defined occurrence in time. As before, the longest chain from start to completion in the project network is the earliest project completion time.

The method of project planning using activity network representations is called the *critical path method* (CPM), originally developed by Du Pont and Remington Rand to control the maintenance of chemical plants. The longest chain is called a critical path since the delay for any task on that path will result in a delay in the project completion. The same approach was developed independently for the Navy Polaris program as a stochastic duration model for the control of the research and development project. This model is called the *program evaluation and review technique* (PERT) and provides an estimate of the mean project duration and the variance of this estimate. CPM/PERT solution procedures are discussed in Chapter 4.

2.1.3 Scheduling Tanker Voyages

Suppose that you have a tramp steamship and you wish to determine the best schedule of ports to visit. Consider a network defined with each port as a node and the arcs (i, j) having costs c_{ij} resulting from transporting a shipload from port i to port j with transit times t_{ij}. The transit times include the time necessary to load the ship at node i, travel from node i to node j, and unload at node j. If profits are treated as negative costs, the directed cycle with a minimum $(\sum c_{ij})/(\sum t_{ij})$ will maximize the average profit per unit time.

In Figure 2.4 the optimal cycle, which represents the desired shipping schedule, is (1, 3), (3, 4), (4, 1), with a ratio of

$$\frac{3+2-4}{1+2+3}=\frac{1}{6}$$

The numbers adjacent to the arcs are the cost and time c_{ij}, t_{ij}. If the t_{ij}'s are all nonnegative, this problem is solved by methods similar to those employed in the shortest-route algorithms.

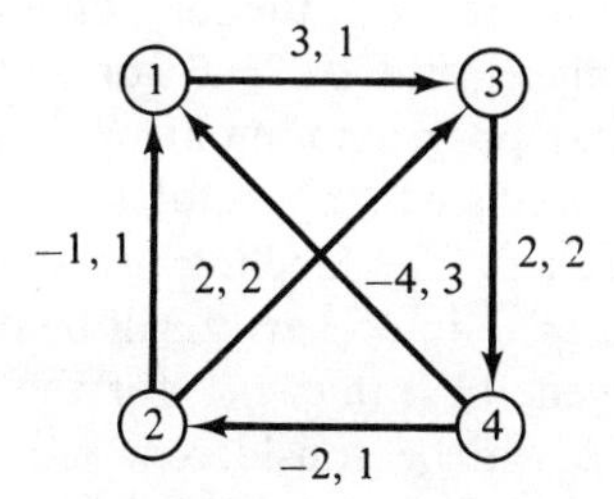

Figure 2.4. Steamship-port scheduling network.

2.1.4 The Maximum-Flow and Minimum-Cost Flow Problems

The preceding examples have been reduced to the selection of a single chain or cycle that optimizes a suitable objective function. In these problems arcs can be interpreted as a device to ship 1 unit of flow from the source to the sink through the nodes of a network. The arc parameter or distance is the generalized cost of using the arc. Usually, these models are referred to as *distance networks*.

A more general interpretation of an arc is that of a transportation link used for shipping multiple units of flow. In this case, the arc parameter represents the maximum amount of flow per unit time that can be shipped along the arc at any time. The parameter is called the *capacity* of the arc, and the network is said to be a *capacitated* flow network.

Let $G = (\mathbf{N}, \mathbf{A})$ be a directed network with $\mathbf{N} = \{1, 2, \ldots, n\}$, arcs in set $\mathbf{A}$, and let U_{ij} be the capacity of arc (i, j). For notational convenience, it will be assumed that node 1 is the source and node n the terminal.

In Section 1.3, an integer-valued function f_{ij} defined on $\mathbf{A}$ was said to be a flow for network G if it satisfies the constraints (1-1), (1-2), and (1-3).

This flow is said to have a value V if $\sum_j f_{sj} = \sum_j f_{jt} = V$. The maximum-flow problem consists of finding the largest feasible value of V, and can be formulated as follows:

$$\text{maximize } V \tag{2-1}$$

subject to

$$\sum_j f_{ij} - \sum_j f_{ji} = \begin{cases} V, & i = 1 \\ O, & i \neq 1, \quad i \neq n \\ -V, & i = n \end{cases} \tag{2-2}$$

$$0 \leq f_{ij} \leq U_{ij}, \qquad (i, j) \in \mathbf{A} \tag{2-3}$$

In Eqs. (2-2), the summations are for all the nodes for which f_{ij} is defined. For integer-valued arc parameters U_{ij}, the foregoing model is totally unimodular, and hence f_{ij} is integer in the optimal solution. Although this model can be solved using a linear programming algorithm, it is more efficient to solve it using a labeling procedure that generates a sequence of flow increases until the maximum is reached. The labeling procedure will be discussed in Section 2.14. The concept of unimodularity will be discussed in Section 2.2.

To provide a more general framework for the applications to follow, several additional features will be considered. A linear cost coefficient c_{ij} will be defined for each arc (i, j), and, in addition to the upper bound U_{ij}, a lower bound L_{ij} will be introduced. A supply $b_i \geq 0$ will be defined for each node i, with the understanding that a negative value for b_i will be interpreted as the flow demand associated with node i. In this model, a node i can be considered as a source node if $b_i > 0$, as an intermediate node if $b_i = 0$, and as a terminal node if $b_i < 0$. With these generalizations, the minimum-cost flow problem can be defined as finding a set of integer-valued arc flows f_{ij} that will

$$\text{minimize } \sum_i \sum_j c_{ij} f_{ij} \tag{2-4}$$

subject to

$$\sum_j f_{ij} - \sum_j f_{ji} = b_i, \qquad i \in \mathbf{N} \tag{2-5}$$

$$L_{ij} \leq f_{ij} \leq U_{ij}, \qquad (i, j) \in \mathbf{A} \tag{2-6}$$

Although this model can be solved by the simplex algorithm, a more efficient solution procedure has been developed by exploiting the network structure of the model and its relationship to the complementary slackness condition of linear programs. The procedure, well known by the name "out-of-kilter method" [15], is discussed in Chapter 3.

2.1.5 The Transportation Model

The *transportation* problem was one of the first network flow problems. It was first solved by Hitchcock [29] in 1941, and since then has been applied to many types of shipping and distribution problems. It is undoubtedly the most widely used network model. The model and its solution methodology are discussed in Section 2.10.

The problem can best be described in terms of shipments from plants to warehouses. Suppose that there are m plants and n warehouses. Each plant has a supply s_i, $i = 1, 2, \ldots, m$; and each warehouse creates a demand d_j, $j = 1, 2, \ldots, n$. The problem of minimizing total shipping costs is equivalent to finding x_{ij} values such that the following model will be optimized:

$$\text{minimize} \sum_i \sum_j c_{ij} x_{ij} \tag{2-7}$$

subject to

$$\sum_j x_{ij} = s_i, \qquad i = 1, 2, \ldots, m \tag{2-8}$$

$$\sum_i x_{ij} = d_j, \qquad j = 1, 2, \ldots, n \tag{2-9}$$

$$x_{ij} \geq 0, \qquad \begin{array}{l} i = 1, 2, \ldots, m; \\ j = 1, 2, \ldots, n \end{array} \tag{2-10}$$

In Eqs. (2-8) and (2-9) it is assumed that $\sum_i s_i = \sum_j d_j$ in order for the constraints to be consistent. This model can be described in terms of a network by letting the plants and warehouses be the nodes and the feasible shipping routes be the arcs. Under this interpretation, a route linking plant i with warehouse j is represented by a directed arc (i, j). This is depicted in Figure 2.5 for $m = 2$ and $n = 3$. Note that the transportation model is a special

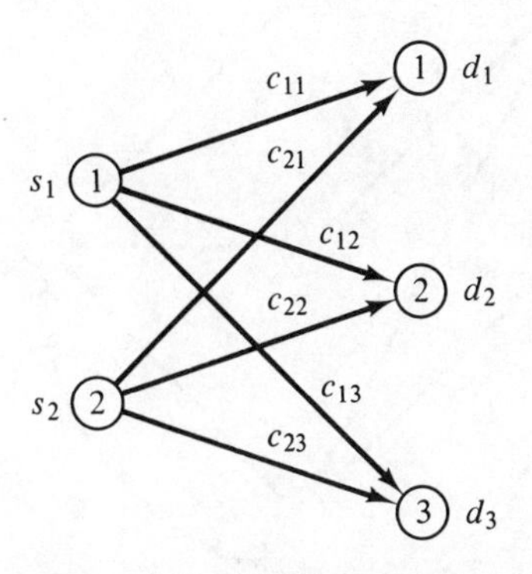

Figure 2.5. Network of a transportation problem.

case of the minimum-cost flow problem formulated in Section 2.1.4. In this case, $b_j = -d_j$, $b_i = s_i$, $L_{ij} = 0$, and $U_{ij} = \infty$.

A special case of the transportation problem where $m = n$, $s_i = 1$, and $d_j = 1$ is called the *assignment* problem, because it is typically described in terms of finding an optimal assignment of workers to jobs. The assignment model and its solution methodology are discussed in Section 2.12.

2.1.6 A Caterer Problem

A caterer must provide d_j napkins on each of n successive days. He can buy new napkins at α cents each, or have his dirty napkins laundered. He has two types of laundry available, slow and fast. For purposes of illustration, assume that the slow laundry takes 2 days and costs b cents per napkin and the fast laundry takes 1 day and costs c cents per napkin, where $c > b$. The problem is to select a purchasing and laundry policy to meet the demands at a minimal cost. This problem has a network formulation if we define

p_j as the number of napkins purchased for day j.
s_j as the number of napkins sent to the slow laundry on day j.
q_j as the number of napkins sent to the fast laundry on day j.
h_j as the number of soiled napkins held over on day j.

A network for this problem with $n = 3$ is given in Figure 2.6, where the variables are indicated adjacent to their respective arcs and the supplies are given adjacent to the nodes. The dashed arc is used to complete the supply–demand balance. The unique aspect of this problem is that the demand for clean napkins d_j also works as a supply of dirty napkins at

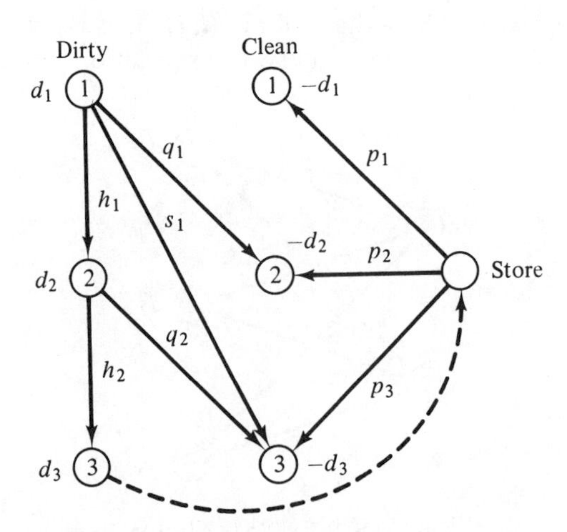

Figure 2.6. Napkin caterer problem network.

the end of each period. After some further manipulation, this problem can be solved as a transportation problem.

Although this problem seems frivolous when described in terms of napkins, the original application was the scheduling of equipment overhaul in a military repair facility. In that context the demands d_j are the known or forecast requirements to repair a particular type of equipment. The slow and fast service might correspond to doing the repairs at the local repair facility and sending the equipment to a central repair facility, respectively.

2.1.7 Employment Scheduling

The following model can be used to determine an employment policy that properly balances the costs of hiring and firing, with the expense of having some idle employees on the payroll for a short period of time. It is assumed that the demand for labor is deterministic but not uniform throughout a known planning horizon.

Assume that the minimum labor requirement D_j is known for each period j. Let x_{ij} be the number of people hired at the beginning of period i and fired at the end of period $j - 1$, and let c_{ij} be the cost per employee associated with x_{ij}. With these definitions, the available force in period j is

$$\sum_{r \leq j} \sum_{t > j} x_{rt}$$

or

$$\sum_{r=1}^{j} \sum_{t=j+1}^{n} x_{rt}$$

If the excess available manpower is defined to be s_j, then the work-force scheduling problem for $n - 1$ periods is to find nonnegative integers x_{ij} and s_j that optimize the mathematical model formulated in Eqs. (2-11) to (2-14):

$$\text{minimize} \sum_{r=1}^{n-1} \sum_{t=r+1}^{n} c_{rt}\, x_{rt} \qquad \text{(2-11)}$$

subject to

$$\sum_{r=1}^{j} \sum_{t=j+1}^{n} x_{rt} - s_j = D_j, \qquad j = 1, \ldots, n-1 \qquad \text{(2-12)}$$

$$s_j \geq 0 \qquad \text{(2-13)}$$

$$x_{ij} \geq 0, \text{ and integer} \qquad \text{(2-14)}$$

This problem does not have an obvious network interpretation and it is only after some manipulation that the network structure becomes evident. The procedure will be illustrated with the following example for $n = 4$. The corresponding constraints are given in Eqs. (2-15) to (2-17):

$$x_{12} + x_{13} + x_{14} \qquad - s_1 = D_1 \quad (2\text{-}15)$$

$$x_{13} + x_{14} + x_{23} + x_{24} \qquad - s_2 = D_2 \quad (2\text{-}16)$$

$$x_{14} \qquad + x_{24} + x_{34} \qquad - s_3 = D_3 \quad (2\text{-}17)$$

Subtracting Eq. (2-15) from Eq. (2-16), Eq. (2-16) from Eq. (2-17), and writing Eq. (2-17) as a redundant equation after multiplying by -1, we obtain the following system of equations:

$$\begin{aligned}
x_{12} + x_{13} + x_{14} \qquad\qquad\qquad - s_1 \qquad &= D_1 \\
-x_{12} \qquad + x_{23} + x_{24} \qquad + s_1 - s_2 \qquad &= D_2 - D_1 \\
- x_{13} \qquad - x_{23} \qquad + x_{34} \qquad + s_2 - s_3 &= D_3 - D_2 \\
- x_{14} \qquad - x_{24} - x_{34} \qquad + s_3 &= -D_3
\end{aligned}$$

An examination of this set of equations reveals that each variable appears exactly twice in the equations, once with a $+1$ coefficient and once with *a* $-$ 1 coefficient. This pattern of constraint coefficients has special significance, and the implications of this structure are discussed fully in Section 2.2. The network for this example is given in Figure 2.7, where the variables are shown next to the arcs and the supplies are shown next to the nodes. This structure is very much like that of the equipment replacement problem, with the exception of the arcs s_j.

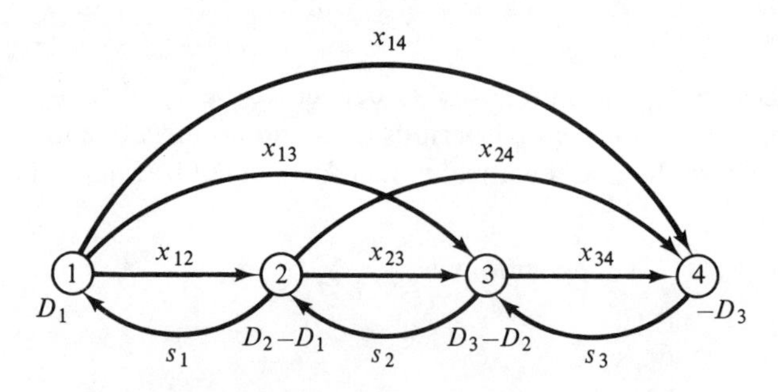

Figure 2.7. Employment scheduling network.

2.1.8 A Fleet Scheduling Problem

A problem that often arises is the scheduling of vehicles to accomplish the shipment of some commodity from points of supply to points of demand. One specific case, the tanker scheduling problem, is concerned with determining the minimum size and optimal routing of a fleet of tankers to achieve a prescribed schedule of deliveries.

Suppose that you are the owner of a tramp steamship company dealing with the delivery of some perishable goods. Several potential customers have announced that they have shipments to be made. Because the cargo is perishable, the customers have given precise shipment dates. If your company is unable to deliver them on the dates desired, your competitor will carry the cargo.

Four shipments (each is a full shipload) are being considered. The information for each shipment is shown in Table 2.1. This is to be interpreted to

Table 2.1. DATA FOR A FLEET SCHEDULING PROBLEM

Shipment	*Origin*	*Destination*	*Delivery Date*	*Profit*
1	Port *A*	Port *C*	3	10
2	Port *A*	Port *C*	8	10
3	Port *B*	Port *D*	3	3
4	Port *B*	Port *C*	6	4

mean that there is a shipload available at port A destined for port C that must be delivered (unloaded) on day 3. The profit associated with each shipment is determined from the revenue and the operating cost directly attributable to that shipment. In addition to the operating cost, there is a fixed charge of 5 to bring a ship into service. The fixed charge might reflect the company's overhead or the costs of recruiting a new crew.

The transit times (including allowance for loading and unloading) are as follows:

Origin \ *Destination*	*C*	*D*
A	3	2
B	2	3

The return times (traveling in ballast without a cargo) are:

Destination \ *Origin*	*A*	*B*
C	2	1
D	1	2

There are several questions of interest for this problem. Which shipments should we make and how many ships should we use? How many ships are necessary to make all of the deliveries at the prescribed times? These questions can be answered by constructing a network to represent all of the possible shipping schedules. Each of the nodes of this network corresponds to a port and a date. Using this convention, there will be a node for each time and place a shipment is delivered and for each time and place a ship starts to load cargo.

For convenience in constructing the network, a time scale has been placed below the network given in Figure 2.8. The darkened arcs correspond to the four shipments listed earlier. The darkened arcs have a capacity of 1, and all others have an infinite capacity. The numbers adjacent to the arcs are the nonzero costs.

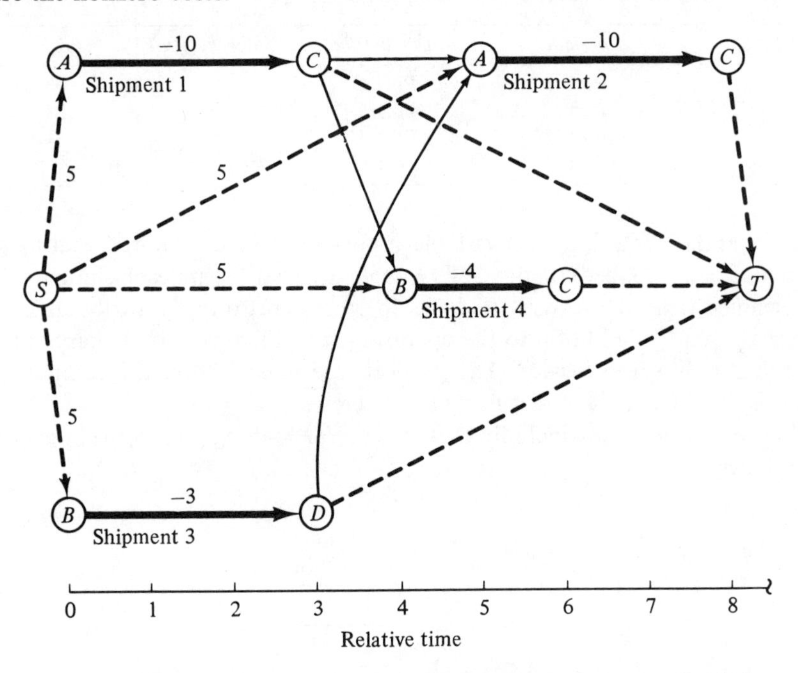

Figure 2.8. Tanker ship problem. Heavy arrows represent loaded trips to ports (nodes). Solid light arrows represent empty trips to pick up loads. Dashed lines represent the cost of putting ships into service (node *S*), and taking them out (node *T*).

Once a ship has completed unloading a shipment, it is free to go to some other port and commence loading the next shipment. Since the return time from port *C* to port *B* is one day, an arc is included from port *C* at time 3 to port *B* at time 4. The remaining arcs from the end of one darkened shipment arc to the beginning of another darkened shipment arc correspond to the other routings of an empty ship from a delivery to a pickup point for another shipment.

To complete the network, two additional nodes are added. A source node S is included at the beginning for the ships to start from, and a sink node T is included at the end of the time scale to provide a place for every ship to return to. The dashed arcs from S correspond to the first use of a ship, and the dashed arcs into T correspond to removing a ship from service.

Each chain from source to sink corresponds to a feasible schedule for a single ship. The earlier questions can be answered by solving a minimal-cost flow problem. A cost of 5 units is placed on each arc out of S to include the fixed charge for the first ship usage, and the shipment arcs have a cost equal to the negative of the profit from that shipment. All other arc costs are zero. The shipment arcs have a capacity of 1, since there is only one shipment of each type. All other arcs have an infinite capacity.

If a fleet of size M is to be used, the minimal-cost flow with a total of M units of flow will give the optimal ship schedules and also determine which shipments should be carried. If the fleet size is to be selected to maximize profit, the problem can be solved parametrically to give a profit/fleet size profile. This is tabulated in Table 2.2 for our example.

Table 2.2. PARAMETRIC SOLUTIONS FOR THE FLEET SCHEDULING PROBLEM

Fleet Size	*Shipment Schedule*	*Cost*	*Total Profit*
1	Ship 1: 1, 2	−15	15
2	Ship 1: 1, 4	−9	
	Ship 2: 3, 2	−8	17

The shipping schedule can be constructed from the network or from the original data. In the example the second ship delivering shipments 3 and 2 starts to load shipment 3 at port B at time zero and finishes unloading the delivery at D at time 3; travels empty to port A; starts to load shipment 2 at time 5; and finishes unloading the delivered shipment at port C at time 8.

The problem described here has also been used to determine optimal commercial bus routing once a given set of revenue-producing trips has been determined. The costs correspond to the "deadhead" trips when a bus is empty and is being relocated to the beginning of a revenue-producing trip.

2.1.9 A Production Planning Model (Smythe and Johnson)

Consider a manufacturer faced with the problem of meeting the demand for a product over the next T periods [49]. We shall denote the expected demands during this interval by $d_1, d_2, \ldots, d_T$, where d_j is the expected demand in the jth period. Suppose that the manufacturer can obtain this item from L different production or procurement sources. Let B_{ij} be the maximum number of units that can be obtained from source i in period j.

Let x_{ij}, $i = 1, 2, \ldots, L$, $j = 1, \ldots, T$, be the number of units to be obtained from source i in period j. Thus the x_{ij} are the decision variables. Let c_{ij} be the cost of obtaining one unit of product from source i in period j, and use c_s to denote the cost of carrying 1 unit of product in inventory from one period to the next. It is assumed that the capacity of the warehouse poses no restrictions on the number of items that can be held in inventory at any time. Finally, we define I_0 to be the initial inventory at the start of the planning horizon.

In formulating the model, it will be helpful to have an expression for the inventory at the end of any period t. Denoting this inventory by I_t, we have the following:

$$I_t = I_0 + \sum_{j=1}^{t} \sum_{i=1}^{L} x_{ij} - \sum_{j=1}^{t} d_j, \qquad t = 1, 2, \ldots, T$$

At this point we might well note that if we choose the x_{ij} such that $I_t \geq 0$ for all t, we shall ensure that the demand is met in each period.

The objective is stated as follows. Choose nonnegative x_{ij} such that the function

$$Z = \sum_{j=1}^{T} \sum_{i=1}^{L} c_{ij}x_{ij} + c_s \sum_{t=1}^{T} \left(I_0 + \sum_{j=1}^{t} \sum_{i=1}^{L} x_{ij} - \sum_{j=1}^{t} d_j\right) \tag{2-18}$$

is minimized. The first term in the objective function (2-18) represents the cost of obtaining the product, and the last term denotes the inventory carrying costs $c_s \sum_{t=1}^{T} I_t$ which are incurred during the planning interval.

The constraints are of two types. The first type ensures that the available supplies are not exceeded. Thus, we have LT constraints of the form

$$x_{ij} \leq B_{ij}, \qquad \text{for } i = 1, 2, \ldots, L \text{ and } j = 1, 2, \ldots, T \tag{2-19}$$

The second type of constraint ensures that the demand in each period will be met. We have T constraints of the form

$$I_0 + \sum_{j=1}^{t} \sum_{i=1}^{L} x_{ij} - \sum_{j=1}^{t} d_j \geq 0, \qquad \text{for } t = 1, 2, \ldots, T \tag{2-20}$$

As a numerical illustration, suppose that three periods are under consideration and that two production alternatives, regular time and overtime, are available in each period. The data are provided in Table 2.3.

In order to represent this problem as a flow network, the network components must be identified. Assume that the nodes of a network represent the sources of the product, the period in which the product is needed, and the demand required. There are seven possible sources: (a) initial inventory, (b) three regular time production runs, and (c) three overtime production runs. Any arc of the network will represent four characteristics: (1) the direction

Table 2.3. PRODUCTION PLANNING NUMERICAL EXAMPLE

Period	Capacities (units)		Unit Production Cost		Anticipated Demand
	Regular Time	Overtime	Regular Time	Overtime	
1	100	20	$14	$18	60
2	100	10	17	22	80
3	60	20	17	22	140

of allowable flow, (2) the minimum amount of flow required through the arc, (3) the maximum amount of flow possible through the arc, and (4) the cost per unit flow through the arc. Each arc will connect a pair of nodes i and j and will be labeled as follows:

$$(i) \xrightarrow{(L,\,U,\,C)} (j)$$

where $L =$ minimum amount of allowable flow
$U =$ maximum allowable flow
$C =$ cost per unit flow

The network optimization problem is to determine the minimum-cost flow pattern which will guarantee that all demands in periods 1, 2, and 3 will be met from the limited sources of supply.

The network representation of this problem is shown in Figure 2.9. Consider arc (I, P_1). The arc parameter (0, 15, 0) indicates that no use of the initial inventory in period 1 is required ($L = 0$), but 15 units are available ($U = 15$) at zero cost ($C = 0$). Similarly, the arc (I, P_2) indicates that the initial inventory of 15 units need not be used, but it will cost one dollar ($C = 1$) per unit if it is used in period 2. All arcs from the source nodes (I, $R_1, O_1, R_2, O_2, R_3, O_3$) to the period nodes ($P_1, P_2, P_3$) have similar interpretations. Finally, the arcs (P_1, D_1), (P_2, D_2), and (P_3, D_3) force the required demands to be satisfied. For example, arc (P_2, D_2) specifies that a minimum flow of 80 units is required ($L = 80$), a maximum flow of 80 units is allowed ($U = 80$), and the unit cost is zero ($C = 0$). Hence, arc (P_2, D_2) forces 80 units to be supplied at node D_2. Arcs (P_1, D_1) and (P_3, D_3) place similar requirements on production in periods 1 and 3, respectively.

This problem can be solved as a minimum-cost network flow problem using either the transportation algorithm of Section 2.10 or the out-of-kilter algorithm in Chapter 3. A complete treatment of this class of problems is contained in Chapter 3.

2.1.10 Conclusions

The formulations previously described were selected to illustrate the diversity of problems that have been modeled using network structures. Primary emphasis has been on network flow representations that are deter-

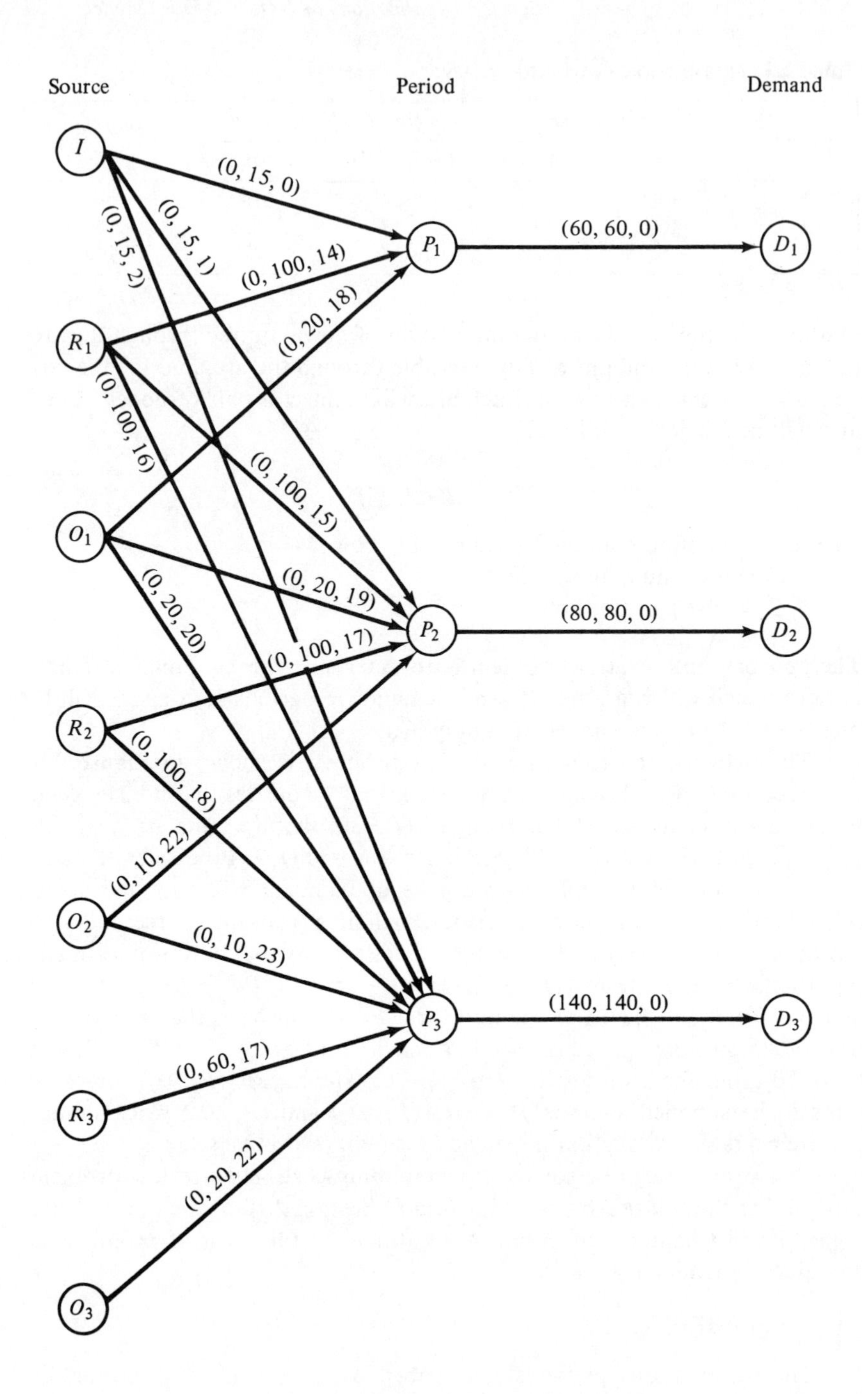

Figure 2.9. Production planning model.

ministic, or stochastic models that have a deterministic equivalent. The formulation work was done using static models. The dynamic characteristics were generally included by representing the time periods as nodes, or by letting each time–location combination be a node. Models of this type are very common, and can be used even if the resultant problem is too large to be solved via conventional linear programming methods.

The shortest-chain models and single-commodity flow models are easily solved. These models have a very special structure that reduces the problem to finding chains in a network. This has undoubtedly contributed to their popularity.

The problems described here were intended to provide an introduction to network models. They are admittedly vast simplifications of the real problems. However, they are useful because they are easily understood and solved. In addition, they provide a visual aid to the solution of each problem. In the sections that follow, we concern ourselves with techniques and algorithmic procedures that can be used to solve these and other problems.

2.2 Linear Programming and Its Relationship to Network Flows

A fundamental fact which will be established at this point is that most network-flow problems studied in the remainder of this book can be formulated as linear programming (LP) models. A natural extension to this fact is why we wish to proceed further than an LP formulation. First, network-flow models of real-world problems are of a special structure which can often be exploited to derive efficient solution procedures rather than seeking a solution via the simplex mechanism. Second, these specialized algorithms often enable us to solve extremely large linear programs in a fraction of the normal computation time. Third, the very fact that we can address mathematical formulations through node–arc representations often aid us in both the presentation and solution of real-world problems.

Since the special structure of network-flow representations as linear programming problems plays a central role in efficient algorithmic development, it is well worth our time to explore this structure. In particular, it is advantageous to predict in advance when linear programming machinery will yield an optimal solution that is all-integer. This section will provide several useful theorems which can be used to predict when linear programs will yield integer solutions. If we knew this in advance, one of two solution procedures could be employed: (1) specialized linear programming machinery could be developed, which is guaranteed to yield an optimal integer solution; and (2) iterative procedures using the network structure in a direct manner could be developed, using additive arithmetic, whose optimal integer solutions would correspond to those of linear programming.

The following discussion is based on a presentation due to Garfinkel and Nemhauser [19]. Consider a network with α source nodes, β sink nodes, and ϕ intermediate nodes through which a single commodity is to be shipped from sources to sinks. Each arc in the network is characterized by an upper bound U_{ij} on the arc flow f_{ij}, and a cost per unit flow c_{ij}. The network can be represented as a directed graph $G = (\mathbf{N}, \mathbf{A})$, where $\mathbf{N} = \mathbf{N}_\alpha \cup \mathbf{N}_\beta \cup \mathbf{N}_\phi$, and $\mathbf{N}_\alpha$, $\mathbf{N}_\beta$, and $\mathbf{N}_\phi$ are the sets of source nodes, terminal nodes, and intermediate nodes, respectively. The minimum-cost flow problem, introduced in Section 2.1.4, can be reformulated mathematically as follows:

$$\text{minimize} \sum_i \sum_j c_{ij} f_{ij} \tag{2-21}$$

subject to

$$\sum_j f_{ij} - \sum_j f_{ji} \leq a_i \qquad i \in \mathbf{N}_\alpha \tag{2-22}$$

$$\sum_j f_{ij} - \sum_j f_{ji} = 0 \qquad i \in \mathbf{N}_\phi \tag{2-23}$$

$$\sum_j f_{ji} - \sum_j f_{ij} \geq b_i \qquad i \in \mathbf{N}_\beta \tag{2-24}$$

$$0 \leq f_{ij} \leq U_{ij} \qquad (i, j) \in \mathbf{A} \tag{2-25}$$

The problem can be stated as minimizing the total cost of commodity flows [Eq. (2-21)] through the network. Equations (2-22) specify that no more than $\sum_i a_i$ units are available, and equations (2-24) specify that at least $\sum_i b_i$ units must be delivered. Equations (2-25) bound arc flows. Equations (2-23) represent the conservation-of-flow requirements discussed in Chapter 1 for all intermediate nodes. Four important problems introduced in Section 2.1 can be represented as special cases of the general formulation given by Eqs. (2-21) to (2-25). The four special cases are (1) the transportation problem (Section 2.1.5), (2) the assignment problem (Section 2.1.5), (3) the maximum-flow problem (Section 2.1.4), and (4) the shortest-path problem (Section 2.1.1).

Case 1: The transportation problem. In the special case where commodities are shipped directly from source nodes to sink nodes, there are no intermediate nodes. If a fixed supply is to satisfy a forced demand at minimum cost, the problem is a classical capacitated transportation problem stated as follows:

$$\text{minimize} \sum_i \sum_j c_{ij} f_{ij} \tag{2-26}$$

subject to

$$\sum_j f_{ij} \leq a_i \qquad i \in \mathbf{N}_\alpha \tag{2-27}$$

$$\sum_j f_{ji} \geq b_i \qquad i \in \mathbf{N}_\beta \tag{2-28}$$

$$0 \leq f_{ij} \leq U_{ij} \qquad (i, j) \in \mathbf{A} \tag{2-29}$$

If there are no forced upper bounds of zero on one or more arcs connecting sources to sources or sinks to sinks, this is the *transshipment* problem.

Case 2: The assignment problem. In the special case where all upper bounds are relaxed such that $U_{ij} = \infty$, for $i, j \in \mathbf{N}$, and $a_i = 1$ for $i \in \mathbf{N}_\alpha$ and $b_i = 1$ for $i \in \mathbf{N}_\beta$, the formulation given above represents an assignment problem:

$$\text{minimize} \sum_i \sum_j c_{ij} f_{ij} \tag{2-30}$$

subject to

$$\sum_j f_{ij} \leq 1 \qquad i \in \mathbf{N}_\alpha \tag{2-31}$$

$$\sum_j f_{ji} \geq 1 \qquad i \in \mathbf{N}_\beta \tag{2-32}$$

$$f_{ij} \geq 0 \qquad (i, j) \in \mathbf{A} \tag{2-33}$$

Note that the constraints will be satisfied as strict equalities if there are as many sink nodes in the set $\mathbf{N}_\beta$ as source nodes in the set $\mathbf{N}_\alpha$. A necessary condition for a feasible solution is that the number of sources must be equal to or greater than the number of sinks.

Case 3: The maximum-flow problem. Consider the case where there is only a single source node, $i = 1$; a single sink node, $i = n$; unlimited supply, $a_1 = \infty$; and no forced demands, $b_n = 0$. If there is a capacity constraint on each arc and the costs per unit flow are defined as $c_{in} = -1$ for $i \neq n$ and zero otherwise, the problem discussed in case 1 becomes that of maximizing the total amount of flow into the sink node n:

$$\text{maximize} \sum_{i \neq n} f_{in} \tag{2-34}$$

subject to

$$\sum_j f_{ij} - \sum_j f_{ji} = 0 \qquad i \neq 1, i \neq n \tag{2-35}$$

$$0 \leq f_{ij} \leq U_{ij} \qquad (i, j) \in \mathbf{A} \tag{2-36}$$

Case 4: The shortest-path problem. Assume that arc parameter c_{ij} represents cost or time required to transmit 1 unit of flow from node i to node j. As mentioned in Section 2.1.1, the problem is to determine the minimum cost (time) incurred in shipping 1 unit of flow from a source node to a designated sink node. Define $\alpha = 1$, $\beta = 1$, $\mathbf{N}_\alpha = \{1\}$ and $\mathbf{N}_\beta = \{n\}$. Further, define $a_1 = 1$ and $b_n = 1$. Since the total supply is exactly 1 unit, the arc bounds are $0 \leq f_{ij} \leq 1$ and need not be considered since their violation is impossible. Finally, assume that there are no negative cycles in the network.

If all arcs are directed, the problem is one of determining the shortest chain from node 1 (source) to node n (sink). Mathematically, this problem can be formulated as follows:

$$\text{minimize} \sum_i \sum_j c_{ij} f_{ij} \tag{2-37}$$

subject to

$$\sum_{j \in \mathbf{N}} f_{1j} \leq 1 \tag{2-38}$$

$$\sum_{j \in \mathbf{N}} f_{jn} \geq 1 \tag{2-39}$$

$$\sum_j f_{ij} - \sum_j f_{ji} = 0 \qquad i \neq 1, i \neq n \tag{2-40}$$

$$f_{ij} \geq 0 \qquad (i, j) \in \mathbf{A} \tag{2-41}$$

It should be noted that the inequality constraints will hold as strict equalities in the optimal solution.

The previous discussion has shown that the formulation of Eqs. (2-21) to (2-25) encompasses a wide variety of important network-flow problems. Algorithms to solve these and other relevant problems will be developed later in this chapter. It is now our purpose to show that any basic, feasible solution to the general formulation must be integer, given that a_i, b_i, and U_{ij} are integer. To prove this result, two important *definitions* will now be presented.

***Definition 1*:** Any square, integer matrix whose determinant is 0, +1, or −1 is called *unimodular*.

***Definition 2*:** If all square submatrices of an integer matrix **M** are unimodular, **M** is said to be *totally unimodular*.

The following system of linear inequality constraints (2-42) and (2-43) defines the feasible region for the elements of a nonnegative vector **F**. It will be assumed that the matrix **D** and the vector **B** are integer:

$$\mathbf{DF} \leq \mathbf{B} \tag{2-42}$$

$$\mathbf{F} \geq \mathbf{0} \tag{2-43}$$

These definitions relate to the following two theorems.

***Theorem 1*:** If the matrix **D** is totally unimodular, *every* basic solution to $\mathbf{DF} = \mathbf{B}$ and $\mathbf{F} \geqslant \mathbf{0}$ is an integer solution.

***Theorem 2*:** Consider an arbitrary integer matrix **D**. The following two conditions are equivalent with respect to matrix **D**.

a. **D** is totally unimodular.
b. All linear programming basic solutions to the set of constraints $\mathbf{DF} \leq \mathbf{B}$; $\mathbf{F} \geq \mathbf{0}$ are integer.

These two theorems are of central importance to our discussion, since they imply that if the constraint set of any linear programming problem is of the form $\mathbf{DF} \leq \mathbf{B}$, $\mathbf{F} \geq \mathbf{0}$, **D** totally unimodular, and **B** integer-valued, then any extreme point (basic) solution generated by linear programming will be integer. Of particular interest is the optimal solution, whose variables must also be integer. We will now show that the network-flow representation of Eqs. (2-21) to (2-25) has an optimal integer solution, when the parameters a_i, b_i, and U_{ij} are integer.

For convenience and clarity of exposition, consider an arbitrary network representation of Eqs. (2-21) to (2-25) with a single source node: $\alpha = 1$, $\mathbf{N}_\alpha = \{1\}$; a single sink node: $\beta = 1$, $\mathbf{N}_\beta = \{6\}$; and six intermediate nodes: $\phi = 6$, $\mathbf{N}_\phi = \{2, 3, 4, 5, 7, 8\}$.

The graphical representation of the sample problem to be considered is given in Figure 2.10.

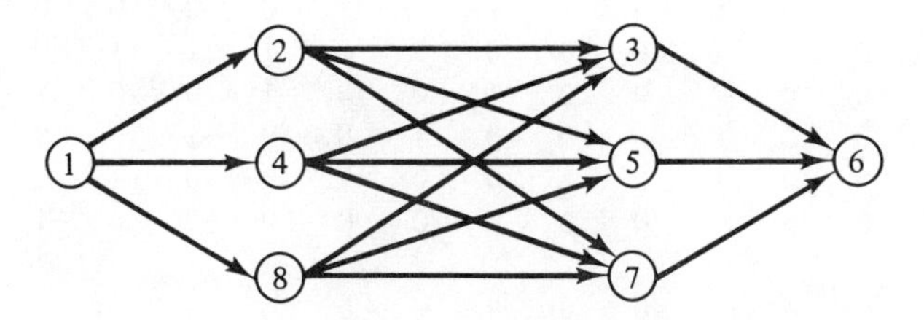

Figure 2.10. Flow network.

The problem can be mathematically formulated as follows:

$$\text{minimize} \sum_i \sum_j c_{ij} f_{ij}$$

subject to

$$\begin{aligned}
&f_{12} + f_{14} + f_{18} \leq a_1 \\
&f_{23} + f_{25} + f_{27} - f_{12} = 0 \\
&f_{43} + f_{45} + f_{47} - f_{14} = 0 \\
&f_{83} + f_{85} + f_{87} - f_{18} = 0 \\
&f_{36} - f_{23} - f_{43} - f_{83} = 0 \\
&f_{56} - f_{25} - f_{45} - f_{85} = 0 \\
&f_{76} - f_{27} - f_{47} - f_{87} = 0 \\
&f_{36} + f_{56} + f_{76} \geq b_6 \\
&0 \leq f_{ij} \leq U_{ij} \qquad (i, j) \in \mathbf{A}
\end{aligned}$$

To prove that any basic, feasible solution to this problem is all-integer, it is sufficient to show that the coefficient matrix for the general problem is totally unimodular. The constraint matrix can be represented by the following partition of the constraint coefficients:

$$\mathbf{D} = \left[\frac{\mathbf{Z}}{\mathbf{I}}\right] \tag{2-44}$$

The **Z** matrix is generated from the coefficients of the conservation of flow constraints and the supply–demand constraints. The **I** matrix is an identity matrix formed from the coefficients of the single-term bounding constraints. The objective is to prove the total unimodularity of the matrix **D**.

The submatrices **Z** and **I** of the coefficient matrix **D** defined in Eq. (2-44) can be written as follows:

$$\begin{array}{cc} & \begin{array}{ccccccccccccccc} f_{12} & f_{14} & f_{18} & f_{23} & f_{25} & f_{27} & f_{43} & f_{45} & f_{47} & f_{83} & f_{85} & f_{87} & f_{36} & f_{56} & f_{76} \end{array} \\ \mathbf{Z} = \begin{array}{c} 1\\2\\4\\8\\3\\5\\7\\6 \end{array} & \left[\begin{array}{ccccccccccccccc} 1 & 1 & 1 & 0 & 0 & 0 & 0 & 0 & 0 & 0 & 0 & 0 & 0 & 0 & 0 \\ -1 & 0 & 0 & 1 & 1 & 1 & 0 & 0 & 0 & 0 & 0 & 0 & 0 & 0 & 0 \\ 0 & -1 & 0 & 0 & 0 & 0 & 1 & 1 & 1 & 0 & 0 & 0 & 0 & 0 & 0 \\ 0 & 0 & -1 & 0 & 0 & 0 & 0 & 0 & 0 & 1 & 1 & 1 & 0 & 0 & 0 \\ 0 & 0 & 0 & -1 & 0 & 0 & -1 & 0 & 0 & -1 & 0 & 0 & 1 & 0 & 0 \\ 0 & 0 & 0 & 0 & -1 & 0 & 0 & -1 & 0 & 0 & -1 & 0 & 0 & 1 & 0 \\ 0 & 0 & 0 & 0 & 0 & -1 & 0 & 0 & -1 & 0 & 0 & -1 & 0 & 0 & 1 \\ 0 & 0 & 0 & 0 & 0 & 0 & 0 & 0 & 0 & 0 & 0 & 0 & -1 & -1 & -1 \end{array}\right] \end{array}$$

$$\mathbf{I} = \left[\begin{array}{ccccccccccccccc} 1 & 0 & 0 & 0 & 0 & 0 & 0 & 0 & 0 & 0 & 0 & 0 & 0 & 0 & 0 \\ 0 & 1 & 0 & 0 & 0 & 0 & 0 & 0 & 0 & 0 & 0 & 0 & 0 & 0 & 0 \\ 0 & 0 & 1 & 0 & 0 & 0 & 0 & 0 & 0 & 0 & 0 & 0 & 0 & 0 & 0 \\ 0 & 0 & 0 & 1 & 0 & 0 & 0 & 0 & 0 & 0 & 0 & 0 & 0 & 0 & 0 \\ 0 & 0 & 0 & 0 & 1 & 0 & 0 & 0 & 0 & 0 & 0 & 0 & 0 & 0 & 0 \\ 0 & 0 & 0 & 0 & 0 & 1 & 0 & 0 & 0 & 0 & 0 & 0 & 0 & 0 & 0 \\ 0 & 0 & 0 & 0 & 0 & 0 & 1 & 0 & 0 & 0 & 0 & 0 & 0 & 0 & 0 \\ 0 & 0 & 0 & 0 & 0 & 0 & 0 & 1 & 0 & 0 & 0 & 0 & 0 & 0 & 0 \\ 0 & 0 & 0 & 0 & 0 & 0 & 0 & 0 & 1 & 0 & 0 & 0 & 0 & 0 & 0 \\ 0 & 0 & 0 & 0 & 0 & 0 & 0 & 0 & 0 & 1 & 0 & 0 & 0 & 0 & 0 \\ 0 & 0 & 0 & 0 & 0 & 0 & 0 & 0 & 0 & 0 & 1 & 0 & 0 & 0 & 0 \\ 0 & 0 & 0 & 0 & 0 & 0 & 0 & 0 & 0 & 0 & 0 & 1 & 0 & 0 & 0 \\ 0 & 0 & 0 & 0 & 0 & 0 & 0 & 0 & 0 & 0 & 0 & 0 & 1 & 0 & 0 \\ 0 & 0 & 0 & 0 & 0 & 0 & 0 & 0 & 0 & 0 & 0 & 0 & 0 & 1 & 0 \\ 0 & 0 & 0 & 0 & 0 & 0 & 0 & 0 & 0 & 0 & 0 & 0 & 0 & 0 & 1 \end{array}\right]$$

Observe that the **Z** matrix is simply the node–arc incidence matrix defined and discussed in Chapter 1. In order to prove total unimodularity, the following theorems will be utilized.

Theorem 3 [28]: An integer matrix **M** is totally unimodular provided that the following conditions are satisfied:

a. Each entry in the matrix is either 0, +1, or −1.
b. Each column of the matrix contains no more than two nonzero entries.
c. The rows of matrix **M** can be partitioned into two sets, $\mathbf{E}_1$ and $\mathbf{E}_2$, such that:

   i. If any column contains two nonzero elements of the same sign, one element can be placed in set $\mathbf{E}_1$ and the other in set $\mathbf{E}_2$.
   ii. If any column contains two nonzero elements of opposite sign, both elements can be placed in the same set.

Theorem 4 [19]: If a totally unimodular matrix **M** is combined with an identity matrix **I** such that $\mathbf{P} = [\mathbf{M} \mid \mathbf{I}]$, then **P** is also totally unimodular.

Corollary: The transpose of a totally unimodular matrix is also totally unimodular.

Referring to either the Eqs. (2-21) to (2-25) or the numerical example, it is clear that the **Z** matrix contains only two nonzero entries in each column, and these entries are either +1, or −1, thus satisfying conditions (a) and (b) of Theorem 3. Condition (c) is satisfied by defining set $\mathbf{E}_1$ as containing all rows and set $\mathbf{E}_2$ empty. Hence, the matrix **Z** is totally unimodular. It follows from Theorem 4 that the matrix $\left[\frac{\mathbf{Z}}{\mathbf{I}}\right]$ is also totally unimodular; hence the existence of an all integer optimal solution is guaranteed by Theorems 1 and 2.

The significance of this development has undoubtedly been masked by the mathematical statements embodied in its presentation. The fundamental result in this development is that when a linear program assumes the form of Eqs. (2-21) to (2-25) and the elements of the constraint matrix exhibit the property of total unimodularity, then the *optimal* solution to the corresponding linear program is *integer*. In actual practice, one would rarely solve the linear programming problem directly, but rather would employ efficient, specialized network-flow algorithms or modified primal simplex rules to solve the equivalent network formulation. Hence, many linear programming problems can be solved as a network-flow problem. When this is possible, the computational advantages and simplicity of network solution procedures will often be spectacular. Such network formulations of real-world problems and the corresponding solution procedures motivate the developments that follow throughout the remainder of this book.

2.3 The Shortest-Route Problem—Dijkstra's Algorithm

A network-flow problem that is important from an applied standpoint is that of finding a path from the source node to the sink node which minimizes the total cost or time spent in the shipment of a given amount of flow along that path. Let us assume in a general formulation that associated with each arc (i, j) in a directed network is a generalized cost parameter, c_{ij}. Dummy or "free" arcs would carry a cost $c_{ij} = 0$, while impossible arcs might carry a cost $c_{ij} = \infty$. The problem discussed in this section can be interpreted as that of shipping 1 unit of flow at minimum cost from the source node s to the sink node t of a given network. Mathematically, this problem can be formulated as a linear program as indicated in Eqs. (2-45)–(2-49):

$$\text{minimize} \sum_i \sum_j c_{ij} f_{ij} \tag{2-45}$$

subject to

$$\sum_j f_{sj} - \sum_j f_{js} = 1 \tag{2-46}$$

$$\sum_j f_{ij} - \sum_j f_{ji} = 0, \qquad i \neq s, i \neq t \tag{2-47}$$

$$\sum_j f_{tj} - \sum_j f_{jt} = -1 \tag{2-48}$$

$$f_{ij} \geq 0 \tag{2-49}$$

The first constraint guarantees that 1 unit of flow leaves the source. The second set of constraints ensures the conservation of this unit of flow as it moves through the network. The third constraint specifies that 1 unit of flow arrives at the terminal node. The minimal path can be identified as the connected sequence of arcs (i, j) such that $f_{ij} = 1$. Rather than solve this linear programming problem we will use a special method, known as *Dijkstra's algorithm* [11], designed to take advantage of the particular structure of the model.

Let us assume that the cost or distance between any two nodes is a nonnegative number represented by c_{ij}. The algorithm assigns to all nodes a *label* which is either *temporary* or *permanent*. Initially, each node except the source node receives a temporary label which represents the direct distance between the source node and that node. The source node is assigned a permanent label of zero. Any node that cannot be reached directly from the source node is assigned a temporary label of ∞, while all others receive temporary labels of $c_{sj}, j \neq s$. When it is determined that a node must belong to the minimal path, its label becomes permanent. The algorithm operates on the simple logic that if a shortest path from node s to node j is known and node k belongs to this path, then the minimal path from s to k is the portion of the original path ending at node k. The algorithm starts with $j = s$ and successively resets j until $j = t$ and then the process is stopped.

2.3.1 An Iterative Procedure

For a given node j, the symbol δ_j is used to represent an estimate of the length of the shortest path from the source node s to node j. When this estimate cannot be further improved, we call it a "permanent label" and represent it by the symbol $\boxed{\delta_j}$. Otherwise, it is referred to as a "temporary label." Initially, only the source node is permanently labeled. Every other label is temporary and contains the direct distance from the source to the corresponding node. To identify the node that is closest to the source node, we select the minimum of the temporary labels and declare it permanent. From this point on the algorithm cycles in two phases until the sink node is permanently labeled.

1. Inspect the remaining temporary labeled nodes. Compare each *temporary label* already assigned with the *sum* of the last *permanent* label and the *direct distance* from the node with the permanent label to the node under consideration. The minimum of these two distances is defined as the new temporary label for that node. Note that if the old temporary label is still minimal, it will remain unchanged.
2. Select the minimum temporary label and declare it permanent. If node t is given a permanent label, the algorithm is terminated. Otherwise, return to step 1. The entire procedure can easily be done by employing a decision table where the columns represent the nodes in the network, the rows the steps in the iterative process, and the entries the permanent and temporary labels.

2.3.2 Example of Dijkstra's Algorithm

The purpose of this section is to illustrate the application of Dijkstra's algorithm on the network given in Figure 2.11. In this figure, s represents the source, t the sink, and the number c_{ij} on each arc (i, j) is the "length" or cost associated with the arc.

The algorithm is started by assigning the permanent label $\boxed{0}$ to the source node s, and temporary labels $\delta_j = c_{sj}$ to nodes $j = 1, 2, \ldots, 6, t$. That is, $\delta_1 = 0$, and $\delta_j = \infty$ for $j = 2, \ldots, t$. Since $\delta_1 = 0$ is the minimum of all temporary labels, node 1 receives a permanent label $\boxed{0}$.

Nodes 2 and 3 are directly connected to node 1, the last node given a permanent label. Note that $\delta_1 + c_{12} = 0 + 3 < \infty$ and $\delta_1 + c_{13} = 0 + 7 < \infty$; therefore, nodes 2 and 3 are given new temporary labels $\delta_2 = 3$ and $\delta_3 = 7$, respectively. Since $\delta_2 < \delta_3$, node 2 is permanently labeled with $\delta_2 = \boxed{3}$.

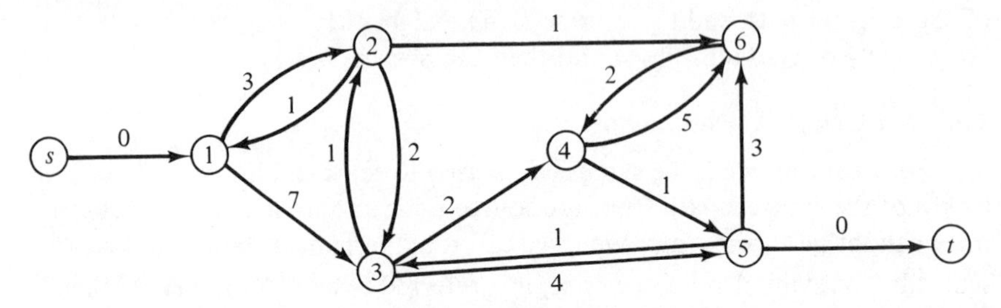

Figure 2.11. Sample problem network for Dijkstra's algorithm.

Nodes 3 and 6 are directly connected to node 2. Also, $\delta_2 + c_{23} = 3 + 2 = 5 < 7$ and $\delta_2 + c_{26} = 3 + 1 = 4 < \infty$. Therefore, the new temporary labels for nodes 3 and 6 are $\delta_3 = 5$ and $\delta_6 = 4$, respectively. Node 6 becomes permanently labeled since $\delta_6 < \delta_3$. That is, $\delta_6 = \boxed{4}$.

At this point, the temporary labels are $\delta_3 = 5$ and $\delta_4 = \delta_5 = \delta_t = \infty$. The last node given a permanent label is node 6, which is directly connected to only node 4. Note that $\delta_6 + c_{64} = 4 + 2 = 6 < \infty$. Hence, the new temporary label for node 4 is $\delta_4 = 6$. Since $\delta_3 = \min\{\delta_3, \delta_4, \delta_5, \delta_t\}$, node 3 receives a permanent label equal to $\boxed{5}$. Continuing in this fashion, the problem is terminated when node t is permanently labeled. The intermediate results for this example are summarized in Table 2.4.

Table 2.4. SUMMARY OF COMPUTATIONAL WORK FOR THE EXAMPLE USING DIJKSTRA'S METHOD

Step \ *Node*	s	1	2	3	4	5	6	t
0	$\boxed{0}$	∞	∞	∞	∞	∞	∞	∞
1	$\boxed{0}$	0	∞	∞	∞	∞	∞	∞
2	$\boxed{0}$	$\boxed{0}$	∞	∞	∞	∞	∞	∞
3	$\boxed{0}$	$\boxed{0}$	3	7	∞	∞	∞	∞
4	$\boxed{0}$	$\boxed{0}$	$\boxed{3}$	7	∞	∞	∞	∞
5	$\boxed{0}$	$\boxed{0}$	$\boxed{3}$	5	∞	∞	4	∞
6	$\boxed{0}$	$\boxed{0}$	$\boxed{3}$	5	∞	∞	$\boxed{4}$	∞
7	$\boxed{0}$	$\boxed{0}$	$\boxed{3}$	5	6	∞	$\boxed{4}$	∞
8	$\boxed{0}$	$\boxed{0}$	$\boxed{3}$	$\boxed{5}$	6	∞	$\boxed{4}$	∞
9	$\boxed{0}$	$\boxed{0}$	$\boxed{3}$	$\boxed{5}$	6	9	$\boxed{4}$	∞
10	$\boxed{0}$	$\boxed{0}$	$\boxed{3}$	$\boxed{5}$	$\boxed{6}$	9	$\boxed{4}$	∞
11	$\boxed{0}$	$\boxed{0}$	$\boxed{3}$	$\boxed{5}$	$\boxed{6}$	7	$\boxed{4}$	∞
12	$\boxed{0}$	$\boxed{0}$	$\boxed{3}$	$\boxed{5}$	$\boxed{6}$	$\boxed{7}$	$\boxed{4}$	∞
13	$\boxed{0}$	$\boxed{0}$	$\boxed{3}$	$\boxed{5}$	$\boxed{6}$	$\boxed{7}$	$\boxed{4}$	7
14	$\boxed{0}$	$\boxed{0}$	$\boxed{3}$	$\boxed{5}$	$\boxed{6}$	$\boxed{7}$	$\boxed{4}$	$\boxed{7}$

As can be seen in Table 2.4, the permanent label of node t is $\delta_t = \boxed{7}$. Therefore, the length of the shortest path from node s to node t is equal to 7. This path is found by identifying arcs for which the difference in the permanent labels of two directly connected nodes is exactly equal to the length of the arc connecting the nodes. In symbols, if $\boxed{\delta_i}$ and $\boxed{\delta_j}$ are permanent labels for nodes i and j, respectively, the condition for these nodes to be on the shortest path can be written as

$$\delta_j = \delta_i + c_{ij} \tag{2-50}$$

The relationship given in Eq. (2-50) can be used recursively in a backward fashion starting with node t. Once the node connected to node t is identified, we repeat the procedure to identify the next node on the shortest path. This tracing procedure is terminated when node s is reached. It can be verified that the shortest path for the example under consideration corresponds to the sequence of nodes s–1–2–6–4–5–t.

2.3.3 Shortest-Route Problem: Buying a New Car

In deciding when to buy a new car, capital costs plus increasing maintenance costs should be considered. In this section we use the shortest-route algorithm to decide how often to replace a car over a period of 8 years so as to minimize total costs. It is assumed that a decision is made at the beginning of each year on the basis of the purchase cost of a new car, the maintenance cost for the period the car would be kept, and the salvage value of the car when it is replaced. Assume that a person starts without a car and wishes to buy a new car at least every 4 years.

The network representation for the system under consideration is given in Figure 2.12. The beginning of each year is represented by a node. If a car is bought at the beginning of year i and replaced by a new one at the begin-

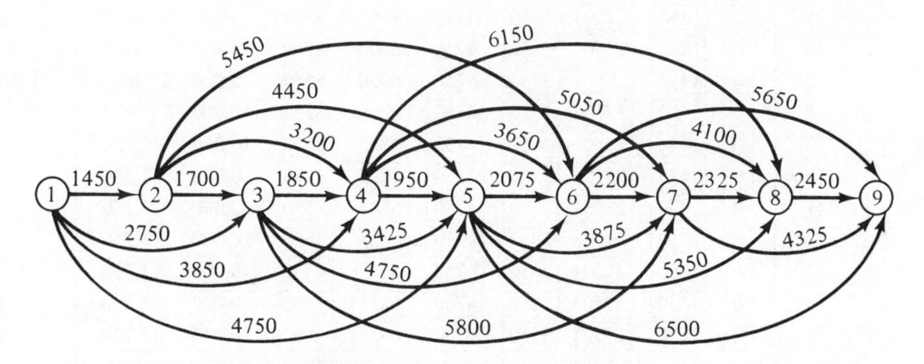

Figure 2.12. Buying a new car.

ning of year j, this replacement alternative is represented by arc (i, j). Let c_{ij} be the total cost associated with this alternative. Hence, the quantitative parameter of arc (i, j) is given by

$$c_{ij} = P_i + \sum_{k=1}^{j-1} m_k - S_j \tag{2-51}$$

where P_i = purchase price at the beginning of year i
m_k = maintenance cost during year k
S_j = salvage value at the beginning of year j

Projected costs c_{ij} are shown on each arc (i, j) of the network given in Figure 2.12. Total costs over planning periods of the same length are not constant but vary due to inflationary effects and increasing purchase and maintenance costs. The optimal solution to this problem is given by the shortest path from node $s = 1$ to node $t = 9$. The path can be interpreted as the route followed by 1 unit of flow going from node 1 to node 9, under the assumption that c_{ij} is the shipping cost along arc (i, j). A summary of the computations performed to obtain the shortest chain using Dijkstra's algorithm is shown in Table 2.5. The minimum-cost policy is to buy a car in periods 1, 5, and 9. The corresponding total cost is $11,250.

Table 2.5. CALCULATIONS

Step \ Node	1	2	3	4	5	6	7	8	9
0	[0]	∞	∞	∞	∞	∞	∞	∞	∞
1	[0]	1450	2750	3850	4750	∞	∞	∞	∞
2	[0]	[1450]	2750	3850	4750	∞	∞	∞	∞
3	[0]	[1450]	2750	3850	4750	6900	∞	∞	∞
4	[0]	[1450]	[2750]	3850	4750	6900	∞	∞	∞
5	[0]	[1450]	[2750]	3850	4750	6900	8550	∞	∞
6	[0]	[1450]	[2750]	[3850]	4750	6900	8550	∞	∞
7	[0]	[1450]	[2750]	[3850]	4750	6900	8550	10,000	∞
8	[0]	[1450]	[2750]	[3850]	[4750]	6900	8550	10,000	∞
9	[0]	[1450]	[2750]	[3850]	[4750]	6825	8550	10,000	11,250
10	[0]	[1450]	[2750]	[3850]	[4750]	[6825]	8550	10,000	11,250
11	[0]	[1450]	[2750]	[3850]	[4750]	[6825]	[8550]	10,000	11,250
12	[0]	[1450]	[2750]	[3850]	[4750]	[6825]	[8550]	10,000	11,250
13	[0]	[1450]	[2750]	[3850]	[4750]	[6825]	[8550]	[10,000]	11,250
14	[0]	[1450]	[2750]	[3850]	[4750]	[6825]	[8550]	[10,000]	11,250
15	[0]	[1450]	[2750]	[3850]	[4750]	[6825]	[8550]	[10,000]	[11,250]

2.3.4 Computer Code for Dijkstra's Algorithm

```
C***********************************************************************
C                                                                      *
C     PURPOSE:                                                         *
C                                                                      *
C        TO OBTAIN THE SHORTEST CHAIN IN A NETWORK FROM AN ORIGIN      *
C        TO ALL OTHER NODES IN THE NETWORK                             *
C                                                                      *
C     LOCATION:  SUBROUTINE DIJKA IN NETWORK OPTIMIZATION              *
C                                                                      *
C     USAGE:                                                           *
C                                                                      *
C        AT PRESENT THIS PROGRAM HANDLES UP TO 50 NODES AND 50 ARCS.   *
C        THIS NUMBER MAY BE INCREASED BY INCREASING THE DIMENSION      *
C        STATEMENTS IN SUBROUTINE DIJKA AND THE MAIN PROGRAM.          *
C                                                                      *
C     INPUT DATA:                                                      *
C                                                                      *
C        SET 1:  SINGLE CARD WITH THE CALL NAME OF THE ALGORITHM.      *
C                -FORMAT(A4)                                           *
C                                                                      *
C        SET 2:  SINGLE CARD WITH THE NUMBER OF NODES AND ARCS IN      *
C                THE PROBLEM.                                          *
C                -FORMAT(2I10)                                         *
C                                                                      *
C        SET 3:  THE TOTAL NUMBER OF CARDS IN  THIS SET IS EQUAL TO    *
C                THE NUMBER OF ARCS IN THE NETWORK.                    *
C                THE FOLLOWING QUANTITIES ARE READ FROM EACH CARD:     *
C                1) START NODE                                         *
C                2) END NODE                                           *
C                3) LENGTH OF THE ARC                                  *
C                  -FORMAT(4X,I6,I10,F10.2)                            *
C                                                                      *
C        SET 4:  THIS SET CONSISTS OF ONLY ONE CARD WHICH INDICATES    *
C                THE END OF A PROBLEM. THE USER SHOULD PUNCH THE WORD  *
C                'PEND' -FORMAT(A4)                                    *
C                                                                      *
C        SET 5:  THIS SET CONSISTS OF ONLY ONE CARD WHICH INDICATES    *
C                THE END OF THE INPUT DECK. THE USER SHOULD PUNCH THE  *
C                WORD 'EXIT'  -FORMAT(A4)                              *
C                                                                      *
C     PROGRAM CONTENTS:                                                *
C                                                                      *
C        THE SUBROUTINES WHICH COMPRISE THE PROGRAM ARE:               *
C            DIJKA: INPUT AND CALCULATIONS                             *
C            TRACED: PRINTS THE NODES AND DISTANCES ON                 *
C                    THE OPTIMAL CHAIN.                                *
C                                                                      *
C     DEFINITION OF TERMS USED:                                        *
C        I - STARTING NODE                                             *
C        J - ENDING NODE                                               *
C        VAL - ARC DISTANCE                                            *
C        D - WORKING ARRAY OF ARC DISTANCES                            *
C        OPTPAT - OPTIMAL SOLUTION ARRAY                               *
C                                                                      *
C     TECHNIQUE USED:                                                  *
C        THE TECHNIQUE USED IN THIS ALGORITHM IS EXPLAINED IN          *
C        SECTION 2.3.1                                                 *
C                                                                      *
C     REFERENCES:                                                      *
C                                                                      *
C        E.W.DIJKSTRA, A NOTE ON TWO PROBLEMS IN CONNECTION WITH       *
C        GRAPHS, NUM. MATH., 1,269-271,1959.                           *
C                                                                      *
C***********************************************************************
```

2.3.5 A Problem in Oil Transport Technology (Phillips)

The Black Gold Petroleum Company has recently found large deposits of oil on the North Slope of Alaska. In order to develop this field, a transportation network must be developed from the North Slope to one of six possible shipping points in the United States. The total transmission line will require seven pumping stations between a North Slope ground oil storage plant and the shipping points. A number of sites are possible for each substation, but not every site in one area can be reached from every site in the next area for a number of reasons. These include (1) geographic inaccessibility due to terrain features such as mountains or lakes, (2) the inability to purchase lease rights-of-way, and (3) restricted wildlife areas. Associated with each pair of pumping stations is the cost of constructing the connecting transmission line. The construction cost between any two points depends on the location of the given origin and the given destination. The problem is to determine a feasible pumping configuration for crude oil, while minimizing the attendant manufacturing costs. Figure 2.13 represents the distribution network for this problem. Each site is represented by a node, and accessible sites are connected by arcs labeled with the corresponding construction costs.

In solving practical network-flow problems, it becomes advantageous to utilize the digital computer to perform necessary calculations. A computer

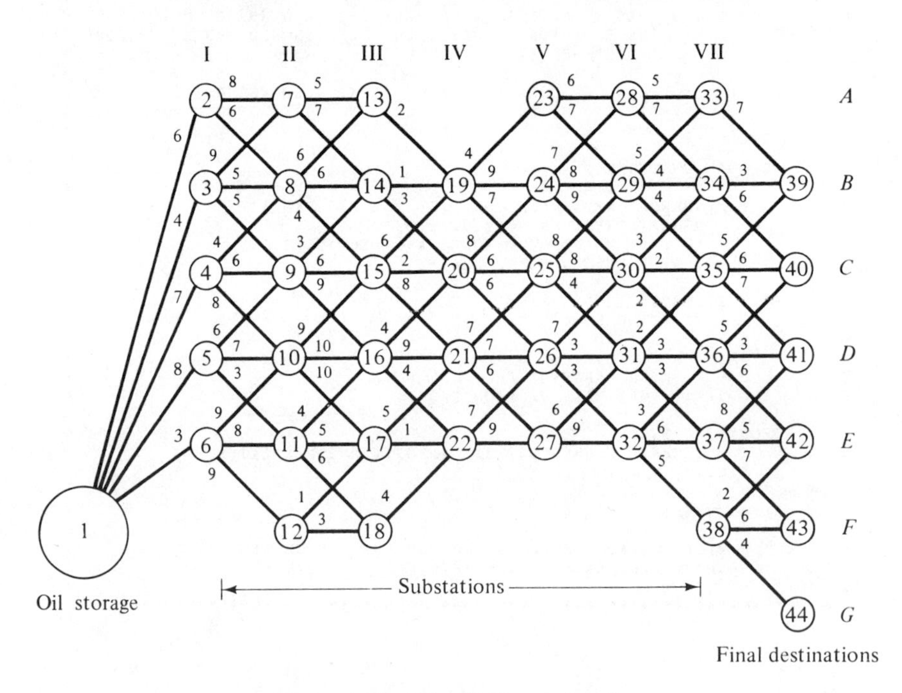

Figure 2.13. Distribution network.

code has been constructed with this objective in mind. The program is written in ANSI FORTRAN with standard input/output formats. Program usage and output interpretation are illustrated through the example given above.

The optimal route from node 1 to node 44 represents a solution with "length" or cost equal to 33. The optimal construction policy is attained by the following chain and corresponding "distances". The solution was obtained using the computer code described in Section 2.3.4.

From node 1 to node 3 is a distance of 4.00
From node 3 to node 9 is a distance of 5.00
From node 9 to node 14 is a distance of 3.00
From node 14 to node 20 is a distance of 3.00
From node 20 to node 26 is a distance of 6.00
From node 26 to node 32 is a distance of 3.00
From node 32 to node 38 is a distance of 5.00
From node 38 to node 44 is a distance of 4.00

In a similar fashion, it is possible to find the shortest chain between the source node 1 and each of the other terminal nodes 39, 40, 41, 42, and 43.

2.4 Multiterminal Shortest-Chain Route Problems

In this section we are concerned with finding shortest chains between all pairs of nodes of a network $G = (\mathbf{N}, \mathbf{A})$. The shortest chain between any two nodes represents the minimal-cost route to ship 1 unit of flow, where arc lengths are interpreted as distance or distribution costs per unit. The arcs in set **A** may be directed or undirected. However, since the direction of flow along the undirected arcs (if any) of a route cannot be anticipated, each undirected arc must be replaced by two directed arcs with opposite orientations and lengths equal to that of the undirected arc.

In the developments that follow, the arc lengths may be positive or negative. However, the total distance or cost of any cycle or circuit must be nonnegative.

The algorithm to be discussed in this section is due to Floyd [12], and our presentation follows the development by Hu [31]. Let the set of nodes **N** be defined as $\mathbf{N} = \{1, 2, \ldots, n\}$, and c_{ij} be the length or quantitative parameter of arc (i, j) directed from node i to node j. Additionally, let d^*_{ik} represent the length of the shortest chain from node i to node k.

Assume that the best estimate of the shortest chain connecting nodes i and k is given by d_{ik}. We know that the length of any chain passing through an intermediate node j, $\bar{d}_{ik} = d_{ij} + d_{jk}$ must be greater than d_{ik}, or else the current shortest chain would be of length $\bar{d}_{ik}$. However, if the length of the new chain $\bar{d}_{ik}$ is less than d_{ik}, the current value of d_{ik} should be changed to $\bar{d}_{ik}$. Floyd's algorithm operates on this basic logic. The shortest chain between any pair of nodes i and k is originally defined to be the length of the connecting arc (i, k). The algorithm successively examines all possible intermediate nodes between i and k, and if the length of a chain through any intermediate node is shorter than the current distance d_{ik}, then d_{ik} is changed to the new value. This procedure is repeated for all possible pairs of nodes until all d^*_{ik} values are obtained.

For any three distinct nodes i, j, and k, the conditions previously discussed can be summarized by Eq. (2-52):

$$d_{ij} + d_{jk} \geq d_{ik}, \qquad i \neq j \neq k \tag{2-52}$$

since, otherwise, the shortest chain from node i to node k should contain node j, and then d_{ik} would not be the length of the shortest chain. Initially, Floyd's algorithm starts with c_{ik} being the estimate of d_{ik}, and then iteratively improves the estimate until the shortest chain between nodes i and k is obtained. Figure 2.14 illustrates the procedure followed by the algorithm for a given pair of nodes i and k.

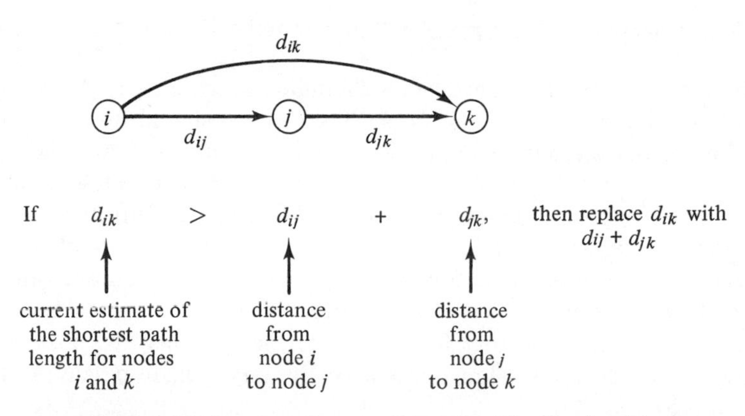

Figure 2.14. Pattern of comparisons for Floyd's algorithm.

Floyd's algorithm solves the multiterminal shortest-chain/route problem in exactly n iterations for a network with n nodes. In order to proceed with the formal development of the iterative solution procedure, we will define the symbol d^j_{ik} as the length of the shortest chain from node i to node k at the jth iteration of the algorithm.

The mechanism to obtain d^j_{ik} at the jth iteration of the algorithm can be summarized by means of the "triple operation" defined below:

$$d^j_{ik} = \min\,[d^{j-1}_{ik};\, d^{j-1}_{ij} + d^{j-1}_{jk}], \qquad i \neq j \neq k \tag{2-53}$$

If the triple operation defined in (2-53) is performed for a given pair of nodes i and k, considering all nodes j, $i \neq j \neq k$, in the order given by the sequence $j = 1, 2, \ldots, n$, the final value d_{ik}^n would be the length of the shortest chain from node i to node k.

We use a matrix method to keep track of the lengths of the shortest chains and of the arcs in those chains. At each iteration of the algorithm, we construct two matrices. The first matrix, referred to as the *shortest-path matrix*, contains the current estimates of the lengths of the shortest chains. That is, at the jth iteration the matrix is defined as $\mathbf{D}^j = [d_{ik}^j]$. The algorithm starts with $\mathbf{D}^0 = [d_{ik}^0]$, where $d_{ik}^0 = c_{ik}$. Then $\mathbf{D}^1$ is obtained by performing the triple operation (2-53) on all entries of $\mathbf{D}^0$, and so on, until the optimality criterion defined in Eq. (2-52) is satisfied for all pairs of nodes.

The purpose of the second matrix, referred to as the *route matrix*, is to identify the intermediate nodes (if any) of the shortest chains. The route matrix at the jth iteration is defined as $R^j = [r_{ik}^j]$ where r_{ik}^j is the first intermediate node of the shortest chain from node i to node k, using intermediate nodes from the set $\{1, 2, \ldots, j\}$ and such that $i \neq j \neq k$. The algorithm starts with $\mathbf{R}^0 = [r_{ik}^0]$, where $r_{ik}^0 = k$. At the jth iteration, r_{ik}^j can be obtained by using the following relationship:

$$r_{ik}^j = \begin{cases} j, & \text{if } d_{ik}^{j-1} > d_{ij}^{j-1} + d_{jk}^{j-1} \\ r_{ik}^{j-1}, & \text{otherwise} \end{cases} \tag{2-54}$$

After initialization, the following procedure is repeated successively for $j = 1, 2, \ldots, n$ using elements of the $(j - 1)$st matrix for all calculations.

***Step 1*:** Block out all elements in the jth row and jth column. Call these collections of elements the *pivotal row* and the *pivotal column*, respectively.

***Step 2*:** Starting with the first element of the distance matrix (top-left), which is not in the pivotal row or pivotal column (d_{ik}^{j-1}, $k \neq j$), compare this element to the *sum* of the elements d_{jk}^{j-1} and d_{ij}^{j-1} in the pivotal row and the pivotal column, respectively.

If the sum of $d_{ij}^{j-1} + d_{jk}^{j-1}$ is greater than or equal to d_{ik}^{j-1}, choose new values for i and k, proceed to the next element d_{ik}^{j-1}, and repeat step 2. If $d_{ik}^{j-1} > d_{ij}^{j-1} + d_{jk}^{j-1}$, then replace d_{ik}^{j-1} in the *shortest path matrix* with $d_{ik}^j = d_{ij}^{j-1} + d_{jk}^{j-1}$ and replace the corresponding element in the *route matrix* with the number j. Once all elements have been examined, return to step 1 with $j = j + 1$ and start anew. Proceed until $j = n$, at which point the optimal distance matrix and route matrix have been generated.

By Eq. (2-53), to obtain $\mathbf{D}^j$ from $\mathbf{D}^{j-1}$, only the entries of $\mathbf{D}^{j-1}$ that do not lie in either the pivot row or the pivot column are investigated to perform the triple operations required to determine if the use of the pivot node would result in shorter chains.

Note that a simplex-like procedure can be employed when examining each element d_{ik}^{j-1}, $i, j \neq k$. Once d_{ik}^{j-1} is chosen, if $d_{ik}^{j-1} = \infty$ and *either* d_{ij}^{j-1} or $d_{jk}^{j-1} = \infty$, a change should not be made. If $d_{ik}^{j-1} \neq \infty$ and *either* d_{ij}^{j-1} or d_{jk}^{j-1} is greater than d_{ik}^{j-1}, a change should *not* be made. Hence, the only case that remains to be considered is when $d_{ik}^{j-1} \neq \infty$ and both $d_{ij}^{j-1}, d_{jk}^{j-1} \neq \infty$, with each being less than d_{ik}^{j-1}. Hence, the following rules will lead to computational simplifications prior to each iteration.

1. $d_{ij}^{j-1} = \infty$; that is, the ith element of the pivot column is ∞. By using Eq. (2-53), we conclude that in this case d_{ik}^{j} should remain d_{ik}^{j-1}. Therefore, the triple operation is not needed.
2. $d_{jk}^{j-1} = \infty$; that is, the kth element of the pivot row is ∞. By using Eq. (2-53), we conclude that in this case d_{ik}^{j} should remain d_{ik}^{j-1}. Therefore, the triple operation is not needed.

The algorithm under consideration can be summarized as a sequence of two simple computational steps. As preparation for the execution of these steps in an iterative fashion, $\mathbf{D}^0$ and $\mathbf{R}^0$ are defined as $\mathbf{D}^0 = [c_{ik}]$, and $\mathbf{R}^0 = [k]$. Also, the parameter j is set equal to zero. This parameter is used to indicate the pivot node at each iteration.

Step 1: $j = j + 1$. Define node j as the pivot node, and delete the pivot row and the pivot column of matrix $\mathbf{D}^{j-1}$. Also, delete all rows and columns of matrix $\mathbf{D}^{j-1}$ with ∞-entries in either the pivot row or the pivot column.

Step 2: Compare each element d_{ik}^{j-1} that has not been deleted versus the sum of the following two elements:

a. The element d_{ij}^{j-1} in the pivot column and in the row of d_{ik}^{j-1}.
b. The element d_{jk}^{j-1} in the pivot row and in the column of d_{ik}^{j-1}.

If the sum of the two elements is less than d_{ik}^{j-1}, then set $d_{ik}^{j} = d_{ij}^{j-1} + d_{jk}^{j-1}$ and set $r_{ik}^{j} = j$. Otherwise, set $d_{ik}^{j} = d_{ik}^{j-1}$ and $r_{ik}^{j} = r_{ik}^{j-1}$. Once all nondeleted elements of D^{j-1} have been investigated, check if $j = n$. If so, stop. Otherwise, go to step 1.

2.4.1 Example of the Multiterminal Shortest-Chain Problem

To illustrate the application of Floyd's algorithm, let us find the shortest chain between every pair of nodes in the network of Figure 2.15. The shortest-path and the route matrices can be initialized as indicated below. Since $n = 8$, the number of iterations for the algorithm is equal to 8.

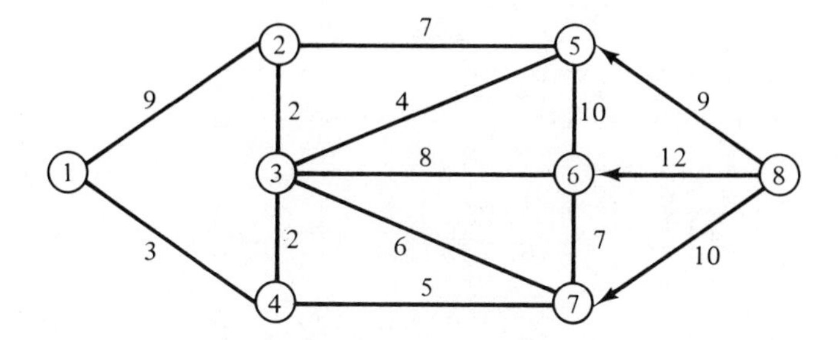

Figure 2.15. Multiterminal shortest-route network.

$$
\mathbf{D}^0 = \begin{array}{c} \\ 1 \\ 2 \\ 3 \\ 4 \\ 5 \\ 6 \\ 7 \\ 8 \end{array}
\begin{array}{c}
\begin{array}{cccccccc} 1 & 2 & 3 & 4 & 5 & 6 & 7 & 8 \end{array} \\
\left[\begin{array}{cccccccc}
0 & 9 & \infty & 3 & \infty & \infty & \infty & \infty \\
9 & 0 & 2 & \infty & 7 & \infty & \infty & \infty \\
\infty & 2 & 0 & 2 & 4 & 8 & 6 & \infty \\
3 & \infty & 2 & 0 & \infty & \infty & 5 & \infty \\
\infty & 7 & 4 & \infty & 0 & 10 & \infty & \infty \\
\infty & \infty & 8 & \infty & 10 & 0 & 7 & \infty \\
\infty & \infty & 6 & 5 & \infty & 7 & 0 & \infty \\
\infty & \infty & \infty & \infty & 9 & 12 & 10 & 0
\end{array}\right]
\end{array}
$$

$$
\mathbf{R}^0 = \begin{array}{c} \\ 1 \\ 2 \\ 3 \\ 4 \\ 5 \\ 6 \\ 7 \\ 8 \end{array}
\begin{array}{c}
\begin{array}{cccccccc} 1 & 2 & 3 & 4 & 5 & 6 & 7 & 8 \end{array} \\
\left[\begin{array}{cccccccc}
1 & 2 & 3 & 4 & 5 & 6 & 7 & 8 \\
1 & 2 & 3 & 4 & 5 & 6 & 7 & 8 \\
1 & 2 & 3 & 4 & 5 & 6 & 7 & 8 \\
1 & 2 & 3 & 4 & 5 & 6 & 7 & 8 \\
1 & 2 & 3 & 4 & 5 & 6 & 7 & 8 \\
1 & 2 & 3 & 4 & 5 & 6 & 7 & 8 \\
1 & 2 & 3 & 4 & 5 & 6 & 7 & 8 \\
1 & 2 & 3 & 4 & 5 & 6 & 7 & 8
\end{array}\right]
\end{array}
$$

***Iteration 1*:** Here $j = 1$ is defined as the pivot node. Hence, we can delete the first row and the first column of matrix $\mathbf{D}^0$. Additionally, columns 3, 5, 6, 7, and 8 have ∞ entries in the pivot row; and rows 3, 5, 6, 7, and 8 have ∞ entries in the pivot column. Therefore, only the entries of $\mathbf{D}^0$ which are shown in Figure 2.16 need to be analyzed with the triple operation to investigate if use of node 1 would result in shorter chains.

Ignoring the diagonal elements of $\mathbf{D}^0$, the estimates that must be inves-

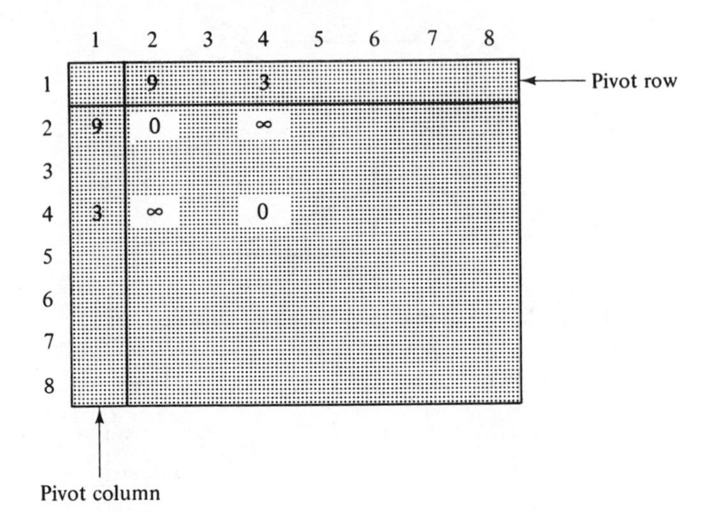

Figure 2.16. Entries to be investigated by triple operations at Iteration 1.

tigated are only d^0_{24} and d^0_{42}. The corresponding triple operations yield the following results:

$$d^1_{24} = \min\,[d^0_{24}; d^0_{21} + d^0_{14}] = \min\,[\infty; 9 + 3] = 12$$
$$d^1_{42} = \min\,[d^0_{42}; d^0_{41} + d^0_{12}] = \min\,[\infty; 3 + 9] = 12$$

Note that $d^1_{24} = 12$ and $d^1_{42} = 12$ are better estimates than $d^0_{24} = \infty$ and $d^0_{42} = \infty$, respectively. Since the use of the pivot node 1 has resulted in shorter chains from node 2 to node 4, and from node 4 to node 2, we set $r^1_{24} = 1$ and $r^1_{42} = 1$. All other entries of $\mathbf{D}^0$ and $\mathbf{R}^0$ remain unchanged, and then we can write for $\mathbf{D}^1$ and $\mathbf{R}^1$:

$$\mathbf{D}^1 = \begin{array}{c} \\ 1 \\ 2 \\ 3 \\ 4 \\ 5 \\ 6 \\ 7 \\ 8 \end{array} \begin{array}{c} \begin{array}{cccccccc} 1 & 2 & 3 & 4 & 5 & 6 & 7 & 8 \end{array} \\ \begin{bmatrix} 0 & 9 & \infty & 3 & \infty & \infty & \infty & \infty \\ 9 & 0 & 2 & 12 & 7 & \infty & \infty & \infty \\ \infty & 2 & 0 & 2 & 4 & 8 & 6 & \infty \\ 3 & 12 & 2 & 0 & \infty & \infty & 5 & \infty \\ \infty & 7 & 4 & \infty & 0 & 10 & \infty & \infty \\ \infty & \infty & 8 & \infty & 10 & 0 & 7 & \infty \\ \infty & \infty & 6 & 5 & \infty & 7 & 0 & \infty \\ \infty & \infty & \infty & \infty & 9 & 12 & 10 & 0 \end{bmatrix} \end{array}$$

$$\mathbf{R}^1 = \begin{array}{c|cccccccc} & 1 & 2 & 3 & 4 & 5 & 6 & 7 & 8 \\ \hline 1 & 1 & 2 & 3 & 4 & 5 & 6 & 7 & 8 \\ 2 & 1 & 2 & 3 & 1 & 5 & 6 & 7 & 8 \\ 3 & 1 & 2 & 3 & 4 & 5 & 6 & 7 & 8 \\ 4 & 1 & 1 & 3 & 4 & 5 & 6 & 7 & 8 \\ 5 & 1 & 2 & 3 & 4 & 5 & 6 & 7 & 8 \\ 6 & 1 & 2 & 3 & 4 & 5 & 6 & 7 & 8 \\ 7 & 1 & 2 & 3 & 4 & 5 & 6 & 7 & 8 \\ 8 & 1 & 2 & 3 & 4 & 5 & 6 & 7 & 8 \end{array}$$

***Iteration 2*:** Now $j = 2$ is defined as the pivot node. Hence, we delete the second row and the second column of matrix $\mathbf{D}^1$. Additionally, columns 6, 7, and 8 have ∞ entries in the pivot row; and rows 6, 7, and 8 have ∞ entries in the pivot column. Therefore, only the entries of $\mathbf{D}^1$ are shown in Figure 2.17 need to be analyzed with the triple operation to investigate if use of node 2 would result in shorter chains.

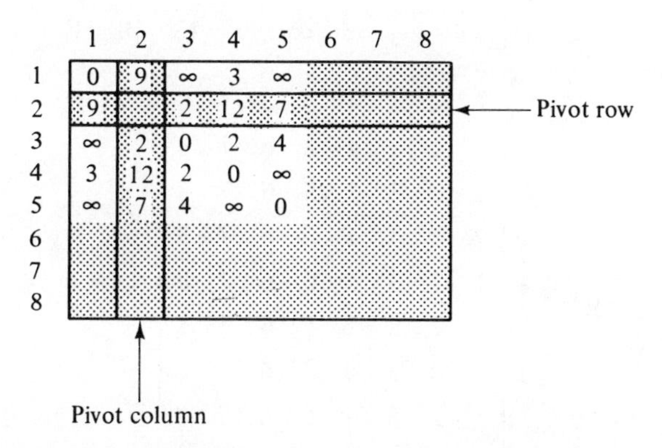

Figure 2.17. Entries to be investigated by triple operations at Iteration 2.

Ignoring the diagonal elements of $\mathbf{D}^1$, the elements that must be investigated are d^1_{13}, d^1_{14}, d^1_{15}, d^1_{31}, d^1_{34}, d^1_{35}, d^1_{41}, d^1_{43}, d^1_{45}, d^1_{51}, d^1_{53}, and d^1_{54}. It can be verified that only the estimates d^1_{13}, d^1_{15}, d^1_{31}, d^1_{45}, d^1_{51}, and d^1_{54} can be improved. The new estimates are given by

$$d^2_{13} = \min\,[d^1_{13};\, d^1_{12} + d^1_{23}] = \min\,[\infty;\, 9 + 2] = 11$$

$$d^2_{15} = \min\,[d^1_{15};\, d^1_{12} + d^1_{25}] = \min\,[\infty;\, 9 + 7] = 16$$

$$d^2_{31} = \min\,[d^1_{31};\, d^1_{32} + d^1_{21}] = \min\,[\infty;\, 2 + 9] = 11$$

$$d_{45}^2 = \min [d_{45}^1; d_{42}^1 + d_{25}^1] = \min [\infty; 12 + 7] = 19$$
$$d_{51}^2 = \min [d_{51}^1; d_{52}^1 + d_{21}^1] = \min [\infty; 7 + 9] = 16$$
$$d_{54}^2 = \min [d_{54}^1; d_{52}^1 + d_{24}^1] = \min [\infty; 7 + 12] = 19$$

Accordingly, $r_{13}^2 = r_{15}^2 = r_{31}^2 = r_{45}^2 = r_{51}^2 = r_{54}^2 = 2$ and $r_{ik}^2 = r_{ik}^1$, for all other entries such that $d_{ik}^2 = d_{ik}^1$. The new matrices $\mathbf{D}^2$ and $\mathbf{R}^2$ can be written as follows:

$$\mathbf{D}^2 = \begin{array}{c} \\ 1 \\ 2 \\ 3 \\ 4 \\ 5 \\ 6 \\ 7 \\ 8 \end{array} \begin{array}{c} \begin{array}{cccccccc} 1 & 2 & 3 & 4 & 5 & 6 & 7 & 8 \end{array} \\ \begin{bmatrix} 0 & 9 & 11 & 3 & 16 & \infty & \infty & \infty \\ 9 & 0 & 2 & 12 & 7 & \infty & \infty & \infty \\ 11 & 2 & 0 & 2 & 4 & 8 & 6 & \infty \\ 3 & 12 & 2 & 0 & 19 & \infty & 5 & \infty \\ 16 & 7 & 4 & 19 & 0 & 10 & \infty & \infty \\ \infty & \infty & 8 & \infty & 10 & 0 & 7 & \infty \\ \infty & \infty & 6 & 5 & \infty & 7 & 0 & \infty \\ \infty & \infty & \infty & \infty & 9 & 12 & 10 & 0 \end{bmatrix} \end{array}$$

$$\mathbf{R}^2 = \begin{array}{c} \\ 1 \\ 2 \\ 3 \\ 4 \\ 5 \\ 6 \\ 7 \\ 8 \end{array} \begin{array}{c} \begin{array}{cccccccc} 1 & 2 & 3 & 4 & 5 & 6 & 7 & 8 \end{array} \\ \begin{bmatrix} 1 & 2 & 2 & 4 & 2 & 6 & 7 & 8 \\ 1 & 2 & 3 & 1 & 5 & 6 & 7 & 8 \\ 2 & 2 & 3 & 4 & 5 & 6 & 7 & 8 \\ 1 & 1 & 3 & 4 & 2 & 6 & 7 & 8 \\ 2 & 2 & 3 & 2 & 5 & 6 & 7 & 8 \\ 1 & 2 & 3 & 4 & 5 & 6 & 7 & 8 \\ 1 & 2 & 3 & 4 & 5 & 6 & 7 & 8 \\ 1 & 2 & 3 & 4 & 5 & 6 & 7 & 8 \end{bmatrix} \end{array}$$

Continuing in this fashion, we obtain the results shown in Table 2.6 for iterations $j = 3, 4, 5, 6, 7, 8$. As can be verified, no improvements were possible in iterations 6, 7, and 8. Hence, the optimal solution corresponds to matrices $\mathbf{D}^5$ and $\mathbf{R}^5$. As an illustration of the results shown in Table 2.6 let us analyze the shortest chain from node 1 to node 5. The length of this chain is equal to $d_{15}^5 = 9$. In order to find the corresponding sequence of nodes, refer to matrix $\mathbf{R}^5$ and proceed as follows. The value of r_{15}^5 is equal to 4, indicating that node 4 is the first intermediate node in the shortest path from node 1 to node 5. Now we obtain the value of r_{45}^5 to identify the node that follows node

Table 2.6. SUMMARY OF RESULTS FOR SAMPLE PROBLEM

Iteration j	$\mathbf{D}^j$	$\mathbf{R}^j$
3	$\begin{bmatrix} 0 & 9 & 11 & 3 & 15 & 19 & 17 & \infty \\ 9 & 0 & 2 & 4 & 6 & 10 & 8 & \infty \\ 11 & 2 & 0 & 2 & 4 & 8 & 6 & \infty \\ 3 & 4 & 2 & 0 & 6 & 10 & 5 & \infty \\ 15 & 6 & 4 & 6 & 0 & 10 & 10 & \infty \\ 19 & 10 & 8 & 10 & 10 & 0 & 7 & \infty \\ 17 & 8 & 6 & 5 & 10 & 7 & 0 & \infty \\ \infty & \infty & \infty & \infty & 9 & 12 & 10 & 0 \end{bmatrix}$	$\begin{bmatrix} 1 & 2 & 2 & 4 & 2 & 2 & 2 & 8 \\ 1 & 2 & 3 & 3 & 3 & 3 & 3 & 8 \\ 2 & 2 & 3 & 4 & 5 & 6 & 7 & 8 \\ 1 & 3 & 3 & 4 & 3 & 3 & 7 & 8 \\ 2 & 3 & 3 & 3 & 5 & 6 & 3 & 8 \\ 3 & 3 & 3 & 3 & 5 & 6 & 7 & 8 \\ 3 & 3 & 3 & 4 & 3 & 6 & 7 & 8 \\ 1 & 2 & 3 & 4 & 5 & 6 & 7 & 8 \end{bmatrix}$
4	$\begin{bmatrix} 0 & 7 & 5 & 3 & 9 & 13 & 8 & \infty \\ 7 & 0 & 2 & 4 & 6 & 10 & 8 & \infty \\ 5 & 2 & 0 & 2 & 4 & 8 & 6 & \infty \\ 3 & 4 & 2 & 0 & 6 & 10 & 5 & \infty \\ 9 & 6 & 4 & 6 & 0 & 10 & 10 & \infty \\ 13 & 10 & 8 & 10 & 10 & 0 & 7 & \infty \\ 8 & 8 & 6 & 5 & 10 & 7 & 0 & \infty \\ \infty & \infty & \infty & \infty & 9 & 12 & 10 & 0 \end{bmatrix}$	$\begin{bmatrix} 1 & 4 & 4 & 4 & 4 & 4 & 4 & 8 \\ 4 & 2 & 3 & 3 & 3 & 3 & 3 & 8 \\ 4 & 2 & 3 & 4 & 5 & 6 & 7 & 8 \\ 1 & 3 & 3 & 4 & 3 & 3 & 7 & 8 \\ 4 & 3 & 3 & 3 & 5 & 6 & 3 & 8 \\ 4 & 3 & 3 & 3 & 5 & 6 & 7 & 8 \\ 4 & 3 & 3 & 4 & 3 & 6 & 7 & 8 \\ 1 & 2 & 3 & 4 & 5 & 6 & 7 & 8 \end{bmatrix}$
5	$\begin{bmatrix} 0 & 7 & 5 & 3 & 9 & 13 & 8 & \infty \\ 7 & 0 & 2 & 4 & 6 & 10 & 8 & \infty \\ 5 & 2 & 0 & 2 & 4 & 8 & 6 & \infty \\ 3 & 4 & 2 & 0 & 6 & 10 & 5 & \infty \\ 9 & 6 & 4 & 6 & 0 & 10 & 10 & \infty \\ 13 & 10 & 8 & 10 & 10 & 0 & 7 & \infty \\ 8 & 8 & 6 & 5 & 10 & 7 & 0 & \infty \\ 18 & 15 & 13 & 15 & 9 & 12 & 10 & 0 \end{bmatrix}$	$\begin{bmatrix} 1 & 4 & 4 & 4 & 4 & 4 & 4 & 8 \\ 4 & 2 & 3 & 3 & 3 & 3 & 3 & 8 \\ 4 & 2 & 3 & 4 & 5 & 6 & 7 & 8 \\ 1 & 3 & 3 & 4 & 3 & 3 & 7 & 8 \\ 4 & 3 & 3 & 3 & 5 & 6 & 3 & 8 \\ 4 & 3 & 3 & 3 & 5 & 6 & 7 & 8 \\ 4 & 3 & 3 & 4 & 3 & 6 & 7 & 8 \\ 5 & 5 & 5 & 5 & 5 & 6 & 7 & 8 \end{bmatrix}$
6	Unchanged	Unchanged
7	Unchanged	Unchanged
8	Unchanged	Unchanged

4 on the chain to node 5. This value is equal to 3. Similarly, the value $r_{35}^5 = 5$ indicates that node 5 is next to node 3. Therefore, the shortest path from node 1 to node 5 corresponds to the sequence of nodes 1, 4, 3, 5.

2.4.2 Multiterminal Shortest-Route Problem: Design of a Mail Distribution System

A major department store plans to establish an internal mail distribution system which will utilize pneumatic tubes to distribute the mail among eight different departments. However, some departments will only dispatch incoming mail within the store, and will not have the capability to receive internal mail. All other departments will be able to dispatch and receive mail without restrictions.

The arrangement of the tubes has been previously decided and is depicted in the network shown in Figure 2.18. Each department is represented by a node, and each tube by an arc. The arcs with specified orientations correspond to distribution links which can be used only in those directions. The number on each arc indicates the distance between the corresponding departments.

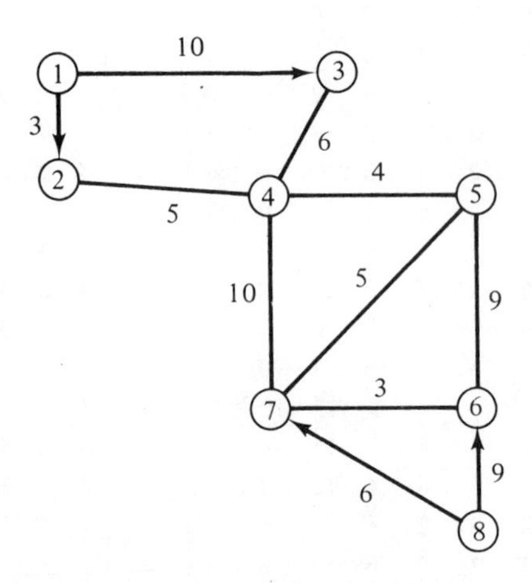

Figure 2.18. Mail distribution system.

In order to enable each department to determine the shortest route to use when sending mail to another department, it is desirable to provide the sending units with a routing sheet showing the shortest route between any pair of departments. This objective can be achieved by applying Floyd's algorithm to the network of Figure 2.18. In this case, the shortest-path and the route matrices can be initialized as follows:

$$\mathbf{D}^0 = \begin{bmatrix} 0 & 3 & 10 & \infty & \infty & \infty & \infty & \infty \\ \infty & 0 & \infty & 5 & \infty & \infty & \infty & \infty \\ \infty & \infty & 0 & 6 & \infty & \infty & \infty & \infty \\ \infty & 5 & 6 & 0 & 4 & \infty & 10 & \infty \\ \infty & \infty & \infty & 4 & 0 & 9 & 5 & \infty \\ \infty & \infty & \infty & \infty & 9 & 0 & 3 & \infty \\ \infty & \infty & \infty & 10 & \infty & 3 & 0 & \infty \\ \infty & \infty & \infty & \infty & \infty & 9 & 6 & 0 \end{bmatrix}$$

$$
\mathbf{R}^0 = \begin{bmatrix}
1 & 2 & 3 & 4 & 5 & 6 & 7 & 8 \\
1 & 2 & 3 & 4 & 5 & 6 & 7 & 8 \\
1 & 2 & 3 & 4 & 5 & 6 & 7 & 8 \\
1 & 2 & 3 & 4 & 5 & 6 & 7 & 8 \\
1 & 2 & 3 & 4 & 5 & 6 & 7 & 8 \\
1 & 2 & 3 & 4 & 5 & 6 & 7 & 8 \\
1 & 2 & 3 & 4 & 5 & 6 & 7 & 8 \\
1 & 2 & 3 & 4 & 5 & 6 & 7 & 8
\end{bmatrix}
$$

The reader may verify that the optimal solution corresponds to the following matrices:

$$
\mathbf{D}^8 = \begin{bmatrix}
0 & 3 & 10 & 8 & 12 & 20 & 17 & \infty \\
\infty & 0 & 11 & 5 & 9 & 17 & 14 & \infty \\
\infty & 11 & 0 & 6 & 10 & 18 & 15 & \infty \\
\infty & 5 & 6 & 0 & 4 & 12 & 10 & \infty \\
\infty & 9 & 10 & 4 & 0 & 8 & 5 & \infty \\
\infty & 17 & 18 & 12 & 9 & 0 & 3 & \infty \\
\infty & 15 & 15 & 19 & 15 & 3 & 0 & \infty \\
\infty & 20 & 21 & 15 & 11 & 9 & 6 & 0
\end{bmatrix}
$$

$$
\mathbf{R}^8 = \begin{bmatrix}
1 & 2 & 3 & 2 & 4 & 7 & 5 & 8 \\
1 & 2 & 4 & 4 & 4 & 7 & 5 & 8 \\
1 & 4 & 3 & 4 & 4 & 7 & 5 & 8 \\
1 & 2 & 3 & 4 & 5 & 7 & 5 & 8 \\
1 & 4 & 4 & 4 & 5 & 7 & 7 & 8 \\
1 & 7 & 7 & 7 & 7 & 6 & 7 & 8 \\
1 & 5 & 5 & 5 & 5 & 6 & 7 & 8 \\
1 & 7 & 7 & 7 & 7 & 6 & 7 & 8
\end{bmatrix}
$$

As an illustration, the shortest chain from node 8 to node 2 has a length equal to $d^8_{82} = 20$; in order to find the corresponding node sequence, proceed as follows. From $\mathbf{R}^8$, $r^8_{82} = 7$, $r^8_{72} = 5$, $r^8_{52} = 4$, and $4^8_{42} = 2$. Therefore, the shortest path from node 8 to node 2 corresponds to the sequence of nodes 8, 7, 5, 4, 2.

2.4.3 Computer Code for the Multiterminal Shortest-Route Algorithm

```
C***********************************************************************
C                                                                      *
C     PURPOSE:                                                         *
C        TO OBTAIN OPTIMAL CHAINS OR ROUTES FOR THE MULTITERMINAL      *
C        SHORTEST CHAIN PROBLEM                                        *
C                                                                      *
C     LOCATION: SUBROUTINE MULTSH IN NETWORK OPTIMIZATION              *
C                                                                      *
C     USAGE:                                                           *
C                                                                      *
C        AT PRESENT THIS PROGRAM HANDLES UP TO 50 NODES AND 50 ARCS.   *
C        THIS NUMBER MAY BE INCREASED BY INCREASING DIMENSION          *
C        STATEMENTS IN SUBROUTINE MULTSH AND THE MAIN PROGRAM          *
C                                                                      *
C     INPUT DATA:                                                      *
C                                                                      *
C        SET 1:  SINGLE CARD WITH THE CALL NAME OF THE ALGORITHM:MTSC *
C                 -FORMAT(A4)                                          *
C                                                                      *
C        SET 2:  SINGLE CARD WITH THE NUMBER OF NODES AND ARCS IN THE *
C                PROBLEM.                                              *
C                 -FORMAT(2I10)                                        *
C                                                                      *
C        SET 3:  THE TOTAL NUMBER OF CARDS FOR THIS SET IS EQUAL TO    *
C                THE NUMBER OF ARCS FOR THE PROBLEM.                   *
C                THE FOLLOWING QUANTITIES ARE READ FROM EACH CARD:     *
C                   1) START NODE                                      *
C                   2) END NODE                                        *
C                   3) LENGTH OF THE ARC                               *
C                      -FORMAT(4X,I6,I10,F10.2)                        *
C                                                                      *
C        SET 4:  THIS SET CONSISTS OF ONLY ONE CARD WHICH INDICATES    *
C                THE END OF A PROBLEM. THE USER SHOULD PUNCH THE       *
C                WORD 'PEND' -FORMAT(A4)                               *
C                                                                      *
C        SET 5:  THIS SET CONSISTS OF ONLY ONE CARD WHICH INDICATES    *
C                THE END OF THE INPUT DECK. THE USER SHOULD PUNCH THE *
C                WORD 'EXIT'    -FORMAT(A4)                            *
C                                                                      *
C     PROGRAM CONTENTS:                                                *
C                                                                      *
C        THE SUBROUTINES WHICH COMPRISE THIS PROGRAM ARE:              *
C            MULTSH: INPUT AND CALCULATIONS                            *
C            PRNTRX: PRINTS THE ORIGINAL MATRIX                        *
C            PRNTRY: PRINTS THE OPTIMAL SOLUTION                       *
C                                                                      *
C                                                                      *
C     DEFINITION OF TERMS USED:                                        *
C                                                                      *
C        I - STARTING NODE                                             *
C        J - ENDING NODE                                               *
C        B - ARC DISTANCE                                              *
C        D - WORKING ARRAY OF ARC DISTANCES                            *
C                                                                      *
C     TECHNIQUE USED:                                                  *
C        THE TECHNIQUE USED IN THIS ALGORITHM CONSIDERS THE DIRECT     *
C        DISTANCE BETWEEN NODE I AND NODE K AS THE INITIAL ESTIMATE    *
C        OF THE LENGTH OF THE SHORTEST PATH FROM I TO K.AT THE J-TH    *
C        ITERATION, THE METHOD COMPARES THE LENGTH OF A PATH FROM      *
C        NODE I TO NODE K HAVING INTERMEDIATE NODES FROM THE SET       *
C        [1,2,..,J] VERSUS THE LENGTH OF A PATH THROUGH NODE J WITH    *
C        INTERMEDIATE NODES FROM THE SET [1,2,..,J-1]. IF THE USE OF   *
C        NODE J RESULTS IN A SHORTER PATH FROM NODE I TO K, THE LENGHT*
C        OF THIS PATH BECOMES THE NEW ESTIMATE OF THE MINIMAL PATH     *
C        LENGTH.                                                       *
C                                                                      *
C     REFERENCE:                                                       *
C        T.C.HU, INTEGER PROGRAMMING AND NETWORK FLOWS,                *
C        ADDISON-WESLEY PUBLISHING CO., MASS, 1969.                    *
C                                                                      *
C***********************************************************************
```

2.4.4 Routing of Rail Cars

Switching charges for rail cars constitute a significant percentage of the total transportation costs for moving cargo in and out of the Port of Houston. In the network of Figure 2.19, node 1 represents the incoming yard, where the inbound trains are broken down, and the other nodes represent different destinations within the port complex. The arc values indicate the corresponding switching costs.

As an illustration, the following situation can be considered. A 200-car train might contain 100 covered hopper cars loaded with grain. These grain hoppers must be switched from node 1 to grain elevators located at nodes 3, 6, and 10. Once empty, the hopper cars can either be switched to fertilizer plants to be loaded with bulk nitrogen at nodes 2, 5 and 8, or be switched directly to node 11, the outgoing yard. Other types of rail units, such as box cars, flat cars, or tank cars might be routed to other terminals located at nodes 4, 7, and 9, where the units would be unloaded and then transfered to the outgoing yard.

The application of the multiterminal shortest-path algorithm to the port rail network of Figure 2.19 would yield the minimal switching cost route between any two terminals of the port complex. For covered hopper cars, for example, the optimal solution to this problem would provide paths connecting the incoming yard, the grain elevators, the fertilizer plants, and the outgoing yard. This problem was solved using the computer code described in Section 2.4.3, and the results are given below. A value of 9999999.00 has

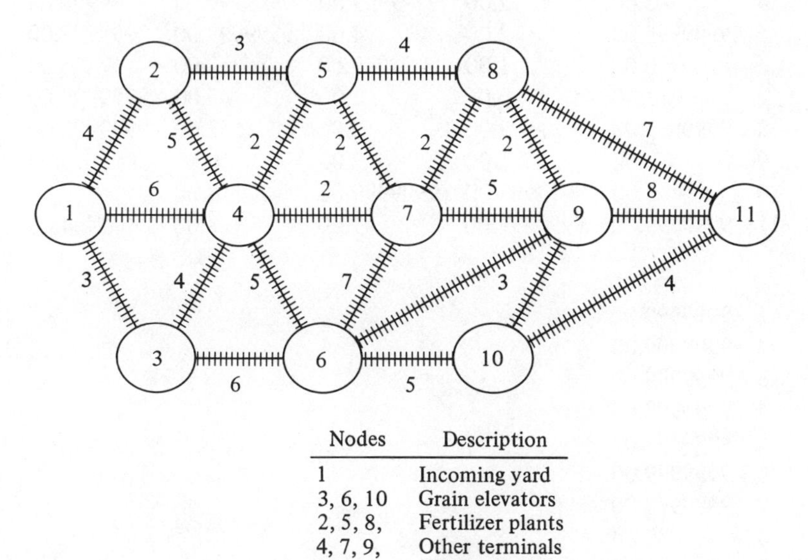

Nodes	Description
1	Incoming yard
3, 6, 10	Grain elevators
2, 5, 8,	Fertilizer plants
4, 7, 9,	Other terminals
11	Outgoing yard

Figure 2.19. Port-rail distribution network.

been defined as the length of those arcs which actually do not exist in the distribution network of Figure 2.19.

MULTITERMINAL SHORTEST CHAINS PROBLEM

ORIGINAL MATRIX

	1	2	3	4	5
1	0.0	4.00	3.00	6.00	9999999.00
2	4.00	0.0	9999999.00	5.00	3.00
3	3.00	9999999.00	0.0	4.00	9999999.00
4	6.00	5.00	4.00	0.0	2.00
5	9999999.00	3.00	9999999.00	2.00	0.0
6	9999999.00	9999999.00	6.00	5.00	9999999.00
7	9999999.00	9999999.00	9999999.00	2.00	2.00
8	9999999.00	9999999.00	9999999.00	9999999.00	4.00
9	9999999.00	9999999.00	9999999.00	9999999.00	9999999.00
10	9999999.00	9999999.00	9999999.00	9999999.00	9999999.00
11	9999999.00	9999999.00	9999999.00	9999999.00	9999999.00

	6	7	8	9	10
1	9999999.00	9999999.00	9999999.00	9999999.00	9999999.00
2	9999999.00	9999999.00	9999999.00	9999999.00	9999999.00
3	6.00	9999999.00	9999999.00	9999999.00	9999999.00
4	5.00	2.00	9999999.00	9999999.00	9999999.00
5	9999999.00	2.00	4.00	9999999.00	9999999.00
6	0.0	1.00	9999999.00	2.00	5.00
7	1.00	0.0	2.00	5.00	9999999.00
8	9999999.00	2.00	0.0	2.00	9999999.00
9	2.00	5.00	2.00	0.0	3.00
10	5.00	9999999.00	9999999.00	3.00	0.0
11	9999999.00	9999999.00	7.00	8.00	4.00

	11
1	9999999.00
2	9999999.00
3	9999999.00
4	9999999.00
5	9999999.00
6	9999999.00
7	9999999.00
8	7.00
9	8.00
10	4.00
11	0.0

SHORTEST DISTANCE MATRIX

	1	2	3	4	5
1	0.0	4.00	3.00	6.00	7.00
2	4.00	0.0	7.00	5.00	3.00
3	3.00	7.00	0.0	4.00	6.00
4	6.00	5.00	4.00	0.0	2.00
5	7.00	3.00	6.00	2.00	0.0
6	9.00	6.00	6.00	3.00	3.00
7	8.00	5.00	6.00	2.00	2.00
8	10.00	7.00	8.00	4.00	4.00
9	11.00	8.00	8.00	5.00	5.00
10	14.00	11.00	11.00	8.00	8.00
11	17.00	14.00	15.00	11.00	11.00

	6	7	8	9	10	11
1	9.00	8.00	10.00	11.00	14.00	17.00
2	6.00	5.00	7.00	8.00	11.00	14.00
3	6.00	6.00	8.00	8.00	11.00	15.00
4	3.00	2.00	4.00	5.00	8.00	11.00
5	3.00	2.00	4.00	5.00	8.00	11.00
6	0.0	1.00	3.00	2.00	5.00	9.00
7	1.00	0.0	2.00	3.00	6.00	9.00
8	3.00	2.00	0.0	2.00	5.00	7.00
9	2.00	3.00	2.00	0.0	3.00	7.00
10	5.00	6.00	5.00	3.00	0.0	4.00
11	9.00	9.00	7.00	7.00	4.00	0.0

MATRIX REPRESENTING THE NODES ON THE SHORTEST CHAINS

	1	2	3	4	5	6	7	8	9	10	11
1	1	2	3	4	2	3	4	4	3	3	4
2	1	2	1	4	5	5	5	5	5	5	5
3	1	1	3	4	4	6	4	4	6	6	4
4	1	2	3	4	5	7	7	7	7	7	7
5	2	2	4	4	5	7	7	8	7	7	8
6	3	7	3	7	7	6	7	7	9	10	10
7	4	5	4	4	5	6	7	8	6	6	8
8	7	5	7	7	5	7	7	8	9	9	11
9	6	6	6	6	6	6	6	8	9	10	10
10	6	6	6	6	6	6	6	9	9	10	11
11	8	8	8	8	8	10	8	8	10	10	11

2.5 Shortest-Path Models with Fixed Charges

In previous sections, algorithms have been presented which can be utilized to determine the minimum-cost route from a given source node s to a given sink node t. In addition, the algorithm of Dijkstra will produce the shortest route from an arbitrary source node s to all other nodes in the network. These algorithms, and indeed all shortest-route algorithms, are based upon the following fundamental observation: "If the shortest route between nodes s and t passes through node k, then that segment of the route from node s to node k is also the shortest route to node k. In addition, the route from node k to node t is the shortest route between those two nodes."

We will now investigate a particular class of problems for which this fundamental observation will not be true. This class of problems is known as the shortest-route model with fixed charges or simply the *fixed-charge problem.*

Fixed-charge problems arise when certain additional penalties or costs are incurred in traversing through one or more nodes in a network. As an illustration of this type of problem, let us consider the shipping of commodities from known storage locations to several centers of consumption, under the assumption that there is a fixed cost for the initial use of the transshipment points of a distribution network. Another example would be the routing of freighters on sea voyages between ports of call, under the consideration of port entry charges. These charges vary from port to port, depending upon the size of the freighters, the nature of the shipment, and government control policies. In general, such problems occur when transshipments facilities must be built, bought, or rented for interim usage.

Several iterative procedures based upon Lagrangian multipliers and/or duality theory have been proposed to solve fixed-charge network models of the type previously described [23, 37, 36]. Although the analysis of these approaches is beyond the scope of this book, we would like to address a particular class of transportation problems which can be readily solved using the mechanics presented in earlier sections of the book.

Many transportation planning problems are based upon the optimal routing of traffic through a network of one-way and/or two-way streets. In the analysis of traffic-flow problems, it is common to include delays at intersections or "turn penalties." Such delays or penalties can be expressed as fixed costs. These turn penalties are usually dependent upon the direction of entry and the direction of exit through an intersection. A prohibitive turn can be treated as one with infinite costs or penalty. To illustrate the computational aspects of this problem, let us consider the rectangular grid network of Figure 2.20. This network represents a segment of city traffic intersections leading from an entry point at node 1 to an exit point at node 8. The cost of traversing each arc is given beside each arc. Additionally, it is assumed that

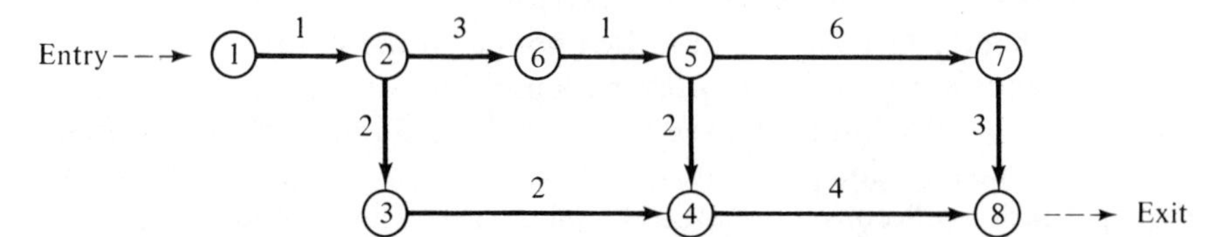

Figure 2.20. Sample network with turn penalties.

any turn will incur a penalty equal to 3. It is desired to identify the most economical route to travel from node 1 to node 8. The following theorem will be used.

Theorem [18]: In a network with turn penalties, the shortest route from node s to node t through an intermediate node k may not include the shortest route from node s to node k, or from node k to node s.

This theorem can be verified by examining the traffic network given in Figure 2.20. It is easy to obtain the most economical route by enumerating all possibilities and identifying the one with the lowest total cost. The corresponding results are summarized in Table 2.7. From the results given in the table it can be seen that the most economical path from node 1 to node 8 is P_3 with a total cost equal to 15. Note that the cost of the subpath of P_3 which connects node 1 with node 4 is equal to 11. From the inspection of the arcs of P_1 and P_2, it is concluded that the other alternative route from node 1 to node 4 is a subpath of P_2 consisting of arcs (1, 2), (2, 6), (6, 5), and (5, 4).

Table 2.7. PATHS FOR EXAMPLE WITH TURN PENALTIES

Path	*Arc Sequence in the Path*	*Travel Cost*	*Turn Cost*	*Total Cost*	
P_1	(1, 2)	1	0	1	
	(2, 6)	3	0	3	
	(6, 5)	1	0	1	
	(5, 7)	6	0	6	
	(7, 8)	3	3	6	17
P_2	(1, 2)	1	0	1	
	(2, 6)	3	0	3	
	(6, 5)	1	0	1	
	(5, 4)	2	3	5	
	(4, 8)	4	3	7	17
P_3	(1, 2)	1	0	1	
	(2, 3)	2	3	5	
	(3, 4)	2	3	5	
	(4, 8)	4	0	4	15

The total cost of this subpath is 10. Therefore, this subpath is the shortest path from node 1 to node 4, and it does not lie on the shortest route from node 1 to node 8. Hence, the shortest-path problem with turn penalties cannot be directly solved using conventional shortest-path network procedures. However, a modification to the basic problem structure can be made to allow the application of conventional techniques [18]. The procedure can be described as follows. A network $G = (\mathbf{N}, \mathbf{A})$ with β nodes in set $\mathbf{N}$ and α arcs in set $\mathbf{A}$ will be considered.

***Step 1*:** Add a pseudo node $\bar{s}$ and connect it to the source node s by defining a direct arc $(\bar{s}, s)$; add a pseudo node $\bar{t}$ and connect the terminal node t to $\bar{t}$ by defining a direct arc $(t, \bar{t})$. For the numerical example under consideration, we obtain the result shown in Figure 2.21(a).

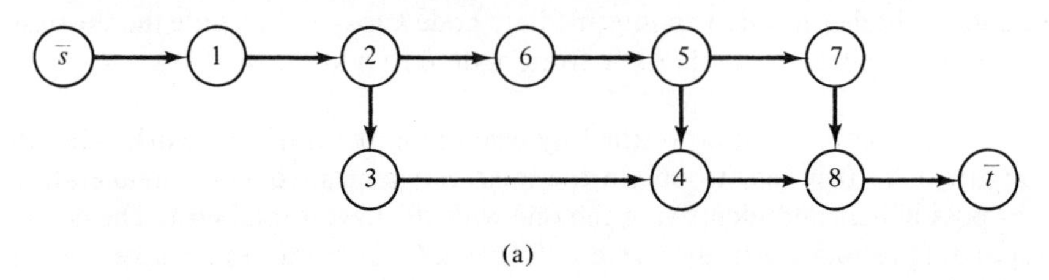

(a)

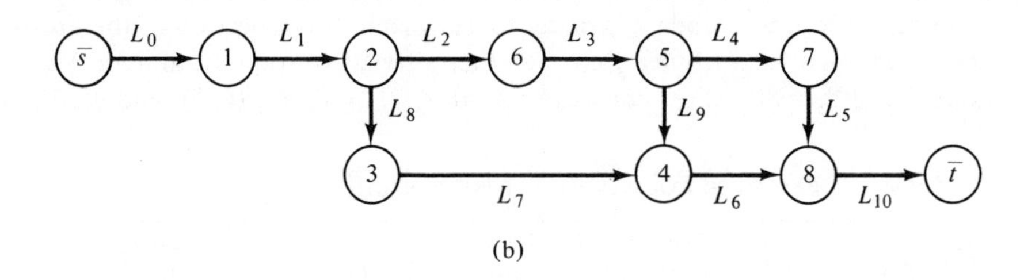

(b)

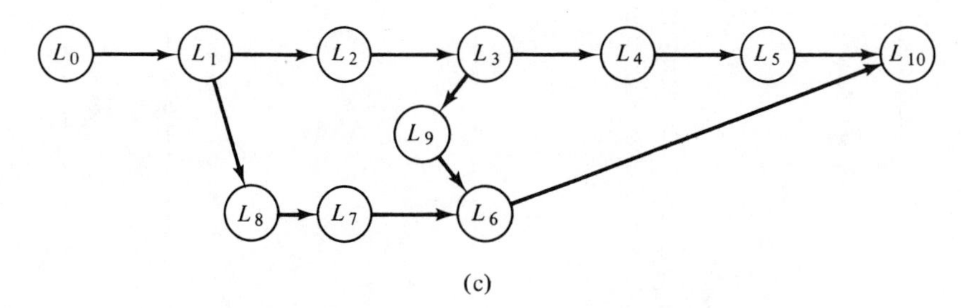

(c)

Figure 2.21. Turn penalty algorithm: (a) step 1; (b) step 2; (c) step 3.

***Step 2*:** Assign a pseudo label L_k to each arc of the network. Let $L_0, L_1, \ldots, L_\alpha, L_{\alpha+1}$ be the pseudo labels for the $\alpha + 2$ arcs of the enlarged network. For the numerical example being discussed, we would obtain the results shown in Figure 2.21(b).

***Step 3*:** Create a pseudo network consisting of $\alpha + 2$ nodes $L_0, L_1, \ldots, L_{\alpha+1}$ such that the directed arc (L_i, L_j) in the pseudo network is defined if the arc with label L_i immediately precedes the arc with label L_j in the labeled network of step 2. The arc parameter of the branch connecting node L_i to node L_j is given by $c(L_i) + p(L_i, L_j)$, where $c(L_i)$ is the original cost of L_i and $p(L_i, L_j)$ is the turn penalty associated with the branch (L_i, L_j). By definition, the costs of original arcs $(\bar{s}, s)$ and $(t, \bar{t})$ are equal to zero, and nodes $\bar{s}$ and $\bar{t}$ are never turned. For the sample problem of this section, the pseudo network given in Figure 2.21(c) is appropriate. The arc costs are calculated in Table 2.8.

Table 2.8. ARC COSTS FOR PSEUDO NETWORK OF STEP 3

From Node	*To Node*	*Cost*
L_0	L_1	0
L_1	L_2	1
L_1	L_8	4
L_8	L_7	5
L_2	L_3	3
L_3	L_9	4
L_9	L_6	5
L_3	L_4	1
L_4	L_5	9
L_5	L_{10}	3
L_6	L_{10}	4
L_7	L_6	2

Using any shortest-path algorithm, the most economical route from node L_0 to node L_{10} can be obtained. The reader can verify that the most economical route corresponds to the sequence of arcs (L_0, L_1), (L_1, L_8), (L_8, L_7), (L_7, L_6), (L_6, L_{10}). The cost of the route is equal to 15, as indicated in Table 2.9(a).

Translating this solution back into the original network formulation, we obtain the results given in Table 2.9(b). Therefore, the minimal-cost route in the original network corresponds to the sequence of nodes 1, 2, 3, 4, 8, and has a total cost equal to 15.

Table 2.9(a). OPTIMAL PSEUDO ROUTE

From Node	*To Node*	*Cost*
L_0	L_1	0
L_1	L_8	4
L_8	L_7	5
L_7	L_6	2
L_6	L_{10}	4

Table 2.9(b). OPTIMAL SOLUTION

Arc	*Original Node*	*Cost*
(L_0, L_1)	1	—
(L_1, L_8)	2	1
(L_8, L_7)	3	5
(L_7, L_6)	4	5
(L_6, L_{10})	8	4

2.6 The K-Shortest-Path Problem

In applications such as communication and transportation networks, it is sometimes desirable to have knowledge of several shortest paths, arranged in increasing order according to their lengths, in contrast to simply a shortest path. As an example, the availability of good alternatives can be used by transportation planners to model more realistically the flow of vehicular traffic on road networks. As a second example, the routing of messages through a communication network when some routes are temporarily obstructed can be based on the best alternatives available.

Thus, the identification of additional solutions provides an alternative approach for planning when the best solution is not available or is infeasible. Also, the knowledge of next-best solutions allows for a sensitivity analysis of the solutions with respect to external factors not included in the network model.

The general version of the K-shortest-path problem admits cycles in the paths and ties between the path lengths. Several relaxations of the problem are possible in order to provide alternative solution procedures to specific network problems of interest. Traditional approaches have been developed by several analysts and have been summarized and discussed in review papers, such as the one by Dreyfus [12]. Some new methods exploit a fairly strong analogy between the solution to the general K-shortest-path problem and the solution of a system of ordinary linear equations. One of these methods, known as the "double-sweep method" [47], simultaneously calculates the K shortest path lengths from a particular source node to all other nodes in the network. This method will be the subject of our present discussion.

2.6.1 The Double-Sweep Method

Consider a directed network with nodes numbered 1 through n. Suppose that for each node there is a vector with estimates of the K shortest path lengths from a given source node. Under the assumption that the initial estimates do not underestimate the path lengths, the double-sweep method successively reduces the estimates until the optimal vector of estimates is achieved in a finite number of iterations.

Each iteration consists of two passes. In the forward pass, the nodes are considered in increasing numerical order (i.e., $j = 1, 2, \ldots, n$). After identifying the list of nodes i incident to node j, such that $i < j$, the K shortest path lengths from the source to node j are successively examined to verify if shorter path lengths are possible through the incident nodes. If such path lengths exist, they will be used as new estimates in further iterations. A similar procedure is performed during the backward pass of the algorithm, but in this case the nodes are considered in decreasing numerical order (i.e., $j = n, n - 1, \ldots, 1$) and only incident nodes $i > j$ are investigated.

The solution procedure of the double-sweep method involves the use of two special algebraic operations developed in reference [47]. These operations are performed on vectors rather than on single numbers, and for that reason are called "generalized operations." The vectors to be considered must all have the same dimension, and can have finite as well as infinite elements, but it is required that the finite elements be in *strictly* increasing order and precede the infinity elements. A result of this requirement is that all the finite elements of the vector are numerically different. The following are examples of acceptable vectors:

$$\mathbf{A} = [-4, 0, 7, \infty]$$

$$\mathbf{B} = [3, 4, \infty, \infty]$$

On the other hand, the following vectors are not acceptable:

$$\mathbf{C} = [-4, 3, 3, 9]$$

$$\mathbf{D} = [-9, 0, \infty, 9]$$

The two generalized operations are now discussed. The first operation is referred to as a *generalized minimization* and the second one as a *generalized addition.* They are binary operations; that is, they can be performed on only two vectors at a time.

The generalized minimization simply defines a set formed with the elements of two given vectors of equal dimension, and then constructs a third vector, of same dimension, by putting elements of the set in strictly increasing order of magnitude. If the number of finite elements in the new vector is less than its dimension, the vector is completed with infinite values. As an illustration, let us consider the vectors $\mathbf{A} = [-4, 0, 1, \infty]$ and $\mathbf{B} = [1, 7, 8, 9]$. The set formed with the elements of vectors **A** and **B** is given by $\{-4, 0, 1, \infty, 1, 7, 8, 9\}$. The elements of this set can be arranged in strictly increasing order as follows: $-4, 0, 1, 7, 8, 9, \infty$. Since the vectors under consideration have dimension equal to 4, the generalized minimization of **A** and **B** is given by the vector $\mathbf{C} = [-4, 0, 1, 7]$, which contains the four minimal elements of the set under consideration.

The generalized addition defines a set formed with the cross sums of the elements of two given vectors of equal dimension, and then constructs a third vector, of same dimension, with the minimal element of the set in its first position. All the remaining elements are arranged as explained above. As an illustration, let us consider the two vectors **A** and **B** used to explain the generalized addition. The cross sums of the elements of these two vectors are as follows:

		Elements of **A**			
		−4	0	1	∞
Elements of **B**	1	−3	1	2	∞
	7	3	7	8	∞
	8	4	8	9	∞
	9	5	9	10	∞

The set consisting of cross sums is therefore defined as $\{-3, 1, 2, \infty, 3, 7, 8, \infty, 4, 8, 9, \infty, 5, 9, 10, \infty\}$. Also, the elements of this set can be arranged in strictly increasing order as follows: $-3, 1, 2, 3, 4, 5, 7, 8, 9, 10, \infty$. Then, the generalized addition of **A** and **B** is equal to $\mathbf{C} = [-3, 1, 2, 3]$.

A formal definition of these operations is now presented in order to support further discussion of the double-sweep algorithm. The following notation will be used:

$\mathbf{R}$	the set of real numbers
$\mathbf{R}_\infty$	the set consisting of the elements of **R** plus the element ∞
K	the desired number of path lengths
$\mathrm{S}(K)$	the set of vectors of dimension K with elements from $\mathbf{R}_\infty$ arranged in strictly increasing order
Min_j	the operation of identifying the jth minimal element of a given set

Since by definition the elements of $\mathbf{S}(K)$ are vectors that have their entries in *strictly* increasing order, we must adopt the convention that $\infty < \infty$, when there are several infinite entries in the vectors.

Generalized minimization. Let A, B, and C be three vectors from $S(K)$. That is,

$$\begin{aligned}
\mathbf{A} &= [a_1, a_2, \ldots, a_K], & a_1 &< a_2 < \ldots < a_K;\ a_i \in \mathbf{R}_\infty \\
\mathbf{B} &= [b_1, b_2, \ldots, b_K], & b_1 &< b_2 < \ldots < b_K;\ b_i \in \mathbf{R}_\infty \\
\mathbf{C} &= [c_1, c_2, \ldots, c_K], & c_1 &< c_2 < \ldots < c_K;\ c_i \in \mathbf{R}_\infty
\end{aligned}$$

Let $\mathbf{T}_+$ be the set formed with the elements of the vectors $\mathbf{A}$ and $\mathbf{B}$. The generalized minimization $\oplus$ is defined by $\mathbf{A} \oplus \mathbf{B} = \mathbf{C}$, where $c_j = \min_j [\mathbf{T}_+]$, for $j = 1, \ldots, K$.

Generalized addition. Let $\mathbf{A}$, $\mathbf{B}$, and $\mathbf{C}$ be three vectors from $\mathbf{S}(K)$. Let $\mathbf{T}_x$ be the set formed with the cross sums of the elements of the vectors $\mathbf{A}$ and $\mathbf{B}$. The generalized addition $\otimes$ is defined by $\mathbf{A} \otimes \mathbf{B} = \mathbf{C}$, where $c_j = \min_j [\mathbf{T}_x]$, for $j = 1, \ldots, K$.

Computational methodology. Let $\mathbf{A}$ be any vector in $\mathbf{S}(K)$, and define $\mathbf{F} = [0, \infty, \infty, \ldots, \infty]$ and $\mathbf{V} = [\infty, \infty, \ldots, \infty]$, both in $\mathbf{S}(K)$; then it is obvious that the following relations always hold:

$$\mathbf{A} \oplus \mathbf{V} = \mathbf{A} \tag{2-55}$$

$$\mathbf{A} \otimes \mathbf{V} = \mathbf{V} \tag{2-56}$$

$$\mathbf{A} \otimes \mathbf{F} = \mathbf{A} \tag{2-57}$$

The computational aspects of the double-sweep method developed by Shier [47] can be summarized as follows. Assume that all nodes in the network are numbered 1 through n, and that the length of each arc (i, j) is equal to d_{ij}. Now, define:

$$\begin{aligned}
\mathbf{D}_{ij} &= [d_{ij}, \infty, \infty, \ldots, \infty] \in \mathbf{S}(K) \\
\mathbf{D} &= [\mathbf{D}_{ij}] \\
\mathbf{L} &= [\mathbf{L}_{ij}], \qquad \text{where } \mathbf{L}_{ij} = \mathbf{D}_{ij} \text{ for } i > j;\quad \mathbf{L}_{ij} = \mathbf{V} \text{ for } i \leq j \\
\mathbf{U} &= [\mathbf{U}_{ij}], \qquad \text{where } \mathbf{U}_{ij} = \mathbf{D}_{ij} \text{ for } i < j;\quad \mathbf{U}_{ij} = \mathbf{V} \text{ for } i \geq j
\end{aligned}$$

In these definitions $\mathbf{D}_{ij}$, $\mathbf{L}_{ij}$, and $\mathbf{U}_{ij}$ are vectors from $\mathbf{S}(K)$, and $\mathbf{D}$, $\mathbf{L}$, and $\mathbf{U}$ are matrices whose entries are vectors from $\mathbf{S}(K)$. If the network under consideration has more than one arc directly joining two nodes i and j, we can redefine $\mathbf{D}_{ij}$ as follows:

$$\mathbf{D}_{ij} = [d_{ij}^1, d_{ij}^2, \ldots, d_{ij}^t, \infty, \infty, \ldots, \infty] \in \mathbf{S}(K)$$

where $d_{ij}^1, d_{ij}^2, \ldots, d_{ij}^t$ are the lengths of the arcs joining i and j. Note that if $t > k$, no ∞ entries will be in $\mathbf{D}_{ij}$.

Let $\mathbf{E}_{0m}$ be a vector in $\mathbf{S}(K)$ containing the initial estimates of the K shortest path lengths from the origin to node m. It is assumed that the first element of $\mathbf{E}_{0m}$ is equal to zero when m is the source node. All the vectors $\mathbf{E}_{0m}$, $m = 1, 2, \ldots, n$, can be arranged in an array $\mathbf{E}(0)$ as follows:

$$\mathbf{E}(0) = [\mathbf{E}_{01}, \mathbf{E}_{02}, \ldots, \mathbf{E}_{0n}]$$

Note that $\mathbf{E}(0)$ is actually a vector whose entries are elements of $\mathbf{S}(K)$. At the wth single sweep, the double-sweep algorithm constructs a vector $\mathbf{E}(w)$, defined as follows:

$$\mathbf{E}(w) = [\mathbf{E}_{w1}, \mathbf{E}_{w2}, \ldots, \mathbf{E}_{wn}]$$

where $\mathbf{E}_{wm}$ is an element of $\mathbf{S}(K)$ containing the current estimates of the K shortest path lengths from the origin to node m. The generation of vectors of estimates is accomplished by the following pair of recursive relationships:

$$\mathbf{E}(2r + 1) = \mathbf{E}(2r) \oplus \mathbf{E}(2r + 1) \otimes \mathbf{L} \quad (2\text{-}58)$$

$$\mathbf{E}(2r + 2) = \mathbf{E}(2r + 1) \oplus \mathbf{E}(2r + 2) \otimes \mathbf{U} \quad (2\text{-}59)$$

These relationships are used in alternating fashion, for each value $r = 0, 1, 2, \ldots$. The solution converges to an optimum when the estimates remain unchanged after two successive applications of the relationships. The operation defined in Eq. (2-58) is called a *backward sweep* and the one defined in Eq. (2-59) is called a *forward sweep*. In both of these equations, the generalized addition is performed first.

Once the path lengths are obtained, a tracing procedure is used to identify the corresponding paths. In order to describe the tracing procedure, suppose that it is desired to find the path (or paths) from the origin to node i, corresponding to the m-shortest path length. Let H_{mi} be this path length, and let j be a node incident to node i. Then,

$$H_{mi} = H_{tj} + d_{ji} \quad (2\text{-}60)$$

where, as defined before, d_{ji} is the length of the arc (j, i) and H_{tj} is the t-shortest path length, $t \leq m$, corresponding to node j. Thus the tracing procedure performs a node search at each node i in order to identify the node

j for which Eq. (2-60) is satisfied. Once j is found, the same procedure is repeated until the origin node is reached.

If only paths without cycles are desired, the previous tracing procedure should be modified. Essentially, the modification consists of a test to inspect if a node has already been considered by the backward relationship of Eq. (2-60) when the node becomes a candidate to belong in the path under consideration.

2.6.2 SAMPLE PROBLEM FOR THE DOUBLE-SWEEP METHOD

Consider the network represented in Figure 2.22. The following hypothetical input information is assumed. It is desired to investigate the three shortest path lengths from node 1 to all the other nodes.

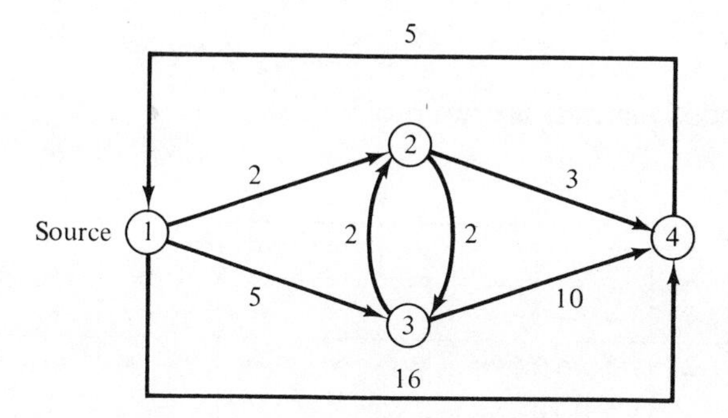

Figure 2.22. Network for sample problem.

For this example, $K = 3$ and $n = 4$. An acceptable arbitrary choice for the vectors of initial estimates is given by $\mathbf{E}_{01} = [0, 21, 22]$, $\mathbf{E}_{02} = \mathbf{E}_{03} = \mathbf{E}_{04} = [20, 21, 22]$. If choosing finite estimates is difficult or impossible, all the entries can be set equal to ∞, with the exception of the estimate of the shortest path length for the source node, which must be always equal to zero.

The distance matrix $\mathbf{D}$, and the matrices $\mathbf{L}$ and $\mathbf{U}$ consist of 16 elements, each element being a vector from S(3), arranged as follows:

$$\mathbf{D} = [\mathbf{D}_{ij}] = \begin{bmatrix} \mathbf{D}_{11} & \mathbf{D}_{12} & \mathbf{D}_{13} & \mathbf{D}_{14} \\ \mathbf{D}_{21} & \mathbf{D}_{22} & \mathbf{D}_{23} & \mathbf{D}_{24} \\ \mathbf{D}_{31} & \mathbf{D}_{32} & \mathbf{D}_{33} & \mathbf{D}_{34} \\ \mathbf{D}_{41} & \mathbf{D}_{42} & \mathbf{D}_{43} & \mathbf{D}_{44} \end{bmatrix}$$

$$\mathbf{L} = [\mathbf{L}_{ij}] = \begin{bmatrix} \mathbf{V} & \mathbf{V} & \mathbf{V} & \mathbf{V} \\ \mathbf{D}_{21} & \mathbf{V} & \mathbf{V} & \mathbf{V} \\ \mathbf{D}_{31} & \mathbf{D}_{32} & \mathbf{V} & \mathbf{V} \\ \mathbf{D}_{41} & \mathbf{D}_{42} & \mathbf{D}_{43} & \mathbf{V} \end{bmatrix}$$

$$\mathbf{U} = [\mathbf{U}_{ij}] = \begin{bmatrix} \mathbf{V} & \mathbf{D}_{12} & \mathbf{D}_{13} & \mathbf{D}_{14} \\ \mathbf{V} & \mathbf{V} & \mathbf{D}_{23} & \mathbf{D}_{24} \\ \mathbf{V} & \mathbf{V} & \mathbf{V} & \mathbf{D}_{34} \\ \mathbf{V} & \mathbf{V} & \mathbf{V} & \mathbf{V} \end{bmatrix}$$

The $\mathbf{D}_{ij}$ vectors can be obtained by enlarging the entries of $[d_{ij}]$ to accommodate $K - 1 = 2$ infinite values. The array $[d_{ij}]$ is obtained directly from the network:

$$[d_{ij}] = \begin{bmatrix} 0 & 2 & 5 & 16 \\ \infty & 0 & 2 & 3 \\ \infty & 2 & 0 & 10 \\ 5 & \infty & \infty & 0 \end{bmatrix}$$

The **D**, **L**, and **U** matrices are given as follows:

D

0 ∞ ∞	2 ∞ ∞	5 ∞ ∞	16 ∞ ∞
∞ ∞ ∞	0 ∞ ∞	2 ∞ ∞	3 ∞ ∞
∞ ∞ ∞	2 ∞ ∞	0 ∞ ∞	10 ∞ ∞
5 ∞ ∞	∞ ∞ ∞	∞ ∞ ∞	0 ∞ ∞

L

∞ ∞ ∞	∞ ∞ ∞	∞ ∞ ∞	∞ ∞ ∞
∞ ∞ ∞	∞ ∞ ∞	∞ ∞ ∞	∞ ∞ ∞
∞ ∞ ∞	2 ∞ ∞	∞ ∞ ∞	∞ ∞ ∞
5 ∞ ∞	∞ ∞ ∞	∞ ∞ ∞	∞ ∞ ∞

U

∞ ∞ ∞	2 ∞ ∞	5 ∞ ∞	16 ∞ ∞
∞ ∞ ∞	∞ ∞ ∞	2 ∞ ∞	3 ∞ ∞
∞ ∞ ∞	∞ ∞ ∞	∞ ∞ ∞	10 ∞ ∞
∞ ∞ ∞	∞ ∞ ∞	∞ ∞ ∞	∞ ∞ ∞

As an illustration, consider the case $r = 0$. The backward sweep produces a vector of estimates, $\mathbf{E}(1)$, given by

$$\mathbf{E}(1) = \mathbf{E}(0) \oplus \mathbf{E}(1) \otimes \mathbf{L}$$

The generalized addition $\mathbf{E}(1) \otimes \mathbf{L}$ is equal to

$$\begin{bmatrix} 25 & 22 & \infty & \infty \\ 26 & 23 & \infty & \infty \\ 27 & 24 & \infty & \infty \end{bmatrix}$$

After performing the backward sweep, the result obtained is $\mathbf{E}(1) = \mathbf{E}(0)$. The computational details of the evaluation of $\mathbf{E}(1)$ will now be discussed in detail. The steps involved in the operation $\mathbf{E}(1) \otimes \mathbf{L}$ resemble those of the right-multiplication of a vector by a matrix, with the elements of the product being identified in a backward fashion. In our case, however, the arrays consist of K-tuples from $\mathbf{S}(K)$ instead of single elements from $\mathbf{R}_{\infty}$, and the $\oplus$ and $\otimes$ operations are used instead of the regular $+$ and $\times$ operations, respectively. Once a given element of $\mathbf{E}(1) \otimes \mathbf{L}$ is computed, it is compared against the corresponding element of $\mathbf{E}(0)$ by means of a generalized minimization, in order to produce the corresponding element of $\mathbf{E}(1)$. In Figure 2.23 we display the arrays $\mathbf{E}(0)$, $\mathbf{L}$, $\mathbf{E}(1) \otimes \mathbf{L}$, and indicate how to obtain the elements of $\mathbf{E}(1)$ in the backward sweep under consideration.

The first forward sweep is performed next. The corresponding operation is indicated by

$$\mathbf{E}(2) = \mathbf{E}(1) \oplus \mathbf{E}(2) \otimes \mathbf{U}$$

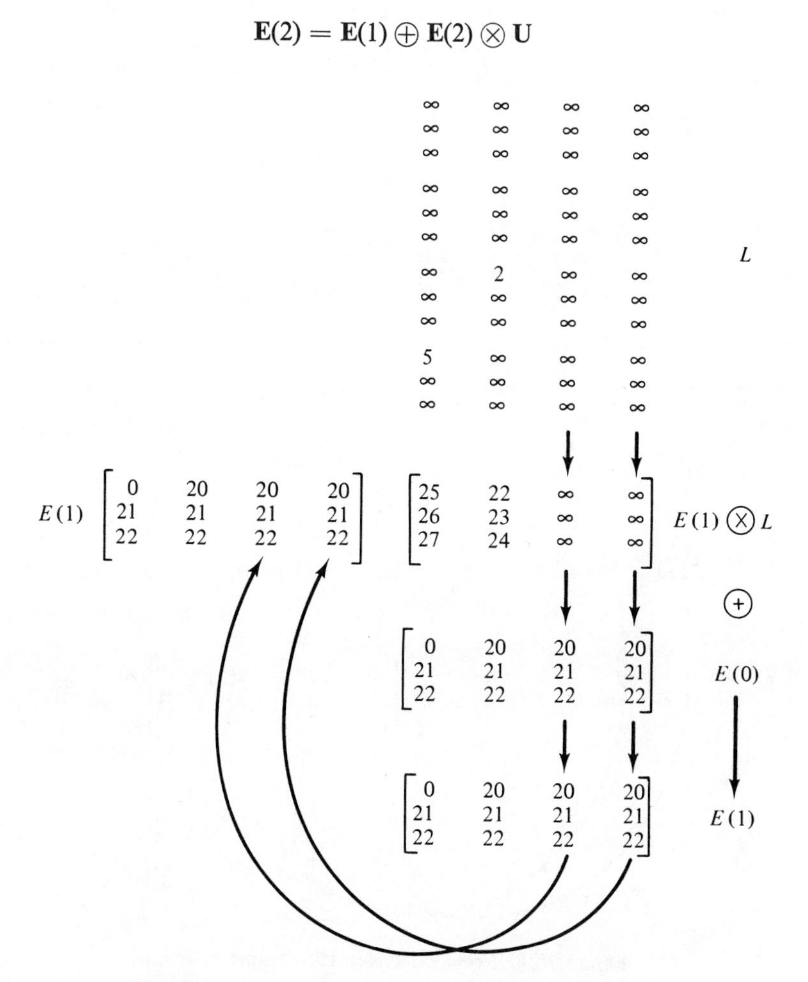

Figure 2.23. Backward sweep for sample problem.

Similarly, the generalized addition $\mathbf{E}(2) \otimes \mathbf{U}$ is given by the following vector of elements in $\mathbf{S}(3)$

$$\begin{bmatrix} \infty & 2 & 4 & 5 \\ \infty & 23 & 5 & 14 \\ \infty & 24 & 22 & 15 \end{bmatrix}$$

Then, the new vector of estimates, $\mathbf{E}(2)$, is given by

$$\begin{bmatrix} 0 & 2 & 4 & 5 \\ 21 & 20 & 5 & 14 \\ 22 & 21 & 20 & 15 \end{bmatrix}$$

In this case, we perform a sequence of steps similar to the ones described for the generalized addition $\mathbf{E}(1) \otimes \mathbf{L}$, but considering $\mathbf{E}(2) \otimes \mathbf{U}$ instead, and computing the elements in a forward fashion, as shown in Figure 2.24.

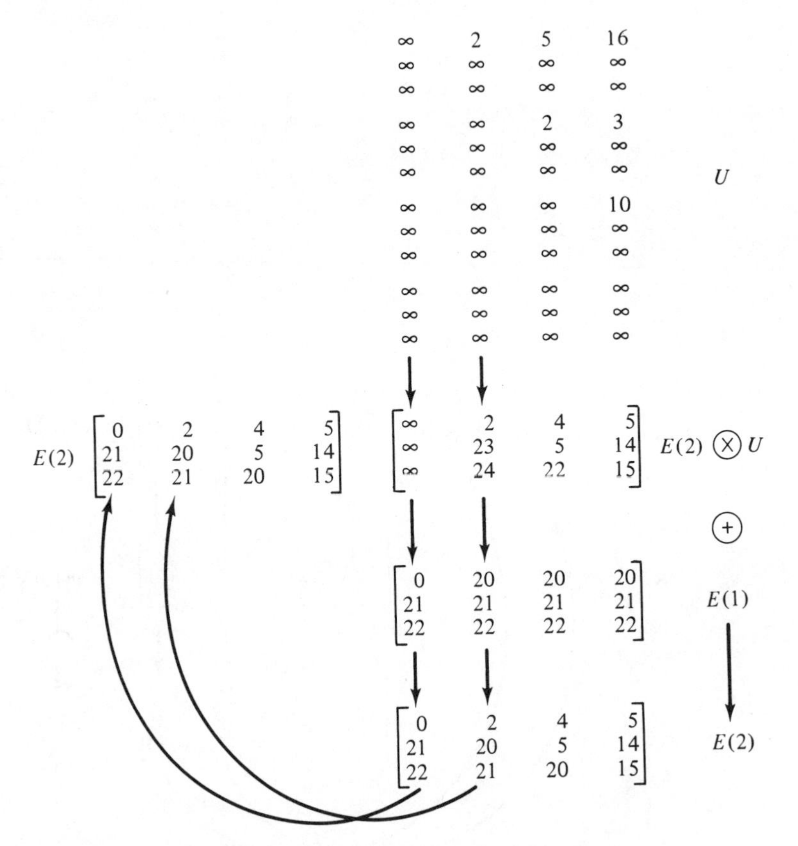

Figure 2.24. Forward sweep for sample problem.

Proceeding in this fashion, it is possible to obtain the results summarized in Table 2.10. For $r = 2$, both the backward and forward sweeps yield the same solution. As previously explained, this indicates that the solution is optimal.

Table 2.10. RESULTS FOR SAMPLE PROBLEM—DOUBLE-SWEEP METHOD

r	*Type of Pass*	*Vector of Estimates*
0	Backward	$\begin{bmatrix} 0 & 20 & 20 & 20 \\ 21 & 21 & 21 & 21 \\ 22 & 22 & 22 & 22 \end{bmatrix}$
0	Forward	$\begin{bmatrix} 0 & 2 & 4 & 5 \\ 21 & 20 & 5 & 14 \\ 22 & 21 & 20 & 15 \end{bmatrix}$
1	Backward	$\begin{bmatrix} 0 & 2 & 4 & 5 \\ 10 & 6 & 5 & 14 \\ 19 & 7 & 20 & 15 \end{bmatrix}$
1	Forward	$\begin{bmatrix} 0 & 2 & 4 & 5 \\ 10 & 6 & 5 & 9 \\ 19 & 7 & 8 & 10 \end{bmatrix}$
2	Backward	$\begin{bmatrix} 0 & 2 & 4 & 5 \\ 10 & 6 & 5 & 9 \\ 14 & 7 & 8 & 10 \end{bmatrix}$
2	Forward	$\begin{bmatrix} 0 & 2 & 4 & 5 \\ 10 & 6 & 5 & 9 \\ 14 & 7 & 8 & 10 \end{bmatrix}$

The results from the application of the tracing procedure described recursively by Eq. (2-60) are shown in Figure 2.25, for the shortest, 2nd shortest, and 3rd shortest paths from node 1 to node 4. Notice that the second-shortest path length corresponds to a path with a cycle.

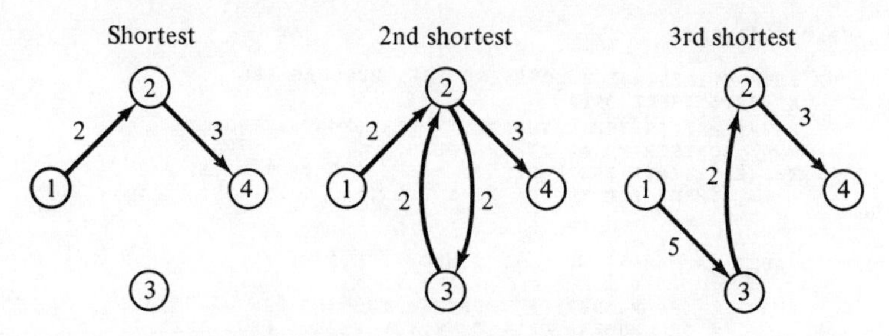

Figure 2.25. Paths for sample problem.

2.6.3 Computer Code for the Double-Sweep Algorithm

```
C***********************************************************************
C                                                                      *
C PURPOSE:                                                             *
C                                                                      *
C      THE ALGORITHM COMPUTES THE K SHORTEST PATH LENGTHS IN A NETWORK *
C      FROM AN ORIGIN TO ALL OTHER NODES IN THE NETWORK.               *
C                                                                      *
C LOCATION: SUBROUTINE KSHORT IN NETWORK OPTIMIZATION                  *
C                                                                      *
C USAGE:                                                               *
C                                                                      *
C      AT THE PRESENT THE PROGRAM HANDLES NETWORKS WITH UP TO 50 NODES *
C      AND 50 ARCS. THE CAPACITY CAN BE INCREASED BY INCREASING THE    *
C      DIMENSION STATEMENTS IN SUBROUTINE KSHORT AND MAIN PROGRAM. ONE *
C      OR MORE PROBLEMS CAN BE HANDLED AT A TIME.                      *
C                                                                      *
C                                                                      *
C INPUT DATA:                                                          *
C      SET 1:  SINGLE CARD WITH THE CALL NAME OF THE ALGORITHM:KSRT    *
C              -FORMAT(A4)                                             *
C      SET 2:  SINGLE CARD WHICH SPECIFIES THE NUMBER OF NODES AND     *
C              ARCS IN THE PROBLEM.                                    *
C              -FORMAT(2I10)                                           *
C                                                                      *
C      SET 3:  THERE ARE THREE DISTINCT SECTIONS OF INPUT DATA IN      *
C              THIS SET:                                               *
C              SECTION 1:                                              *
C                1) START NODE                                         *
C                2) END NODE                                           *
C                3) ARC LENGTH                                         *
C                -FORMAT(3I10)                                         *
C                  A SINGLE CARD IS READ FOR EACH ARC IN THE NETWORK.  *
C                  THE DATA IN THIS SECTION SHOULD BE PUNCHED IN       *
C                  INCREASING ORDER OF ENDING NODE.                    *
C                                                                      *
C              SECTION 2:                                              *
C                1) K,NS,IMAX                                          *
C                -FORMAT(4X,3I5)                                       *
C                                                                      *
C              SECTION 3:                                              *
C                1) NF,PMAX                                            *
C                -FORMAT(4X,2I5)                                       *
C                  NOTE: IF THIS PROBLEM IS TO BE SEQUENTIALLY REPEATED *
C                        FOR DIFFERENT SOURCES AND/OR TERMINALS, THEN  *
C                        ONLY SECTIONS 2 AND 3 SHOULD BE REPEATED.     *
C                                                                      *
C      SET 4:  THIS SET CONSISTS OF ONLY ONE CARD WHICH INDICATES      *
C              THE END OF A PROBLEM. THE USER SHOULD PUNCH THE WORD    *
C              'PEND'  - FORMAT(A4)                                    *
C                                                                      *
C      SET 5:  THIS SET CONSISTS OF ONLY ONE CARD WHICH INDICATES      *
C              THE END OF THE INPUT DECK. THE USER SHOULD PUNCH THE    *
C              WORD 'EXIT'  -FORMAT(A4)                                *
C                                                                      *
C PROGRAM CONTENTS:                                                    *
C                                                                      *
C      THE SUBROUTINES WHICH COMPRISE THIS PROGRAM ARE:                *
C          KSHORT: INPUT DATA                                          *
C          DSWP: PRINTS THE K SHORTEST PATH LENGTHS FROM THE           *
C                ORIGIN TO AL OTHER NODES.                             *
C          TRACE: PRINTS THE NODES IN THE K SHORTEST PATHS             *
C                 FROM THE SOURCE TO A SPECIFIC TERMINAL.              *
C                                                                      *
C                                                                      *
C THE VARIABLE AND THE ARRAYS IN COMMON USE ARE:                       *
C                                                                      *
C        N       = THE NUMBER OF NODES IN THE NETWORK                  *
C        MU      = THE NUMBER OF ARCS (I,J) WITH I LESS THAN J.        *
C        ML      = THE NUMBER OF ARCS (I,J) WITH J LESS THAN I.        *
```

```
C       LLEN     = AN ARRAY WHOSE J-TH ENTRY IS THE NUMBER OF ARCS            *
C                  (I,J) WITH J LESS THAN I.                                  *
C       LINC     = AN ARRAY CONTAINING THE NODES I INCIDENT TO NODE J         *
C                  WITH I GREATER THAN J, LISTED IN ORDER OF INCREASINGJ.*
C       LVAL     = AN ARRAY CONTAINING THE ARC LENGTH VALUES CORROSPON-      *
C                  DING TO ARCS IN LINC.                                      *
C       ULEN     = AN ARRAY WHOSETJ-TH ENTRY IS THE NUMBER OF ARCS (I,J)     *
C                  WITH I LESS THAN J.                                        *
C       UINC     = AN ARRAY CONTAINING THE NODES I INCIDENT TO NODE J         *
C                  WITH I LESS THAN J, LISTED IN ORDER OF INCREASING J.      *
C       UVAL     = AN ARRAY CONTAINING THE ARC LENGTH VALUES CORROSPON-      *
C                  DING TO ARCS IN UINC.                                      *
C       INF      = A NUMBER LARGER THAN ANY OTHER PATH LENGTH. A NON          *
C                  EXISTING PATH IS ASSIGNED THE LENGTH INF.                  *
C       START    = AN ARRAY WHOSE J-TH ELEMENT INDICATES THE FIRST            *
C                  POSITION ON  INC WHERE NODES INCIDENT TO NODE J ARE       *
C                  LISTED.                                                    *
C       INC      = AN ARRAY CONTAINING NODES I WHICH ARE INCIDENT TO NODE*
C                  J,LISTED IN ORDER OF INCREASING J.                         *
C       VAL      = AN ARRAY CONTAINING THE ARC LENGTH VALUES CORROSPON-      *
C                  DING TO ARCS IN INC.                                       *
C                                                                             *
C ADDITIONAL VARIABLES WHOSE VALUES MUST BE SPECIFIED BY THE USER ARE:       *
C                                                                             *
C       K        = THE NUMBER OF DISTINT PATH LENGTHS REQUIRED.               *
C       IMAX     = THE MAXIMUM NUMBER OF DOUBLE SWEEP ITERATIONS ALLOWED.*
C       NS,NF    = THE INITIAL AND FINAL NODES OF ALL K SHORTEST PATHS       *
C                  TO BE GENERATED.                                           *
C       PMAX     = THE MAXIMUM NUMBER OF PATHS TO BE GENERATED BETWEEN        *
C                  NODES NS AND NF.                                           *
C                                                                             *
C TECHNIQUE USED:                                                             *
C      THE TECHNIQUE USED IN THIS ALGORITHM IS EXPLAINED IN                   *
C      SECTION 2.6.1                                                          *
C                                                                             *
C REFERENCES:                                                                 *
C                                                                             *
C      DOUGLAS R. SHIER, 'COMPUTATIONAL EXPERIENCE WITH AN ALGORITHM          *
C      FOR FINDING THE K SHORTEST PATHS IN A NETWORK',J.OF RESEARCH           *
C      OF THE NATIONAL BUREAU OF STANDARDS-VOL.78B,NO. 3, JULY-SEPT,          *
C      1974.                                                                  *
C                                                                             *
C                                                                             *
C                                                                             *
C******************************************************************************
```

If only paths without cycles are of interest in the context of a given problem, the TRACE subroutine can be modified to stop the identification of a path as soon as a cycle is encountered. The proper modification of the subroutine is as follows:

```
        DO 60 J = 1, K
        IF(X(NI, J) - LT) 60, 180, 70
    180 DO 181 IA = 1, KK
        IF(NI - P(IA)) 181, 70, 181
    181 CONTINUE
        GO TO 80
     60 CONTINUE
```

These instructions replace the instructions in lines 52 through 54 of the TRACE subroutine.

2.6.4 Computational Results

An excellent discussion of the computational efficiency of the double-sweep method is published in [48]. Also, in [48] it is established that the method has several theoretical and computational advantages over other algorithms designed to solve the K-shortest path problem.

In this section we consider networks having a rectangular grid structure, such as the rural road transportation networks in many agricultural areas of the United States. An interesting characteristic of these networks is the availability of several routes with a given length. If it is desired to find a shortest path satisfying one or more external conditions, such as load-carrying capacities of bridges in the routes, we would need to identify all the paths corresponding to a given length. The purpose of this section is to present some computational results related to both the identification of path lengths and the path-tracing procedure.

For a given network with NS representing the source node and NF representing the terminals, we will report on (1) number of paths traced, (2) maximal path length investigated for tracing, (3) total execution time, and

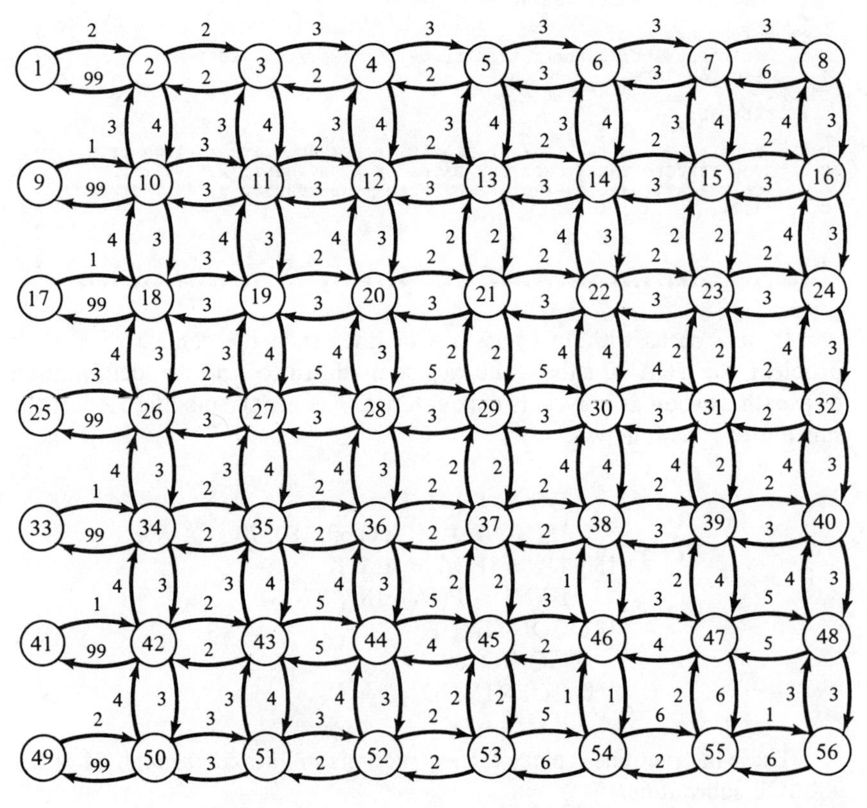

Figure 2.26. Sample network for computational results.

(4) tracing execution time. A network with 56 nodes and 182 arcs as shown in Figure 2.26 will be used. Two cases are examined for this network; in Case 1, NS = 1, NF = 56, $K = 10$; in Case 2, NS = 56, NF takes on seven different values, each value corresponding to a user for which routes between NS and NF are being searched, $K = 5$, and the maximal number of paths between NS and NF is 30. The computational results are given in Tables 2.11(a) and 2.11(b).

In order to obtain the results of Table 2.11(a), the parameter PMAX,

Table 2.11(a). RESULTS FOR CASE 1 (IBM 360)

Number of Paths Traced	*Maximum Path Length*	*Total Execution Time (sec)*	*Trace Execution Time (sec)*
10	3rd	4.62	0.80
20	3rd	4.83	1.01
40	4th	4.88	1.06
80	5th	5.80	1.98
160	7th	7.29	3.47
250	7th	9.22	5.40
285	8th	10.00	6.15

Table 2.11(b). RESULTS FOR CASE 2

Runs	*User Number*	*NF Node*	*Number of Paths Generated*	*jth Path Length without Cycles*	*IBM 360 Execution Time (sec)*	*CYBER 175 Execution Time Total (sec)*	*CYBER 175 Execution Time Trace (sec)*
I	1	4	20	5			
	2	8	17	4			
	3	19	30	5			
	4	29	10	5	3.77	0.271	0.108
	5	33	9	5			
	6	39	8	4			
	7	52	6	4			
II	1	5	10	5			
	2	23	12	5			
	3	11	21	5			
	4	34	9	5	3.89	0.281	0.118
	5	40	5	5			
	6	9	30	5			
	7	50	7	4			
III	1	1	27	5			
	2	9	30	5			
	3	17	30	5			
	4	25	28	5	4.62	0.367	0.204
	5	33	9	5			
	6	41	30	5			
	7	49	7	5			

representing the maximum number of paths to be traced, was set equal to 10, 20, 40, 80, 160, 250, and 285, as indicated in column 1. In each case, $K = 10$ path lengths were evaluated, and the order of the path length corresponding to the PMAX-th path is given in column 2. Column 3 records the total execution time in seconds, and column 4 the execution time for the TRACE subroutine. The difference between these two estimates is equal to the time actually needed for the execution of the double-sweep algorithm.

Table 2.11(b) records the execution times for three runs on an IBM 360 machine and a CYBER 175 machine. Column 1 corresponds to the computer runs. Column 2 corresponds to the users of the system. Column 3 contains the terminal node NF corresponding to each user. Column 4 records the number of paths from NS = 56 to NF for each user. For the case under discussion, this number must be less than or equal to PMAX = 30. Column 5 indicates the order of the maximum cycleless path length traced; this number must be less than or equal to $K = 5$. Columns 6 and 7 show total execution times for the IBM 360 and the CYBER 175, respectively. Finally, column 8 gives the execution time for the TRACE subroutine for the CYBER 175.

2.6.5 Computer Example for a Four-Shortest Path Problem

It is desired to find the paths from node 1 to nodes 10, 11, 12 corresponding to the four shortest path lengths in the network given in Figure 2.27. The code uses starting solutions equal to infinity values. The input data for this problem should be arranged as shown in Table 2.12. Note that the end

Table 2.12. INPUT DATA FOR SAMPLE PROBLEM

SET 1: KSRT
SET 2: 12 31
SET 3:

Card Number				*Card Number*				*Card Number*			
1	2	1	4	2	3	1	4	3	1	2	4
4	3	2	5	5	4	2	2	6	5	2	3
7	1	3	4	8	2	3	5	9	5	3	2
10	6	3	2	11	2	4	2	12	5	4	10
13	7	4	5	14	2	5	3	15	3	5	2
16	4	5	10	17	6	5	5	18	8	5	4
19	3	6	2	20	5	6	5	21	9	6	3
22	4	7	5	23	8	7	8	24	5	8	4
25	7	8	8	26	9	8	12	27	6	9	3
28	8	9	12	29	7	10	2	30	8	11	2
31	9	12	2	32	4	1	30	33	11	10	
34	10	10		35	12	10					

SET 4: PEND
SET 5: EXIT

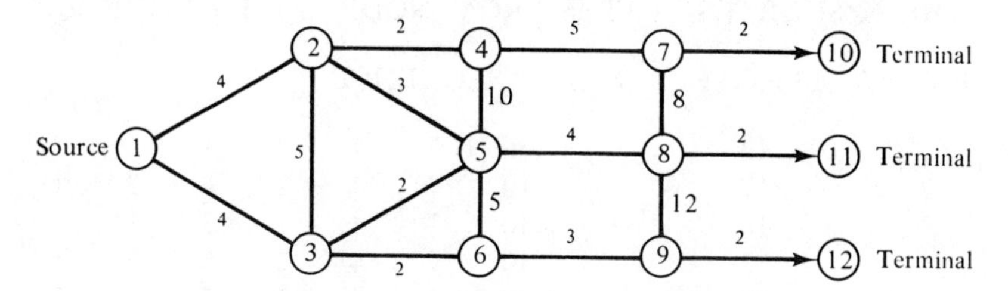

Figure 2.27. Network for computer sample problem.

nodes are listed in *strictly increasing order* from card 1 to card 31. The computer output for this example lists the four shortest path lengths and is shown below.

K = 4 SHORTEST PATH LENGTHS FROM NODE 1

TO NODE				
1	0	8	12	13
2	4	8	9	10
3	4	8	9	12
4	6	10	11	12
5	6	7	10	11
6	6	10	11	12
7	11	15	16	17
8	10	11	14	15
9	9	13	14	15
10	13	17	18	19
11	12	13	16	17
12	11	15	16	17

NUMBER OF ITERATIONS REQUIRED FOR CONVERGENCE = 5

THE K SHORTEST PATHS FROM NODE 1 TO NODE 11

PATH	LENGTH	NODE SEQUENCE						
1	12	11	8	5	3	1		
2	13	11	8	5	2	1		
3	16	11	8	5	3	5	3	1
4	16	11	8	5	3	6	3	1
5	17	11	8	5	2	4	2	1
6	17	11	8	5	3	2	1	
7	17	11	8	5	3	5	2	1
8	17	11	8	5	6	3	1	
9	17	11	8	7	4	2	1	

THE K SHORTEST PATHS FROM NODE 1 TO NODE 10

PATH	LENGTH	NODE SEQUENCE						
1	13	10	7	4	2	1		
2	17	10	7	4	2	4	2	1
3	18	10	7	4	2	3	1	
4	18	10	7	4	2	5	3	1
5	19	10	7	4	2	5	2	1

THE K SHORTEST PATHS FROM NODE 1 TO NODE 12

PATH	LENGTH	NODE SEQUENCE						
1	11	12	9	6	3	1		
2	15	12	9	6	3	5	3	1
3	15	12	9	6	3	6	3	1
4	16	12	9	6	3	2	1	
5	16	12	9	6	3	5	2	1
6	16	12	9	6	5	3	1	
7	17	12	9	6	5	2	1	
8	17	12	9	6	9	6	3	1

2.7 Analysis and Complexity of Shortest-Path Algorithms

The reasons for analyzing algorithms can be classified into two important categories: practical reasons and theoretical reasons. Practical reasons can be summarized as a need to obtain estimates or upper bounds on the computer memory and/or execution-time requirements involved in the implementation of algorithms. Perhaps the most important theoretical reason for the analysis of algorithms is the desirability of quantitative standards that would permit the comparison of two or more algorithms designed to solve the same problem.

The purpose of this section is to provide upper bounds on the amount of computational work involved in the application of Dijkstra's method, Floyd's algorithm, and the double-sweep method. In the algorithms by Dijkstra and by Floyd there are only two types of elementary operations: additions and comparisons. It is usually assumed that an addition and a comparison require approximately the same amount of time. By worst-case conditions for the execution of an algorithm, it is meant that the required number of elementary operations to terminate the algorithm is maximum. This upper bound on amount of computation is a function of both the size of the network and the desired number of solutions (paths), and is referred to as the "computational complexity" of the algorithm. In the case of the double-sweep method, the computational complexity will be given in terms of the number of generalized operations defined in Section 2.6.1.

2.7.1 Computational Complexity of Dijkstra's Method

Let $G = (\mathbf{N}, \mathbf{A})$ be the network under consideration, with n nodes in set $\mathbf{N}$. In the worst-case condition the terminal node is the nth node to receive a permanent label. Assume that at any given time of the execution of the algorithm there are m nodes with permanent labels, and $n - m$ nodes with temporary labels. To identify the $(m + 1)$st node that must be given a permanent label, we update the $n - m$ temporary labels by performing an addition and then one comparison for each label. Once the temporary labels are updated, one more minimization is needed to identify the label that becomes permanent. This minimization operation implies $n - m - 1$ comparisons. Therefore, the number of elementary operations required to assign a permanent label to one more node, given that there are m nodes with permanent labels, is equal to $3(n - m) - 1 \simeq 3(n - m)$.

The total number of elementary operations needed for the termination of the algorithm, under worst-case conditions, is given by

$$\begin{aligned}\sum_{m=1}^{n} 3(n - m) &= 3\sum_{m=1}^{n} (n - m) = 3\sum_{m=1}^{n-1} (n - m) \\ &= 3[(n - 1) + (n - 2) + \ldots + 1] \\ &= \frac{3n(n - 1)}{2}\end{aligned}$$

2.7.2 Computational Complexity of Floyd's Algorithm

Let $G = (\mathbf{N}, \mathbf{A})$ be the network under consideration, with $\mathbf{N}$ defined as the set $\mathbf{N} = \{1, 2, \ldots, n\}$. At any iteration of the algorithm, the total number of entries to be evaluated using the triple operation defined in Eq. (2-53) can be obtained by reasoning as follows:

1. The total number of entries in the matrix is equal to n^2.
2. The number of entries in either the pivot row and/or the pivot column is equal to $2n - 1$.
3. The number of zero entries in the main diagonal is equal to n and need not be reevaluated.
4. One element of the pivot row and one element of the pivot column lie in the main diagonal.
5. The analysis of each entry implies one addition and one comparison according to the triple operation.
6. The maximum total number of operations at a given iteration of the algorithm is then approximately equal to $2n(n - 3)$.

Since the number of iterations is equal to the number of nodes, n, the total number of elementary operations required to terminate the algorithm is at most equal to $2n^2(n - 3)$.

2.7.3 Computational Complexity of the Double-Sweep Method

Let $G = (\mathbf{N}, \mathbf{A})$ be the network under consideration, with $\mathbf{N}$ defined as the set $\mathbf{N} = \{1, 2, \ldots, n\}$. As an illustration of the procedure that can be followed to find the number of elementary operations, let us consider a backward sweep. As indicated in Section 2.6.1, the nodes in a backward sweep are considered according to the order in the sequence $n, n - 1, \ldots, 1$. The number of generalized additions $\otimes$ and generalized minimizations $\oplus$ required to perform a backward sweep is shown in Table 2.13.

Table 2.13. NUMBER OF GENERALIZED OPERATIONS NEEDED TO PERFORM A BACKWARD SWEEP

Node	*Number of Generalized Additions* $\otimes$	*Number of Generalized Minimizations* $\oplus$
n	0	0
$n - 1$	1	1
$n - 2$	2	2
$n - 3$	3	3
.	.	.
.	.	.
.	.	.
1	$n - 1$	$n - 1$
Total	$\frac{n}{2}(n - 1)$	$\frac{n}{2}(n - 1)$

Therefore, according to the results in Table 2.13, the number of generalized operations for a backward sweep is equal to $n(n - 1)$. It can be shown, as we did for the backward sweep, that the total number of generalized operations in a forward sweep is also equal to $n(n - 1)$. Then the total number of generalized operations in a double sweep is given by $2n(n - 1)$.

If K is the number of path lengths to be investigated, the total number of estimates produced at each single sweep is equal to nK. Since at least one estimate is improved at each single sweep, under worst-case conditions the number of double sweeps is equal to $nK/2$. Therefore, the total number of generalized operations needed for the termination of the double-sweep method is given by

$$\frac{nK}{2} 2n(n - 1) = Kn^2(n - 1) \simeq Kn^3$$

2.8 Minimal Spanning Tree Problems

Before describing this network-flow problem, let us first review some basic definitions presented in Chapter 1 which directly relate to minimal spanning trees. In this section, an undirected network is considered.

Define a *tree* as a set of connected undirected edges (arcs) that contains no cycles. Therefore, given a set of m nodes connected by undirected edges, a subset of exactly $(m - 1)$ arcs is needed to form a *tree*. In other words, each node is connected to another node by a unique path.

Suppose that we are given a network containing n nodes comprising a complete set S. Then a *spanning tree* is a set of $(n - 1)$ arcs (edges) and n connected nodes. Any proper subset of S can also form a tree, but this would not be a spanning tree for the set S. As before, suppose that we define a cost or distance c_{ij} for every arc connecting nodes i and j contained in S. Then we can develop the notion of a *minimal* (or maximal) spanning tree. A *minimal* spanning tree is that spanning tree such that the sum of all the c_{ij} for each arc in the tree is a minimum with respect to all spanning trees for the network.

The minimal spanning tree problem is found to be useful in many real-life applications. For instance, suppose that you are a natural-gas supplier and you wish to supply K distinct user groups from a raw material supply. The minimal spanning tree for the supply network would be that distribution system that would connect all users in a minimum total cost (or distance). Other uses include minimal-cost solutions to specialized transportation networks.

Minimal spanning trees are also often used in suboptimization or decomposition of larger, more complex network algorithms. In addition to the practical and operational significance of this problem, it is relatively unique in the operations research field in terms of solution procedures.

2.8.1 Minimal Spanning Trees—Computational Procedure

Preceding algorithms dealing with commodity flows have utilized recursive computational schemes that iterate to an optimal solution. As it happens, the minimal spanning tree problem is one of the few problems in operations research that can be solved via a "greedy approach" which requires a minimal amount of effort. Using the "greedy" rationale, we present the following algorithm.

As previously defined, the minimal spanning tree problem is concerned with selecting the arcs for a given network that have the minimum total value while providing a path (or route) between each pair of nodes. We achieve this by selecting the arcs in such a way that the resulting network forms a tree (as defined earlier) that spans or connects all the given nodes. Hence, in short, we are interested in finding the spanning tree with a minimum total arc value.

The minimal spanning tree problem can be solved in a straightforward way. Starting with any node in the network, the procedure initiates by identifying the shortest possible arc to any other node. Join these two nodes with the chosen arc, and find another node that is closest to either of these connected nodes. Add that arc and node to the network. This process is repeated until all the nodes have been connected. An algorithm based upon this "greedy" selection of candidate arcs can be described in the following manner.

An algorithm to determine the minimal spanning tree

Step 1: Define two sets of nodes using the original list of nodes.

a. S = a set of *connected nodes.*
b. $\bar{S}$ = a set of *nonconnected nodes.*

Initially, *all nodes* will be placed in the set $\bar{S}$.

Step 2: Choose *any* node in $\bar{S}$ and connect it to its nearest neighbor. (This step will place two nodes initially in the set S.)

Step 3: Connect the node in $\bar{S}$ closest to a connected node in set S; call the node that has been selected from set $\bar{S}$, δ. Transfer δ from set $\bar{S}$ to set S.

Step 4: Repeat Step 3 until *all* nodes are in set S.

2.8.2 EXAMPLE OF A GREEDY SOLUTION (FIGURE 2.28)

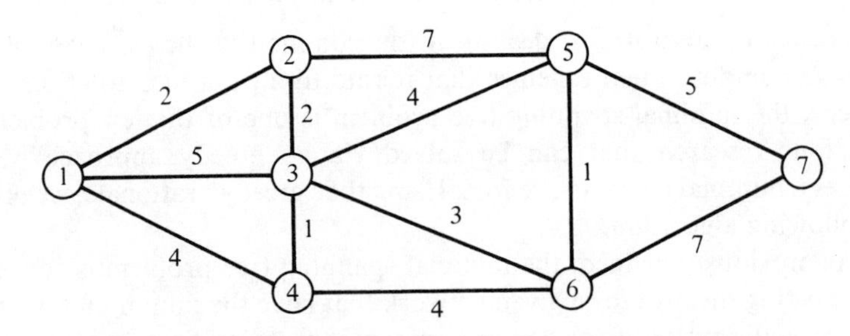

Figure 2.28. Sample network for a minimum spanning tree.

Step 1: $\bar{S} = (1, 2, 3, 4, 5, 6, 7)$

$S = \varnothing$

***Step 2*:** Choose node 6.

$\mathbf{S} = (6, 5)$ Cost = 1 unit

$\bar{\mathbf{S}} = (1, 2, 3, 4, 7)$ ⑥—1—⑤

***Step 3*:**

a. Choose node 3.

$$\mathbf{S} = (6, 5, 3)$$
$$\bar{\mathbf{S}} = (1, 2, 4, 7)$$

b. Choose node 4.

$$\mathbf{S} = (6, 5, 3, 4)$$
$$\bar{\mathbf{S}} = (1, 2, 7)$$

Cost = 4 units

⑥—1—⑤, ⑥—3—③

Cost = 4 units
Step 3, a

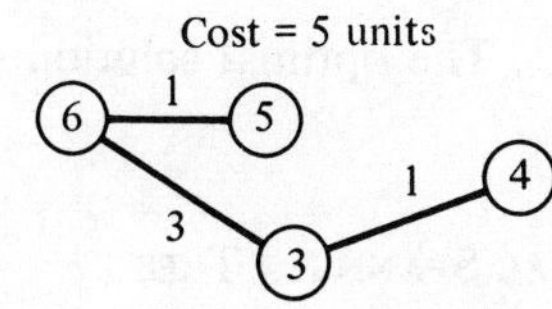

Cost = 5 units
Step 3, b

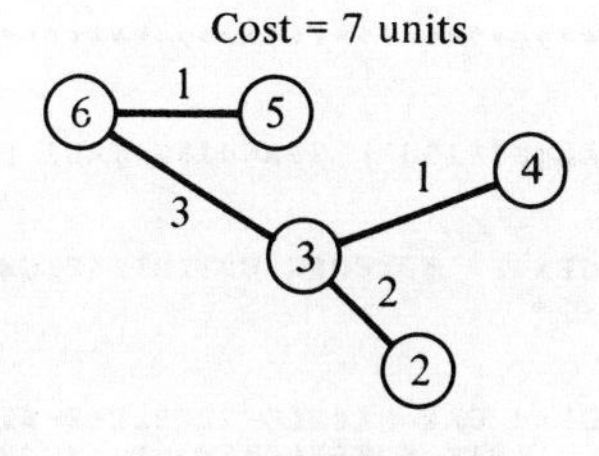

Cost = 7 units
Step 3, c

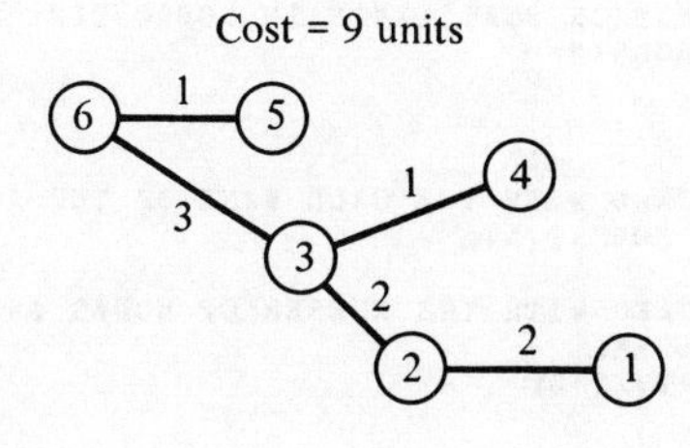

Cost = 9 units
Step 3, d

c. Choose node 2.

$$\mathbf{S} = (6, 5, 3, 4, 2)$$
$$\bar{\mathbf{S}} = (1, 7)$$

d. Choose node 1.

$$\mathbf{S} = (6, 5, 3, 4, 1, 2)$$
$$\bar{\mathbf{S}} = (7)$$

e. Choose node 7.

$$\mathbf{S} = (6, 5, 3, 4, 1, 2, 7)$$
$$\bar{\mathbf{S}} = \varnothing$$

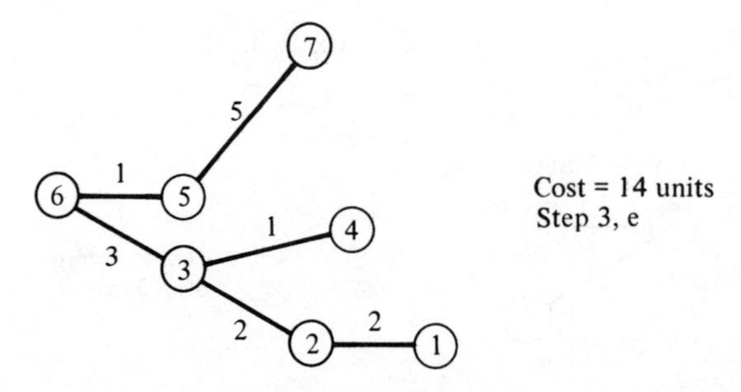

Cost = 14 units
Step 3, e

The algorithm terminates since $\bar{\mathbf{S}} \equiv \varnothing$. The optimal solution yields a total cost of 14.

2.8.3 Computer Code for the Minimal Spanning Tree Algorithm

```
C***********************************************************************
C*                                                                     *
C*       PURPOSE:                                                      *
C*            TO OBTAIN THE OPTIMAL(MINIMUM) SPANNING TREE FOR AN      *
C*            UNDIRECTED NETWORK.                                      *
C*                                                                     *
C*       LOCATION: SUBROUTINE MINSPA IN NETWORK OPTIMIZATION           *
C*                                                                     *
C*       USAGE:                                                        *
C*                                                                     *
C*            AT PRESENT THIS PROGRAM CAN HANDLE PROBLEMS WITH UP TO   *
C*            50 NODES AND 50 ARCS. THIS NUMBER MAY BE INCREASED  BY   *
C*            INCREASING DIMENSION STATEMENTS IN SUBROUTINE MINSPA     *
C*            AND THE MAIN PROGRAM                                     *
C*                                                                     *
C*       INPUT DATA:                                                   *
C*                                                                     *
C*            SET 1: SINGLE CARD WITH THE CALL NAME OF THE ALGORITHM:  *
C*                   MSTR    -FORMAT(A4)                               *
C*                                                                     *
C*            SET 2: SINGLE CARD WITH THE NUMBER OF NODES AND ARCS IN  *
C*                   THE PROBLEM.                                      *
C*                     -FORMAT(2I10)                                   *
C*                                                                     *
```

```
C*          SET 3: THE TOTAL NUMBER OF CARDS FOR THIS SET IS EQUAL  *
C*                 TO THE TOTAL NUMBER OF ARCS IN THE NETWORK .     *
C*                 THE FOLLOWING QUANTITIES ARE READ FROM EACH CARD:*
C*                 1) START NODE                                    *
C*                 2) END NODE                                      *
C*                 3) ARC VALUE                                     *
C*                   -FORMAT(4X,I6,I10,F10.2)                       *
C*                                                                  *
C*          SET 4: THIS SET CONSISTS OF ONLY ONE CARD WHICH         *
C*                 INDICATES THE END OF A PROBLEM. THE USER SHOULD  *
C*                 PUNCH THE WORD 'PEND'  -FORMAT(A4)               *
C*                                                                  *
C*          SET 5: THIS SET CONSISTS OF ONLY ONE CARD WHICH         *
C*                 INDICATES THE END THE INPUT DECK. THE USER SHOULD*
C*                 PUNCH THE WORD 'EXIT'  -FORMAT(A4)               *
C*                  -FORMAT(A4)                                     *
C*                                                                  *
C*     DEFINITION OF THE TERMS USED:                                *
C*                                                                  *
C*          I - STARTING NODE                                       *
C*          J - ENDING NODE                                         *
C*          A1 - ARC DISTANCE                                       *
C*          D - WORKING ARRAY OF ARC DISTANCES                      *
C*                                                                  *
C*     TECHNIQUE USED:                                              *
C*                                                                  *
C*          SELECT NODE 1 AND CONNECT IT TO THE NEAREST DISTINCT    *
C*          NODE. THEN IDENTIFY THE UNCONNECTED NODE THAT IS NEAREST*
C*          TO A CONNECTED NODE AND CONNECT THESE TWO NODES. REPEAT *
C*          THIS PROCEDURE UNTIL ALL NODES HAVE BEEN CONNECTED.     *
C*                                                                  *
C*     REFERENCE:                                                   *
C*                                                                  *
C*          HILLIER,S.F.,LIEBERMAN,J.G.,OPERATIONS RESEARCH,2ND ED.,*
C*          HOLDEN-DAY,INC.,CALIFORNIA,1974.                        *
C*                                                                  *
C*******************************************************************
```

2.8.4 Allocation of Highway Maintenance Funds

The highway department has a limited amount of funding available with which to provide asphalt topping and labor for several unimproved roads in a particular county. These roads connect a series of 10 small farming communities. It is the intention of the highway department to provide the county residents with the capability to be able to travel between any community on a year-round basis. The present situation with the unimproved roads makes them impassable during a moderate to heavy rainfall. Additionally, the roads must be maintained on a frequent basis because of the formation of chuckholes.

To minimize costs, the highway department must come up with a plan that will provide the improved interconnection capability between all 10 communities with the minimum amount of expenditures on materials and labor. Figure 2.29 represents the 10 communities and the unimproved road network. The "distances" are actually construction costs for each road segment. The problem is solved using the minimal-spanning-tree computer code. Figure 2.30 represents the solution that interconnects all 10 cities and shows the portions of the road network to be paved. The computer solution is shown as it would appear from the program.

MINIMAL SPANNING TREE PROBLEM
THE MINIMAL SPANNING TREE CONSISTS OF THE FOLLOWING ARCS

STARTING NODE	ENDING NODE	DISTANCE
1	2	7.00
2	3	4.00
3	7	2.00
3	8	6.00
4	5	6.00
5	6	4.00
6	7	5.00
6	10	4.00
6	9	3.00

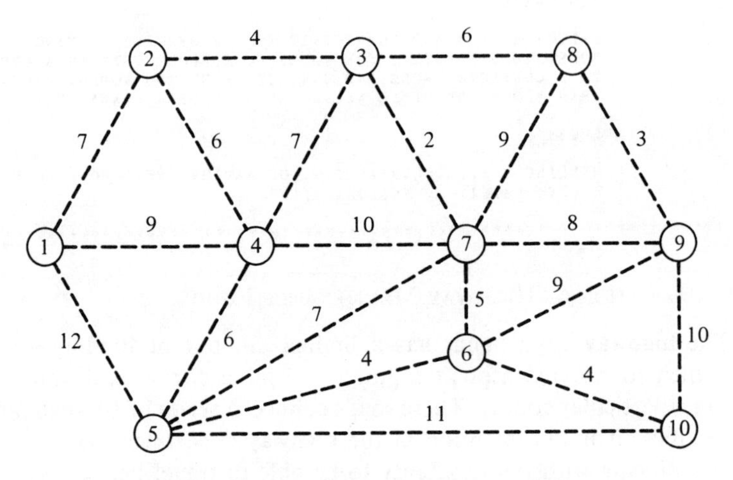

Figure 2.29. Minimal spanning tree problem.

2.8.5 Applications of Minimal Spanning Trees

The minimal spanning tree (MST) is useful in a wide variety of applications. Many apparently dissimilar problems have been cleverly modeled as MST problems. MST modeling has been used by Dei Rossi, Heiser, and King [10] in the design of a television cable linking all stations in a network, and Saltman, Bolotsky, and Ruthberg [46] have shown its usefulness in the design of a leased-line telephone network. Perhaps the most interesting and important application is in cluster analysis, where MST has been proved to be efficient in solving problems that few clustering methods could handle [53]. Another application is in network reliability, where the weight of an

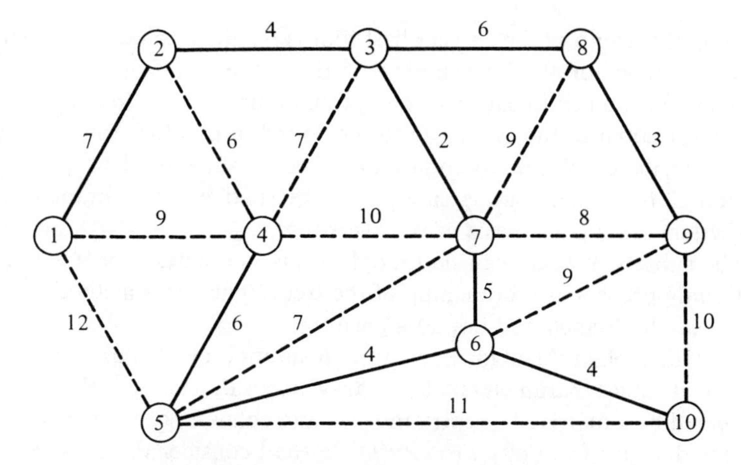

Figure 2.30. Minimal spanning tree solution.

MST represents the minimum probability that the tree will fail at one or more arcs. Gomory and Hu [25] have used MST calculations as subproblems for solving multiterminal flow problems. A similar application is proposed by Held and Karp [26, 27] for solving traveling salesman problems.

2.9 The Traveling Salesman Problem

The traveling salesman problem can be stated as follows. A salesman, starting from a city, intends to visit each of $(n - 1)$ other cities once and only once and return to the start. The problem is to determine the order in which he should visit the cities to minimize the total distance traveled, assuming that the direct distances between all city pairs are known. Instead of "distance," any other measure of effectiveness may be substituted, such as cost, time, and so on.

The structure of the problem shows that there are $(n - 1)!$ possible tours, of which one or more should be optimal. However, if some of the cities are inaccessible, the optimal value (or minimum) could be infinity. In most cases, one can assume that the distance between a city i and another city j is symmetric; that is, the distance from city i to city j is the same as the distance from city j to city i. However, in the algorithm that follows, this requirement is not necessary.

The algorithm presented is called a *branch-and-bound* algorithm, and was first developed by Little et al. [41]. The method is to first identify a feasible solution and then to decompose the set of all remaining feasible tours into smaller and smaller subsets. At each step of the decomposition, a lower bound on the length of the current best tour is readily available. The bounds provide a guide for the partitioning of the subsets of feasible tours and eventually for the identification of an optimal tour. When a tour with length less than or equal to the minimum lower bound of all other tours is found, this intermediate solution becomes the "best" available. This process of "bounding" tours, eliminating suboptimal alternatives, and "branching" to new (better) tours is the basis of the algorithm.

The subsets of tours can be thought of as the nodes of a tree and the partitioning process as a branching of the tree. Hence, this method is called a "tree-search" branch-and-bound algorithm.

We will explain the algorithm through solution of an illustrative example. The distance parameters of the traveling salesman problem will be represented by a matrix. The entry in row i and column j of the matrix represents the distance from city i to city j. We shall consider the distance from city i to city j as the length of a *link* (i, j). Define $\mathbf{D}$ as the *distance matrix* which represents all distances involved in traveling from city i to city j. If there are n cities (stops) in the problem, $\mathbf{D}$ will be an $(n \times n)$ matrix with nonnegative entries, d_{ij}. $\mathbf{D}$ will initially be the original distance matrix of the problem, but will undergo various transformations as the algorithm proceeds. A tour, $\mathbf{T}$, can be represented as a set of n ordered city pairs:

$$\mathbf{T} = \{(i_1, i_2), (i_2, i_3), \ldots, (i_{n-1}, i_n), (i_n, i_1)\}$$

Each feasible tour forms a cycle, visiting each city once and only once and returning to the starting city. Each ordered pair (i, j) represents an arc or a link in the trip. The length of a tour $\mathbf{T}$ is the sum of the matrix elements identified in $\mathbf{T}$ and will be denoted by $Z(\mathbf{T})$.

$$Z(\mathbf{T}) = \sum_{(i, j) \in \mathbf{T}} d_{ij}$$

It is clear that a feasible tour always includes one and only one entry in each row and each column of the matrix $\mathbf{D}$. Note that a $Z(\mathbf{T})$ is determined for *any* feasible tour, and that the *optimal* tour cannot have a length exceeding the current value of $Z(\mathbf{T})$. Hence, the current value of $Z(\mathbf{T})$ constitutes an *upper bound* on the length of the tour corresponding to the optimal solution. This upper bound will be represented by $Z_U(\mathbf{T})$.

2.9.1 Construction of a Lower Bound

A useful concept in constructing lower bounds will be that of reduction. If a constant, C, is subtracted from each element of a row of the distance matrix, the length of any tour under the new matrix is C units less than under the old matrix. This is so because every tour must contain one and only one element from that row. The relative lengths of all tours are unchanged, however, and hence any tour optimal under the original distance matrix is also optimal under the new matrix. It is clear that the same argument applies to *column elements* as well.

The process of subtracting the smallest element from the entries of each row and each column is called *row reduction* and *column reduction*, respectively. A matrix with nonnegative elements and at least one zero in *each* row and *each* column is called a *reduced matrix* and may be obtained by successively reducing the rows and the columns of the matrix. If $Z(\mathbf{T})$ is the length of a tour $\mathbf{T}$ under a matrix before reduction, $Z_1(\mathbf{T})$ the length under the matrix after reduction, and H the sum of the constants used in forming the reduced matrix of $\mathbf{T}$, then

$$Z(\mathbf{T}) = H + Z_1(\mathbf{T})$$

Since a reduced matrix contains only nonnegative elements, H constitutes a lower bound on the length of $\mathbf{T}$ under the matrix before reduction.

Let us consider a six-city problem [41] characterized by the distance matrix of Table 2.14. We assume that $d_{ij} = \infty$ for all $i = j$.

Table 2.14. INITIAL DISTANCE MATRIX

	1	2	3	4	5	6
1	∞	27	43	16	30	26
2	7	∞	16	1	30	30
3	20	13	∞	35	5	0
4	21	16	25	∞	18	18
5	12	46	27	48	∞	5
6	23	5	5	9	5	∞

We will first arbitrarily choose a feasible tour, such as the one consisting of the links (1, 4), (4, 5), (5, 3), (3, 6), (6, 2), and (2, 1). The length of this feasible tour is $Z_U(\mathbf{T}) = 16 + 18 + 27 + 0 + 5 + 7 = 73$. The value of $Z_U(\mathbf{T})$ sets an upper bound on the optimal tour; that is, an optimal tour cannot have a $Z(\mathbf{T})$ greater than $Z_U(\mathbf{T})$. It is not essential to identify a feasible solution initially to evaluate the optimal solution. However, the evaluation of an upper bound often helps in the reduction of the computational effort involved.

Next we proceed with the reduction of the distance matrix, first by rows and then by columns. This is done by first identifying the smallest element in *each* row, and subtracting that number from the corresponding row elements. The result is shown in Table 2.15, where the C_i column contains the reducing constants.

Table 2.15. ROW REDUCTION

	1	2	3	4	5	6	C_i
1	∞	11	27	0	14	10	16
2	6	∞	15	0	29	29	1
3	20	13	∞	35	5	0	0
4	5	0	9	∞	2	2	16
5	7	41	22	43	∞	0	5
6	18	0	0	4	0	∞	5

Column reduction is now attempted. Upon inspection of Table 2.15, we see that each column except column 1 contains a zero entry. Hence, only column 1 can be reduced. We reduce column 1 by 5 units to yield the results shown in Table 2.16, where the Q_j row contains the reducing constants for each column. The intersection of the Q_j row and the C_i column shows the sum of the reducing constants. The total reduction is given by $H = \sum_{\text{all } i, j} (C_i + Q_j)$, and is equal to 48 for Table 2.16. This is a lower bound $Z_L^0(\mathrm{T})$ on *all* tours for the problem.

Table 2.16. COLUMN REDUCTION

	1	2	3	4	5	6	C_i
1	∞	11	27	0	14	10	16
2	1	∞	15	0	29	29	1
3	15	13	∞	35	5	0	0
4	0	0	9	∞	2	2	16
5	2	41	22	43	∞	0	5
6	13	0	0	4	0	∞	5
Q_j	5	0	0	0	0	0	48

The basic task now becomes that of identifying the optimal (minimal length) tour. Note that an ideal solution at this point would be the selection of exactly one zero-length cell in each row and column. In other words, if such a zero-length tour could be found, the optimal solution would be one with a length of 48 units. Rather than try to identify simultaneously all the links of an optimal tour from the current distance matrix, the following algorithm will construct the optimal solution one link at a time from the distance matrix. The final tour will form a minimal-length solution which includes whatever links are necessary to achieve that goal. Fundamentally, it would seem to be advantageous to start the solution with a link that possesses zero length, and subsequently add links that possess zero or minimal lengths. The procedure is based upon two fundamental observations. First, if a link (i, j) is *selected*, it is impossible to include another link corresponding to an entry from row i or column j. Second, if a link (i, j) can be *excluded* from the final solution, that link need not be considered in further calculations. Hence, it is sufficient to consider only two possible results for any one link: it will either be included in the current (partial) solution and all other subsequent (partial) solutions until a tour is obtained, or it will be excluded from further consideration.

2.9.2 Branching

The partitioning of the set of all tours into disjoint subsets will be represented by the branching of a tree as shown in Figure 2.31. The node labeled "all tours" represents the set of all tours. The node labeled (i, j) represents the subset of "all tours" that *include* the link (i, j). The node labeled $\overline{(i, j)}$ represents the subset of "all tours" that *do not include* the link (i, j). In Figure 2.31 no further branching is indicated at node $\overline{(i, j)}$. The node labeled (k, l) represents all tours that include *both* (i, j) and (k, l). As before, the node labeled $\overline{(k, l)}$ represents all tours that include (i, j) *but not* (k, l). The reason for branching can easily be explained; if we partition the set of all tours into smaller and smaller subsets, we should eventually have a subset containing a single tour. The links or the city pairs that make up a single tour can then be recovered by tracing backward to the initial (all-tours) node. The branching process is controlled by the lower-bound criterion. If at any iteration of the algorithm the lower bound for a subset of tours is larger than the lower bound for another subset, the first subset can be deleted from further consideration. That is, no additional branching is performed from the corresponding node.

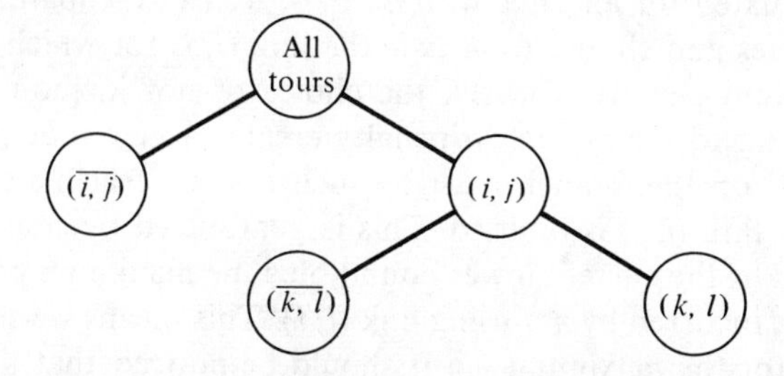

Figure 2.31. Partitioned subsets.

In general, when a node, say X, branches into two subsequent nodes, the node that *includes* a new link will be represented by Y and the node that *does not include* the new link will be represented by $\bar{Y}$.

2.9.3 Computational Procedure

Considering the numerical example, we have the initial node representing "all tours," with the lower and upper bounds as shown in Figure 2.32.

The next step in the procedure involves selecting the link on which to

$$Z_L^0(T) = 48 \leqslant \left(\begin{array}{c}\text{All}\\ \text{tours}\end{array}\right) \leqslant 73 = Z_U^0(T)$$

Figure 2.32. Bounds on the initial solution.

base the branching. The purpose of branching is to partition the tours of X into a subset Y that is quite *likely* to include the best tour of the node X and another $\bar{Y}$ that is quite *unlikely* to include it. As previously indicated, the most advantageous tours to be considered for Y are those containing a link (i, j) for which $d_{ij} = 0$. However, we have at least one zero element in each row and each column. The question to be answered at this point is which one of these links is to be chosen for inclusion in Y? To arrive at an answer, let us consider the possible tours for $\bar{Y}$ [tours not including link (i, j)]. Since city i must connect to some other city, these tours must include the link with length equal to the smallest element in row i, excluding d_{ij}. Let us denote this distance by A_i. Since city j must be reached from some other city, the tours must also include the link with length equal to the smallest element in column j *other* than d_{ij}. Let us denote this distance by B_j, and the sum of A_i and B_j by ϕ_{ij}. The values of ϕ_{ij} will be called *secondary penalties*. If an element (i, j) is *not* included in a possible tour, then another element in the ith row and another element in the jth column *must* be in the final tour. The ϕ_{ij} calculation yields the (minimum) penalty which would be incurred if we are forced to *omit* link (i, j) from the optimal tour. If the penalty for *not* using link (i, j) is calculated for all links with $d_{ij} = 0$, we can compare the corresponding ϕ_{ij} values and choose to *include* the link (i, j) for which we would incur the maximum penalty. Clearly, the choice of not including the link (i, j) corresponding to the *maximum* penalty creates a new tour possibility. The lower bound on this branch must be such that no feasible tour which does not contain link (i, j) is omitted. This is guaranteed by computing the new lower bound as the current lower bound plus the maximum penalty that could possibly be incurred by *not* using link (i, j). This means we shall search over all $d_{ij} = 0$ for the maximum ϕ_{ij}. It should be noticed that $\phi_{ij} = 0$ for all (i, j) such that $d_{ij} \neq 0$. This is true because if we set $d_{ij} = \infty$ and then reduce row i and column j, the sum of the reducing constants is ϕ_{ij}.

For our example, the A_i and B_j values are written in a separate column and a separate row, respectively. Matrix **D** with these entries is shown in Table 2.17. The ϕ_{ij} values for the different links or city pairs (excess distance involved in not selecting the particular link with zero element in the reduced matrix) are given in Table 2.18.

We observe that $\phi_{ij} = 10$ is the maximum value. By using link (1, 4) we can save 10 units of distance, which is a greater saving than that achieved by selecting any other link. Hence, we will base our branching on the link (1, 4).

Table 2.17. FIRST DECISION MATRIX

	1	2	3	4	5	6	C_i	A_i
1	∞	11	27	0	14	10	16	10
2	1	∞	15	0	29	29	1	1
3	15	13	∞	35	5	0	0	5
4	0	0	9	∞	2	2	16	0
5	2	41	22	43	∞	0	5	2
6	13	0	0	4	0	∞	5	0
Q_j	5	0	0	0	0	0	48	
B_j	1	0	9	0	2	0		

Table 2.18. SECONDARY PENALTIES

Link (i, j)	*Distance* $\phi_{ij} = A_i + B_j$
(1, 4)	$10 + 0 = 10$
(2, 4)	$1 + 0 = 1$
(3, 6)	$5 + 0 = 5$
(4, 1)	$0 + 1 = 1$
(4, 2)	$0 + 0 = 0$
(5, 6)	$2 + 0 = 2$
(6, 2)	$0 + 0 = 0$
(6, 3)	$0 + 9 = 9$
(6, 5)	$0 + 2 = 2$

At this point, $Y = (1, 4)$ and $\bar{Y} = \overline{(1, 4)}$. The lower bound on the subsets of $\bar{Y}$ is $Z_L^0(\mathbf{T}) + \phi_{14}$, and for our example it is equal to $48 + 10 = 58$. Before we can determine the new lower bound on Y, which includes the link (1, 4), certain modifications are to be performed on matrix **D**. Note that if we include any link (k, l) in the tours we do not need to consider row k and column l any further. Hence, for this example, row 1 and column 4 will be *deleted* from further consideration. Further, if link (k, l) is in a tour, that is, (k, l) is a link in a directed cycle generated by the links of Y, then link (l, k) cannot be in a tour generated by Y. This can be guaranteed by setting $d_{lk} = \infty$ in all further iterations. Finally, there may be other links which if included for further consideration might also create subtours. (A subtour is a cycle that does not include all the cities.) These links are referred to as *forbidden links* and can be eliminated from consideration by setting their lengths to ∞ in matrix **D**. After these modifications, the matrix **D** is a candidate for further reductions in row l and column k, since any column that had a single zero in row k and any row that had a zero in column l can be further reduced. In addition, any other rows and columns that contain forbidden links might also be candidates for reduction. The lower bound on Y can now be computed as the sum of the new reducing constants and the previous lower bound. Modifications to Table 2.17, along with new C_i, Q_j, A_i, and B_j, are given in Table 2.19. Note that row 1 and column 4 have been deleted, and there are no forbidden links.

The new lower bound for Y in this example is $48 + 1 = 49$. The tree can now be expanded as indicated in Figure 2.33.

Table 2.19. SECOND DECISION MATRIX

	1	2	3	5	6	C_i	A_i
2	0	∞	14	28	28	1	14
3	15	13	∞	5	0	0	5
4	∞	0	9	2	2	0	2
5	2	41	22	∞	0	0	2
6	13	0	0	0	∞	0	0
Q_j	0	0	0	0	0	1	
B_j	2	0	9	2	0		

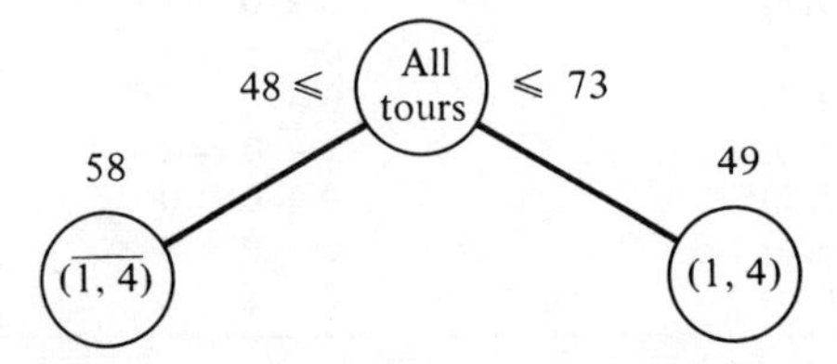

Figure 2.33. Bounds on the first iteration.

The secondary reductions are given by the following values: $\phi_{21} = 16$, $\phi_{36} = 5$, $\phi_{42} = 2$, $\phi_{56} = 2$, $\phi_{62} = 0$, $\phi_{63} = 9$, and $\phi_{65} = 2$. The maximum value is clearly 16, and the new lower bound for $\bar{Y}$ is equal to $49 + 16 = 65$. Consider now the set Y, which *includes* link (2, 1) in the partial tour. First delete row 2 and column 1 from further consideration. The length of link (1, 2) is now ∞, but is not in the next tableau. However, note that link (4, 2) is now forbidden, since it would create a subtour. To prevent this, set $d_{42} = \infty$. The matrix **D** after reduction is given in Table 2.20. The new lower bound on Y is $49 + 2 = 51$, and the tree at this point is given in Figure 2.34.

Table 2.20. THIRD DECISION MATRIX

	2	3	5	6	C_i	A_i
3	13	∞	5	0	0	5
4	∞	7	0	0	2	0
5	41	22	∞	0	0	22
6	0	0	0	∞	0	0
Q_j	0	0	0	0	2	
B_j	13	7	0	0		

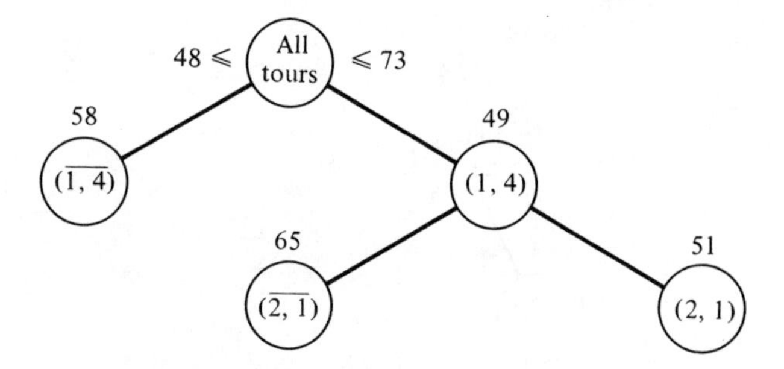

Figure 2.34. Bounds on the second iteration.

The minimum lower bound for the sets of tours represented by the tree of Figure 2.34 is equal to 51 and corresponds to node (2, 1). Therefore, in the next iteration, branching will be based on node (2, 1). Proceeding in this fashion, we would continue branching out of the node associated with the minimum lower bound. For discussion purposes, however, let us assume that we branch as indicated in Figure 2.35 in order to obtain a complete tour. The corresponding decision matrix is given in Table 2.21.

Table 2.21. FINAL DECISION MATRIX

	2	3	C_i	A_i
4	∞	0	7	∞
6	0	∞	0	0
Q_j	0	0	7	
B_j	∞	0		

If the length of this complete tour is less than that of any partial tour represented by other nodes on the tree, the tour is optimal. However, we see that node $(\overline{1, 4})$ has a lower bound of 58, which is less than the tour value of 63 which we just obtained. Hence, we have to explore the subset of tours that *does not* contain the link (1, 4). The matrix **D** at this point will be the old matrix **D** at the time of branching based upon link (1, 4). We will set $d_{14} = \infty$ in the matrix **D** to exclude all tours that include the link (1, 4). For our example, this would be the matrix **D** of Table 2.14 with ∞ substituted for the value 16 of link (1, 4). The operations previously described for branching and bounding are now performed on this new matrix and the tree obtained is given in Figure 2.36. This procedure of exploring previous intermediate branching points which might generate a better tour is called *backtracking*.

Proceeding as before, we partition the subset of tours containing link (6, 3). The tree shown in Figure 2.37 is subsequently obtained.

Observe that the original tour containing link (1, 4) now has the smallest lower bound. Hence, it must be the optimal tour. The links or city pairs in

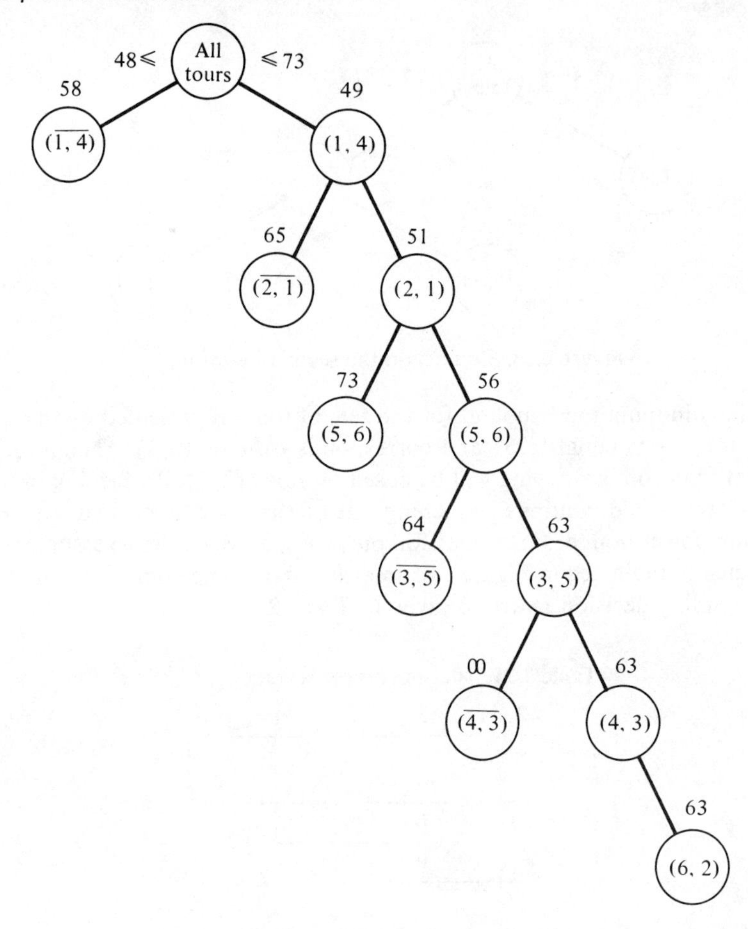

Figure 2.35. Intermediate solution.

the optimal tour are (4, 3), (6, 2), (3, 5), (5, 6), (2, 1), and (1, 4). This is a directed cycle with a total length of 63.

2.9.4 Final Remarks

For completeness, it should be noted that the traveling salesman problem is combinatorial in nature because of the exponentially increasing number of possible solutions as the number of cities increases. Further, the problem belongs to that class of problems known as *NP-complete*. It has been shown that a large class of combinatorial problems are equivalent in the sense that either all or none can be solved by algorithms with polynomial computational complexity (see Section 2.7 for a discussion of this topic). This group of problems constitutes the NP-complete class. Finally, the traveling salesman problem cannot be directly formulated and solved as a linear programming problem. It is included as a network-flow algorithm

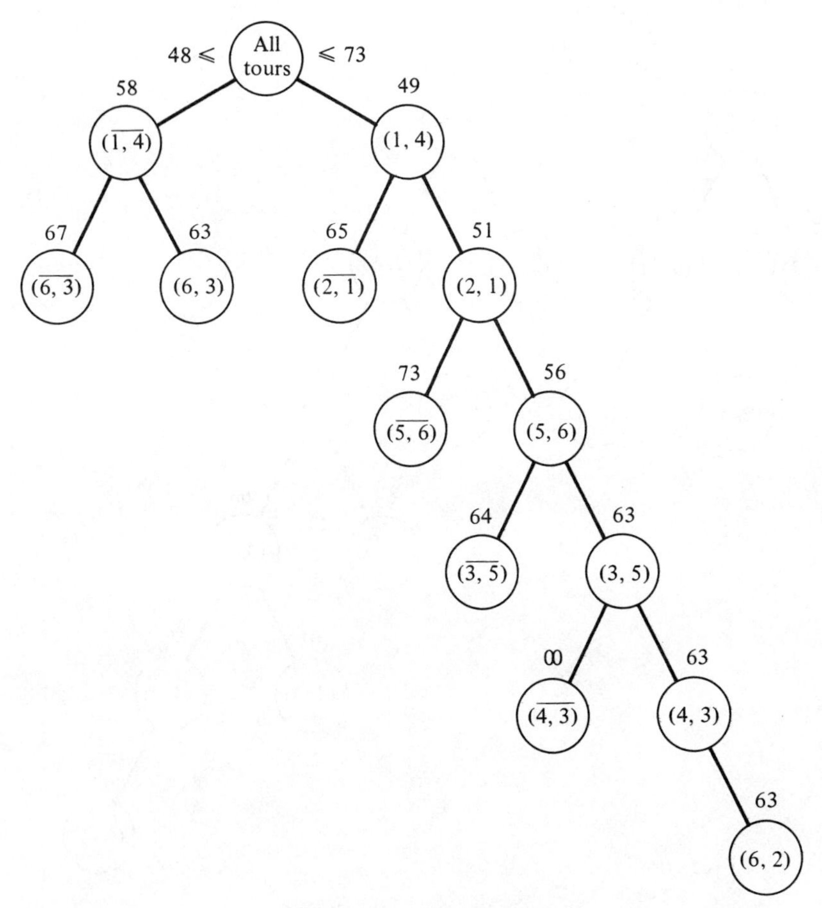

Figure 2.36. Backtracking.

because of its historical association with this class of techniques, and its graphical representation as an optimal tour.

Many traveling salesman problems have been formulated and solved to investigate movements of people or goods, but these applications obscure the important structure of the problem. The main difference between the traveling salesman problem and the other shortest-distance problems is the requirement of having a directed cycle that contains all the nodes (cities or certain milestones) of the network once and only once. Assigning class loads to teachers, scheduling different models on an assembly line, or sequencing NC/DNC machining operations are other important applications of the traveling salesman problem model. In addition to these applications, important results relating to machine scheduling, optimal changeover rates, and sequencing of conveyor-serviced production systems have also appeared.

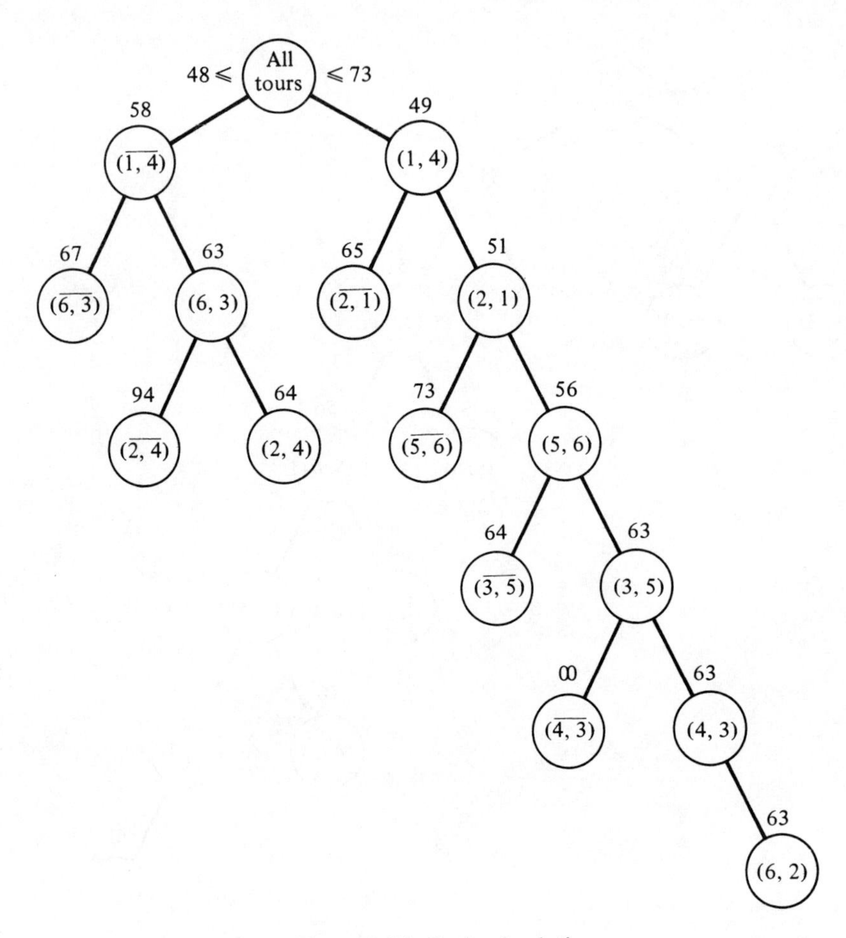

Figure 2.37. Optimal solution.

2.9.5 Computer Code for the Traveling Salesman Algorithm

```
C***********************************************************************
C                                                                      *
C     PURPOSE:                                                         *
C                                                                      *
C       THIS PROGRAM OBTAINS OPTIMAL SOLUTION TO THE TRAVELING         *
C       SALESMAN PROBLEM.                                              *
C                                                                      *
C     LOCATION: SUBROUTINE TRASAL IN NETWORK OPTIMIZATION              *
C                                                                      *
C     USAGE:                                                           *
C                                                                      *
C       THIS PROGRAM CAN HANDLE PROBLEMS WITH UP TO 50 NODES AND 50    *
C       ARCS. THIS NUMBER CAN BE INCREASED BY INCREASING THE           *
C       DIMENSION STATEMENTS IN SUBROUTINE TRASAL AND MAIN PROGRAM.    *
C                                                                      *
C     INPUT DATA:                                                      *
C                                                                      *
C       SET 1:  SINGLE CARD WITH THE CALL NAME OF THE ALGORITHM:TSPR   *
C               -FORMAT(A4)                                            *
```

```
C                                                                          *
C         SET 2:  SINGLE CARD WITH THE NUMBER OF NODES AND ARCS IN         *
C                 THE PROBLEM.                                             *
C                 -FORMAT(2I10)                                            *
C                                                                          *
C         SET 3:  THIS SET CONSISTS OF AS MANY CARDS AS THE NUMBER         *
C                 OF ARCS IN  THE PROBLEM. THE FOLLOWING QUANTITIES        *
C                 ARE READ FROM EACH CARD:                                 *
C                 1) STARTING NODE                                         *
C                 2) ENDING NODE                                           *
C                 3) ARC VALUE                                             *
C                 -FORMAT(4X,I6,I10,F10.2)                                 *
C                                                                          *
C         SET 4:  THIS SET CONSISTS OF ONLY ONE CARD WHICH INDICATES       *
C                 THE END OF A PROBLEM. THE USER SHOULD PUNCH THE WORD     *
C                 'PEND'  -FORMATA(A4)                                     *
C                                                                          *
C         SET 5:  THIS SET CONSISTS OF ONLY ONE CARD WHICH INDICATES       *
C                 THE END OF THE INPUT DECK. THE USER SHOULD PUNCH THE     *
C                 WORD 'EXIT' -FORMAT(A4)                                  *
C                                                                          *
C        PROGRAM CONTENTS:                                                 *
C                                                                          *
C           THE SUBROUTINES WHICH COMPRISE THIS PROGRAM ARE:               *
C               SUBROUTINE TRASAL: INPUT/OUTPUT/TREE CONSTRUCTION/         *
C                                  BACKTRACKING/FATHOMING                  *
C               FUNCTION MINCOL: COLUMN REDUCTION                          *
C               FUNCTION MINROW: ROW REDUCTION                             *
C               SUBROUTINE PRTRXS: PRINTS THE INITIAL SOLUTION             *
C                                                                          *
C        DEFINITION OF THE TERMS USED:                                     *
C                                                                          *
C           I - STARTING NODE                                              *
C           J - ENDING NODE                                                *
C           VAL - ARC VALUE                                                *
C                                                                          *
C        TECHNIQUE USED:                                                   *
C                                                                          *
C          THE TECHNIQUE USED IS BASICALLY THE BRANCH AND BOUND            *
C          ALGORITHM AS GIVEN BY LITTLE,J.D.C.,ET AL.                      *
C                                                                          *
C        REFERENCE:                                                        *
C                                                                          *
C          LITTLE J.D.C,ET AL, AN ALGORITHM FOR THE TRAVELLING             *
C          SALESMAN PROBLEM, OPERATIONS RESEARCH, 11(5),972-978,1963.      *
C                                                                          *
C***************************************************************************
```

2.9.6 Scheduling International Travel

A Texas A&M graduate student recently won third prize in the Irish Sweepstakes. After having spent five consecutive years at College Station studying industrial engineering, he decided it was time to break away and see what was going on in the rest of the world. He decided that there were seven places that might be interesting to visit during the Christmas break. He took his list over to the local travel service and got a price quotation for air fares between the various places he planned to visit. The prices in dollars are shown in Table 2.22. Since the fares remain the same in either direction, the cost matrix is symmetric.

Before committing himself to reservations, he wants to figure out the cheapest round-robin route to take him through all the cities. This presents no problem since he has studied the traveling salesman algorithm.

The corresponding traveling salesman model was solved using the computer code described in Section 2.9.5. It was concluded that the cheapest way

Table 2.22. IRISH SWEEPSTAKES

From	To							
	College Station	*Acapulco*	*Honolulu*	*Tokyo*	*Hong Kong*	*London*	*Sydney*	*Rome*
College Station	—	190	210	680	690	460	780	750
Acapulco	190	—	380	760	790	610	670	450
Honolulu	210	380	—	890	900	340	410	600
Tokyo	680	760	890	—	480	760	510	250
Hong Kong	690	790	900	480	—	890	490	560
London	460	610	340	760	890	—	720	600
Sydney	780	670	410	510	490	720	—	500
Rome	750	450	600	250	560	600	500	—

to visit all of the locations would be to go from College Station to London to Honolulu to Sydney to Hong Kong to Tokyo to Rome to Acapulco and then back to College Station, all for the price of $3070.

TRAVELLING SALESMAN PROBLEM****BY BRANCH AND BOUND

ORIGINAL MATRIX

	1	2	3	4	5
1	99999999	19	21	68	69
2	19	99999999	38	76	79
3	21	38	99999999	89	90
4	68	76	89	99999999	48
5	69	79	90	48	99999999
6	46	61	34	76	89
7	78	67	41	51	49
8	75	45	60	25	56

	6	7	8
1	46	78	75
2	61	67	45
3	34	41	60
4	76	51	25
5	89	49	56
6	99999999	72	60
7	72	99999999	50
8	60	50	99999999

```
**WE HAVE A FEASIBLE SOLUTION**                    307

            1    2    8    4    5    7    3    6

WE HAVE AN OPTIMAL SOLUTION OF                     307

                          BY

GO FROM NODE 1  TO NODE 2  AT A DISTANCE OF  19
GO FROM NODE 2  TO NODE 8  AT A DISTANCE OF  45
GO FROM NODE 8  TO NODE 4  AT A DISTANCE OF  25
GO FROM NODE 4  TO NODE 5  AT A DISTANCE OF  48
GO FROM NODE 5  TO NODE 7  AT A DISTANCE OF  49
GO FROM NODE 7  TO NODE 3  AT A DISTANCE OF  41
GO FROM NODE 3  TO NODE 6  AT A DISTANCE OF  34
GO FROM NODE 6  TO NODE 1  AT A DISTANCE OF  46
```

2.10 The Transportation Problem

In the standard interpretation of this model, there are m supply centers with items available to be shipped to n demand centers. More specifically, source i can ship a_i items, and destination j requires b_j items. The parameters a_i and b_j are fixed with respect to a specified planning horizon. The cost of shipping 1 unit from center i to center j is assumed to be independent of the number of units shipped, and will be represented by c_{ij}. The objective of the model is to select, for the duration of the given planning horizon, a routing plan that minimizes total transportation costs. Only one single type of commodity is considered. However, in Chapter 5 we show that a special class of multicommodity transportation problems can be transformed to a single commodity model.

The basic elements required to formulate a transportation problem are the unit shipping costs, the availability figures of the production centers, and the requirements of the demand centers. The graphical structure of the model to be formulated is shown in Figure 2.38.

Usually, the input information required for the formulation of the model is displayed in a rectangular array which contains the cost coefficients for all possible routes, and the availability and demand figures for the production and demand centers, respectively. As an illustration, such an arrangement is shown in Table 2.23 for the case corresponding to Figure 2.38.

For a situation where there are two sources and three destinations, the following hypothetical example will be considered. The values c_{ij} may include

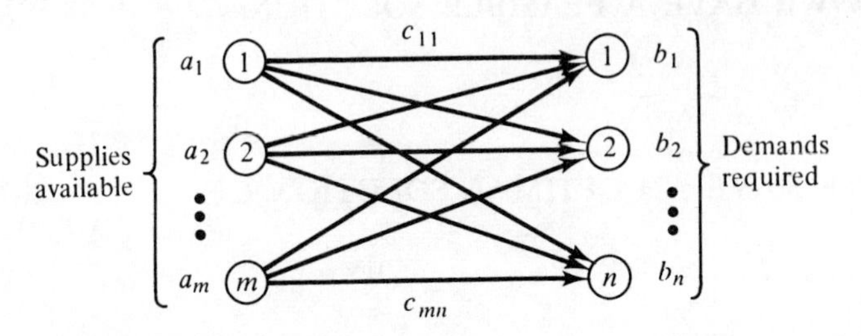

Figure 2.38. Graphical structure of the transportation problem.

Table 2.23. INPUT INFORMATION FOR THE TRANSPORTATION MODEL

	$j=1$	$j=2$	$j=3$	$\cdots$	$j=n$	
$i=1$	c_{11}	c_{12}	c_{13}	$\cdots$	c_{1n}	a_1
$i=2$	c_{21}	c_{22}	c_{23}	$\cdots$	c_{2n}	a_2
$i=3$	c_{31}	c_{32}	c_{33}	$\cdots$	c_{3n}	a_3
$\vdots$	$\vdots$	$\vdots$	$\vdots$		$\vdots$	$\vdots$
$i=m$	c_{m1}	c_{m2}	c_{m3}	$\cdots$	c_{mn}	a_m
	b_1	b_2	b_3	$\cdots$	b_n	

the unit costs of producing the items if such costs differ from source to source:

	1	2	3	a_i
1	3	4	7	20
2	6	3	2	10
b_j	10	12	8	

If for physical or economic reasons a certain route must not be used, its transportation cost per unit is set equal to infinity.

2.10.1 Mathematical Model

Let x_{ij} be the number of units to be shipped from source i to destination j. The objective function of the model minimizes the total transportation cost. The constraints ensure that all supplies are used and all demands are satisfied. It is assumed that all the parameters of the model are integer.

$$\text{minimize } Z = \sum_i \sum_j c_{ij} x_{ij}$$

subject to

$$\sum_j x_{ij} = a_i, \qquad i = 1, \ldots, m$$
$$\sum_i x_{ij} = b_j, \qquad j = 1, \ldots, n$$
$$x_{ij} \geq 0, \qquad i = 1, \ldots, m; j = 1, \ldots, n$$

The previous model satisfies some necessary and sufficient conditions for an LP model with integer data to have an optimal solution that is integer. For this model to be feasible, the supply and demand parameters must satisfy the relationship

$$\sum_i a_i = \sum_j b_j$$

which implies that at least one constraint is redundant. An equivalent formulation of the model is given as follows, where the objective function maximizes $-Z$.

$$\text{maximize } \sum_i \sum_j (-c_{ij}) x_{ij}$$

subject to

$$\sum_j x_{ij} = a_i, \qquad i = 1, \ldots, m$$
$$\sum_i x_{ij} = b_j, \qquad j = 1, \ldots, n$$
$$x_{ij} \geq 0, \qquad i = 1, \ldots, m; j = 1, \ldots, n.$$

In order to support later discussion of the transportation simplex algorithm, the dual problem of the previous linear program is formulated:

$$\text{minimize } \{\sum_i a_i R_i + \sum_j b_j K_j\}$$

subject to

$$R_i + K_j + c_{ij} \geq 0, \qquad \text{for all } i \text{ and all } j$$

In the foregoing model, R_i is the dual variable corresponding to the ith availability constraint, and K_j the dual variable corresponding to the jth demand constraint. Both R_i and K_j are unrestricted in sign because they are associated with equality constraints in the primal problem.

For the previous example, the primal and dual linear programs are formulated as follows.

Primal problem

$$\text{Maximize}\,\{-3x_{11} - 4x_{12} - 7x_{13} - 6x_{21} - 3x_{22} - 2x_{23}\}$$

subject to

$$\begin{array}{llllllll}
x_{11} + x_{12} + x_{13} & & & & & & = 20 \\
& & & x_{21} + & x_{22} + & x_{23} & = 10 \\
x_{11} & & & + \; x_{21} & & & = 10 \\
& x_{12} & & & + \; x_{22} & & = 12 \\
& & x_{13} & & & + \; x_{23} & = 8
\end{array}$$

$$x_{ij} \geq 0$$

Dual problem

$$\text{Minimize}\,\{20R_1 + 10R_2 + 10K_1 + 12K_2 + 8K_3\}$$

subject to

$$\begin{array}{lllll}
R_1 & + \; K_1 & & & + 3 \geq 0 \\
R_1 & & + \; K_2 & & + 4 \geq 0 \\
R_1 & & & + \; K_3 & + 7 \geq 0 \\
R_2 & + \; K_1 & & & + 6 \geq 0 \\
R_2 & & + \; K_2 & & + 3 \geq 0 \\
R_2 & & & + \; K_3 & + 2 \geq 0
\end{array}$$

R_i and K_j unrestricted in sign

A feasible solution to the transportation problem can be displayed in a rectangular arrangement of m rows and n columns, with x_{ij} being the number of units that are shipped from the ith source to the jth terminal:

$$\begin{bmatrix}
x_{11} & x_{12} & x_{13} & \cdots & x_{1n} \\
x_{21} & x_{22} & x_{23} & \cdots & x_{2n} \\
\cdot & \cdot & \cdot & & \cdot \\
\cdot & \cdot & \cdot & & \cdot \\
\cdot & \cdot & \cdot & & \cdot \\
x_{m1} & x_{m2} & x_{m3} & \cdots & x_{mn}
\end{bmatrix}$$

A feasible solution, the corresponding transportation costs, and the supply-demand requirements can be simultaneously displayed in a single array as indicated in Figure 2.39.

c_{11} x_{11}	c_{12} x_{12}	c_{13} x_{13}	$\cdots$	c_{1n} x_{1n}	a_1
c_{21} x_{21}	c_{22} x_{22}	c_{23} x_{23}	$\cdots$	c_{2n} x_{2n}	a_2
$\vdots$	$\vdots$	$\vdots$	$\vdots$	$\vdots$	$\vdots$
c_{m1} x_{m1}	c_{m2} x_{m2}	c_{m3} x_{m3}	$\cdots$	c_{mn} x_{mn}	a_m
b_1	b_2	b_3	$\cdots$	b_n	

Figure 2.39. Transportation tableau.

As an illustration, the following is a feasible solution for the numerical example under consideration:

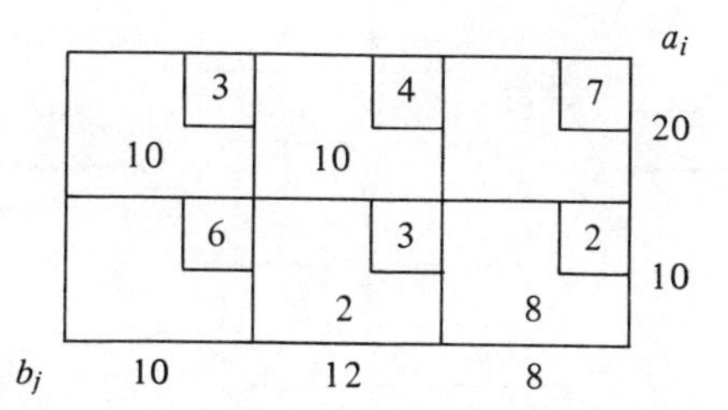

The total transportation cost of this shipping policy is equal to $(10)(3) + (10)(4) + (2)(3) + (8)(2) = 92$.

2.10.2 The Transportation Simplex Algorithm

Before describing the algorithm, some important concepts will be defined. A basic feasible solution for a transportation problem with m sources and n destinations is a feasible solution containing $m + n - 1$ or fewer variables at positive levels. As discussed in Section 2.10.1, at most $m + n - 1$ constraints are independent in the linear program of the transportation problem. The mn variables in a basic solution can be classified into two groups, one with $m + n - 1$ variables and a second one with $mn - (m + n - 1)$ variables. A basic feasible solution can be obtained by setting the variables in the second group equal to zero and then solving the independent constraints of the model for the variables in the first group. The variables in the first group are called *basic* and the variables in the second group *nonbasic*. Accordingly, a route will be called basic if it corresponds to a basic variable (i.e., its flow is not set equal to 0) and nonbasic if it corresponds to a nonbasic variable. A basic solution is said to be degenerate if at least one of its basic variables is equal to zero.

Starting with a nondegenerate basic feasible solution, the transportation simplex algorithm examines if it is profitable to replace one of the current basic routes with another (nonbasic) route not yet in use, performing a test that is equivalent to the optimality condition for linear programs. If, as a result of the test, the route substitution operation should be effected, all flows must be properly adjusted to preserve the feasibility of the new basic solution.

The algorithm consists of four steps. The purpose of each step and its interaction with the other steps are shown in the flowchart given in Figure 2.40. To facilitate our discussion we will refer to the set of basic routes as a *basis*.

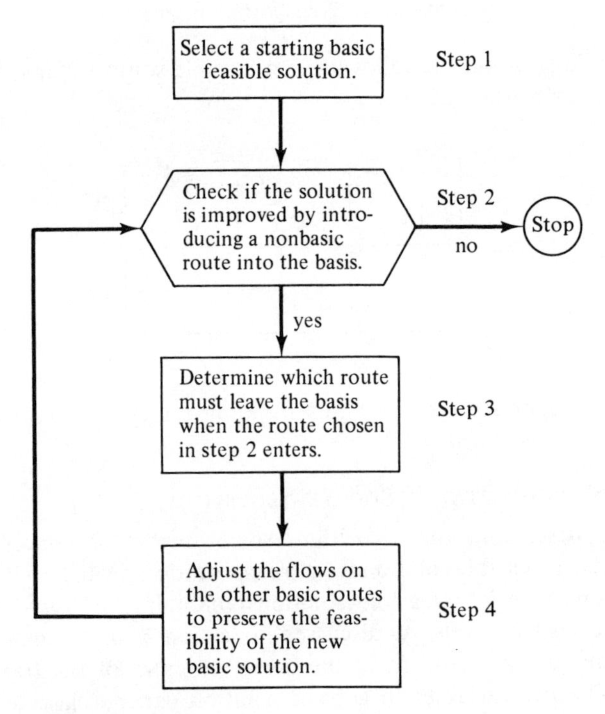

Figure 2.40. Flowchart for the transportation simplex algorithm.

Let $\mathbf{Y}$ be defined as the vector of dual variables, that is, $\mathbf{Y} = [R_1, R_2, \ldots, R_m, K_1, K_2, \ldots, K_n]$, and let $\mathbf{A}_{ij}$ be the vector of constraint coefficients of x_{ij} in the primal program. As can be verified in the example of Section 2.10.1, $\mathbf{A}_{ij}$ can be written as

$$\mathbf{A}_{ij} = \begin{bmatrix} 0 \\ \cdot \\ \cdot \\ \cdot \\ 1 \\ 0 \\ \cdot \\ \cdot \\ \cdot \\ 1 \\ 0 \\ \cdot \\ \cdot \\ \cdot \\ 0 \end{bmatrix} \begin{matrix} \\ \\ \\ \\ \leftarrow i\text{th position} \\ \\ \\ \\ \\ \leftarrow j\text{th position} \\ \\ \\ \\ \\ \\ \end{matrix}$$

The optimality condition for the primal can be written as $\mathbf{Y}\,\mathbf{A}_{ij} + c_{ij} \geq 0$. Since $\mathbf{Y}\,\mathbf{A}_{ij} = R_i + K_j$, the condition can be rewritten as $R_i + K_j + c_{ij} \geq 0$, which is the same condition for the *dual* to be feasible. This implies that the dual is infeasible and the primal nonoptimal, if the route linking the ith source with the jth terminal is nonbasic and has $R_i + K_j + c_{ij} < 0$. Also, the most negative $R_i + K_j + c_{ij}$ value indicates the route with the most potential improvement. Now the algorithm will be described.

***Step 1*:** Selection of a starting feasible solution. There are several methods to obtain this solution. The two most popular methods are probably the *northwest corner method* and the *Vogel approximation method* (VAM). A short description of each method follows.

The northwest corner method. Begin by assigning as much flow as possible to the route linking the first source with the first terminal. Let x_{11} be the number of units assigned. If $a_1 = b_1$, delete both source 1 and terminal 1, and select the route corresponding to x_{22}. Otherwise, delete the source or terminal whose supply or demand is satisfied, and select either x_{12} or x_{21}, depending on which one corresponds to a route that is feasible. Assign as much flow as possible to the selected route, and remove the source or terminal whose supply or demand is also satisfied. If the supply and the demand conditions are simultaneously satisfied, both the source and the terminal are removed. Continuing in this fashion, if x_{ij} is the last basic variable selected, the next one to be considered is $x_{i,j+1}$ if the ith source has any supply available, $x_{i+1,j}$ if the jth terminal has any unfilled demand, or $x_{i+1,j+1}$ otherwise.

As an illustration, a starting solution by the northwest corner rule will be obtained for the transportation problem defined in Table 2.24. Here M represents an arbitrarily large number.

Table 2.24. STARTING SOLUTION BY THE NORTHWEST CORNER RULE

	1	2	3	4	5	6	a_i
1	19	49	43	55	47	53	150
2	20	37	28	42	30	38	200
3	19	33	32	40	70	35	300
4	20	22	20	30	25	M	350
5	20	32	29	48	37	41	250
b_j	100	150	250	250	300	200	

(a) Row 1 assignment

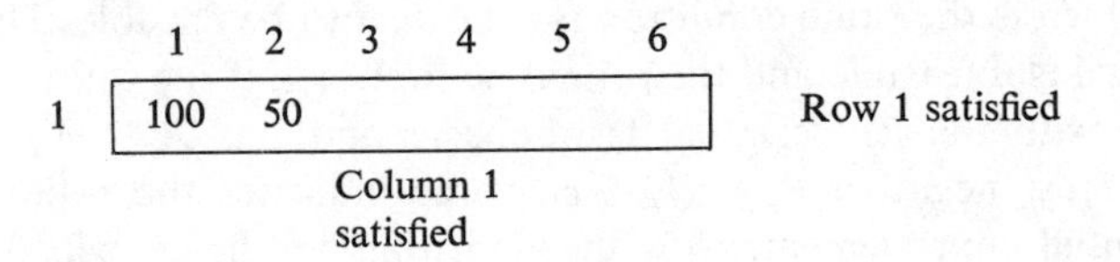

	1	2	3	4	5	6
1	100	50				

Row 1 satisfied

Column 1 satisfied

(b) Row 2 assignment

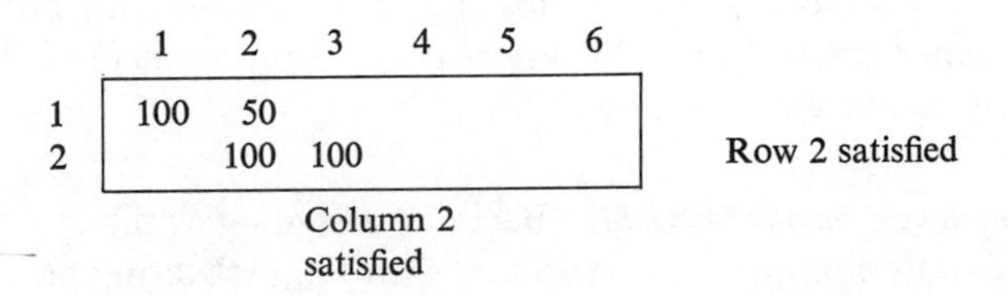

	1	2	3	4	5	6
1	100	50				
2		100	100			

Row 2 satisfied

Column 2 satisfied

(c) Row 3 assignment

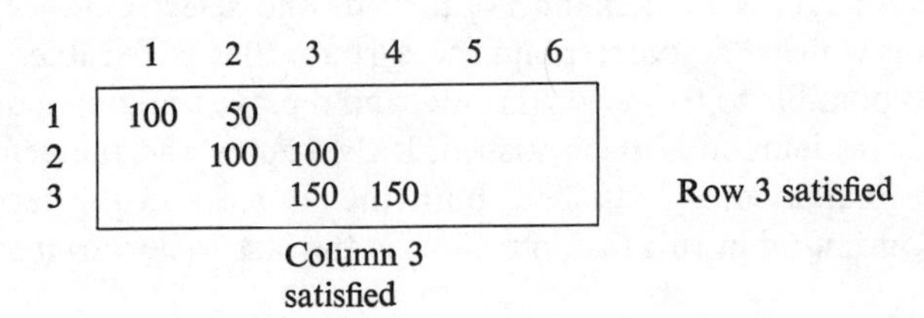

	1	2	3	4	5	6
1	100	50				
2		100	100			
3			150	150		

Row 3 satisfied

Column 3 satisfied

(d) Row 4 assignment

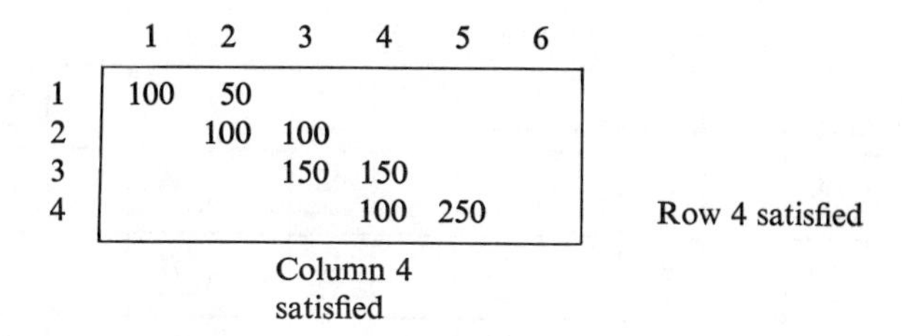

	1	2	3	4	5	6	
1	100	50					
2		100	100				
3			150	150			
4				100	250		Row 4 satisfied
				Column 4 satisfied			

(e) Starting basic feasible solution by the northwest corner rule:

	1	2	3	4	5	6	a_i	
1	100	50					150	
2		100	100				200	
3			150	150			300	
4				100	250		350	
5					50	200	250	$Z = 40{,}950$
b_j	100	150	250	250	300	200		

Vogel approximation method. This heuristic procedure is superior to the northwest corner method because it incorporates the cost information into the route selection process. Essentially, the method computes the minimal penalty incurred if the most economical route is not chosen to leave each source or to go to each terminal, and then selects the source or terminal associated with the largest minimal penalty. Once the source or terminal with the largest minimal penalty is chosen, the method assigns as many units of flow as possible to the most economical route leaving the source or going into the terminal, whichever was chosen. The implication of this assignment is that the largest minimal penalty is avoided. After the chosen route is "saturated," the source and the terminal linked by the route are checked to see what supply–demand condition becomes satisfied. The source or terminal whose condition is satisfied is removed, new penalties are computed, and the procedure is repeated until all sources and terminals are removed. The penalty for any source or terminal is easily computed and is equal to the difference between the smallest and the next-to-the-smallest costs. If both costs are equal, the penalty is then equal to zero. Also, when the largest minimal penalty is not unique for a set of sources and or terminals, the ties can be broken arbitrarily.

Refer to the preceding numerical example. Details of the computations involved in the determination of a starting basic feasible solution by the VAM algorithm are shown in Table 2.25. The penalty values for sources and terminals are written in parentheses, and the largest values are identified by

Table 2.25. STARTING SOLUTION BY THE VAM ALGORITHM

	(0)	(10)	(8)	(10)	(5)	(3)	
★(24)	100 [19]	[49]	[43]	[55]	[47]	[53]	150
(8)	[20]	[37]	[28]	[42]	[30]	[38]	200
(13)	[19]	[33]	[32]	[40]	[70]	[35]	300
(0)	[20]	[22]	[20]	[30]	[25]	[M]	350
(9)	[20]	[32]	[29]	[48]	[37]	[41]	250
	100	150	250	250	300	200	

		★(10)	(8)	(10)	(5)	(3)	
(4)		[49]	[43]	[55]	[47]	[53]	50
(2)		[37]	[28]	[42]	[30]	[38]	200
(1)		[33]	[32]	[40]	[70]	[35]	300
(2)		150 [22]	[20]	[30]	[25]	[M]	350
(3)		[32]	[29]	[48]	[37]	[41]	250
		150	250	250	300	200	

			(8)	★(10)	(5)	(3)	
(4)			[43]	[55]	[47]	[53]	50
(2)			[28]	[42]	[30]	[38]	200
(3)			[32]	[40]	[70]	[35]	300
(5)			[20]	200 [30]	[25]	[M]	200
(8)			[29]	[48]	[37]	[41]	250
			250	250	300	200	

Table 2.25. (cont.)

			(1)	(2)	(7)	(3)	
(4)			43	55	47	53	50
(2)			28	42	30	38	200
(3)			32	40	70	35	300
★ (8)			29 (250)	48	37	41	250
			250	50	300	200	

				(2)	★ (17)	(3)	
(6)				55	47	53	50
(8)				42	30 (200)	38	200
(5)				40	70	35	300
				50	300	200	

				(15)	★ (23)	(18)	
(6)				55	47 (50)	53	50
(5)				40	70	35	300
				50	100	200	

				40 (50)	70 (50)	35 (200)	300
				50	50	200	

asterisks. The modified supply–demand parameters are also shown at each iteration of the method.

Therefore, as a result of the steps shown in Table 2.25, the following starting basic feasible solution is obtained by the VAM algorithm:

	1	2	3	4	5	6	a_i
1	100				50		150
2					200		200
3				50	50	200	300
4		150		200			350
5			250				250
b_j	100	150	250	250	300	200	

$Z = 39{,}300$

The total transportation cost associated with the distribution policy given by the northwest corner solution is equal to 40,950. The distribution cost of the solution given by the VAM method is equal to 39,300. This illustrates the general superiority of the VAM algorithm over the northwest corner method.

In the numerical example, $m + n - 1 = 10$. Therefore, the VAM solution is degenerate but that of the northwest corner rule is not.

***Step 2*:** Identification of the nonbasic route to be brought into the basis. Since the number of dual values is equal to $m + n$ and we have $m + n - 1$ equations $R_i + K_j + c_{ij} = 0$ to solve, it is possible to set $R_1 = 0$ and solve for the remaining values. After setting $R_1 = 0$, the K_j values for terminals with basic variables in the first row can be determined. Once these K_j values are available, we can find the R_i values of those rows with basic variables in the columns corresponding to the K_j values just obtained. This procedure is then repeated until all dual values are obtained. After computing the $R_i + K_j + c_{ij}$ values for all nonbasic routes, the optimality condition is verified for each one. Recall that an optimal solution is obtained when all $R_i + K_j + c_{ij} \geq 0$. If all these values are not nonnegative, the route with the most negative value has the largest potential for improvement and should be brought into the basis.

Note that when the number of basic routes is less than $m + n - 1$, it is not possible to find the values R_i and K_j, and therefore we cannot measure the potential improvements of the nonbasic routes. A simple way to find the dual values consists of assigning infinitesimal flows to as many routes as necessary to complete a nondegenerate solution. The infinitesimal flows, however, must be assigned to independent routes. An independent route is defined as one for which the flow cannot be obtained as a linear combination of basic flows. In the $m \times n$ tableau used to record the input data and the solutions, a route corresponds to a cell. An independent route or cell is one for which we cannot construct a loop, starting at that cell, with positive flows at each corner.

As an illustration of step 2, the starting basic feasible solution by VAM will be considered. As previously mentioned, this solution is degenerate because the number of basic routes is less than $m + n - 1 = 10$. In this case, one infinitesimal flow is needed, and it is represented by ϵ in Table 2.26. The reader should verify that a loop cannot be constructed using the ϵ-cell as the starting point.

Table 2.26. VALID POSITION FOR ϵ-CELL

100				50		150
				200		200
			50	50	200	300
	150		200			350
	ϵ	250				250
100	150	250	250	300	200	

Note that in Table 2.27, however, the ϵ-cell is not in a valid position since the four routes corresponding to the corners of the loop shown form a non-independent set of cells.

We will now check the VAM solution to determine if it is optimal or not. In Table 2.28, the symbol ● represents a basic flow (including ϵ). The value in the lower left corner of the nonbasic flow cells is equal to $R_i + K_j + c_{ij}$. The R_i values are shown to the left of each row and the K_j values above

Table 2.27. INVALID POSITION FOR ϵ-CELL

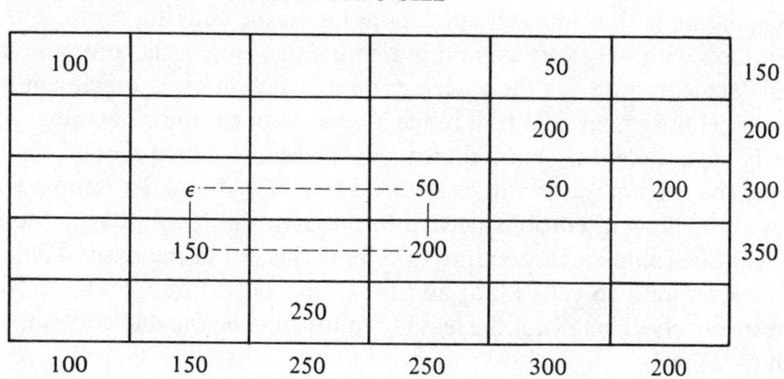

100				50		150
				200		200
	ϵ		50	50	200	300
	150		200			350
		250				
100	150	250	250	300	200	

Table 2.28. ROUTE POTENTIAL IMPROVEMENTS

	$K_1 = -19$	$K_2 = -9$	$K_3 = -6$	$K_4 = -17$	$K_5 = -47$	$K_6 = -12$
$R_1 = 0$	[19] ●	[49] 40	[43] 37	[55] 38	[47] ●	[53] 41
$R_2 = 17$	[20] 18	[37] 45	[28] 39	[42] 42	[30] ●	[38] 43
$R_3 = -23$	[19] −23	[33] 1	[32] 3	[40] ●	[70] ●	[35] ●
$R_4 = -13$	[20] −12	[22] ●	[20] 1	[30] ●	[25] −35	[M] M −25
$R_5 = -23$	[20] −22	[32] ●	[29] ●	[48] 8	[37] −33	[41] 6

each column. As a result of step 2, the nonbasic route represented by cell (4, 5) should enter the basis, since $R_4 + K_5 + c_{45} = -35$ and is the most negative value (maximum potential improvement).

***Step 3*:** Identification of the route that must leave the basis. Once we determine the route that enters the basis, we can find the pattern of alterations of the basic routes when the flow on the entering route is increased from zero. The pattern of alterations is called a *stepping-stone path.*

In the example under consideration, it was determined that the route corresponding to cell (4, 5) should enter the basis. If the flow on this route is increased, some other routes must either increase or decrease their flows to preserve the feasibility of the new solution, since the original supply–demand conditions must be always satisfied. When a route increases its flow, we call it a *recipient route*, and when it decreases its flow we call it a *donor route.* Obviously, the leaving route must be the donor route with the smallest flow in the current solution. This procedure will always drive one or more basic variables to zero. When more than one, the solution becomes degenerate.

In order to identify the recipient and donor routes, we first start the stepping-stone path at the cell corresponding to the entering route, and then close a loop with only vertical and horizontal displacements, and basic route cells in each corner. Since the entering route by definition is a recipient route, we can label the routes next to it in the loop as donor routes. The other routes are assigned recipient (+) and donor (−) labels alternately around the loop.

For the current numerical example, Table 2.29 shows the stepping-stone path with the flow alterations needed to preserve the feasibility of the solution when the route corresponding to cell (4, 5) is put in the basis. The donor routes correspond to cells (3, 5) and (4, 4) and have flows equal to 50 and 200, respectively. Therefore, the leaving route must be the one corresponding to cell (3, 5).

Table 2.29. STEPPING-STONE PATH

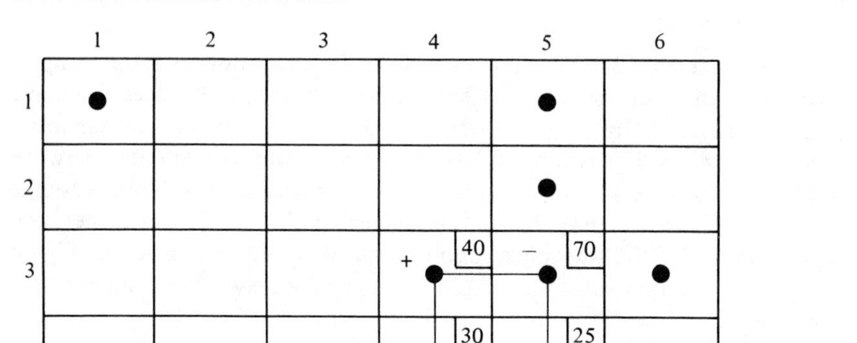

***Step 4*:** Adjustment of flows on basic routes. The amount by which we can modify the flows is given by the flow of the leaving route. To find the new flows, we add that amount to the flows on recipient routes and subtract the same amount from the flows on donor routes.

In summary, this procedure creates a new basic solution for which the supply and demand constraints are satisfied. The stepping-stone path simply "swaps" flow units from one cell to another along the loop. The application of step 4 to the example under study yields the new solution displayed in Table 2.30.

This solution is now considered in step 2 to investigate if it satisfies the optimality condition. It can be verified that the solution is indeed optimal, and the algorithm is then terminated.

Table 2.30. NEW BASIC FEASIBLE SOLUTION (OPTIMAL)

	1	2	3	4	5	6
1	100				50	
2					200	
3				100		200
4		150		150	50	
5		ϵ	250			

2.10.3 A Network Interpretation for the Transportation Simplex Algorithm

The purpose of this section is to illustrate how the transportation simplex algorithm can be described using only network concepts. We feel that some students might find our current presentation more appealing to their intuition, since we will perform the simplex method directly on the network rather than in the standard tableau format of Section 2.10.2. To discuss the present topic we will use the example given in that section. The network representation of the starting solution obtained by VAM is given in Figure 2.41, where the numbers assigned to the arcs of the network are equal to the corresponding flows.

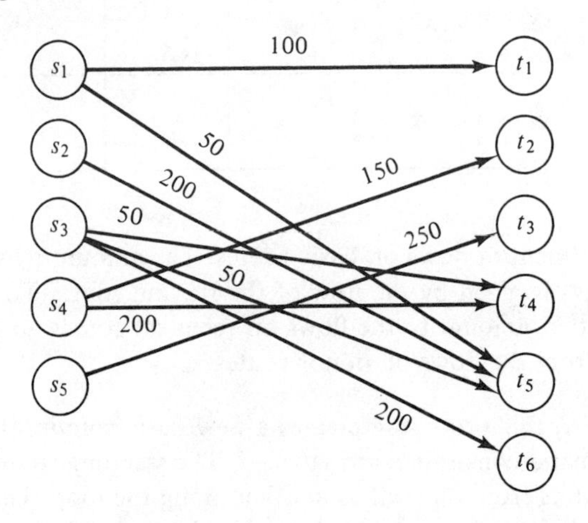

Figure 2.41. Network representation of the starting solution by VAM.

It can be verified that the number of arcs with positive flow is in this case equal to 9 and that there are two unconnected components in this network. The first component corresponds to the subnetwork $G_1 = (\mathbf{N}_1, \mathbf{A}_1)$, where $\mathbf{N}_1 = \{s_1, s_2, s_3, s_4, t_1, t_2, t_4, t_5, t_6\}$ and $\mathbf{A}_1 = \{(s_1, t_1), (s_1, t_5), (s_2, t_5), (s_3, t_4), (s_3, t_5), (s_3, t_6), (s_4, t_2), (s_4, t_4)\}$. The second component corresponds to the subnetwork $G_2 = (\mathbf{N}_2, \mathbf{A}_2)$, where $\mathbf{N}_2 = \{s_5, t_3\}$ and $\mathbf{A}_2 = \{(s_5, t_3)\}$. In order to have exactly one connected component in the network of Figure 2.41, we need one more arc either from source node s_5 to any terminal node $t_j, j \neq 3$, or from any source node $s_i, i \neq 5$, to terminal node t_3. As in the example of Section 2.10.2, we will consider an arc (s_5, t_2) with an infinitesimal flow ϵ in order not to violate the demand-supply constraints of the network model. The connected network is shown in Figure 2.42. In general, $m + n - 1$ arcs are needed to construct a connected (one-component) network and

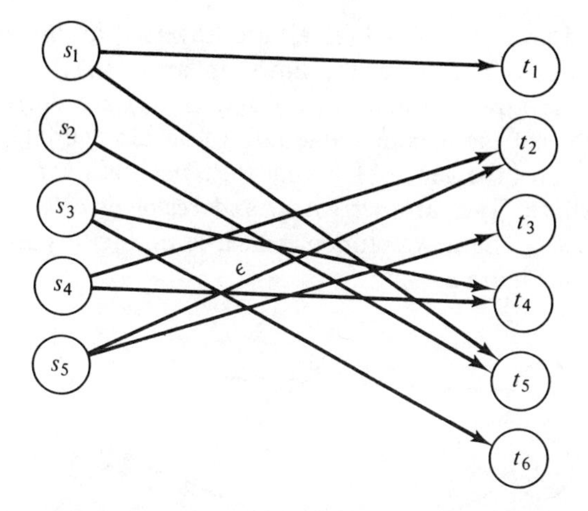

Figure 2.42. Connected network for the starting solution by VAM.

degeneracy is equivalent to two or more unconnected components. The $m + n - 1$ arcs correspond to the "independent cells."

As a result of the analysis of the profitability of nonbasic routes, it was concluded in step 2, Section 2.10.2, that route (s_4, t_5) should be used. The corresponding pattern of alterations, represented as a loop in the tableau format, corresponds to a cycle in the network, as shown in Figure 2.43. If this cycle is traversed following the node sequence s_4, t_5, s_3, t_4, s_4, we observe that arcs (s_4, t_5) and (s_3, t_4) are traversed in the same direction as their

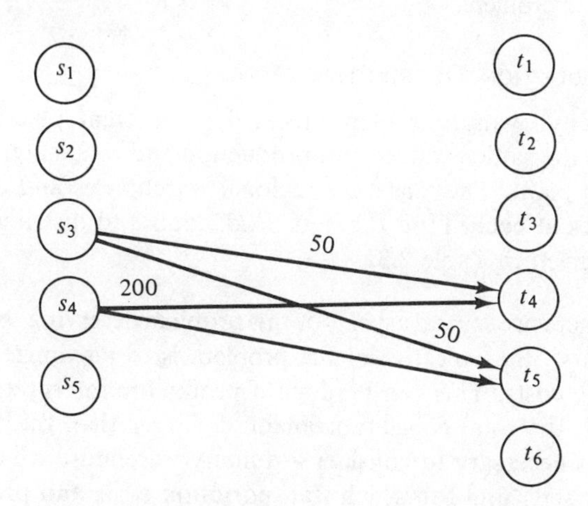

Figure 2.43. Network representation of the pattern of alterations in a transportation network.

orientations, and arcs (s_3, t_5) and (s_4, t_4) are traversed in direction opposite to their orientations. Therefore, the flows on arcs (s_4, t_5) and (s_3, t_4) are increased, and the flows on arcs (s_3, t_5) and (s_4, t_4) are decreased. It can easily be verified that the maximal amount of flow along arc (s_4, t_5) is equal to 50. This illustrates the general fact that the *donor cells* correspond to arcs in the cycles whose flows are *decreased*, and *recipient cells* to arcs whose flows are *increased*. The new solution, which is optimal as seen in Section 2.10.2, is given in Figure 2.44.

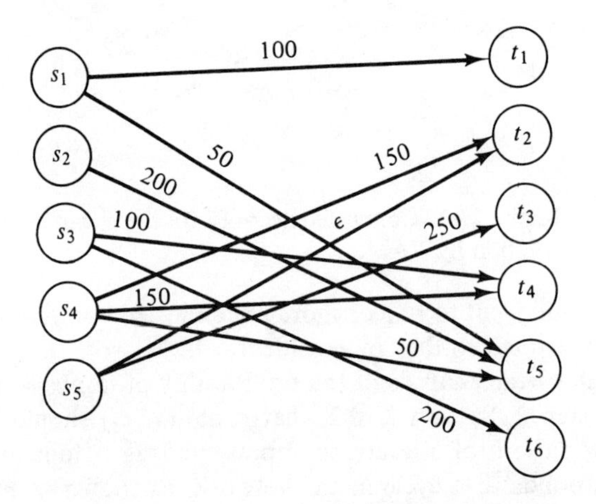

Figure 2.44. Network representation of the optimal solution of the sample transportation problem.

2.10.4 A PRODUCTION–DISTRIBUTION MODEL

A company owns four plants located in different places. Each plant manufactures the same product, but production and raw material costs differ from plant to plant. There are five regional warehouses and customers pay different prices at each. Find the best production and distribution schedule for the data given in Table 2.31.

***Solution*:** There are several aspects of this problem deserving some additional comments. First, the objective of the problem is to maximize profits rather than minimize costs. This can be done if profits are interpreted as negative costs. Second, the total capacity available is larger than the total demand; therefore, it is necessary to consider a dummy warehouse which will absorb the excess capacity and for which transportation costs and profits are equal to zero. A third consideration is that some of the shipments will result in a loss if all the requirements are to be met, as in the standard version of the transportation model. If desired, we can introduce a dummy plant with

Table 2.31. PRODUCTION AND DISTRIBUTION SCHEDULE DATA[a]

Plant:	1	2	3	4
Production (not including raw materials):	20	25	19	20
Raw materials:	10	9	12	8
Transportation costs per unit:				
Warehouse 1	4	10	12	19
Warehouse 2	3	7	5	6
Warehouse 3	5	9	4	7
Warehouse 4	7	5	9	3
Warehouse 5	6	8	7	6

[a]Assume that the capacities of the four plants, for a given planning horizon, are 200, 250, 325, and 150 units. Also, assume that the maximum sales are 130, 160, 200, 150, and 210 for the five warehouses. The sales prices for these warehouses are 40, 42, 41, 42, and 41 dollars per unit.

capacity x and with zero profits and transportation costs, and assign the nonprofitable shipments to that plant. Provided that x is large enough, its value will not affect shipments between real plants and warehouses. A correct choice for x is the minimum between the total capacity of all plants and the total maximum sales of all warehouses. In the example, this value is equal to 850.

Table 2.32. TRANSPORTATION MODEL FOR PROFIT MAXIMIZATION

		Warehouse						
		W_1	W_2	W_3	W_4	W_5	W_6	
	P_1	6	9	6	5	5	0	200
	P_2	−4	1	−2	3	−1	0	250
Plant	P_3	−3	6	6	2	3	0	325
	P_4	−7	8	6	11	7	0	150
	P_5	0	0	0	0	0	0	850
		130	160	200	150	210	925	

Table 2.32 summarizes the profit information and the supply–demand figures for this problem. Here, P_5 and W_6 are a dummy plant and a dummy warehouse, respectively. To convert this maximization problem into an equivalent minimization problem, we can multiply all entries of the profit matrix by -1 and add 11 (the maximal profit entry). The corresponding minimization problem is defined in Table 2.33, where warehouses are defined as sources and plants as terminals.

Table 2.33. TRANSPORTATION MODEL FOR COST MINIMIZATION

		Terminal					
		P_1	P_2	P_3	P_4	P_5	a_i
Source	W_1	5	15	14	18	11	130
	W_2	2	10	5	3	11	160
	W_3	5	13	5	5	11	200
	W_4	6	8	9	0	11	150
	W_5	6	12	8	4	11	210
	W_6	11	11	11	11	11	925
	b_j	200	250	325	150	850	

The optimal solution to this problem is obtained by the transportation simplex algorithm, and it is arranged in the matrix $X = [x_{ij}^*]$ as follows:

$$X = [x_{ij}^*] = \begin{bmatrix} 130 & 0 & 0 & 0 & 0 \\ 70 & 0 & 90 & 0 & 0 \\ 0 & 0 & 200 & 0 & 0 \\ 0 & 0 & 0 & 150 & 0 \\ 0 & 0 & 35 & 0 & 175 \\ 0 & 250 & 0 & 0 & 675 \end{bmatrix}$$

The maximal profit associated with this policy is equal to $4905. According to the solution, no real shipment is sent to terminal 2. That is, no shipment is made from plant 2.

2.11 The Transshipment Problem

In the transportation model it is assumed that no route from one source to a terminal can use other sources or terminals as intermediate points. If transshipment through sources and terminals is allowed, the new problem can still be modeled and solved as a regular transportation problem. In this case, of course, the computational work involved in solving the model is increased as a result of the additional routes that must be considered. These additional routes link each source with all the other sources, and each terminal with all the other terminals.

Assuming that the transportation costs for the additional routes are available, a modified transportation problem can be formulated to solve the transshipment problem if the supply–demand parameters for the additional routes do not influence the selection of routes in the basic algorithm. This condition is required because the supply–demand conditions of the additional routes represent fictitious specifications which are incorporated only for modeling purposes.

Let f be defined as the minimum between $\sum a_i$ and $\sum b_j$. That is, f is the actual number of units being shipped through the modified transportation network. It is, therefore, obvious that we can set a "demand" parameter $\bar{b}_i = f$, for the ith original source, and a "supply" parameter $\bar{a}_j = f$, for the jth original terminal. In other words, it is possible for all flow to be transshipped through only one source or one terminal.

By similar reasoning, we can conclude that the original supply–demand parameters must be increased by f. Note that a value $\bar{f} < f$ would overconstrain the network flows, but a value $\bar{f} \geq f$ would allow sufficient flow for transshipment without adding constraints which are not specified in the transshipment problem. Hence, a reasonable choice for the parameter is f.

As an example of the transshipment problem, assume that in the transportation problem formulated in Section 2.10.1, each source and each terminal can be considered as a transshipment point. According to the original data for the transportation problem, $\sum_i a_i = \sum_j b_j = 30$. Therefore, $f = 30$.

The new transshipment model can be solved as a transportation problem with the following supply–demand parameters:

1. For original sources, $\bar{a}_i = a_i + 30$, $i = 1, 2$.
2. For original sources, $\bar{b}_i = 30$, $i = 1, 2$.
3. For original terminals, $\bar{a}_j = 30$, $j = 1, 2, 3$.
4. For original terminals, $\bar{b}_j = b_j + 30$, $j = 1, 2, 3$.

The equivalent transportation model is defined in Table 2.34, where the shipping costs of the additional routes are assumed known.

Table 2.34. A TRANSSHIPMENT TABLEAU

		Original terminal			Original source		
		1	2	3	1	2	
Original source	1	3	4	7	0	5	50
	2	6	3	2	3	0	40
Original terminal	1	0	5	4	2	5	30
	2	9	0	1	3	2	30
	3	2	4	0	2	6	30
		40	42	38	30	30	

2.12 The Assignment Problem

This problem consists of allocating some resources to some operating patterns in such a way that the costs of the assignments are minimized, each resource is assigned exactly once, and each operating pattern is performed exactly once. Examples of resources and operating patterns along with the corresponding measure of effectiveness or "cost" are given in Table 2.35.

Table 2.35. APPLICATIONS OF THE ASSIGNMENT MODEL

Example	*Resource*	*Operating Pattern*	*Measure of Effectiveness*
1	Workers	Jobs	Time
2	Trucks	Routes	Cost
3	Machines	Locations	Material handling
4	Crews	Flights	Layover time
5	Salesmen	Territories	Volume of sales

The cost matrix **C** is defined as $\mathbf{C} = [c_{ij}]$, where c_{ij} is the cost of assigning the ith resource to the jth operating pattern. In general, i and j take on the values $1, 2, \ldots, m$, where m is the number of operating patterns or resources.

Define x_{ij} as follows:

$$x_{ij} \doteq \begin{cases} 1, & \text{if resource } i \text{ is assigned to operating pattern } j \\ 0, & \text{otherwise} \end{cases}$$

Therefore, a solution to the problem can be arranged in a rectangular array $\mathbf{X} = [x_{ij}]$. A feasible solution is called an *assignment*. There are $m!$ feasible solutions for a given value of m. A feasible solution is produced when one selects one and only one cell from each row of $\mathbf{X} = [x_{ij}]$ and one and only one cell from each column of the same matrix.

Given the matrix $\mathbf{C} = [c_{ij}]$, it is possible to display a feasible solution to the problem by encircling the "costs" corresponding to the entries in **X** such that $X_{ij} = 1$. As an example, for $m = 3$, we can have the following feasible solution:

$$\mathbf{C} = [c_{ij}] = \begin{bmatrix} 4 & 7 & \textcircled{0} \\ \textcircled{0} & 3 & 8 \\ 6 & \textcircled{3} & 9 \end{bmatrix} \qquad \mathbf{X} = [x_{ij}] = \begin{bmatrix} 0 & 0 & 1 \\ 1 & 0 & 0 \\ 0 & 1 & 0 \end{bmatrix}$$

2.12.1 Mathematical Model

The objective function of this model minimizes the total cost of the assignments. The constraints can be classified into two groups. The first group of constraints ensures that each resource is allocated exactly once.

The second group ensures that each operating pattern is used exactly once. The formulation of the model is given as follows:

$$\text{minimize } Z = \sum_i \sum_j c_{ij} x_{ij}$$

subject to

$$\sum_j x_{ij} = 1, \qquad \text{for all } i$$

$$\sum_i x_{ij} = 1, \qquad \text{for all } j$$

$$x_{ij} \geq 0, \qquad \text{for all } i \text{ and all } j$$

It is clear that the assignment problem is a special case of the transportation problem, where each a_i and each b_j are equal to 1. Hence, this model can be solved by any linear programming algorithm or by the transportation simplex algorithm. However, the method discussed in the following section is more efficient, since it exploits the special mathematical structure of the model.

2.12.2 THE HUNGARIAN ALGORITHM

Let us consider an assignment model with cost matrix $\mathbf{C} = [c_{ij}]$. Suppose that all the entries of the ith row are modified by the addition of a real number γ_i, and that all entries of the jth column are modified by the addition of a real number δ_j. Therefore, a transformed cost matrix $\mathbf{D}$ is obtained, for which

$$d_{ij} = c_{ij} + \gamma_i + \delta_j$$

or

$$c_{ij} = d_{ij} - \gamma_i - \delta_j$$

The previous expression implies that $c_{ij}X_{ij} = d_{ij}X_{ij} - \gamma_i X_{ij} - \delta_j X_{ij}$. Therefore,

$$\begin{aligned}\sum_i \sum_j c_{ij} x_{ij} &= \sum_i \sum_j d_{ij} x_{ij} - \sum_i \sum_j \gamma_i x_{ij} - \sum_i \sum_j \delta_j x_{ij} \\ &= \sum_i \sum_j d_{ij} x_{ij} - \sum_i \gamma_i - \sum_j \delta_j\end{aligned}$$

This result implies that minimizing $\sum \sum c_{ij} x_{ij}$ is equivalent to minimizing $\sum \sum d_{ij} x_{ij}$, subject to the constraints of the assignment model. The Hungarian algorithm capitalizes on this result by subtracting the smallest value from each row and each column and trying to identify a feasible solution with zero costs in the modified cost matrix. If such a feasible solution exists, it is an optimal assignment. If it does not exist, the algorithm modifies the matrix once again in order to create more cells with zero costs.

The algorithm consists of three phases: (1) the row–column reduction process, (2) the assignment identification process, and (3) the modification of the reduced matrix. A brief description of each step is given next.

***Phase 1*:** Row–column reduction process. The purpose of this step of the algorithm is to create as many zero-cost cells as possible. To do this, we can subtract the smallest value in each row from all the elements in that row, and then subtract the smallest element in each column from all the elements in that column. Now this reduced-cost matrix can be used instead of the original cost matrix for the purpose of making assignments.

Phase 2: Identification of assignments. If the row–column reduction process allows us to select one zero-cost cell from each row and one from each column, such that the resulting solution is feasible, that assignment will also be optimal. If a zero-cost assignment in the reduced matrix is not possible, the matrix must be further modified, as explained in Phase 3.

***Phase 3*:** Modification of the reduced matrix. If there are insufficient zero-cost cells to select a zero-cost assignment, more zeros can be created by using the following simple procedure. Identify the minimal group of rows and columns containing all the zero-cost entries of the reduced matrix, and find the minimal value not in this group. If this value is subtracted from all the entries of the matrix, the zeros become negatives, and at least one element outside the group becomes equal to zero. However, since the matrix has negative entries, a zero-cost assignment is not necessarily optimal.

To have no negative numbers in the group, choose the absolute value of the most negative number and add it to all the elements in each row and each column of the group. Notice that by this procedure the elements at the intersections of the rows and columns of the group are increased by twice the chosen value, and all the negative elements become zero or positive, as they were before.

The final result is that the new reduced matrix has more zeros in positions outside the rows and columns containing the zeros of the current nonoptimal solution.

The fundamental procedure of the Hungarian algorithm just discussed can be reduced to the following two sets of rules. The first set can be used to search for a feasible solution having all zero-cost assignments in the reduced matrix. The second set of rules can be used to modify the reduced-cost matrix.

1. Rules to Search for a Zero-Cost Feasible Solution (Phase 2):

   a. Identify the rows that have exactly one uncrossed zero. In each of these rows, make an assignment in the uncrossed zero cell and cross out any zeros not yet crossed out in those columns in which assignments were made. Rows are considered in increasing numerical order.

b. Identify the columns that have exactly one uncrossed zero. In each of these columns, make an assignment in the uncrossed zero cell and cross out any zeros not yet crossed out in those rows in which assignments were made. Columns are considered in increasing numerical order.

c. Repeat steps a and b until it is not possible to cross out more zeros. If all zeros were crossed out, the assignment is optimal. If some zeros remained uncrossed, it may still be possible to select a complete assignment, for example by trial and error, among several optimal alternatives. If no complete assignment is possible, the matrix needs to be further modified.

2. Rules to Modify the Reduced Matrix (Phase 3):

   a. Count the number of zeros in each row and each column not yet crossed out.
   b. Cross out the row or column with the maximum number of zeros. In case of ties break them arbitrarily. Rows are crossed out with horizontal lines, and columns with vertical lines.
   c. Repeat steps a and b until all zeros are covered with lines.
   d. Subtract the minimal uncovered number from all uncovered numbers, and add it to the elements at the intersection of any two lines.

Figure 2.45 gives a flowchart for the Hungarian algorithm. The rules for Phase 3 are those previously described. Note that the negative elements mentioned in the justification of the algorithm are not directly considered in these rules.

2.12.3 Example of the Hungarian Algorithm

The Hungarian algorithm will be illustrated on the assignment problem defined by the following cost matrix:

$$\mathbf{C} = [c_{ij}] = \begin{bmatrix} 2 & 10 & 9 & 7 \\ 15 & 4 & 14 & 8 \\ 13 & 14 & 16 & 11 \\ 4 & 15 & 13 & 19 \end{bmatrix}$$

1. Row–Column Reduction Process:

   a. The minimal numbers are 2, 4, 11, and 4 for rows 1, 2, 3, and 4, respectively. Subtracting each minimal value from all entries in its row, we obtain the following matrix:

$$\begin{bmatrix} 0 & 8 & 7 & 5 \\ 11 & 0 & 10 & 4 \\ 2 & 3 & 5 & 0 \\ 0 & 11 & 9 & 15 \end{bmatrix}$$

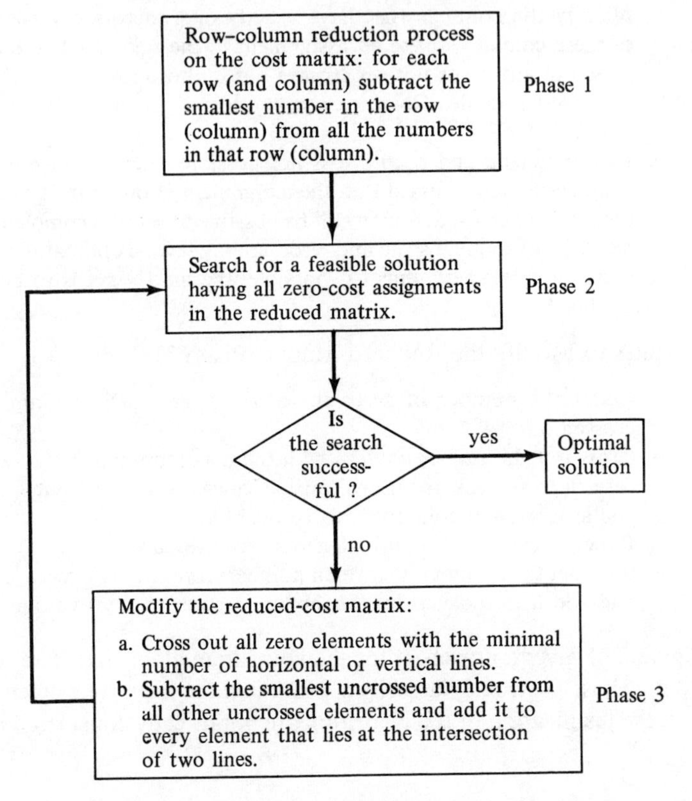

Figure 2.45. Flowchart for the Hungarian algorithm.

b. The minimal numbers are 0, 0, 5, and 0, for columns 1, 2, 3, and 4, respectively. Subtracting each minimal value from all entries in its column, we obtain the following reduced matrix:

0	8	2	5
11	0	5	4
2	3	0	0
0	11	4	15

2. Search for a Feasible Solution with All Zero-Cost Assignments:

a. Rows 1, 2, and 4 have single uncrossed zeros. Considering these rows, in increasing numerical order, we would first make an assignment in the cell (1, 1), and cross out the zero in cell (4, 1). The second assignment can be made in cell (2, 2). Row 4 cannot be used since its zero was crossed out when the assignment in cell (1, 1) was made.

b. Columns 4 and 5 have single uncrossed zeros. Considering these columns in increasing numerical order, we can make the third assignment in cell (3, 3). No assignment in column 4 is possible since its zero is crossed out after making the assignment in cell (3, 3). The results up to this point are summarized as follows:

⓪	8	2	5
11	⓪	5	4
2	3	⓪	0
0	11	4	15

Therefore, no complete assignment was possible, and then we must further modify the reduced-cost matrix.

3. Modification of the Reduced-Cost Matrix:

   a. The number of zeros in rows 1, 2, 3, and 4 of the foregoing matrix are 1, 1, 2, and 1, respectively. For the columns these numbers are 2, 1, 1, and 1.
   b. Row 3 and column 1 have a maximum of two zeros each. If row 3 is chosen, a horizontal line will cover all the elements of this row.
   c. The number of uncovered zeros in rows 1, 2, and 4 are 1, 1, and 1, respectively. The numbers for the columns are 2, 1, 0, and 0. Then we must choose column 2 and cross it out with a vertical line. After this only one uncovered zero remains in cell (2, 2). We can cover it with either a horizontal line through row 2 or a vertical line through column 2. If the vertical line is chosen, the following results are obtained:

	8	2	5
	11	4	15

   d. The minimal uncovered value is 2. Substracting it from all the uncovered values and adding it to the intersection elements, the new reduced-cost matrix is given as follows:

0	6	0	3
13	0	5	4
4	3	0	0
0	9	2	13

Repeating the procedure to obtain a zero-cost feasible solution, we find the following optimal solution:

0	6	⓪	3
13	⓪	5	4
4	3	0	⓪
⓪	9	2	13

All the results are summarized in Figure 2.46.

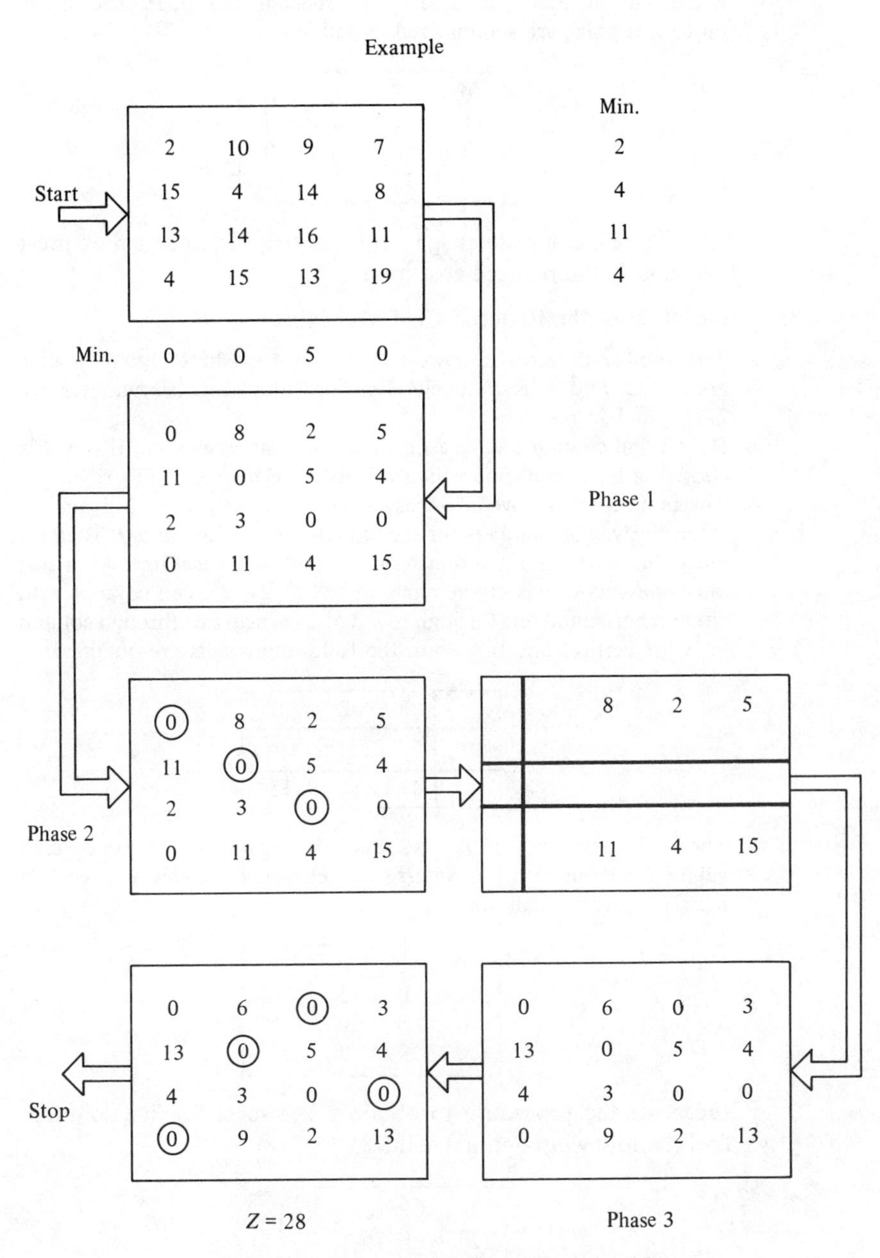

Figure 2.46. Steps in the Hungarian algorithm.

2.12.4 Remarks

1. The total number of feasible assignments is equal to $m!$, where m is the number of rows or columns of the cost matrix (square matrix).

2. For any impossible assignment, the corresponding cost is set equal to M (big positive number in minimization problems).

3. If the original matrix is not square, add as many dummy rows or columns as needed, and set the costs equal to proper values determined from the context of the problem.

4. If the problem is a maximization case, multiply all entries of the cost matrix by -1, and add a suitable number to all entries to remove negative values. Then solve the problem as a minimization problem.

5. In general, if the minimum number of lines needed to cover the zero elements is equal to the number of rows or columns (square matrix), there is a zero-cost assignment possible.

2.12.5 A Product–Plant Allocation Problem

A company needs to develop a production plan to start manufacturing three new products. Assume that the company has five plants and that three of them must be selected to manufacture the new items, one product per plant. The unit production and distribution costs for all the product–plant combinations are given below:

1. Unit production costs (dollars):

		Plant				
		1	2	3	4	5
Product	1	20	23	38	15	35
	2	8	29	6	35	35
	3	5	8	3	4	7

2. Unit distribution costs (dollars):

		Plant				
		1	2	3	4	5
Product	1	20	50	20	10	13
	2	7	90	8	35	60
	3	5	5	4	15	6

According to the current plans, the following annual production for each item is required to meet demands at the prices shown:

Product	*Planned Production*	*Planned Price*
1	35,000	$55
2	160,000	50
3	54,000	30

***Solution*:** The total unit cost is equal to the unit production cost plus the unit distribution cost. Since the unit selling price is known for each product, we can find the profit per unit for each product–plant combination as follows:

		Plant 1	2	3	4	5
Product	1	15	−18	−3	30	7
	2	35	−69	36	−20	−45
	3	20	17	23	11	17

Multiplying the profit per unit by the total volume of annual sales, we can find the total annual profit associated with each product–plant combination. The following table shows the profits in thousands of dollars:

		Plant 1	2	3	4	5
Product	1	525	−630	−105	1050	245
	2	5600	−11040	5760	−3200	−7200
	3	1080	918	1242	594	918

The present maximization problem can be formulated as a minimization assignment model by considering profits as negative costs and by including two dummy products (4 and 5) with zero profits. To remove negative values from the cost matrix, 5760 is added to each entry. The corresponding results are given as follows:

		Plant 1	2	3	4	5
Product	1	5235	6390	5865	4710	5515
	2	160	16800	0	8960	12960
	3	4680	4842	4518	5166	4842
	4	0	0	0	0	0
	5	0	0	0	0	0

The optimal solution to this problem is given below. According to the solution, product 1 is assigned to plant 4, product 2 to plant 1, product 3 to plant 3, product 4 to plant 2, and product 5 to plant 5. Evidently, these two last assignments are fictitious. The total maximal profit per year for this solution is equal to $7,892,000.

		Plant				
		1	2	3	4	5
	1	365	1520	1155	⓪	645
	2	⓪	16640	0	8960	12800
Product	3	2	164	⓪	648	164
	4	0	⓪	160	160	0
	5	0	0	160	160	⓪

This solution is obtained after the reduced matrix is modified by covering the zeros with lines in the third and fourth columns and in the fourth and fifth rows.

2.13 The Assignment Problem and the Traveling Salesman Problem

In Section 2.9, an algorithm was presented to solve the traveling salesman problem. Under a special program structure, the assignment algorithm of Section 2.12 can also be used to solve the traveling salesman problem. In general, it should be clear that the optimal (minimum-cost) solution generated for an $(n \times n)$ assignment problem would not constitute a solution to the corresponding $(n \times n)$ traveling salesman problem. However, for the following formulation, the assignment algorithm can be suitably modified to solve the more complicated problem.

Let $\mathbf{D}$ be an $(n \times n)$ distance (cost) matrix which represents the distance (cost) penalties d_{ij} involved in moving from a particular city or location i to another city or location j, where $i = 1, 2, \ldots, n$, $j = 1, 2, \ldots, n$.

Further, define a special structure of $\mathbf{D}$ wherein $d_{ij} \equiv 0$ for $i \geq j$ and d_{ij} completely arbitrary (positive, negative, or zero) for $i < j$ (Table 2.36). Such a distance (cost) matrix is represented by $\bar{\mathbf{D}}$ and is said to be *upper triangular*.

The solution algorithm is attributed to Lawler [40] and proceeds as follows.

***Step 1*:** Delete column 1 and row n from the $\bar{D}$ matrix and solve the resulting $(n - 1) \times (n - 1)$ assignment problem.

Comment: The result of step 1 will generate a *path* from node 1 to node n. It will (possibly) generate other *disjoint closed subpaths* between other

Table 2.36. AN UPPER TRIANGULAR MATRIX

From (i) \ To (j)	1	2	3	$\cdots$	n
1	0	d_{12}	d_{13}	$\cdots$	d_{1n}
2	0	0	d_{23}	$\cdots$	d_{2n}
3	0		0	$\cdots$	d_{3n}
$\vdots$	$\vdots$	$\vdots$	$\vdots$		$\vdots$
n	0	0	0	$\cdots$	0

$= \bar{D}$

nodes which are not on the main path from node 1 to node n. If the assignment algorithm produces an acyclic path or chain which passes through each node once and only once, the solution is obtained. Hence, we will assume that disjoint closed subpaths do exist.

Before proceeding, note that $d_{ij} \equiv 0$ for all arcs (i, j) such that $i \geq j$. Such arcs will be called *null arcs*. The path from node 1 to node n, and any disjoint closed subpaths, may contain one or more null arcs.

Step 2: Examine successively all disjoint closed subpaths and *remove* from these portions of the original assignment solution all *null arcs*.

Comment: Since all null arcs have been removed from each closed subpath, each subpath is now composed of arcs that connect a lower-numbered node to a higher-numbered node. (Recall the upper triangular structure of the distance matrix.) Suppose that these subpaths are designated as i_1 to j_1, i_2 to j_2, j_2 to j_3, and so on, where $i_1 < i_2 < \cdots < i_m$ and $i_1 \leq j_1, i_2 \leq j_2, \ldots, i_m \leq j_m$.

Step 3: Starting with node n, which completes the path from node 1 to node n, introduce new null arcs into the solution connecting nodes n to i_m, j_m to $i_{m-1}, \ldots, j_2$ to i_1, and j_1 to node 1.

Comment: The result of step 3 is a complete cyclic path from node 1 back to node 1 which passes through all the other $(n - 1)$ nodes once and only once. Note that since we have added only null arcs, the objective function value of the resulting traveling salesman problem is the same as that of the first assignment problem. Since the assignment solution is optimal (minimum-cost), the resultant traveling salesman solution is also optimal.

To illustrate the procedure, consider the following example due to Lawler [40].

Distance matrix (From rows, To columns) $= \bar{D}$

From \ To	1	2	3	4	5	6	7
1	0	−1	7	−20	3	2	5
2	0	0	12	8	16	9	8
3	0	0	0	3	7	6	2
4	0	0	0	0	4	4	9
5	0	0	0	0	0	−18	−1
6	0	0	0	0	0	0	3
7	0	0	0	0	0	0	0

***Step 1*:**
In the following matrix, the circled entries represent the optimal assignment solution.

Reduced assignment matrix $= \bar{D}$

	2	3	4	5	6	7
1	−1	7	(−20)	3	−2	5
2	(0)	12	8	16	9	8
3	0	0	3	7	6	(2)
4	0	(0)	0	4	4	9
5	0	0	0	0	(−18)	−1
6	0	0	0	(0)	0	3

***Step 2*:** A path from node 1 to node 7 is given by

$$\textcircled{1} \xrightarrow{-20} \textcircled{4} \xrightarrow{0} \textcircled{3} \xrightarrow{2} \textcircled{7}$$

There are two disjoint closed subpaths:

1. $② \xrightarrow{0} ②$
2. $⑤ \xrightarrow{-18} ⑥ \xrightarrow{0} ⑤$

Since arcs (2, 2) and (6, 5) are null arcs, remove them from the disjoint closed subpaths. This creates the following disjoint structure:

$$① \xrightarrow{-20} ④ \xrightarrow{0} ③ \xrightarrow{2} ⑦$$
$$②$$
$$⑤ \xrightarrow{-18} ⑥$$

***Step 3*:** Using null arcs, connect disjoint nodes (paths) in the following manner:

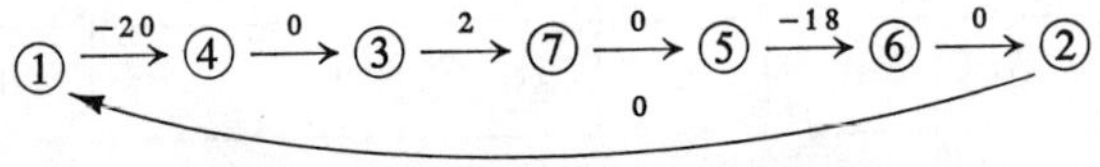

$$① \xrightarrow{-20} ④ \xrightarrow{0} ③ \xrightarrow{2} ⑦ \xrightarrow{0} ⑤ \xrightarrow{-18} ⑥ \xrightarrow{0} ② \xrightarrow{0} ①$$

The optimal traveling salesman solution is now achieved at a cost of -36. In conclusion, note that the following solution is not optimal since the nonnull arc (2, 5) has been added to the sequence of disjoint subpaths.

$$① \xrightarrow{-20} ④ \xrightarrow{0} ③ \xrightarrow{2} ⑦ \xrightarrow{0} ② \xrightarrow{16} ⑤ \xrightarrow{-18} ⑥ \xrightarrow{0} ①$$

2.14 The Maximum-Flow Problem

Let $G = (\mathbf{N}, \mathbf{A})$ be a directed network having a single source $s \in \mathbf{N}$, a single terminal $t \in \mathbf{N}$, and capacitated arcs $(i, j) \in \mathbf{A}$. The maximum-flow problem can be described as that of finding the pattern of flows through the arcs of **A** which will result in a maximum flow shipped from node s to node t. It is assumed that there is an infinite amount of flow available at the source node, and that the flow conservation condition holds for all the intermediate nodes of the network. The problem is nontrivial when the capacity U_{ij} of any arc represents a finite upper bound on the flow f_{ij} along that arc.

The maximum-flow problem was formulated as a linear programming model in Section 2.2, and hence can be solved by the standard simplex method. It is the purpose of the present section to discuss an alternative and more efficient solution procedure. The algorithm starts with a feasible solution and then uses a labeling procedure developed by Ford and Fulkerson [15] in order to produce another feasible solution with a better flow value. In the algorithm, nodes are viewed as flow transshipment points and arcs as distribution channels. Two fundamental concepts are necessary for a formal discussion of the algorithm: the notion of labeling, and the definition of a flow-augmenting path.

The purpose of labeling a node is to indicate both the flow value and the origin of a shipment which causes a change in the current amount of flow on the arc between the origin of the shipment and that node. If a shipment of q_j units of flow is sent from node i to node j and it causes an increase in the flow on the arc, we say that node j is labeled from node i by $+q_j$. In this case, the label $[+q_j, i]$ is assigned to node j. Similarly, if the shipment causes a reduction in the flow on the arc, we say that node j is labeled from node i by $-q_j$. In this case, the label $[-q_j, i]$ is assigned to node j.

The current flow from node i to node j is increased when a flow shipment of q_j units is sent to node j through a directed arc (i, j) which is traversed in the direction of its orientation. In this case, arc (i, j) is referred to as a *forward arc*. This situation is illustrated in Figure 2.47.

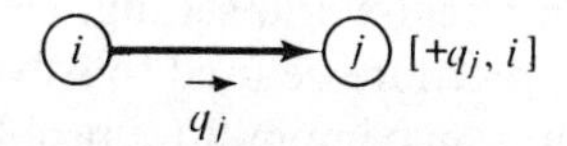

Figure 2.47. The labeling of node j from node i causes an increase in the amount of flow on arc (i, j) because (i, j) is a forward arc.

The current flow from j to i is decreased when a flow of q_j units is sent to node j through a directed arc (j, i) traversed in the direction opposite to that of its orientation. In this case, arc (j, i) is said to be a *reverse arc*. This situation is illustrated in Figure 2.48.

When a given node j is labeled from node i and (i, j) is a forward arc, the flow through this arc is increased and the remaining unused capacity of

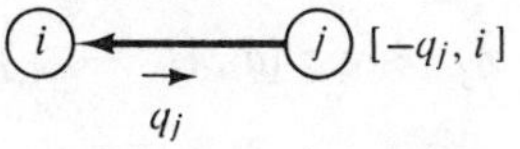

$[-q_j, i]$

$\overrightarrow{q_j}$

Figure 2.48. The labeling of node j from node i causes a decrease in the amount of flow on arc (j, i) because (j, i) is a reverse arc.

the flow must be properly adjusted. This remaining capacity is usually referred to as the "residual capacity" of the arc. The reader can easily realize that only forward branches with strictly positive capacities can be used in labeling. Moreover, node j can be labeled from node i only after node i has been labeled.

A flow-augmenting path from node s to node t is defined as a connected sequence of forward and reverse arcs along which some units of flow can be sent from s to t. The flow on each forward arc is increased without exceeding its capacity, and the flow on each reverse arc is decreased without making it negative. The purpose of the flow-augmenting path is to identify the pattern of flow changes resulting in an increase of the flow at node t, without violating the flow conservation condition for all the intermediate nodes of the network.

2.14.1 A Labeling Procedure for the Maximum-Flow Algorithm

The maximum-flow problem is one that is frequently encountered in practice. However, it is not uncommon to formulate problems with thousands of nodes and arcs in a real-world setting. Hence, it is necessary that a computationally efficient procedure be used to solve such problems. Because of the simple structure of a maximum-flow problem, an efficient solution algorithm can be developed which recursively seeks an optimal (maximum flow) solution through the use of labeling procedures. The methodology will now be presented.

Let (i, j) be a directed arc from node i to node j. Under what circumstances can the flow along this arc be *increased*? According to the previous discussion, it is possible to increase the flow by q_j if (i, j) is a forward arc and node j is labeled with $[+q_j, i]$. The question that should be answered is when can this occur? Suppose that there is a flow $f_{ij} \geq 0$ already assigned to that arc, such that $f_{ij} \leq U_{ij}$. Obviously, q_j can be no more than the residual capacity $U_{ij} - f_{ij}$. Is this sufficient to label node j? The answer is no, because it may not be possible to get exactly $U_{ij} - f_{ij}$ units of flow from node i. Note that we can send only as much to node j as we have surplus at node i, that is, at most q_i. Hence, we can augment the flow across a forward arc (i, j) by q_j, where

$$q_j = \min [q_i, U_{ij} - f_{ij}]$$

The same logic enables us to label node j if arc (j, i) is a reverse arc. Here we ask if a decrease in flow across arc (j, i) is possible. Evidently, it will be possible only if $f_{ji} > 0$. But, how much can we decrease the flow? We can decrease the flow by at most the amount that we could move away from node i, that is, q_i. Hence, we could reduce the flow across a reverse arc (j, i) by q_j, where

$$q_j = \min [q_i, f_{ji}]$$

The algorithmic structure of the method is developed as follows. Initially, the source node is labeled with $[\infty, -]$, which indicates the there is an infinite supply available. We now seek a flow-augmenting path from the source node to the sink node via labeled nodes. All other nodes are initially unlabeled. We proceed across forward and reverse arcs, labeling nodes as we move along, seeking to reach the sink node. One of two events will now occur.

1. The sink node t is labeled $[+q_t, k]$. Therefore, a flow-augmenting path has been found and each arc flow along this path can be increased, or decreased, by an amount q_t. After the flows are changed, the current labels are erased, and the entire procedure is repeated.
2. The sink node t cannot be labeled. This implies that no flow-augmenting path can be found. Hence, the present flow values represent an optimal (maximum-flow) solution.

2.14.2 Example of the Labeling Algorithm

As an illustration of the labeling algorithm, the maximum flow that can be shipped from node s to node t will be found for the network shown in Figure 2.49(a). Each arc (i, j) is assigned a label $[f_{ij}, U_{ij}]$, where f_{ij} is the current value of the arc flow and U_{ij} is the capacity of the arc. Initially, all flows are set equal to zero. At each iteration, the procedure attempts to label the terminal node t. The problem is solved in six iterations. The results of these iterations are summarized in Figure 2.49.

Steps for the 1st iteration	Description
1	Label s with $[\infty, -]$
2	Label 2 with $[+3, s]$
3	Label t with $[+2, 2]$
4	Change flows: $f_{s2} = 2, f_{2t} = 2$

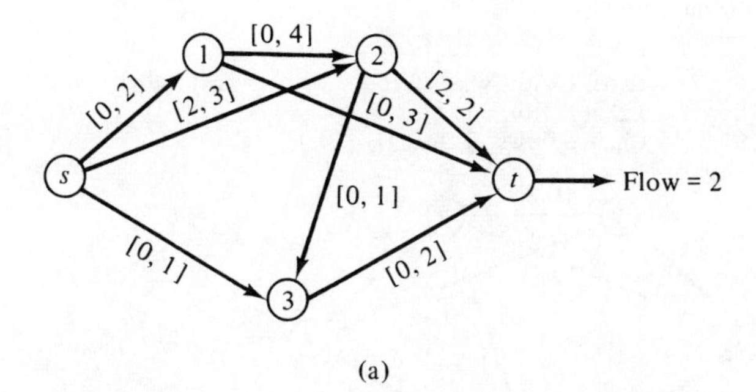

Figure 2.49. Example of the labeling algorithm: (a) first iteration; (b) second iteration; (c) third iteration; (d) fourth iteration; (e) fifth iteration.

Steps for the 2nd iteration	Description
5, 6	Label s with $[\infty, -]$; label 1 with $[+2, s]$
7, 8	Label 2 with $[+2, 1]$; label 3 with $[+1, 2]$
9	Label t with $[+1, 3]$
10	Change flows: $f_{s1} = 1, f_{12} = 1, f_{23} = 1, f_{3t} = 1$

(b)

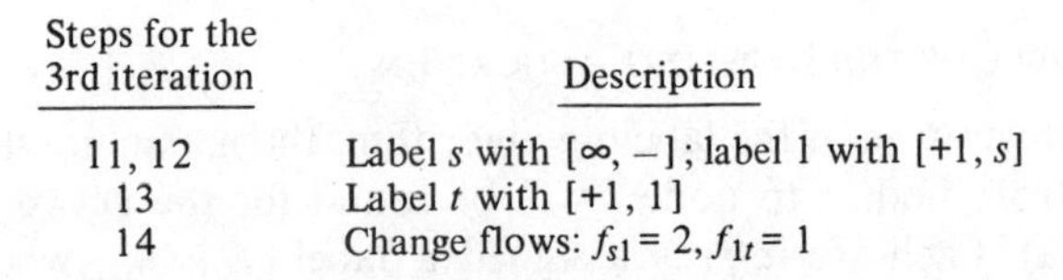

Steps for the 3rd iteration	Description
11, 12	Label s with $[\infty, -]$; label 1 with $[+1, s]$
13	Label t with $[+1, 1]$
14	Change flows: $f_{s1} = 2, f_{1t} = 1$

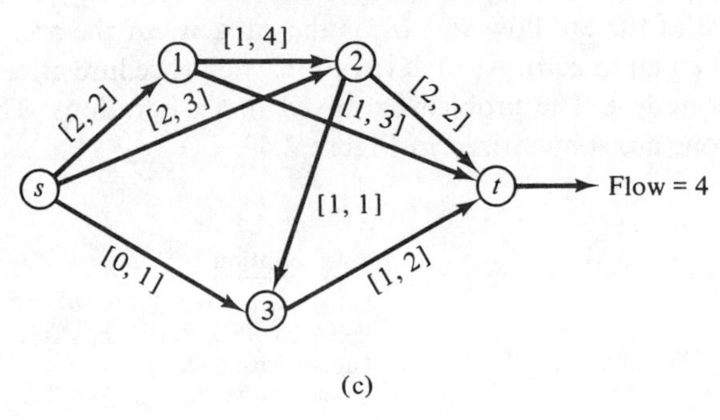

(c)

Steps for the 4th iteration	Description
15, 16, 17	Label s with $[\infty, -]$; label 2 with $[+1, s]$; label 1 with $[-1, 2]$
18	Label t with $[+1, 1]$
19	Change flows: $f_{s2} = 2, f_{12} = 0, f_{1t} = 2$

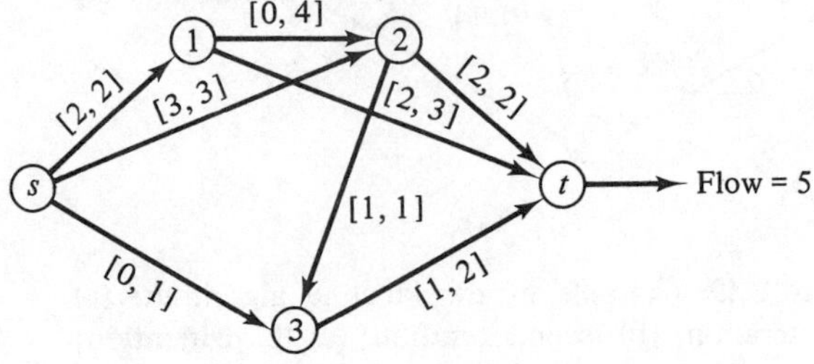

(d)

Figure 2.49. (cont.)

Steps for the 5th iteration	Description
20, 21, 22	Label s with $[\infty, -]$; label 3 with $[+1, s]$; label t with $[+1, 3]$
23	Change flows: $f_{s3} = 1, f_{3t} = 2$

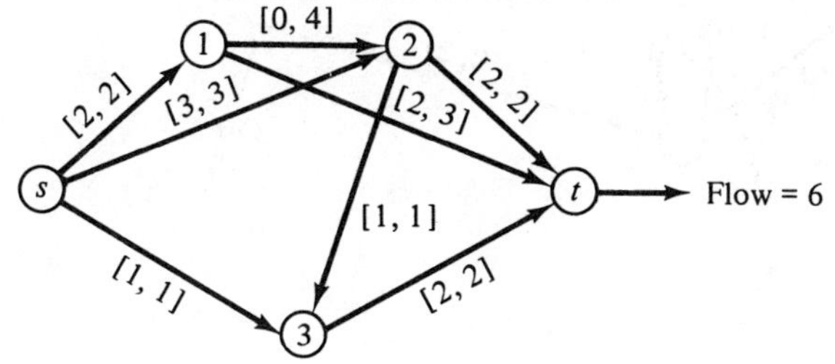

(e)

Steps for the 6th iteration	Description
24	Label s with $[\infty, -]$
25	No other nodes can be labeled; therefore, the maximum flow is equal to 6

Figure 2.49. (cont.)

2.14.3 The Maximum Flow/Minimum Cut Theorem

Suppose that we have a network $G = (\mathbf{N}, \mathbf{A})$. Let us divide the nodes of the set $\mathbf{N}$ into two mutually exclusive subsets, $\mathbf{N}_c$ and $\bar{\mathbf{N}}_c$. These two subsets of nodes are connected by a subset of arcs $\mathbf{A}_c$; other arcs are elements of the subset $\bar{\mathbf{A}}_c$. Further, suppose that the sink node t is contained in the subset $\bar{\mathbf{N}}_c$ and the source node s in the subset $\mathbf{N}_c$. Then the value of any flow from the subset $\mathbf{N}_c$ to $\bar{\mathbf{N}}_c$ through arcs in $\mathbf{A}_c$ can be no more than the sum of all arc capacities in the subset $\mathbf{A}_c$; that is,

$$\sum f_{ij} \leq \sum_{\substack{i \in \mathbf{N}_c \\ j \in \bar{\mathbf{N}}_c}} U_{ij}$$

We denote this "transfer barrier" separating the subset $\mathbf{N}_c$ from $\bar{\mathbf{N}}_c$ as a "cut" $(\mathbf{N}_c, \bar{\mathbf{N}}_c)$. Obviously, the maximum flow that can pass from node s to node t is bounded from above by the cut value. The value of the cut $(\mathbf{N}_c, \bar{\mathbf{N}}_c)$ is calculated by summing all arc capacities for those arcs in the subset $\mathbf{A}_c$ which can be used to ship flow from set $\mathbf{N}_c$ to $\bar{\mathbf{N}}_c$. The maximum flow/minimum cut theorem establishes that the maximum flow from node s to node t is given by the value of the minimum cut that separates node s from node t.

As an illustration of the maximum flow/minimum cut theorem, the network of Figure 2.50 will be considered. Notice that there are several cuts

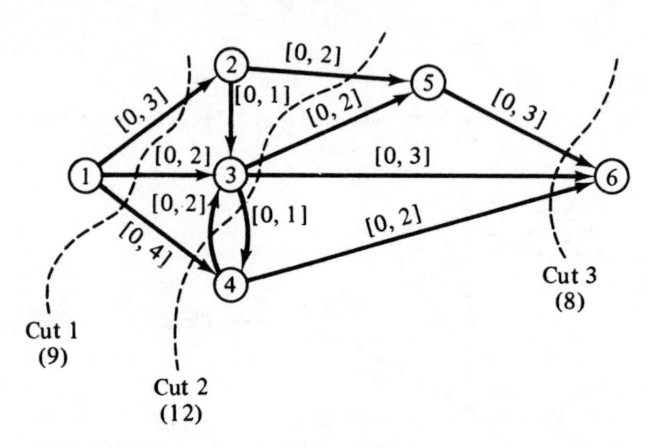

Figure 2.50. Maximum flow/minimum cut theorem.

separating node 6 from node 1, and that the value of the maximum flow is equal to 8. The capacities of the cuts 1, 2, and 3 shown in Figure 2.50 are equal to 9, 12, and 8, respectively. Therefore, cut 3 is a minimal cut.

The implication of the maximum flow/minimum cut theorem is that it is possible to find the maximum flow through a capacitated network by finding the capacities of *all* cuts and choosing the minimum value. Of course, this is of little practical value in solving a maximum-flow problem, since no information is obtained concerning the actual flows (f_{ij}) through the network. However, the result is important from a theoretical standpoint and is most useful in developing sophisticated network flow algorithms or in checking if a given solution is optimal. A proof of the theorem is given as follows. Let F be a feasible value for the flow from node s to node t. For any cut $(\mathbf{N}_c, \bar{\mathbf{N}}_c)$ we can write

$$F = \sum_{j \in \mathbf{N}_c} [\sum_{j \in \mathbf{N}} f_{ij} - \sum_{j \in \mathbf{N}} f_{ji}]$$

Since $\mathbf{N} = \mathbf{N}_c \cup \bar{\mathbf{N}}_c$ and a node cannot be in both $\mathbf{N}_c$ and $\bar{\mathbf{N}}_c$, then

$$F = \sum_{\substack{i \in \mathbf{N}_c \\ j \in \mathbf{N}_c}} f_{ij} + \sum_{\substack{i \in \mathbf{N}_c \\ j \in \bar{\mathbf{N}}_c}} f_{ij} - \sum_{\substack{i \in \mathbf{N}_c \\ j \in \mathbf{N}_c}} f_{ji} - \sum_{\substack{i \in \mathbf{N}_c \\ j \in \bar{\mathbf{N}}_c}} f_{ji}$$

$$F = [\sum_{\substack{i \in \mathbf{N}_c \\ j \in \bar{\mathbf{N}}_c}} f_{ij} - \sum_{\substack{i \in \mathbf{N}_c \\ j \in \bar{\mathbf{N}}_c}} f_{ji}] + [\sum_{\substack{i \in \mathbf{N}_c \\ j \in \mathbf{N}_c}} f_{ij} - \sum_{\substack{i \in \mathbf{N}_c \\ j \in \mathbf{N}_c}} f_{ji}]$$

But the second term in brackets is exactly zero; hence,

$$F = \sum_{\substack{i \in \mathbf{N}_c \\ j \in \bar{\mathbf{N}}_c}} f_{ij} - \sum_{\substack{i \in \mathbf{N}_c \\ j \in \bar{\mathbf{N}}_c}} f_{ji}$$

so it is certainly true that

$$F \leq \sum_{\substack{i \in \mathbf{N}_c \\ j \in \bar{\mathbf{N}}_c}} f_{ij} \tag{2-61}$$

since $f_{ji} \geq 0$. Hence, any flow value F is bounded by the capacity of an arbitrary cut between $\mathbf{N}_c$ and $\bar{\mathbf{N}}_c$, and it follows that the maximum allowable flow is bounded from above by the capacity of the minimum cut.

Using Eq. (2-61) and the fact that $f_{ij} \leq U_{ij}$, it is established that

$$F \leq \sum_{i \in \mathbf{N}_c} \sum_{j \in \bar{\mathbf{N}}_c} U_{ij} \tag{2-62}$$

In Eq. (2-62) the right-hand side is the value of an arbitrary cut $(\mathbf{N}_c, \bar{\mathbf{N}}_c)$ which separates node s from node t. Representing this value by V_{st}, we can write Eq. (2-62) as

$$F \leq V_{st} \tag{2-63}$$

Let $(\mathbf{L}, \bar{\mathbf{L}})$ be a cut such that (a) $s \in \mathbf{L}$, $t \in \bar{\mathbf{L}}$; and (b) $j \in \mathbf{L}$ if there is an $i \in \mathbf{L}$ such that $f_{ij} < U_{ij}$ or $f_{ji} > 0$. From Eq. (2-63),

$$F = \sum_{i \in \mathbf{L}} \sum_{j \in \bar{\mathbf{L}}} f_{ij} - \sum_{i \in \mathbf{L}} \sum_{j \in \bar{\mathbf{L}}} f_{ji} \tag{2-64}$$

But, from the definition of $\mathbf{L}$, if $(i, j) \in (\mathbf{L}, \bar{\mathbf{L}})$, then $f_{ij} = U_{ij}$; and if $(j, i) \in (\bar{\mathbf{L}}, \mathbf{L})$, then $f_{ji} = 0$; hence,

$$\sum_{i \in \mathbf{L}} \sum_{j \in \bar{\mathbf{L}}} f_{ij} = \sum_{i \in \mathbf{L}} \sum_{j \in \bar{\mathbf{L}}} U_{ij} \tag{2-65}$$

and

$$\sum_{i \in \mathbf{L}} \sum_{j \in \bar{\mathbf{L}}} f_{ji} = 0 \tag{2-66}$$

Therefore,

$$F = \sum_{i \in \mathbf{L}} \sum_{j \in \bar{\mathbf{L}}} U_{ij} \geq V_{st} \tag{2-67}$$

From Eqs. (2-63) and (2-67), we finally conclude that

$$F = V_{st} \tag{2-68}$$

2.14.4 Design of a Centralized Sewage Waste Treatment Plant

A waste treatment plant receives input from nine decentralized pumping substations distributed across an urban population. A new urban renewal project, currently being planned, will create a significant amount of new effluence to the existing system. The waste treatment plant is relatively new, but the drainage system is believed to be undersized. The problem is to determine the maximum amount of sewage that can be pumped through the system to the existing processing facility. Figure 2.51 represents the fundamental elements of the system. The nodes correspond to pumping substations and the arcs to pipelines.

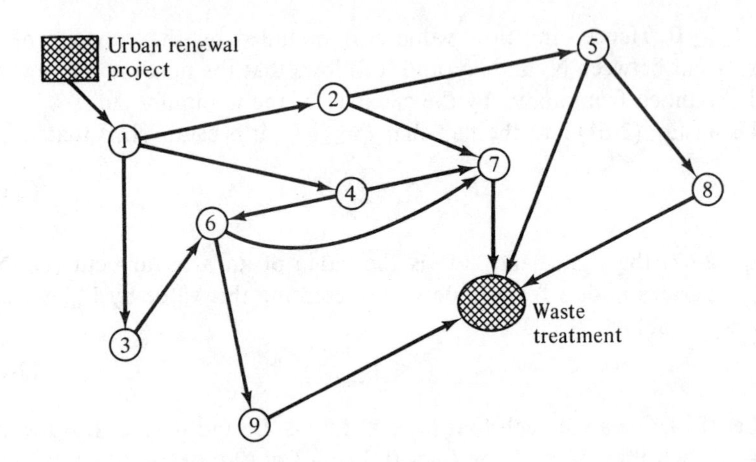

Figure 2.51. Waste sewage problem.

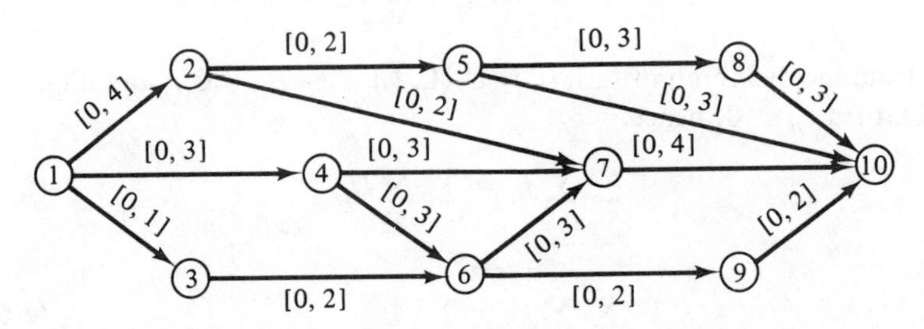

Figure 2.52. Maximum-flow network for waste sewage.

Figure 2.52 depicts the system in conventional network form with upper bounds on all links to represent the maximum flow capacities at any one point in time. In order to obtain a solution using the maximum-flow algorithm, the initial solution of Figure 2.52 will be used. Figure 2.53(a)–(e) shows the sequence of steps followed to arrive at the solution given in Figure 2.53(f). Figure 2.53(f) indicates that a maximum flow rate of 8 is possible, and that pipe sections (6, 7), (5, 8), and (8, 10) have no flow at the optimal solution. This implies that the pumping substation 8 is of no value to the system and can be shut down if desired. Link (6, 7) can likewise be shut off. However, if the maximum attainable flow rate is not great enough for planned expansion, it is clear that new, larger pipelines at links (1, 2) and (2, 5) will increase the system's capacity. Finally, it can be noted that new piping at links (1, 4) and (1, 3) will not increase the capacity of the system without additional modifications.

The results summarized in Figure 2.53(a) are obtained after the following sequence of operations: label 1 with $[\infty, -]$, label 2 with $[+4, 1]$, label 5

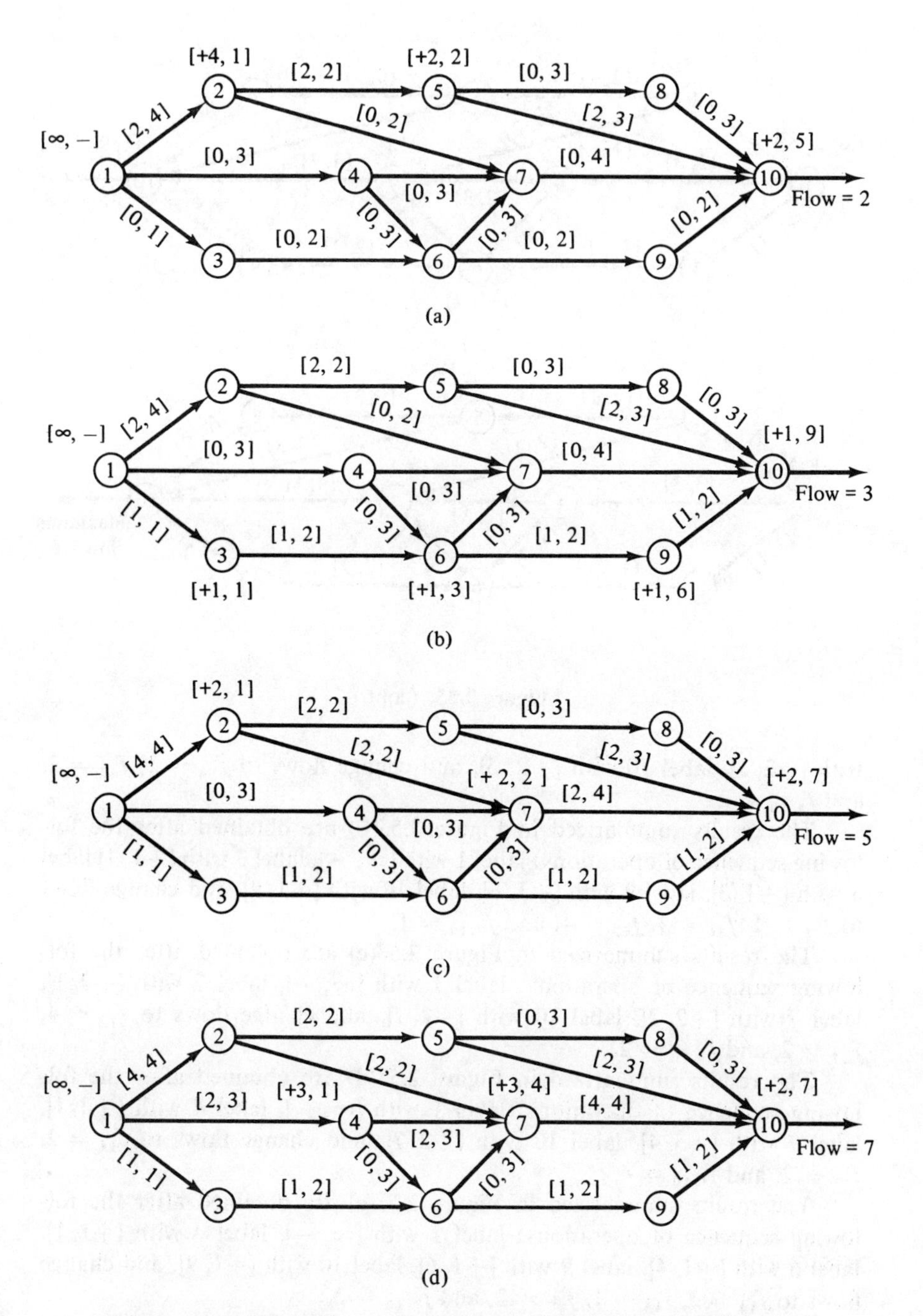

Figure 2.53. Waste sewage problem: (a) first iteration; (b) second iteration; (c) third iteration; (d) fourth iteration; (e) fifth iteration; (f) final solution.

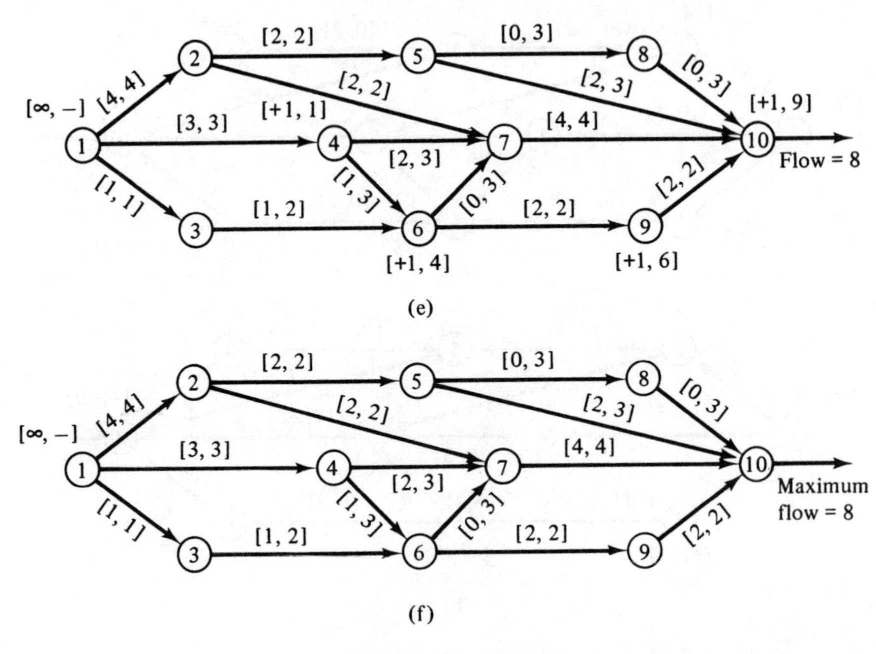

(e)

(f)

Figure 2.53. (cont.)

with [+2, 2], label 10 with [+2, 5], and change flows to $f_{12} = 2$, $f_{25} = 2$, and $f_{5,10} = 2$.

The results summarized in Figure 2.53(b) are obtained after the following sequence of operations: label 1 with [∞, —], label 3 with [+1, 1] label 6 with [+1, 3], label 9 with [+1, 6], label 10 with [+1, 9], and change flows to $f_{13} = 1$, $f_{36} = 1$, $f_{69} = 1$, and $f_{9,10} = 1$.

The results summarized in Figure 2.53(c) are obtained after the following sequence of operations: label 1 with [∞, —], label 2 with [+2, 1], label 7 with [+2, 2], label 10 with [+2, 7], and change flows to $f_{12} = 4$, $f_{27} = 2$, and $f_{7,10} = 2$.

The results summarized in Figure 2.53(d) are obtained after the following sequence of operations: label 1 with [∞, —], label 4 with [+3, 1], label 7 with [+3, 4], label 10 with [+2, 7], and change flows to $f_{14} = 2$, $f_{47} = 2$, and $f_{7,10} = 4$.

The results summarized in Figure 2.53(e) are obtained after the following sequence of operations: label 1 with [∞, —], label 4 with [+1, 1], label 6 with [+1, 4], label 9 with [+1, 6], label 10 with [+1, 9], and change flows to $f_{14} = 3$, $f_{46} = 1$, $f_{69} = 2$, and $f_{9,10} = 2$.

In Figure 2.53(f), no other nodes can be labeled. Therefore, the current solution is optimal with maximum flow value equal to 8 units.

2.14.5 Computer Code for the Maximum-Flow Algorithm

```
C***********************************************************************
C*                                                                     *
C*       PURPOSE:                                                      *
C*                                                                     *
C*         TO OBTAIN THE MAXIMUM FLOW THROUGH A CAPACITATED NETWORK.   *
C*                                                                     *
C*       LOCATION: SUBROUTINE MAXFLOW IN NETWORK OPTIMIZATION          *
C*                                                                     *
C*       USAGE:                                                        *
C*                                                                     *
C*         AT PRESENT THIS PROGRAM CAN HANDLE PROBLEMS WITH UP TO 50   *
C*         NODES AND 50 ARCS. THIS NUMBER MAY BE INCREASED BY          *
C*         INCRESING DIMENSION STATEMENTS IN SUBROUTINE MAXFLOW AND    *
C*         THE MAIN PROGRAM.                                           *
C*                                                                     *
C*       INPUT DATA:                                                   *
C*                                                                     *
C*         SET 1:  SINGLE CARD WITH THE CALL NAME OF THE ALGORITHM:MAXF*
C*                  -FORMAT(A4)                                        *
C*                                                                     *
C*         SET 2:  SINGLE CARD WITH THE NUMBER OF NODES AND ARCS IN    *
C*                 THE PROBLEM.                                        *
C*                  -FORMAT(2I10)                                      *
C*                                                                     *
C*         SET 3:  THE TOTAL NUMBER OF CARDS IN  THIS SET IS  EQUAL    *
C*                 TO THE NUMBER OF DIRECTED ARCS IN THE PROBLEM. THE  *
C*                 FOLLOWING QUANTITIES ARE READ FROM EACH CARD:       *
C*                 1) START NODE                                       *
C*                 2) END NODE                                         *
C*                 3) CAPACITY OF THE ARC                              *
C*                 4) INITIAL  FLOW IN THE ARC (CONSERVATION OF FLOW   *
C*                    SHOULD BE PRESERVED).IF LEFT BLANK, ALL INITIAL  *
C*                    FLOWS WILL BE ZERO.                              *
C*                    -FORMAT(4X,I6,I10,2F10.2)                        *
C*                                                                     *
C*         SET 4: THIS SET CONSISTS OF ONLY ONE CARD WHICH INDICATES   *
C*                THE END OF THE PROBLEM. THE USER SHOULD PUNCH THE    *
C*                WORD 'PEND'  -FORMAT(A4)                             *
C*         SET 5: THIS SET CONSISTS OF ONLY ONE CARD WHICH INDICATES   *
C*                THE END OF THE INPUT DECK. THE USER SHOULD PUNCH THE *
C*                WORD 'EXIT'  -FORMAT(A4)                             *
C*                                                                     *
C*       DEFINITION OF THE TERMS USED:                                 *
C*                                                                     *
C*         I-START NODE                                                *
C*         J-END NODE                                                  *
C*         A1-CAPACITY OF THE ARC                                      *
C*         A2-EXISTING FLOW IN THE ARC                                 *
C*                                                                     *
C*       TECHNIQUE USED:                                               *
C*                                                                     *
C*         STARTING WITH AN ARBITRARY FLOW,WHICH PRESERVES CONSERVATION*
C*         OF FLOW,ALL POSSIBLE FLOW AUGMENTING PATHS FROM THE ORIGIN  *
C*         TO THE DESTINATION ARE SYSTEMATICALLY SEARCHED BY USING A   *
C*         LABELING  PROCEDURE.AS SOON AS A FLOW AUGMENTING PATH IS    *
C*         FOUND,THE FLOW IS INCREASED TO ITS MAXIMUM CAPACITY ALONG   *
C*         THE PATH AND ALL THE LABELS ARE ERASED.THE ALGORITHM TER-   *
C*         MINATES IF FLOW CAN NOT BE INCREASED.                       *
C*                                                                     *
C*       REFERENCE:                                                    *
C*         HU,T.C., INTEGER PROGRAMMING AND NETWORK FLOWS,111-120,     *
C*         ADDISON-WESLEY PUBLISHING COMPANY,INC., MASSACHUSETTES,1969.*
C*                                                                     *
C***********************************************************************
```

2.14.6 A Problem in Grain Shipment and Storage

Many developing countries are investing large sums of money to eliminate bottlenecks and to reduce freight and spoilage costs in agricultural transportation and storage systems. The network approach is an appropriate tool for the study of rapidly developing regions. It incorporates capacity limitations on linkages and can measure the quantitative effects of specific improvements in transportation networks.

The network given in Figure 2.54 is a simplified representation of the transportation system for a coastal region of a developing area. Grains produced at node 1 must meet export demand at port facilities located at node 7. Nodes 2, 3, 4, 5, and 6 are storage and/or transfer locations. The highway system has modern main arteries but is presently incomplete. Additionally, the rail system is antiquated and fragmented. The arc labels of the network of Figure 2.54 indicate a zero-flow starting solution, and the maximum flows allowed on the transportation links of the system.

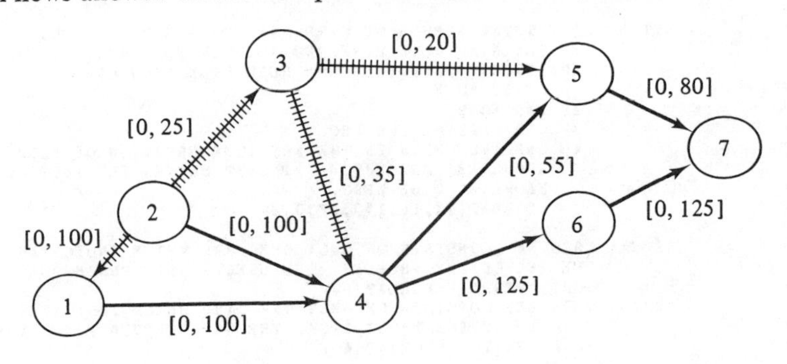

Figure 2.54. Simplified representation of an agricultural transportation network.

This network model can easily be modified to assess the impact of changing capacities of certain arcs or links in the transportation system. Such changes are represented simply by changing the corresponding arc parameters. New facilities are represented by additional arcs and nodes. Likewise, the disappearance of facilities, such as rail abandonment, is represented by the deletion of affected arcs.

The maximum-flow-network approach offers the researcher an effective tool for studying alternative investments in transportation and storage systems, especially when the improvements may significantly affect the entire system. An optimal (maximum-flow) solution is given in Figure 2.55, and the computer solution illustrates the steps to optimality.

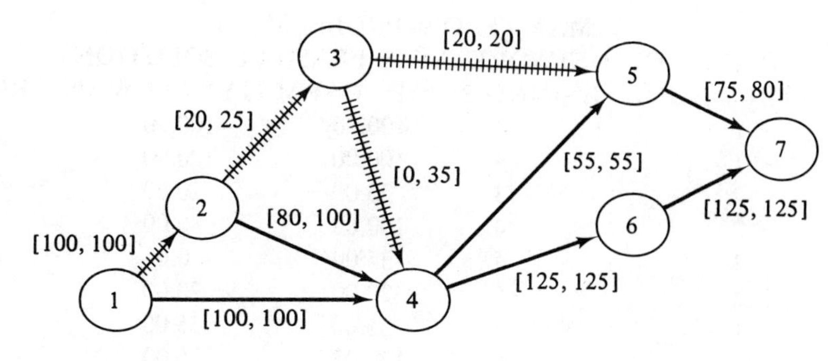

Figure 2.55. Optimal solution for the agricultural transportation network.

MAX/FLOW PROBLEM
STATUS OF NETWORK AT START

STARTING NODE	ENDING NODE	CAPACITY	FLOW IN ARC
1	2	100.00	0.0
1	4	100.00	0.0
2	3	25.00	0.0
2	4	100.00	0.0
3	4	35.00	0.0
3	5	20.00	0.0
4	5	55.00	0.0
4	6	125.00	0.0
5	7	80.00	0.0
6	7	125.00	0.0

MAXIMUM FLOW THROUGH THE ABOVE ARCS IS = 0.0

MAX/FLOW PROBLEM
STATUS OF NETWORK AT A FEASIBLE SOLUTION

STARTING NODE	ENDING NODE	CAPACITY	FLOW IN ARC
1	2	100.00	20.00
1	4	100.00	0.0
2	3	25.00	20.00
2	4	100.00	0.0
3	4	35.00	0.0
3	5	20.00	20.00
4	5	55.00	0.0
4	6	125.00	0.0
5	7	80.00	20.00
6	7	125.00	0.0

MAXIMUM FLOW THROUGH THE ABOVE ARCS IS = 20.00

MAX/FLOW PROBLEM
STATUS OF NETWORK AT A FEASIBLE SOLUTION

STARTING NODE	ENDING NODE	CAPACITY	FLOW IN ARC
1	2	100.00	100.00
1	4	100.00	100.00
2	3	25.00	20.00
2	4	100.00	80.00
3	4	35.00	0.0
3	5	20.00	20.00
4	5	55.00	55.00
4	6	125.00	125.00
5	7	80.00	75.00
6	7	125.00	125.00

MAXIMUM FLOW THROUGH THE ABOVE ARCS IS = 200.00

MAX/FLOW PROBLEM
STATUS OF NETWORK AT OPTIMAL SOLUTION

STARTING NODE	ENDING NODE	CAPACITY	FLOW IN ARC
1	2	100.00	100.00
1	4	100.00	100.00
2	3	25.00	20.00
2	4	100.00	80.00
3	4	35.00	0.0
3	5	20.00	20.00
4	5	55.00	55.00
4	6	125.00	125.00
5	7	80.00	75.00
6	7	125.00	125.00

MAXIMUM FLOW THROUGH THE ABOVE ARCS IS = 200.00

2.15 The Multiterminal Maximal-Flow Problem

There are numerous problems in engineering and business where the systems under consideration can be approximated by deterministic multiterminal flow models. A few examples are (1) transportation systems, where the highways are represented by the arcs of the network model and the capacities represent upper bounds on vehicular traffic intensities; (2) telephone systems, where the lines are represented by the arcs of the network and the capacities indicate the maximal number of calls allowed at any time; and (3) electric power distribution systems, where the transmission elements are represented by arcs and the capacities specify the maximal amounts of energy that can be absorbed by those elements of the system. In all these systems it is assumed that there are several sources at which a given commodity is available, and that the amount that can be transported to several terminals is limited only by the capacities of the distribution links.

Let us consider an undirected capacitated network, that is, one in which the total flow on a given arc must be less than or equal to its capacity. In a previous section, we discussed a labeling procedure to solve the one source/one terminal case, where it is assumed that there is an infinite availability of the commodity at the source. The objective of the problem was to find the maximal amount of commodity that could be shipped from the source to the terminal through the arcs of the network, without violating the capacity conditions.

Several analysts have considered the problem of finding the maximal flow for all pairs of nodes in a given undirected capacitated network [7, 15, 25]. This problem can be viewed as a generalization of the one source/one terminal problem and, accordingly, can be solved by applying the procedure of Section 2.14.1 to each pair of nodes. A more elegant and efficient approach is due to Gomory and Hu [25]. Our presentation uses the fundamental results developed in their paper and will include a justification of the algorithm.

Assuming that the capacity of any arc is the same regardless of the direction in which the arc is traversed, and that each pair of nodes can be defined as a source–destination pair, the total number of maximal-flow problems that would have to be solved is equal to $n(n - 1)/2$, where n is the number of nodes in the network. The Gomory–Hu algorithm solves the same problem in only $n - 1$ maximal-flow determinations.

2.15.1 The Gomory–Hu Algorithm

Let $G = (\mathbf{N}, \mathbf{A})$ be an undirected network, where $\mathbf{N}$ represents the set of nodes and $\mathbf{A}$ the set of arcs. Let c_{ij} be the capacity of arc (i, j) in set $\mathbf{A}$, and let $c_{ij} = c_{ji}$. Also, suppose that the set of nodes is defined as $\mathbf{N} = \{1, 2, \ldots, n\}$. The following notation will be used in the discussion of the algorithm:

v_{ij}	maximal flow between nodes i and j
$(\mathbf{X}, \bar{\mathbf{X}})_{ij}$	minimal cut separating i from j, $i \in \mathbf{X}$ and $j \in \bar{\mathbf{X}}$
$C(\mathbf{X}, \bar{\mathbf{X}})_{ij}$	capacity of the minimal cut separating i from j

Considering any node s as the source and another node t as the terminal, the maximal flow/minimal cut theorem states that $v_{st} = C(\mathbf{X}, \bar{\mathbf{X}})_{st}$. If any other pair of nodes i and j is redefined as a source–terminal pair, the Gomory–Hu approach capitalizes on the solution to the previous maximal-flow problem to determine the desired value v_{ij}, if some rather simple conditions are satisfied. Namely, as proved in reference [25], if nodes i and j are chosen in such a way that both are in $\mathbf{X}$ or both are in $\bar{\mathbf{X}}$, then the other set of nodes defining the cut $(\mathbf{X}, \bar{\mathbf{X}})_{st}$ can be condensed into one node and the maximal value of the flow between i and j will be the same in both the original and the condensed network.

Let $\bar{\mathbf{N}}_{ij}$ be the set of nodes after the condensation of all the nodes on the side of the cut which does not contain nodes i and j. Let $\bar{\mathbf{A}}_{ij}$ be the set of arcs joining the nodes in $\bar{\mathbf{N}}_{ij}$. Therefore, the modified network can be represented by $\bar{G}_{ij} = (\bar{\mathbf{N}}_{ij}, \bar{\mathbf{A}}_{ij})$. The labeling procedure can be applied on $\bar{G}_{ij}$ to obtain the maximal value of the flow between i and j if the capacities of the arcs in $\bar{\mathbf{A}}_{ij}$ are known; we can obtain these capacities following a simple procedure. Let $j_1, j_2, \ldots, j_r$ be the nodes in $\bar{\mathbf{X}}$ directly connected to the same node i in $\mathbf{X}$. If $\bar{\mathbf{X}}$ is to be condensed, all arcs $(i, j_1), (i, j_2), \ldots, (i, j_r)$ are replaced by a single arc-connecting node i with the condensed node $\bar{\mathbf{X}}$. The capacity of this arc is given by

$$c_{i\bar{x}} = \sum_{m=1}^{r} c_{ij_m}$$

As already mentioned, the maximal value of the flow between i and j can be determined by applying the labeling procedure to $\bar{G}_{ij}$. The determination of v_{ij} implies, again, the identification of a minimal cut separating i from j. Let $(X, \bar{X})_{ij}$ be the corresponding cut with minimal capacity. Now another pair of nodes, both in X or both in $\bar{X}$, can be selected and another condensed network can be obtained. The application of the labeling procedure will allow the identification of another cut and another condensed network will be possible. It can be shown that after $n - 1$ pairs are selected, all $n(n - 1)/2$ maximal-flow values of the original network G are available.

The flowchart in Figure 2.56 describes the steps of the Gomory–Hu algorithm. The fundamental idea is to iteratively construct a maximal spanning tree whose branches represent cuts and whose branch parameters represent cut values. A justification of the algorithm and an illustrative example are given afterward.

2.15.2 Justification of the Algorithm

Let $G = (\mathbf{N}, \mathbf{A})$ be an undirected network with arc capacities such that c_{ij} is equal to c_{ji} for all the arcs in $\mathbf{A}$. For i, j, k in $\mathbf{N}$, the maximal flow/minimal cut theorem establishes that $v_{ij} = C(\mathbf{X}, \bar{\mathbf{X}})_{ij}$. If k is in $\bar{\mathbf{X}}$, then $v_{ik} \leqslant C(\mathbf{X}, \bar{\mathbf{X}})_{ij}$, and if k is in $\mathbf{X}$, then $v_{kj} \leqslant C(\mathbf{X}, \bar{\mathbf{X}})_{ij}$. Therefore, $v_{ij} \geqslant v_{ik}$ and $v_{ij} \geqslant v_{kj}$, which jointly imply that

$$v_{ij} \geqslant \min\,[v_{ik}, v_{kj}]$$

If a similar argument is repeated for v_{ik} and v_{kj}, the following results would be obtained: $v_{ik} \geqslant \min\,[v_{ip}, v_{pk}]$ and $v_{kj} \geqslant \min\,[v_{kq}, v_{qj}]$, where $\{i, p, k, q, j\}$ is a connected sequence of nodes in $\mathbf{N}$. Therefore,

$$v_{ij} \geqslant \min\,[v_{ip}, v_{pk}, v_{kq}, v_{qj}]$$

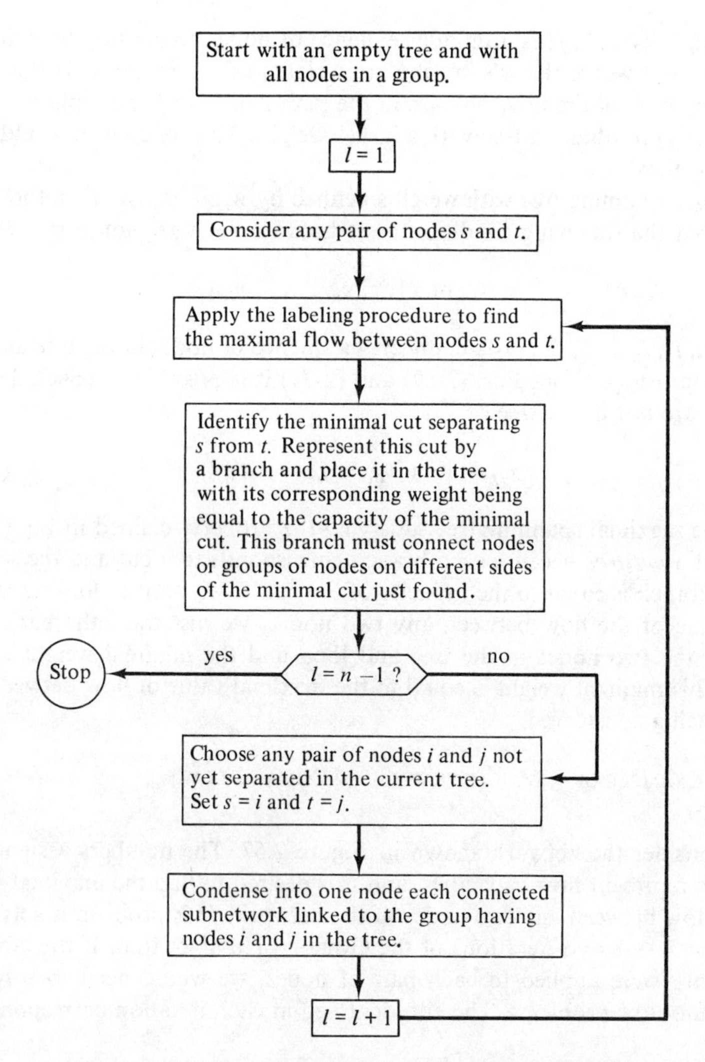

Figure 2.56. Flowchart for the Gomory–Hu algorithm.

In general,

$$v_{ij} \geqslant \min\,[v_{ii_1}, v_{i_1i_2}, v_{i_2i_3}, \ldots, v_{i_rj}] \tag{2-69}$$

where $\{i, i_1, i_2, \ldots, j\}$ is a connected sequence of nodes in **N**.

Before proceeding with our discussion, the following property of a maximal spanning tree will be established for any arc (i, j) *not* in the tree:

$$w_{ij} \leqslant \min\,[w_{ii_1}, w_{i_1i_2}, w_{i_2i_3}, \ldots, w_{i_rj}] \tag{2-70}$$

where $\{i, i_1, i_2, \ldots, j\}$ is a unique sequence of nodes connecting branches of the tree, and where the w's are the arc weights of the network. If this were not true, we could remove any arc in the path from i to j and replace it with the arc (i, j) to obtain a tree with a larger weight. This, of course, would be a contradiction.

For a spanning tree with weights defined by $w_{ij} = v_{ij}$, we therefore conclude that the following condition must hold for each arc not in the tree:

$$v_{ij} \leqslant \min [v_{ii_1}, v_{i_1i_2}, \ldots, v_{i_rj}] \tag{2-71}$$

where $\{i, i_1, i_2, \ldots, i_r, j\}$ is a connected sequence of nodes in the tree along a path from i to j. From Eqs. (2-69) and (2-71) it is possible to conclude that for any arc not in the tree

$$v_{ij} = \min [v_{ii_1}, v_{i_1i_2}, \ldots, v_{i_rj}] \tag{2-72}$$

The maximal spanning tree satisfying the property defined in Eq. (2-72) is called a *cut-tree* because each branch corresponds to a cut and the weight of the branch is equal to the capacity of the cut. If we want to find the maximal value of the flow between any two nodes, we just trace the path connecting the two nodes in the tree and then find the minimal weight in the path. This minimal weight is equal to the maximal value of flow between the nodes being considered.

2.15.3 Example of a Multiterminal Maximal-Flow Problem

Consider the network shown in Figure 2.57. The numbers assigned to the arcs represent flow capacities, and it is desired to find the maximal value of the flow between any two nodes of the network. This problem is solved in $n - 1 = 7 - 1 = 6$ iterations of the Gomory–Hu algorithm. If the labeling procedure were applied to each pair of nodes, we would need to solve 21 maximum-flow problems. The cut identified at each iteration corresponds to

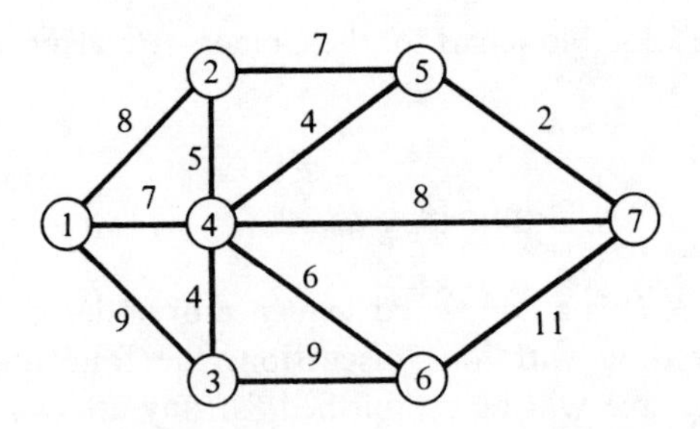

Figure 2.57. Multiterminal maximal-flow sample network.

arcs with residual capacities equal to zero. For presentation purposes, the results from the individual maximum-flow labeling procedures will be omitted.

***Iteration 1*:** Consider $s = 2$ and $t = 5$. The maximal-flow value is equal to 13. Therefore, $v_{25} = v_{52} = 13$. The cut with minimal capacity indicates that we can start building the cut tree with a branch joining node 5 with a condensed node consisting of nodes 1, 2, 3, 4, 6, 7 [Figure 2.58(a)]. The weight of this branch is equal to 13.

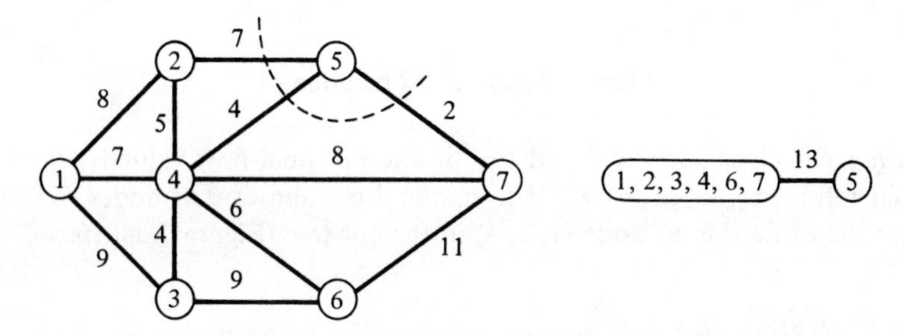

Figure 2.58(a). Maximal-flow problem: first iteration.

***Iteration 2*:** Consider $s = 1$ and $t = 2$. The maximal-flow value is equal to 19. Therefore, $v_{12} = v_{21} = 19$. The minimal cut indicates that nodes 2 and 5 are on one side and the other nodes on the other side [Figure 2.58(b)]. The weight of the branch joining 2 with the condensed node consisting of 1, 3, 4, 6, 7 is equal to 19.

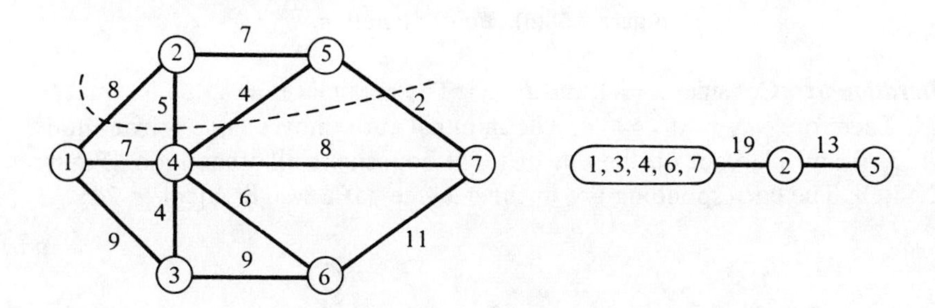

Figure 2.58(b). Second iteration.

***Iteration 3*:** Consider $s = 6$ and $t = 7$. The value of the maximal flow is equal to 21. Therefore, $v_{67} = v_{76} = 21$. The minimal cut indicates that in the cut tree node 7 is connected to the condensed node consisting of 1, 3, 4, 6 by an arc of weight equal to 21. The minimal cut also indicates that nodes

2 and 5 are on one side of the condensed node (1, 3, 4, 6) and node 7 is on the other side [Figure 2.58(c)].

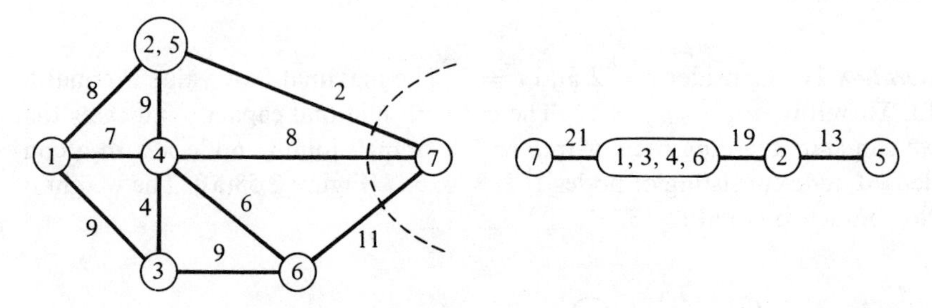

Figure 2.58(c). Third iteration.

Iteration 4: Consider $s = 4$ and $t = 6$. The maximal-flow value is equal to 25. Therefore, $v_{46} = v_{64} = 25$. The minimal cut shows that nodes 6 and 7 are on the same side of node (1, 3, 4) in the cut tree [Figure 2.58(d)].

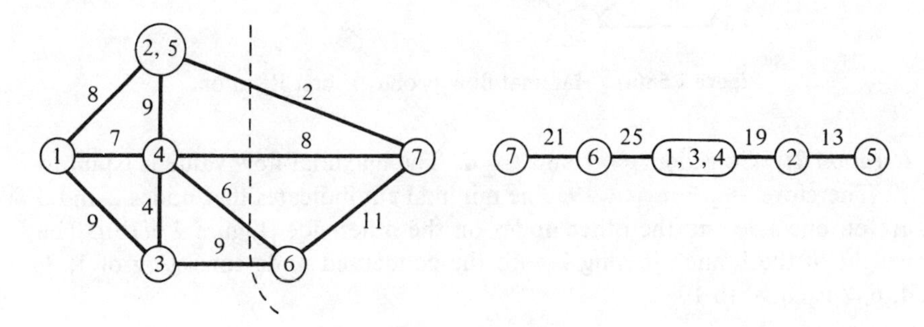

Figure 2.58(d). Fourth iteration.

Iteration 5: Consider $s = 1$ and $t = 4$. The maximal-flow value is equal to 24. Therefore, $v_{14} = v_{41} = 24$. The minimal cut removes node 1 from node (1, 3, 4) and places it on the side of (3, 4) opposite to all other nodes [Figure 2.58(e)]. The corresponding arc in the cut tree has a weight equal to 24:

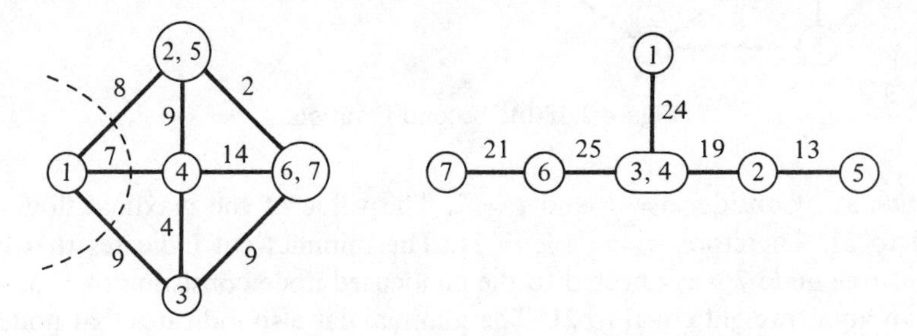

Figure 2.58(e). Fifth iteration.

***Iteration 6*:** Consider $s = 3$ and $t = 4$. The maximal flow value is 22. Therefore, $v_{34} = v_{43} = 22$. The minimal cut removes node 3 and links it to 4 in the cut tree with an arc with weight equal to 22. Now the cut tree is complete, that is, consists of six arcs, and then the procedure is terminated [Figure 2.58(f)].

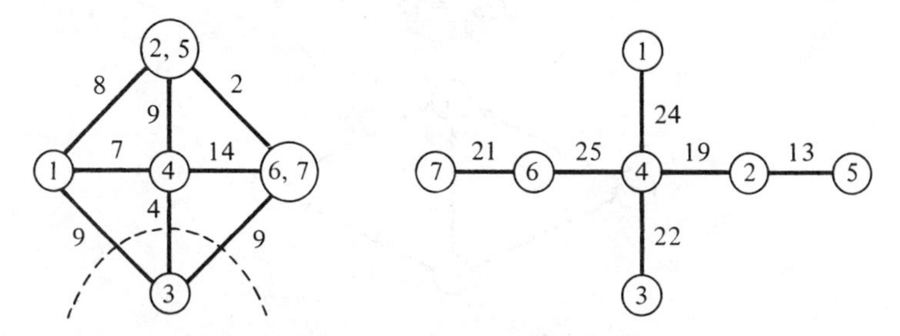

Figure 2.58(f). Sixth iteration.

The maximal-flow values can be arranged in the matrix $\mathbf{V} = [v_{ij}]$ as follows:

$$\mathbf{V} = \begin{bmatrix} — & 19 & 22 & 24 & 13 & 24 & 21 \\ 19 & — & 19 & 19 & 13 & 19 & 19 \\ 22 & 19 & — & 22 & 13 & 22 & 21 \\ 24 & 19 & 22 & — & 13 & 25 & 21 \\ 13 & 13 & 13 & 13 & — & 13 & 13 \\ 24 & 19 & 22 & 25 & 13 & — & 21 \\ 21 & 19 & 21 & 21 & 13 & 21 & — \end{bmatrix}$$

2.16 The Multiterminal Maximum-Capacity Chain Problem

A problem closely related to the multiterminal maximum-flow problem of Section 2.15 is the multiterminal maximum-capacity chain problem. The algorithm of Section 2.15 determines the maximum possible flow between all pairs of nodes. It is clear that the maximum flow between any two nodes might involve flow which forms multiple paths or chains from source to sink. In fact, multiple paths or chains are of no importance to the maximum-flow problem except when they increase flow from one point to another. Consider the simplified network shown in Figure 2.59. (The number on each arc represents the upper bound on the arc flow.) The maximum possible flow between nodes A and D is equal to 40, and corresponds to the following pattern of flows:

$$f_{AB} = 20$$
$$f_{AC} = 20$$
$$f_{CB} = 5$$
$$f_{BD} = 25$$
$$f_{CD} = 15$$

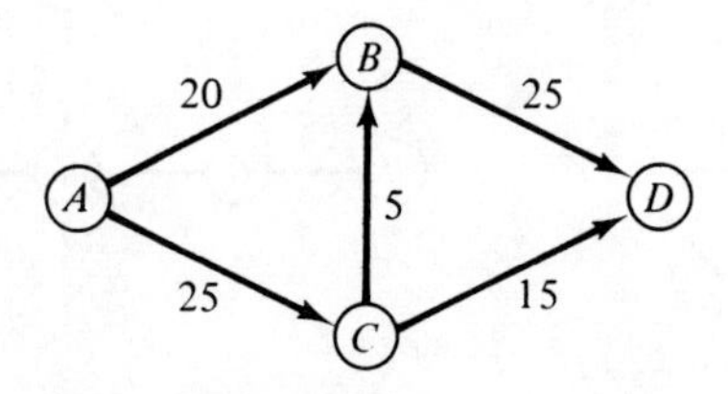

Figure 2.59. Maximum capacity network.

Three chains are formed from A to D:

Ⓐ ⟶ Ⓑ ⟶ Ⓓ

Ⓐ ⟶ Ⓒ ⟶ Ⓑ ⟶ Ⓓ

Ⓐ ⟶ Ⓒ ⟶ Ⓓ

Consider the following problem relevant to the foregoing network. What is the maximum-capacity chain from node A to node D? Clearly, the maximum-capacity chain is the sequence Ⓐ ⟶ Ⓑ ⟶ Ⓓ with a flow value of $F_{\max} = 20$. The problem that we wish to address in this section is the multiterminal maximum-capacity chain problem, that is, the maximum-flow chain between all pairs of nodes.

Hu [31] has developed an efficient computational procedure which is a modification of the triple operation used in the multiterminal shortest-chain (path) problem. The algorithm proceeds as follows.

***Step 1*:** Construct an $(n \times n)$ capacity matrix which represents the capacity between nodes i and j; $i, j = 1, 2, \ldots, n$.

***Step 2*:** a. For $j = 1, 2, \ldots, n$, successively delete the jth row and jth column simultaneously, and perform the following triple operation on each remaining element d_{ik} in the matrix (diagonal elements are also excluded):

$$d_{ik} = \max\{d_{ik}; \min[d_{ij}, d_{jk}]\}, \qquad \text{for all } i, k \neq j, j = 1, 2, \ldots, n$$

A second matrix, referred to as the *route matrix*, must be maintained to keep track of the intermediate nodes of each chain. The route matrix is also an $(n \times n)$ matrix, with the kth entry of the ith row initially equal to k.

b. Simultaneous changes in the route matrix will occur along with those in the capacity matrix using the following rules:

$$r_{ik} = \begin{cases} j; & \text{if } d_{ik} < \min\,[d_{ij}, d_{jk}] \\ \text{unchanged}; & \text{if } d_{ik} \geq \min\,[d_{ij}, d_{jk}] \end{cases}$$

Finally, note that if two nodes are not connected (or a connection is deemed inadmissible), a "$-\infty$" entry for the corresponding value of d_{ij} should be placed in the capacity matrix.

To illustrate the computational procedure, consider the following problem.

2.16.1 Optimal Movement of Bulk Freight

East Texas Freight Company (ETFC) operates a trucking firm that specializes in moving large, bulky equipment. ETFC is bidding a special contract to move bulk freight between seven NASA installations in the Houston, Texas, area. This particular contract calls for the movement of several classes of NASA equipment resulting in extremely high loads when placed on a truck. ETFC must move this equipment, which is highly classified, over only seven approved routes. A major problem is the maximum allowable bridge clearance along each route. ETFC has measured all bridge clearances, and has listed the following maximum clearances between all shipping/delivery points.

In the matrix shown in Table 2.37, each entry is vertical clearance in

Table 2.37. MOVEMENT OF BULK FREIGHT

		To						
		1	2	3	4	5	6	7
	1	0	11	30	–	–	–	–
	2	11	0	–	12	2	–	–
	3	30	–	0	19	–	4	–
From	4	–	12	19	0	11	9	–
	5	–	2	–	11	0	–	–
	6	–	–	4	9	–	0	–
	7	–	–	–	20	1	1	0

excess of 30 feet. ETFC wishes to know the highest load that it can transport between any two shipping/receiving points.

The solution can be generated by finding the multiterminal maximum-capacity route between all pairs of nodes. The results are as follows.

***Iteration 0*:** Initial matrices.

Capacity matrix

	1	2	3	4	5	6	7
1	0	11	30	$-\infty$	$-\infty$	$-\infty$	$-\infty$
2	11	0	$-\infty$	12	2	$-\infty$	$-\infty$
3	30	$-\infty$	0	19	$-\infty$	4	$-\infty$
4	$-\infty$	12	19	0	11	9	$-\infty$
5	$-\infty$	2	0	11	0	$-\infty$	$-\infty$
6	$-\infty$	$-\infty$	4	9	$-\infty$	0	$-\infty$
7	$-\infty$	$-\infty$	$-\infty$	20	1	1	0

Route matrix

	1	2	3	4	5	6	7
1	1	2	3	4	5	6	7
2	1	2	3	4	5	6	7
3	1	2	3	4	5	6	7
4	1	2	3	4	5	6	7
5	1	2	3	4	5	6	7
6	1	2	3	4	5	6	7
7	1	2	3	4	5	6	7

Iteration 1

Capacity matrix

	1	2	3	4	5	6	7
1	0	11	30	$-\infty$	$-\infty$	$-\infty$	$-\infty$
2	11	0	11	12	2	$-\infty$	$-\infty$
3	30	11	0	19	$-\infty$	4	$-\infty$
4	$-\infty$	12	19	0	11	9	$-\infty$
5	$-\infty$	2	$-\infty$	11	0	$-\infty$	$-\infty$
6	$-\infty$	$-\infty$	4	9	$-\infty$	0	$-\infty$
7	$-\infty$	$-\infty$	$-\infty$	20	1	1	0

Route matrix

	1	2	3	4	5	6	7
1	1	2	3	4	5	6	7
2	1	2	1	4	5	6	7
3	1	1	3	4	5	6	7
4	1	2	3	4	5	6	7
5	1	2	3	4	5	6	7
6	1	2	3	4	5	6	7
7	1	2	3	4	5	6	7

Iteration 2

Capacity matrix

	1	2	3	4	5	6	7
1	0	11	30	11	2	$-\infty$	$-\infty$
2	11	0	11	12	2	$-\infty$	$-\infty$
3	30	11	0	19	2	4	$-\infty$
4	11	12	19	0	11	9	$-\infty$
5	2	2	2	11	0	$-\infty$	$-\infty$
6	$-\infty$	$-\infty$	4	9	$-\infty$	0	$-\infty$
7	$-\infty$	$-\infty$	$-\infty$	20	1	1	0

Route matrix

	1	2	3	4	5	6	7
1	1	2	3	2	2	6	7
2	1	2	1	4	5	6	7
3	1	1	3	4	2	6	7
4	2	2	3	4	5	6	7
5	2	2	2	4	5	6	7
6	1	2	3	4	5	6	7
7	1	2	3	4	5	6	7

Iteration 3

Capacity matrix

	1	2	3	4	5	6	7
1	0	11	30	19	2	4	$-\infty$
2	11	0	11	12	2	4	$-\infty$
3	30	11	0	19	2	4	$-\infty$
4	19	12	19	0	11	9	$-\infty$
5	2	2	2	11	0	2	$-\infty$
6	4	4	4	9	2	0	$-\infty$
7	$-\infty$	$-\infty$	$-\infty$	20	1	1	0

Route matrix

	1	2	3	4	5	6	7
1	1	2	3	3	2	3	7
2	1	2	1	4	5	3	7
3	1	1	3	4	2	6	7
4	3	2	3	4	5	6	7
5	2	2	2	4	5	3	7
6	3	3	3	4	3	6	7
7	1	2	3	3	5	6	7

Iteration 4

Capacity matrix

	1	2	3	4	5	6	7
1	0	12	30	19	11	9	$-\infty$
2	12	0	12	12	11	9	$-\infty$
3	30	12	0	19	11	9	$-\infty$
4	19	12	19	0	11	9	$-\infty$
5	11	11	11	11	0	9	$-\infty$
6	9	9	9	9	9	0	$-\infty$
7	19	12	19	20	11	9	0

Route matrix

	1	2	3	4	5	6	7
1	1	4	3	3	4	4	7
2	4	2	4	4	4	4	7
3	1	4	3	4	4	4	7
4	3	2	3	4	5	6	7
5	4	4	4	4	5	4	7
6	4	4	4	4	4	6	7
7	4	4	4	4	4	4	7

Iteration 5

Capacity matrix

	1	2	3	4	5	6	7
1	0	12	30	19	11	9	$-\infty$
2	12	0	12	12	11	9	$-\infty$
3	30	12	0	19	11	9	$-\infty$
4	19	12	19	0	11	9	$-\infty$
5	11	11	11	11	0	9	$-\infty$
6	9	9	9	9	9	0	$-\infty$
7	19	12	19	20	11	9	0

Route matrix

	1	2	3	4	5	6	7
1	1	4	3	3	4	4	7
2	4	2	4	4	4	4	7
3	1	4	3	4	4	4	7
4	3	2	3	4	5	6	7
5	4	4	4	4	5	4	7
6	4	4	4	4	4	6	7
7	4	4	4	4	4	4	7

Iteration 6

Capacity matrix

	1	2	3	4	5	6	7
1	0	12	30	19	11	9	$-\infty$
2	12	0	12	12	11	9	$-\infty$
3	30	12	0	19	11	9	$-\infty$
4	19	12	19	0	11	9	$-\infty$
5	11	11	11	11	0	9	$-\infty$
6	9	9	9	9	9	0	$-\infty$
7	19	12	19	20	11	9	0

Route matrix

	1	2	3	4	5	6	7
1	1	4	3	3	4	4	7
2	4	2	4	4	4	4	7
3	1	4	3	4	4	4	7
4	3	2	3	4	5	6	7
5	4	4	4	4	5	4	7
6	4	4	4	4	4	6	7
7	4	4	4	4	4	4	7

Iteration 7

Capacity matrix

	1	2	3	4	5	6	7
1	0	12	30	19	11	9	$-\infty$
2	12	0	12	12	11	9	$-\infty$
3	30	12	0	19	11	9	$-\infty$
4	19	12	19	0	11	9	$-\infty$
5	11	11	11	11	0	9	$-\infty$
6	9	9	9	9	9	0	$-\infty$
7	19	12	19	20	11	9	0

Route matrix

	1	2	3	4	5	6	7
1	1	4	3	3	4	4	7
2	4	2	4	4	4	4	7
3	1	4	3	4	4	4	7
4	3	2	3	4	5	6	7
5	4	4	4	4	5	4	7
6	4	4	4	4	4	6	7
7	4	4	4	4	4	4	7

The optimal (maximum) route value can be read *directly* from the final capacity matrix for each pair of nodes. The optimal route can be recovered by tracing through the final route matrix. For example, the maximum capacity chain from point 1 to point 4 is given by $d^*_{14} = 19$ (clearance of 49 feet). The route is traced as follows:

$r^*_{14} = 3$; move from node 1 to node 4 through node 3
$r^*_{34} = 4$; move from node 3 to node 4 directly
$r^*_{13} = 3$; move from node 1 to node 3 directly

$$① \xrightarrow{30} ③ \xrightarrow{19} ④ \quad d^*_{14} = \min\,[30, 19] = 19$$

Note that a *direct* link between points 1 and 4 does not exist.

2.16.2 Computer Code for the Multiterminal Maximum-Capacity Route Algorithm

```
C***********************************************************************
C                                                                      *
C        PURPOSE:                                                      *
C           TO OBTAIN THE ROUTE WITH MAXIMUM FLOW CAPACITY FOR EACH   *
C           SOURCE-TERMINAL PAIR OF A DIRECTED NETWORK.                *
C                                                                      *
C        LOCATION: SUBROUTINE MAXCAP IN NETWORK OPTIMIZATION           *
C                                                                      *
C        USAGE:                                                        *
C                                                                      *
C           AT PRESENT THIS PROGRAM HANDLES UP TO 50 NODES AND 50 ARCS *
C           THIS NUMBER MAY BE INCREASED BY INCREASING THE DIMENSION   *
C           STATEMENTS IN SUBROUTINE MAXCAP AND THE MAIN PROGRAM.      *
C                                                                      *
C        INPUT DATA:                                                   *
C                                                                      *
C           SET 1:   SINGLE CARD WITH THE CALL NAME OF THE ALGORITHM:MCRP *
C                    -FORMAT(A4)                                       *
C                                                                      *
C           SET 2:   SINGLE CARD WHICH SPECIFIES THE NUMBER OF NODES AND *
C                    ARCS IN THE PROBLEM.                              *
C                    -FORMAT(2I10)                                     *
C                                                                      *
C           SET 3:   THE TOTAL NUMBER OF CARDS FOR THIS SET IS EQUAL TO *
C                    THE NUMBER OF ARCS FOR THE PROBLEM                *
C                    THE FOLLOWING QUANTITIES ARE READ FROM EACH CARD  *
C                    1) START NODE                                     *
C                    2) END NODE                                       *
C                    3) CAPACITY OF THE ARC                            *
C                    -FORMAT(4X,I6,I10,F10.2)                          *
C                                                                      *
C           SET 4: THIS SET CONSISTS OF ONLY ONE CARD WHICH INDICATES  *
C                  THE END OF A PROBLEM. THE USER SHOULD PUNCH THE WORD *
C                  'PEND'  -FORMAT(A4)                                 *
C                                                                      *
C           SET 5: THIS SET CONSISTS OF ONLY ONE CARD WHICH INDICATES  *
C                  THE END OF THE INPUT DECK. THE USER SHOULD PUNCH THE *
C                  WORD 'EXIT' -FORMAT(A4)                             *
C                                                                      *
```

```
C       PROGRAM CONTENTS:                                               *
C          THE SUBROUTINES WHICH COMPRISE THIS PROGRAM ARE:              *
C              MAXCAP: INPUT DATA                                        *
C              PRNTRX: PRINTS ORIGINAL MATRIX                            *
C              PRNTRY: PRINTS OPTIMAL SOLUTION                           *
C                                                                        *
C       DEFINITION OF TERMS USED:                                        *
C          I - STARTING NODE                                             *
C          J - ENDING NODE                                               *
C          K - INTERMEDIATE NODE                                         *
C          A - ARC |CAPACITY|                                            *
C                                                                        *
C       TECHNIQUE:                                                       *
C          THE TECHNIQUE USED IN THIS ALGORITHM CONSIDERS THE CAPACITY   *
C          OF THE DIRECT ARC BETWEEN NODE I TO NODE K AS THE INITIAL     *
C          ESTIMATE OF THE MAXIMUM CAPACITY FOR FLOW FROM I TO K.        *
C          AT THE J-TH ITERATION, THE METHOD COMPARES THE CAPACITY       *
C          OF A PATH WITH INTERMEDIATE NODES FROM THE SET  1,2,..,J      *
C          VERSUS THE CAPACITY OF A PATH THROUGH NODE J WITH             *
C          INTERMEDIATE NODES FROM THE SET [1,2,...,J-1]. IF THE USE OF  *
C          NODE J RESULTS IN A PATH WITH LARGER CAPACITY FROM NODE I TO  *
C          NODE K, THE CAPACITY OF THIS PATH BECOMES THE NEW ESTIMATE OF*
C          THE MAXIMUM CAPACITY.                                         *
C                                                                        *
C       REFERENCE                                                        *
C          T.C. HU, INTEGER PROGRAMMING AND NETWORK FLOWS,               *
C          ADDISON-WESLEY PUBLISHING CO., MASS, 1969.                    *
C*************************************************************************
```

2.16.3 Transporting the Space Shuttle

The United States space shuttle was recently constructed in California, and subsequently moved to an appropriate launch site. Prior to construction, NASA engineers were assigned the task of determining where the main shuttle body would be constructed, and how it might be moved from the chosen construction site to a feasible launch site. After studying the problem, NASA engineers identified three possible manufacturing locations and three possible launch sites. Because of transportation logistics problems, the only feasible mode of transportation was determined to be by flatbed truck. NASA engineers were concerned about the maximum weight that could be transported from feasible construction sites to launch sites. It was determined that between all site locations there were 28 bridges with dangerous load limits. After checking with the highway department, the network of Figure 2.60 was constructed, which represents bridge crossings between all pairs of points. Each arc represents a bridge crossing, and the number on each arc corresponds to the load limit in tons for each bridge. The problem is to determine the maximum possible load that can be transported from construction sites *A*, *B*, and *C* to launch sites *D*, *E*, and *F*.

The computer solution to this problem is displayed as follows. The *original matrix* consists of all connecting arcs (bridges) and their load limits. Note that 9999999.0 designates a forbidden link with capacity equal to $-\infty$. The *arc capacity matrix* provides the maximum capacity route between all pairs of nodes. The *matrix of nodes* displaying the maximum capacity path is given by the last matrix.

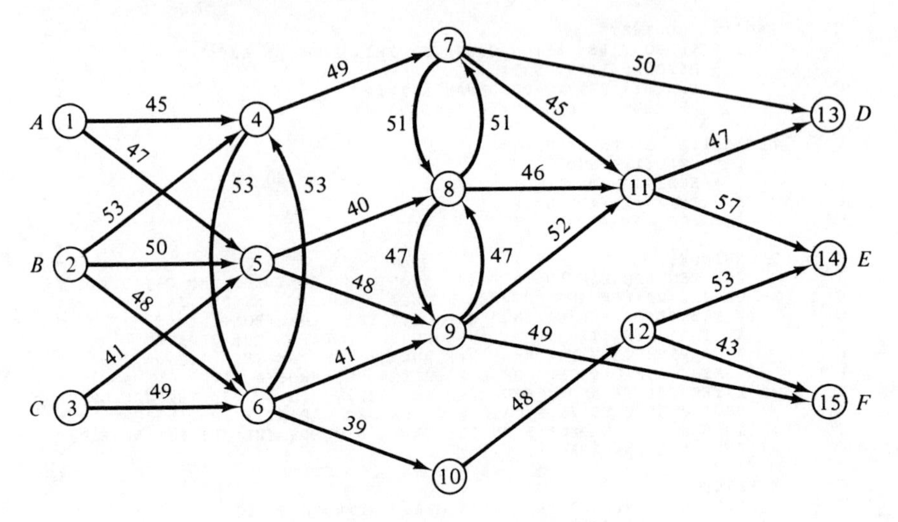

Figure 2.60. Bridge weight restrictions.

ORIGINAL MATRIX

	1	2	3	4	5
1	0.00	9999999.00	9999999.00	45.00	47.00
2	9999999.00	0.00	9999999.00	53.00	50.00
3	9999999.00	9999999.00	0.00	9999999.00	41.00
4	9999999.00	9999999.00	9999999.00	0.00	9999999.00
5	9999999.00	9999999.00	9999999.00	9999999.00	0.00
6	9999999.00	9999999.00	9999999.00	53.00	9999999.00
7	9999999.00	9999999.00	9999999.00	9999999.00	9999999.00
8	9999999.00	9999999.00	9999999.00	9999999.00	9999999.00
9	9999999.00	9999999.00	9999999.00	9999999.00	9999999.00
10	9999999.00	9999999.00	9999999.00	9999999.00	9999999.00
11	9999999.00	9999999.00	9999999.00	9999999.00	9999999.00
12	9999999.00	9999999.00	9999999.00	9999999.00	9999999.00
13	9999999.00	9999999.00	9999999.00	9999999.00	9999999.00
14	9999999.00	9999999.00	9999999.00	9999999.00	9999999.00
15	9999999.00	9999999.00	9999999.00	9999999.00	9999999.00

	6	7	8	9	10
1	9999999.00	9999999.00	9999999.00	9999999.00	9999999.00
2	48.00	9999999.00	9999999.00	9999999.00	9999999.00
3	49.00	9999999.00	9999999.00	9999999.00	9999999.00
4	53.00	49.00	9999999.00	9999999.00	9999999.00
5	9999999.00	9999999.00	40.00	48.00	9999999.00
6	0.00	9999999.00	9999999.00	41.00	39.00
7	9999999.00	0.00	51.00	9999999.00	9999999.00
8	9999999.00	51.00	0.00	47.00	9999999.00
9	9999999.00	9999999.00	47.00	0.00	9999999.00
10	9999999.00	9999999.00	9999999.00	9999999.00	0.00
11	9999999.00	9999999.00	9999999.00	9999999.00	9999999.00
12	9999999.00	9999999.00	9999999.00	9999999.00	9999999.00
13	9999999.00	9999999.00	9999999.00	9999999.00	9999999.00
14	9999999.00	9999999.00	9999999.00	9999999.00	9999999.00
15	9999999.00	9999999.00	9999999.00	9999999.00	9999999.00

	11	12	13	14	15
1	9999999.00	9999999.00	9999999.00	9999999.00	9999999.00
2	9999999.00	9999999.00	9999999.00	9999999.00	9999999.00
3	9999999.00	9999999.00	9999999.00	9999999.00	9999999.00
4	9999999.00	9999999.00	9999999.00	9999999.00	9999999.00
5	9999999.00	9999999.00	9999999.00	9999999.00	9999999.00
6	9999999.00	9999999.00	9999999.00	9999999.00	9999999.00
7	45.00	9999999.00	50.00	9999999.00	9999999.00
8	46.00	9999999.00	9999999.00	9999999.00	9999999.00
9	52.00	9999999.00	9999999.00	9999999.00	49.00
10	9999999.00	48.00	9999999.00	9999999.00	9999999.00
11	0.00	9999999.00	47.00	57.00	9999999.00
12	9999999.00	0.00	9999999.00	53.00	43.00
13	9999999.00	9999999.00	0.00	9999999.00	9999999.00
14	9999999.00	9999999.00	9999999.00	0.00	9999999.00
15	9999999.00	9999999.00	9999999.00	9999999.00	0.00

ARC CAPACITY MATRIX

	1	2	3	4	5
1	0.00	9999999.00	9999999.00	45.00	47.00
2	9999999.00	0.00	9999999.00	53.00	50.00
3	9999999.00	9999999.00	0.00	49.00	41.00
4	9999999.00	9999999.00	9999999.00	0.00	9999999.00
5	9999999.00	9999999.00	9999999.00	9999999.00	0.00
6	9999999.00	9999999.00	9999999.00	53.00	9999999.00
7	9999999.00	9999999.00	9999999.00	9999999.00	9999999.00
8	9999999.00	9999999.00	9999999.00	9999999.00	9999999.00
9	9999999.00	9999999.00	9999999.00	9999999.00	9999999.00
10	9999999.00	9999999.00	9999999.00	9999999.00	9999999.00
11	9999999.00	9999999.00	9999999.00	9999999.00	9999999.00
12	9999999.00	9999999.00	9999999.00	9999999.00	9999999.00
13	9999999.00	9999999.00	9999999.00	9999999.00	9999999.00
14	9999999.00	9999999.00	9999999.00	9999999.00	9999999.00
15	9999999.00	9999999.00	9999999.00	9999999.00	9999999.00

	6	7	8	9	10
1	45.00	47.00	47.00	47.00	39.00
2	53.00	49.00	49.00	48.00	39.00
3	49.00	49.00	49.00	47.00	39.00
4	53.00	49.00	49.00	47.00	39.00
5	9999999.00	47.00	47.00	48.00	9999999.00
6	0.00	49.00	49.00	47.00	39.00
7	9999999.00	0.00	51.00	47.00	9999999.00
8	9999999.00	51.00	0.00	47.00	9999999.00
9	9999999.00	47.00	47.00	0.00	9999999.00
10	9999999.00	9999999.00	9999999.00	9999999.00	0.00
11	9999999.00	9999999.00	9999999.00	9999999.00	9999999.00
12	9999999.00	9999999.00	9999999.00	9999999.00	9999999.00
13	9999999.00	9999999.00	9999999.00	9999999.00	9999999.00
14	9999999.00	9999999.00	9999999.00	9999999.00	9999999.00
15	9999999.00	9999999.00	9999999.00	9999999.00	9999999.00

	11	12	13	14	15
1	47.00	39.00	47.00	47.00	47.00
2	48.00	39.00	49.00	48.00	48.00
3	47.00	39.00	49.00	47.00	47.00
4	47.00	39.00	49.00	47.00	47.00
5	48.00	9999999.00	47.00	48.00	48.00
6	47.00	39.00	49.00	47.00	47.00
7	47.00	9999999.00	50.00	47.00	47.00
8	47.00	9999999.00	50.00	47.00	47.00
9	52.00	9999999.00	47.00	52.00	49.00
10	9999999.00	48.00	9999999.00	48.00	43.00
11	0.00	9999999.00	47.00	57.00	9999999.00
12	9999999.00	0.00	9999999.00	53.00	43.00
13	9999999.00	9999999.00	0.00	9999999.00	9999999.00
14	9999999.00	9999999.00	9999999.00	0.00	9999999.00
15	9999999.00	9999999.00	9999999.00	9999999.00	0.00

MATRIX REPRESENTING THE NODES ON THE MAX-FLO PATH

	1	2	3	4	5	6	7	8	9	10	11	12	13	14	15
1	1	2	3	4	5	4	9	9	5	6	9	10	9	11	9
2	1	2	3	4	5	4	4	7	5	6	9	10	7	11	9
3	1	2	3	6	5	6	6	7	8	6	9	10	7	11	9
4	1	2	3	4	5	6	7	7	8	6	9	10	7	11	9
5	1	2	3	4	5	6	9	9	9	10	9	12	9	11	9
6	1	2	3	4	5	6	4	7	8	10	9	10	7	11	9
7	1	2	3	4	5	6	7	8	8	10	9	12	13	11	9
8	1	2	3	4	5	6	7	8	9	10	9	12	7	11	9
9	1	2	3	4	5	6	8	8	9	10	11	12	8	11	15
10	1	2	3	4	5	6	7	8	9	10	11	12	13	12	12
11	1	2	3	4	5	6	7	8	9	10	11	12	13	14	15
12	1	2	3	4	5	6	7	8	9	10	11	12	13	14	15
13	1	2	3	4	5	6	7	8	9	10	11	12	13	14	15
14	1	2	3	4	5	6	7	8	9	10	11	12	13	14	15
15	1	2	3	4	5	6	7	8	9	10	11	12	13	14	15

Output interpretation of the maximum-capacity computer program. The solutions of interest are summarized in Table 2.38, and the optimal solutions in Table 2.39.

Table 2.38. MAXIMUM-CAPACITY ROUTES

Source	*Terminal*	*Maximum Capacity*
1	13	47
2	13	49
3	13	49
1	14	47
2	14	48
3	14	47
1	15	47
2	15	48
3	15	47

Table 2.39. OPTIMAL SOLUTIONS

Source	*Terminal*	*Path*	*Optimum Capacity*
2	13	2–4–7–13	49
3	13	3–6–4–7–13	49

There are two points which should be discussed. First, two optimal solutions are simultaneously produced, and both provide a path that can accommodate up to a 49-ton payload. Second, because of the manner in which the paths are calculated, the matrix of nodes on the path can reflect the optimum path in a nonserial sequence. This occurs since the length of a path from a node i to another node k is compared to one including an intermediate node j (triple operation calculation). Hence, the path when traced might identify an intermediate node j which either immediately *follows* node i or immediately *precedes* node j, with other nodes on the optimal path between i and k. For example, tracing the path from source node 2 to terminal node 13 one obtains the information from the matrix of nodes given in Table 2.40. Note that the

Table 2.40. TRACING PROCEDURE FOR PATH FROM 2 TO 13

From	*To*	*Through Node*	*Path*
2	13	7	(2) - -→ (7) - -→ (13)
7	13	13	(2) - -→ (7) - -→ (13)
2	7	4	(2) - -→ (4) - -→ (7) ——→ (13)
2	4	4	(2) - -→ (4) - -→ (7) ——→ (13)
4	7	7	(2) - -→ (4) - -→ (7) ——→ (13)

path is identified in an unambiguous fashion. However, consider tracing the path from source node 3 to terminal node 13. One obtains the information given in Table 2.41.

Table 2.41. TRACING PROCEDURE FOR PATH FROM 3 TO 13

From	*To*	*Through Node*	*Path*
3	13	7	(3) ⇢ (7) ⇢ (13)
7	13	13	(3) ⇢ (7) → (13)
3	7	6	(3) ⇢ (6) ⇢ (7) → (13)
6	7	4	(3) ⇢ (6) ⇢ (4) ⇢ (7) → (13)
4	7	7	(3) ⇢ (6) ⇢ (4) → (7) → (13)
3	6	6	(3) → (6) ⇢ (4) → (7) → (13)
6	4	4	(3) → (6) → (4) → (7) → (13)

The reader should easily master the recovery of the optimal solution after a little practice. This example illustrates the basic procedure for all types of solutions.

2.17 Node and Arc Failures in Networks

It is clear that networks can represent a wide range of physical, economic, and operational systems. The majority of those systems discussed to this point have represented physical or economic problem areas. However, network analysis can also play a major role in the study and characterization of operational systems. Within the context of the following discussion, operational problems will be those concerned with the ability of a system to effectively operate or function after one or more of the network components (nodes and arcs) "fail." Recall that a network is said to be connected provided that at least one path exists between every pair of nodes; otherwise, it is said to be disconnected.

To define the nature of the problems we wish to address, it is necessary to precisely define how operational characteristics are directly related to network construction. A network is *disconnected* when a path no longer exists between every pair of nodes. Note that the removal of one or more arcs may not disconnect a network, but successive removal of arbitrary arcs will ultimately result in disconnection. Often one is only interested in whether a finite set of nodes can be disconnected by arc removal. Such a disconnection will be called a *partial disconnection* and will be associated with only two or more nodes. Another important problem which carries the notion of partial disconnection to the extreme is that of determining how many nodes can be removed from a network in order to have no paths between *any* two nodes in the network. Frank and Frisch [18, 36] and Kleitman [36] have studied these two problems extensively. The following discussions are based upon their work.

The first problem that we will discuss is how to find the minimum number of branches that must be removed in order to disconnect a network. The following theorem proved by Frank and Frisch [36] is relevant.

***Theorem*:** If each branch of a connected network is assigned a maximum flow capacity equal to 1, the maximum flow between any two nodes in the network is equal to the minimum number of branches that must be disconnected in order to break all paths from one node to another.

In ordinary terms, this theorem simply states that if all arc capacities are set equal to 1, the maximum flow between any pair of nodes is equal to the number of paths between those two nodes that have no arcs in common. Hence, if the maximum flow between all pairs of nodes is found, the number of branches removed to disconnect the network is the minimum of all maximum flows. Note that if the algorithm of Section 2.15 is used to execute the calculations, an auxiliary benefit is all partial disconnections between all pairs of nodes.

The second problem of this section is that of finding the minimum number of nodes that must be removed from a network to disconnect all pairs of nodes in the network. The problem can be solved via an algorithm based upon the logic of the one presented in Section 2.5. Arbitrarily choose any node i as the source and any other node j as the terminal. Assume that each *node* in the network possesses a flow capacity of 1 unit. This flow capacity corresponds to the turn penalty defined for each node in the algorithm of Section 2.5. In our present discussion, the label for each arc of the pseudo network is equal to 1. Interpret this label as a flow capacity.

Next, determine the maximum flow from the pseudo source to the pseudo terminal. The value of the maximum flow is equal to the maximum number of paths in the original network which have no nodes in common except the source and sink nodes. These paths will be called *node disjoint paths*. This procedure is then repeated recursively for all pairs of nodes in the network. The number of nodes that must be disconnected to break the network is given by the minimum of all the maximum flows.

Although the foregoing procedure is guaranteed to yield the proper solution, the procedure is unwieldy for a large number of nodes. For example, if a network contains 1000 nodes, over 500,000 maximum-flow problems would have to be solved to determine if the network will be totally disconnected if four or less nodes are removed from the network. Such calculations would present themselves when solving electrical distribution problems of only modest size where the nodes represent communication junctions. Clearly, a more efficient procedure is required for real-world problems. Kleitman [36] has shown that the foregoing problem can be solved in less than 4000 maximum-flow problems using the following procedure.

Choose any node in the network and verify that the maximum flow from this node to every other node in the network is 4 or more units. If true, delete this node and all branches connected to that node from the network. Using this reduced network, choose one of the remaining nodes and verify that there exists a flow of 3 or more units between that node and all other nodes in the reduced network. Delete this node and its connecting arcs from the network and verify that a maximum flow of 2 or more units exists between any remaining node and the rest of the remaining network. Choosing a fourth node (after node–arc deletion), verify if there is a flow of 1 unit from this node to all other nodes. Note that by utilizing this procedure, each stage of this process yields the number of node disjoint paths between an (arbitrary) source node and all remaining nodes. If this sequence of operations cannot be executed, the network will fail when fewer than four nodes are deleted. For example, if the number of disjoint paths was found to be only two, then at the next-to-the-last stage of the procedure, it would have been impossible to verify that the required flow existed. This procedure is usually very efficient, since at each stage one node and (possibly) a large number of arcs can be eliminated from the original network. The procedure is illustrated by the following example.

Consider the 8-node, 16-arc network shown in Figure 2.61(a). We wish to determine if the removal of five or fewer nodes will disconnect the network.

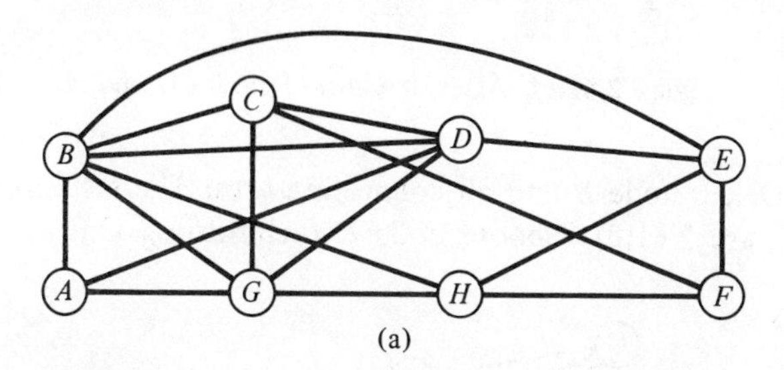

Figure 2.61(a). Sample network for disconnection analysis.

***Stage 1*:** Choose node G (arbitrarily). One can verify that there are five (or more) node disjoint paths between node G and every other node.

***Stage 2*:** Delete node G and all connecting arcs. The resulting network is given in Figure 2.61(b). Choose node A (arbitrarily). One can verify that

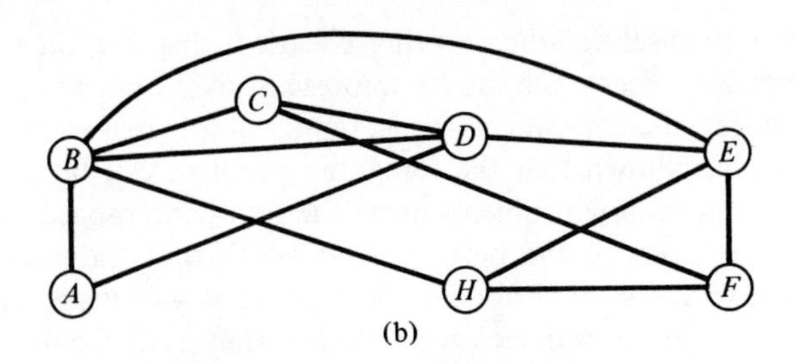

Figure 2.61(b). After deletion of node *G*.

there are at least four node disjoint paths between node *A* and all remaining nodes.

Stage 3: Delete node *A* and all connecting arcs. The resulting network is given in Figure 2.61(c). Choose node *B* (arbitrarily). One can verify that there are at least three node disjoint paths between node *B* and all remaining nodes.

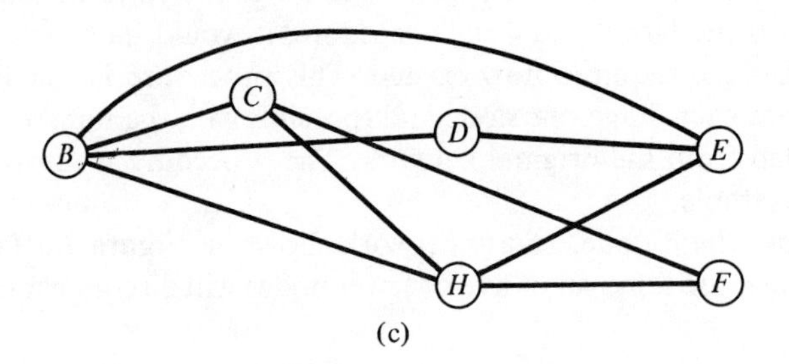

Figure 2.61(c). After deletion of nodes *G* and *A*.

Stage 4: Delete node *B* and all connecting arcs. The resulting network is given in Figure 2.61(d). Choose node *C* (arbitrarily). One can verify that

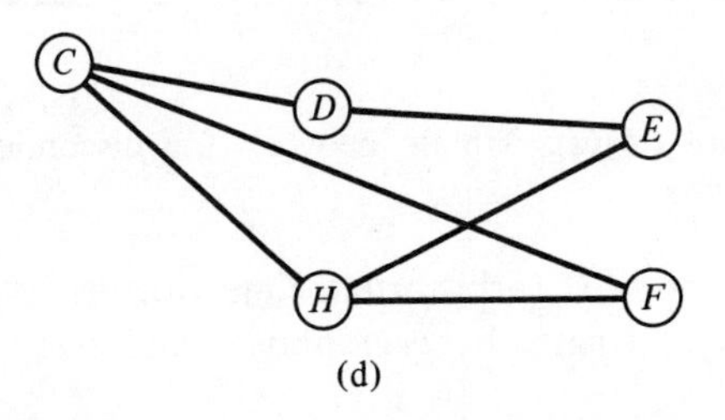

Figure 2.61(d). After deletion of nodes *G*, *A*, and *B*.

there are two or more node disjoint paths between node C and all remaining nodes.

***Stage 5*:** Delete node C and all connecting arcs. This yields the network shown in Figure 2.61(e).

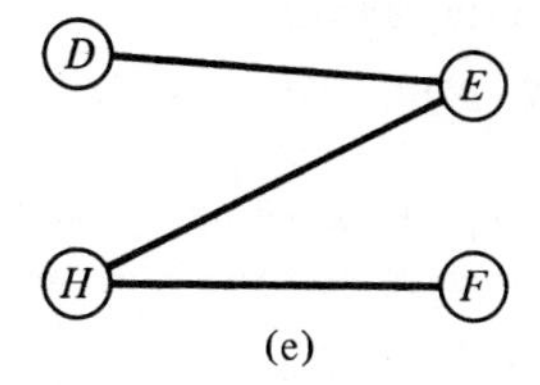

Figure 2.61(e). After deletion of nodes G, A, B, and C.

Note that in the network of Figure 2.61(e) there exists only one node disjoint path between all pairs of nodes. Since stage 5 completes the process, the hypothesis is correct.

EXERCISES

1. Indicate how we can use Dijkstra's method to obtain an arborescence with a given root.
2. Do we have to start the problem anew if a mistake has been made in the evaluation of a permanent label in one intermediate iteration, but the mistake is not detected until after the problem is finished?
3. Find the shortest path from node 1 to node 4 in the network shown (a) by inspection; (b) using Dijkstra's method. Why are the two solutions different?

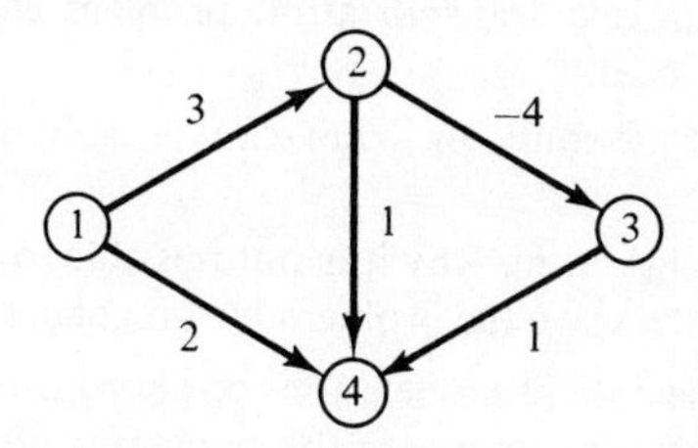

4. Give several practical applications of longest-path problems. Discuss under which conditions a longest-path model can be converted into an equivalent shortest-path model.
5. Which shortest-path algorithms discussed in this chapter find the path lengths first and then use tracing procedures to identify the actual paths? What is the

advantage of such methods over the algorithms which find path lengths and paths simultaneously?

6. The version of Dijkstra's method discussed in this chapter assumes that all arc parameters are nonnegative. Modify the algorithm in order to include arcs with negative parameters. What condition is of critical importance in this case?

7. Explain why the number of iterations of Floyd's algorithm is exactly the same as the number of nodes in the network.

8. Can we have negative arc parameters in Floyd's algorithm and the double-sweep method?

9. Can we obtain cycles in the paths when using the double-sweep method to find the K-shortest paths in a directed network? Discuss what you would do if only cycleless paths are wanted.

10. Explain why the order in which the nodes are numbered may affect the computational efficiency of Floyd's algorithm and the double-sweep method. Discuss each case separately.

11. Demonstrate that in a directed circuitless network with arcs (i, j) such that $i < j$ the double-sweep method reduces itself to one sweep. Which sweep is eliminated? Consider the same exercise for the case where the arcs (i, j) are such that $j < i$.

12. (a) How is it possible to know if two or more solutions are optimal for a traveling salesman problem? (b) How can we force the exclusion of a given city in the optimal tour?

13. Discuss why it is not possible to solve the general traveling salesman problem directly as an LP assignment model.

14. What is the roll of the initial bounds in the traveling salesman algorithm?

15. Under which conditions an undirected arc connecting nodes i and j can be and cannot be replaced by two directed arcs (i, j) and (j, i) with same capacity as that of the original arc. Illustrate your explanations with examples.

16. Why is the starting solution by VAM, in general, superior to that obtained by the northwest corner rule for a transportation problem?

17. Illustrate how a standard transportation problem can be converted into a (large) assignment model.

18. Show how the shortest-route problem can be solved by the transshipment model.

19. In the turn-penalty algorithm, why it is not possible to add the penalty to each arc parameter and then solve the problem by a regular shortest-path algorithm?

20. Formulate the mathematical models for the shortest-path problems with and without turn penalties, and compare the structures of the models.

21. Consider the network shown, in which the arc parameters represent the cost per unit flow across that arc. Find each of the following using hand calculations:
(a) The cheapest route from node 1 to node 13 using Dijkstra's algorithm.
(b) The cheapest route from each node to every other node.
(c) The $k = 3$ cheapest paths from node 2 to node 13.

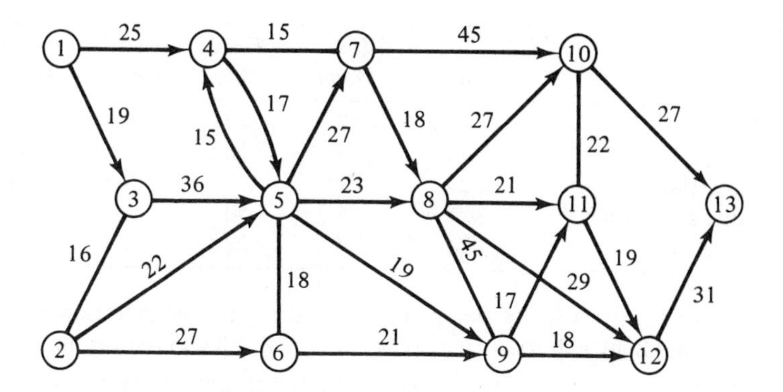

22. Consider the network shown, in which each arc parameter represents the upper bound on arc flow. Find each of the following using hand calculations:
(a) The maximum flow from node 2 to node 10.
(b) The maximum-capacity chain from node 2 to node 10.
(c) The maximum flow between each node and every other node.

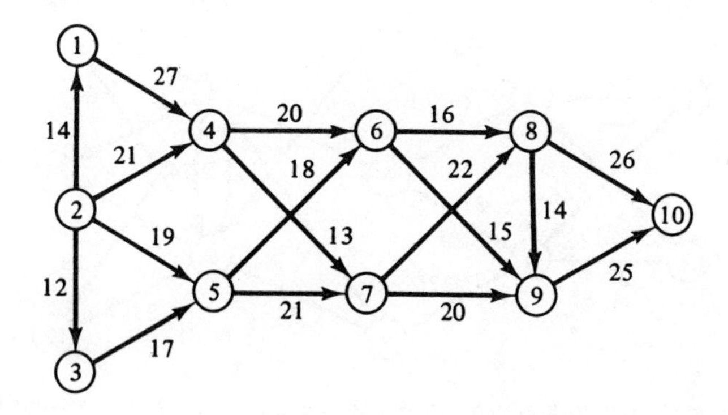

23. Verify the maximum flow found in Exercise 22, part (a), by using the maximum flow/minimum cut theorem.

24. Solve Exercise 21 using the appropriate computer codes.

25. Suppose that you are told that the cost of traversing arc (12, 13) in Exercise 21 is halved. Can you obtain the cheapest route using the previous results?

26. Suppose that in Exercise 21 you are told that arc (4, 5) *must* be included in the cheapest route. Obtain the answer to this new problem. How would this restriction affect your solution procedure if you knew this a priori?

27. Buck Rogers is touring astrospace in the twenty-fifth century, and he is directed to stop at five hostile asteriods to spy on the enemies' space forts. Buck only wants to visit each bastion once, and there are only certain space lanes that connect each asteriod which can be used. The following table lists the megamiles between each asteroid following predetermined space lanes.

	Earth	*Ming's Moon*	*Fatal Fort*	*Dang's Dungeon*	*Awful Asteroid*
Earth	—	52	29	35	29
Ming's Moon	46	—	62	21	44
Fatal Fort	29	62	—	50	27
Dang's Dungeon	37	23	50	—	—
Awful Asteroid	24	44	27	—	—

The distance matrix is not symmetric because of possible detection by space-spears. Find Buck's shortest route if he starts at Earth and returns home safely.

28. Consider the network shown, in which the arc parameters represent the cost per unit flow of using the arcs. Find each of the following using hand calculations:
(a) The cheapest route from node 1 to node 10.
(b) The cheapest route from nodes 1 and 2 to nodes 10, 11, and 12.
(c) The $k = 2$ cheapest routes from node 1 to node 11.

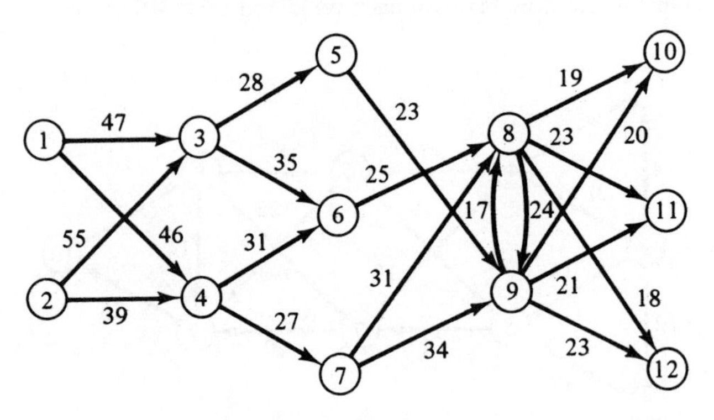

29. Consider the network shown, in which the arc parameters represent the upper bounds on arc flows. Find each of the following using hand calculations:
(a) The maximum flow from node 1 to node 9.
(b) The maximum capacity chain from node 1 to node 8.
(c) The maximum flow between every pair of nodes.

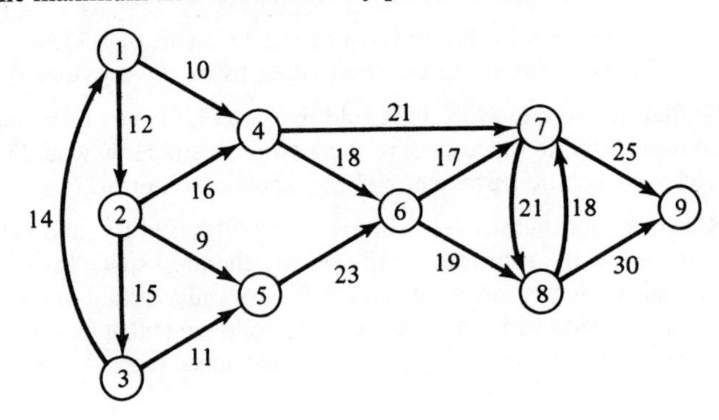

30. Compute the minimal spanning tree for the networks shown.

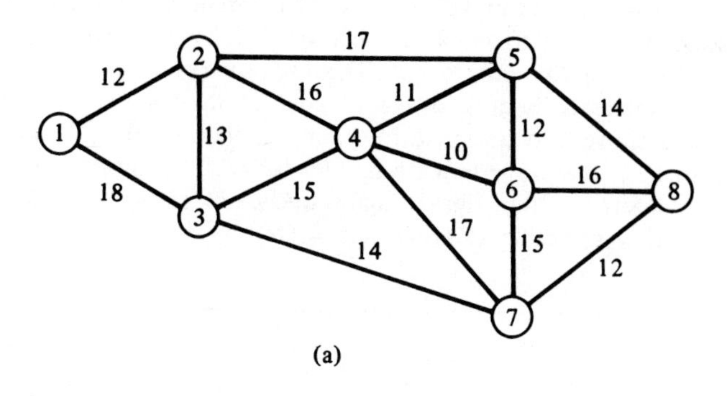

(a)

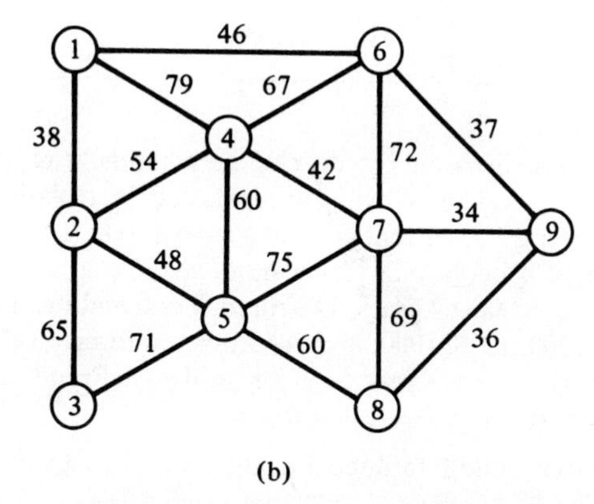

(b)

31. Draw an arborescent spanning tree for the network in Exercise 29.

32. A major conveyor manufacturer is designing a conveyor system from point B to point C. The possible system configuration is as shown. The numbers on each arc represent construction costs from point to point. Every junction point requires extra money to install if a turn is to be made. Points 1, 2, 6, 10, and 9 cost 1 unit extra. Points 3, 5, 8, 7, and 4 cost 2 units extra. Find the least-cost route from point B to point C.

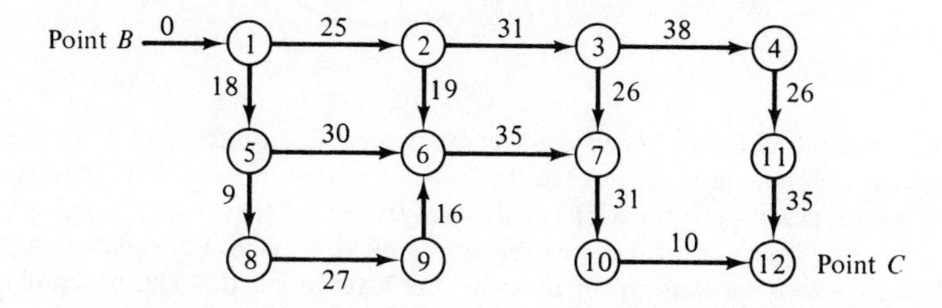

33. The Resto Manufacturing Company produces dining tables and sells them to Texas A&M University. There are three main phases in the production of a table: assembling, painting, and drying. The assembly phase can be done in two different stations, *A* and *B*. Because of differences in the equipment, the total assembly time is different at each station. Similarly, the painting operation can be performed in three stations, *C*, *D*, and *E*, each one consisting of different painting components. Because of plant layout limitations, no tables assembled in station *A* are sent to station *E* for painting, and no tables assembled at *B* are sent to *C*. The average time in minutes for each operation is given in the table. It is desired to find the optimal sequence of stations to minimize the production time per table. (Drying time = 60.)

	Station				
Operation	*A*	*B*	*C*	*D*	*E*
Assembling	120	140	—	—	—
Painting	—	—	50	30	20

34. The ABC Taxi Service Company wants to develop an equipment replacement policy for the upcoming 4 years. The following expenses and depreciation values are anticipated for these years. The current price of a car is \$7000. The car prices are expected to increase 10% every year. The cars depreciate at 25% for the first 2 years and at 15% for the next 2 years. The running and maintenance costs for the next 4 years are 1200, 1500, 1900, and 2400 dollars, respectively. A car can be replaced every year or every 2 years, 3 years, or 4 years. Find the policy that minimizes the total cost for the next 4 years.

35. Find the shortest path from node 1 to node 5 in the network shown by using Dijkstra's algorithm.

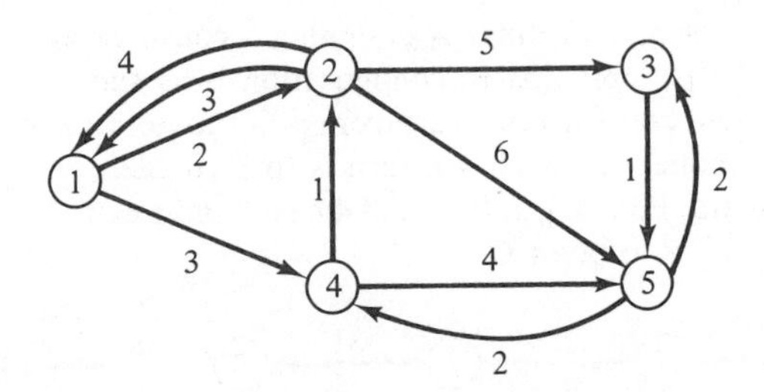

36. A certain device consists of eight components. The probability of failure for each component is as shown. The device will function properly if all the elements in a path from *s* to *t* function satisfactorily. Assume that once block 2 is used, after using block 1, it is not possible to use block 1 again. Formulate this problem as a shortest-path problem under the assumption that the probability of failure is to be minimized.

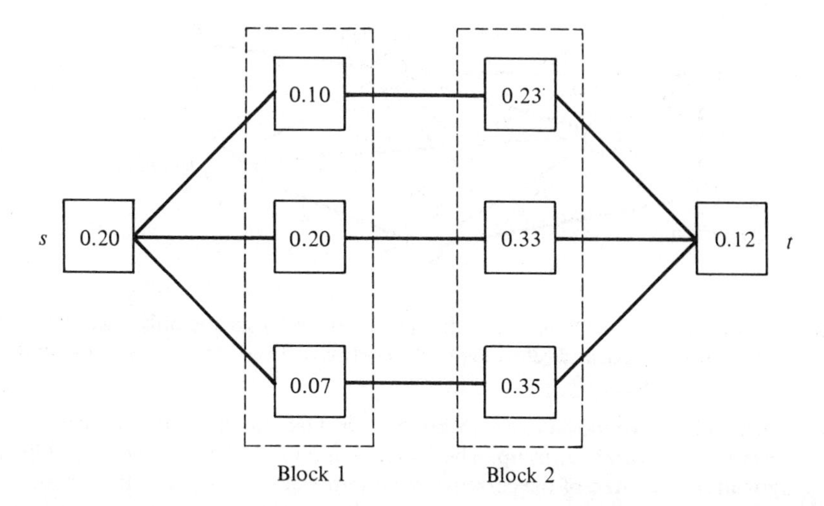

37. A manufacturing company is planning to produce air conditioners involving three main components: cabinets, fans, and motors. It can tool up to fabricate the cabinets for \$20,000, fans for \$50,000, and motors for \$80,000. However, once the fans are in production, the cost of tooling up for the cabinets and motors will be reduced by 5%. If the motors are put into production first, the costs of tooling up for the other units will be reduced by 10%. If the cabinets are in production first, the other costs are reduced by 5%. After 2 units are in production there is an extra reduction of 5% for the third unit. Draw the network, formulate, and solve as a shortest-path problem with one source and one terminal. Define the nodes, the arcs, and the arc parameters.

38. Find the shortest path between each pair of nodes in the network shown.

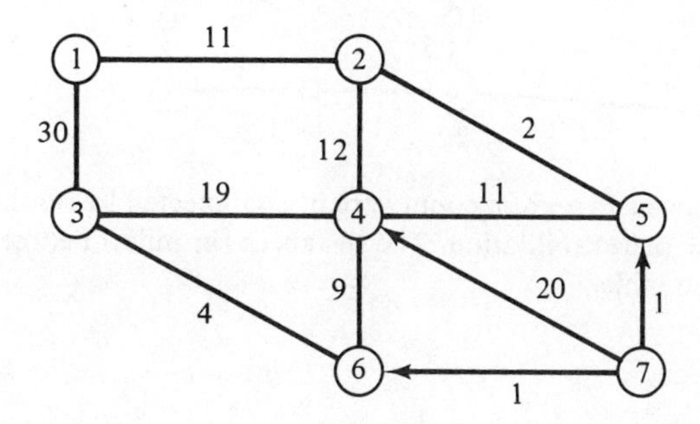

39. The U.S. Postal Service wants to identify the most economical routes to deliver packages from a given city to every other city in the map shown. The approximate transportation cost for each trip is indicated on each link of the network. Find the minimum-cost routes.

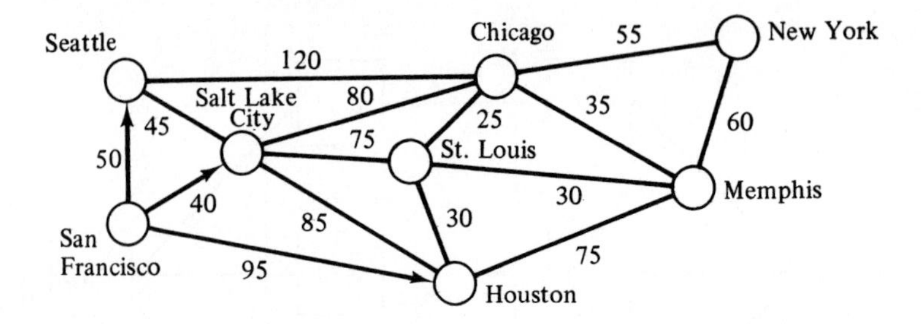

40. Assume in Exercise 39 that each city is to be visited once and only once, starting in Houston, Texas, and returning to the same city. Find the shortest route under these restrictions.

41. The graph shown indicates the location of the farm units and the country grain elevators in a rural township. The major economic activity of the township is agriculture. Because of the current weight limits of the bridges of the township, some of the users of the system might not be able to take the shortest route to the elevators. The arc numbers are road lengths. The capacity of each bridge is shown near the bridge symbol, and the truck capacity is shown next to the symbol of the farm. Given that the available budget is enough to replace two bridges, propose a methodology to determine which ones should be replaced.

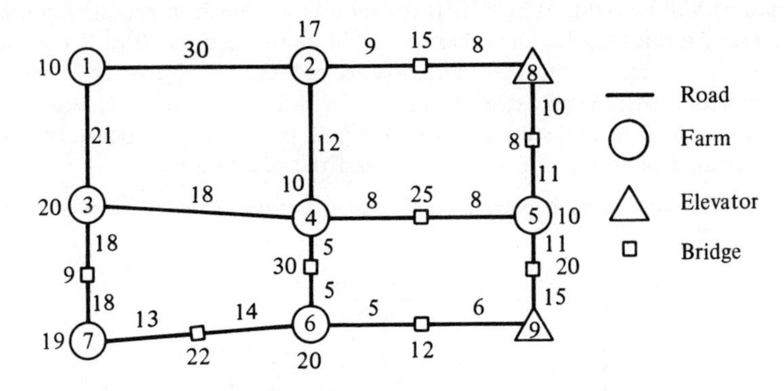

42. Five small towns in a given county are interconnected by rural roads in need of maintenance or rehabilitation. The distances (in miles) between the towns are shown in the table.

		Town j				
		1	2	3	4	5
	1	0	4	2	—	—
	2	—	0	7	5	—
Town i	3	—	5	0	4	6
	4	—	3	—	0	2
	5	—	—	—	—	0

The county highway superintendent wants to know the three roads with the highest priority for maintenance. At present the service condition of the roads is approximately the same. Most of the traffic is generated by round trips between 1 and 5, 2 and 5, and 3 and 4. All trips are distributed as follows:

70% along the shortest path
20% along the second shortest path
10% along the third shortest path

Given that the maintenance costs are identical per linear mile, find the preferred solution.

43. Find the three shortest paths from node 1 to every other node in the network shown.

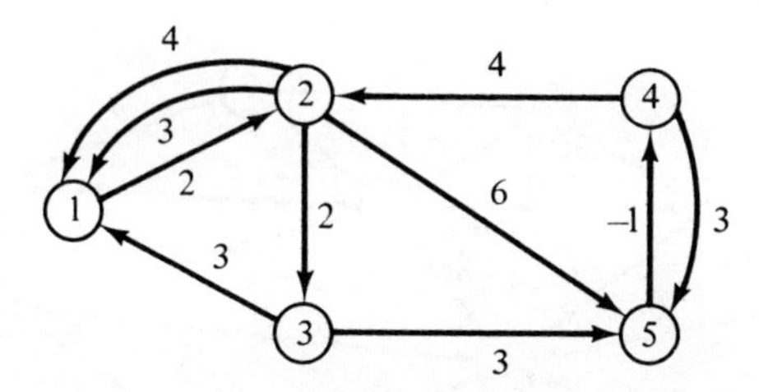

44. A given research and development project consists of four tasks which can be performed at different paces according to the budget allocated to each task, as illustrated. What is the best strategy to finish the project within 18 months without spending more than $35,000? What is the second best strategy?

	Task A		*Task B*		*Task C*		*Task D*	
Pace	*Duration (months)*	*Cost*	*Duration (months)*	*Cost*	*Duration (months)*	*Cost*	*Duration (months)*	*Cost*
Slow	6	$6,000	5	$ 8,000	6	$2,000	7	$10,000
Normal	4	8,000	3	9,000	3	3,000	5	12,000
Fast	3	9,000	2	10,000	1	5,000	3	18,000

45. Find all the paths corresponding to the four shortest path lengths ($k = 4$) by using the double-sweep computer code for the network given below. You will notice that some of the paths have cycles. Modify the TRACE subroutine of the FORTRAN program for the double-sweep algorithm to stop the further identification of a node sequence as soon as a cycle is started. Using the revised program find the cycleless paths. Node 1 is a source and nodes 10, 11, and 12 are terminals.

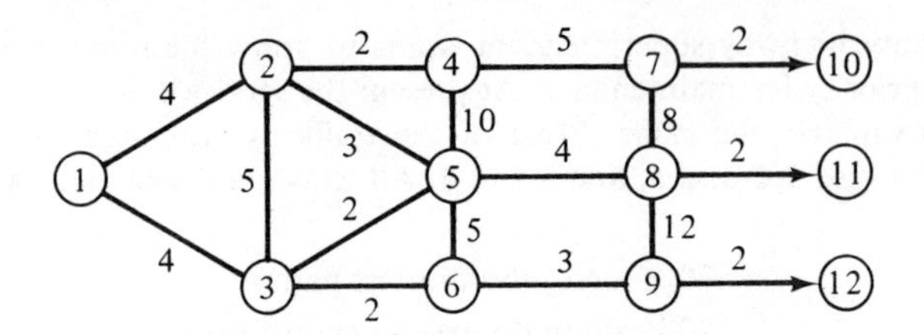

46. The Electrode TV Cable Company desires to establish operation in "Network City." The company has rented space at location 1 for the central operation and has determined satellite centers (nodes 2, 3, . . . , 10) of residential population based on 500 people per square mile. They must run cable to each of the satellite centers and want to use the least amount of cable for economic reasons. Determine the length of cable required and the minimal spanning tree. Distances are given in miles.

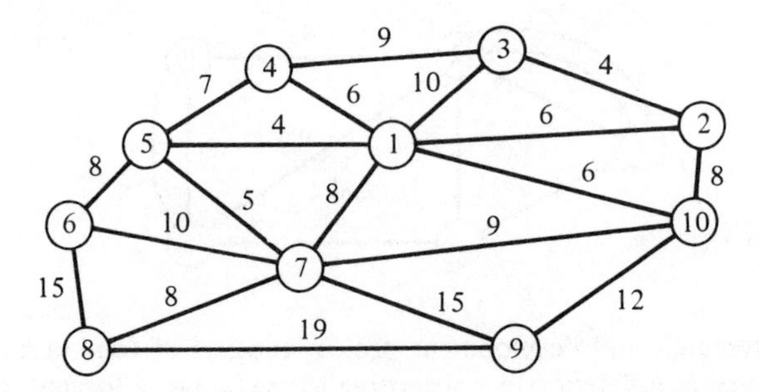

47. The Federal Highway Administration is planning to build roads to connect five major cities in neighboring states. All cities are to be connected with one another either directly or via another city. The road construction costs (in thousands of dollars) are given in the table. Which roads should be constructed?

		To City				
		A	*B*	*C*	*D*	*E*
	A	—	10	40	60	80
	B	10	—	80	70	60
From City	*C*	40	80	—	20	30
	D	60	70	20	—	25
	E	80	60	30	25	—

48. A young engineer wishes to invest in a productive asset for 6 years. He can buy the asset and keep it for 6 years; or replace it at the end of year 5 and keep it for one more year; or keep it for 4 years, then replace it with a new one for the fifth and final year; and so on. He can also replace it at the end of each year. Formulate as an equipment replacement problem and solve for the optimal solution.

49. A manager of small business is interested in investing $12,000 in three different plans, 1, 2, and 3. She has to invest a minimum of $3000, $2000, and $1000 in each plan, respectively. In excess of the minimum amount requirement, any whole-number multiple of $1000 may be invested. Formulate the network representation for this problem.

50. Rail-Riders, Inc., a major manufacturer of materials handling equipment is designing a high-speed people-mover system between six different airline terminals in Space City, U.S.A. The matrix represents the distances in feet between all pairs of terminals via the shortest pairwise construction routes.

	1	*2*	*3*	*4*	*5*	*6*
1	—	870	910	796	860	712
2	870	—	1100	560	430	630
3	900	1100	—	270	600	
4	796	560	270	—	300	850
5	850	430	600	300	—	370
6	712	630	250	850	370	—

(a) OSHA regulations will not allow the tracks in high-speed systems to cross one another. Find the length of track necessary to connect all terminals in a cyclic fashion at minimum cost. (Cost per foot is constant at $100 per foot.)

(b) Rail Riders wishes to design an alternative system such that there are two-way guide paths constructed between a central distribution point and all other terminals. This design would resemble the wheel of a bicycle, with the central station as the "hub." The following table gives the direct distance between the hub and each of the six terminals:

	1	2	3	4	5	6
Hub	150	175	210	180	175	190

Design the new system at minimum cost and compare these results to those of part (a).

51. Coal-allocation problem: Four mines supply coal to five gas-producing plants in the southern United States. The transportation cost includes freight, road haulage, handling, and unloading charges, which are given in the cost matrix (in hundreds of dollars). Find the minimum-cost solution.

		Plant j					
		1	2	3	4	5	*Supply* (*tons*)
Mine i	1	10	9	8	9	7	500
	2	4	8	12	7	9	900
	3	6	3	4	2	8	700
	4	7	6	5	4	3	600
Demand (*tons*)		400	700	400	500	700	

52. Four job orders have been received in a shop. Five machines are available. The cost to make the job at different machines is given in the matrix (in dollars). The manager is interested in scheduling jobs to machines in order to minimize the total cost.

		Job			
		1	2	3	4
Machine	1	12	14	12	18
	2	16	10	10	16
	3	10	8	8	14
	4	8	12	16	12
	5	14	10	18	14

53. A water ski manufacturer has four nationwide distribution centers to distribute water skis to five different retail stores. If the centers have available 100, 200, 300, and 400 units and the retail stores need 200, 250, 300, and 100 units, find the transportation schedule that minimizes cost. The cost matrix is given in hundreds of dollars.

		Retail Store				
		A	*B*	*C*	*D*	*E*
Distribution Center	1	5	7	3	15	9
	2	6	2	8	5	6
	3	3	9	8	2	2
	4	7	6	8	7	4

54. When uranium is mined, approximately 0.711% of it is ^{235}U and the rest is ^{238}U. This is sent to three locations where it is enriched; that is, the percent of ^{235}U is increased. This results in uranium flouride (UF_6), which is highly radioactive. It is next sent to 10 locations to be converted into fuel for nuclear reactors. The government, very concerned with the danger, has developed a scale from 1 to 10 which identifies the danger associated with the various routes, 10 being the most dangerous. The table illustrates this, together with the supply and demand. Identify the transportation schedule that minimizes the total danger.

	1	2	3	4	5	6	7	8	9	10	*Supply* (*lb*)
A	9	8	7	4	6	3	8	6	5	2	500
B	1	7	9	10	8	3	6	7	5	4	500
C	3	3	5	7	7	8	9	1	2	3	250
Demand (*lb*)	100	100	300	150	200	90	110	50	50	100	

55. John Jones went to Puerto Rico last summer. His VISA only allowed him to visit exactly five locations once each. Because he had never been in Puerto Rico before, he went to a tourist information desk and asked them what places they would recommend to visit. They told him not to leave Puerto Rico without visiting the cities of Caguas, Ponce, Mayaguez, and Arecibo. Knowing that gasoline was very expensive, he went to his room in San Juan and made a distance matrix of all the possible routes from San Juan to those cities. He was interested in finding the shortest route to visit all those cities and return to San Juan. Determine his final travel route. The distance matrix is given in miles.

	SJ	*C*	*P*	*A*	*M*
SJ	—	17	64	48	94
C	17	—	47	58	94
P	64	47	—	52	47
A	48	58	52	—	47
M	94	94	47	47	—

56. Consider a production schedule problem at a machine that produces a single product with different colors. There is only one setup cost involved, the cost of changing the color in the machine. The objective is to obtain a production schedule that minimizes the sum of the setup costs in the machine. The cost of a color change is dependent upon an immediately preceding color. The cost of changing from color i to color j is denoted by C_{ij} and can be expressed in the form of a matrix (dollars). Find the optimal production schedule that minimizes changeover costs.

$C = [C_{ij}] =$

	1	2	3	4	5	6	7	8	9	10
1	—	1	2	3	4	5	6	7	8	9
2	2	—	3	4	5	6	7	8	9	1
3	3	4	—	4	5	6	7	8	9	1
4	4	5	6	—	7	9	8	1	2	3
5	5	6	7	8	—	9	1	2	3	4
6	6	7	8	9	1	—	2	3	4	5
7	7	8	9	1	2	3	—	4	5	6
8	8	9	1	2	3	4	5	—	6	7
9	9	1	2	3	4	5	6	7	—	8
10	1	2	3	4	5	6	7	8	9	—

57. Five different processes are to be performed on a material. The cutting tool of the machine to be used will be changed automatically from time to time, depending on the cutting operation called for. Some of the operations must precede others and some operations require the same setups. An arbitrary sequencing of operations will yield repetition in the setups. It is required to minimize the processing time by minimizing the number of setups through optimization of the operation schedule. The operations and their locations are given in Table 1. Their corresponding processing time units for traversing the operation from location i to j and the sequence restrictions are given in Table 2. Operation 1 is the setup operation. Find the minimum-time solution.

Table 1

Operation	*Location*			
	A	*B*	*C*	*D*
2	Mill			
3	Spot face			
4		Spot face		
5		Drill		
6		Tap		
7			Spot face	
8			Drill	
9				Spot face
				Pilot hole

Table 2

	Location								
	A		*B*			*C*		*D*	
Operation	2	3	4	5	6	7	8	9	10
1	20	×	25	×	×	30	×	35	×
2	×	20	20	20	20	40	40	45	45
3	×	×	25	20	25	45	45	30	30
4	20	25	×	30	×	30	30	35	35
5	25	25	×	×	30	20	30	35	35
6	25	25	×	×	×	30	20	35	35
7	50	50	10	20	30	×	20	40	40
8	50	50	20	20	20	×	×	40	40
9	25	25	20	10	20	20	40	×	30
10	25	25	30	30	30	20	25	×	×

58. Solve a traveling salesman problem with six cities where the cost of traveling from city *i* to city *j* is as given in the cost matrix (in dollars).

$C =$

	1	2	3	4	5	6
1	—	27	43	16	30	26
2	7	—	16	1	30	25
3	20	13	—	35	5	0
4	21	16	25	—	18	18
5	12	46	27	48	—	5
6	23	5	5	9	5	—

59. Consider a manufacturing operation located in Michigan. The operation consists of three plants (plastic molding, paint, and assembly) with covered drives and walkways between them. There are six general foremen's offices at the plants to be cleaned on the third shift (nodes 2, 3, 4, 5, 6, and 7). The Clark Scooter with all the cleaning tools is stored at location 1. The indirect cost industrial engineer has estimated the time for the tasks of cleaning and traveling with a standard time manual. However, he is a trained I.E. and desires to find the most efficient route. Distances between locations are shown in Table 1. Determine the most efficient route and the total distance traveled.

Table 1

Undirected Arc (i, j)	*Distance (yds)*
(1, 2)	40
(1, 3)	90
(1, 4)	120
(1, 7)	90
(2, 3)	130
(3, 4)	80
(4, 5)	70
(4, 7)	80
(5, 6)	60
(6, 7)	140

The I.E. decides that Michigan weather is severe in the winter months and outside routes should be considered. Also, the speed of the Clark Scooter is restricted to 2.72 miles/hour inside because of safety regulations, but can travel at 5.09 miles/hour outside. New and alternative outside routes are shown in Table 2 with their respective distances. Determine if the new or alternative route proposal changes the most efficient route obtained above. What is the total distance traveled? What is the total travel time?

Table 2

Alternative Outside Routes		*New Outside Routes*	
Undirected Arc (i, j)	*Distance (yd)*	*Undirected Arc* (i, j)	*Distance (yd)*
(1, 4)	140	(3, 5)	190
(1, 7)	180	(4, 6)	110
(6, 7)	150		

60. Crude oil is pumped from seven different offshore oil platforms to a terminal 50 miles away. The production rate at each wellhead varies from platform to platform. The terminal production manager specifies that he wants all lines coming into the terminal to be at their maximal possible flow at all times. The maximum amount of crude oil, in 10^5 barrels, that can be pumped from platform i to platform j is given in the matrix. Find the maximal flow from oil platform 1 to the terminal T.

	1	2	3	4	5	6	7	T
1	—	6	4	1	—	—	—	—
2	—	—	—	—	4	—	—	—
3	—	—	—	1	1	3	—	—
4	—	—	—	—	—	4	—	—
5	—	—	—	—	—	—	4	—
6	—	—	—	—	—	—	9	4
7	—	—	—	—	—	—	—	6

61. A plumbing contractor has completely "piped" a house for a waste water system. He knows all the capacities of the individual pipes within the house, but he has not yet made the tie in with the main server line. He needs to know the maximum amount of water that can come out of the house so that he can place the best size of pipe and minimize his cost. The flow diagram of the house is shown. The figures shown are the upper-bound flow capacities of the pipes in cubic inches. Determine the maximum flow and the relevant pipe sizes. Discuss the shortcomings of this formulation.

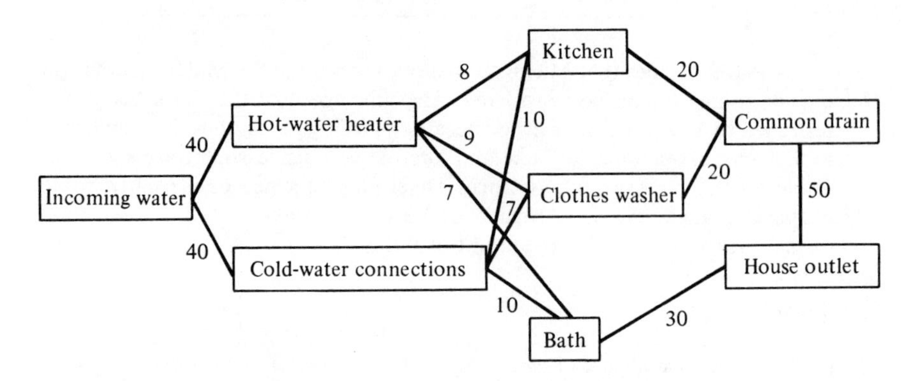

62. Crudeland, an oil-producing country, has recently developed an oil well in the southern part of the country. Accordingly, the country's only refinery was built in a nearby southern town and refined petroleum was distributed across the nation by road transportation in tankers. However, a recent study of alternative means of distribution has shown that installing two other refineries in the north and west of the country, and transporting crude oil via pipelines to these refineries, would have a long-term economic advantage over the existing system. In an attempt to operate refineries at their maximum-capacity levels, it is desired to determine the maximum flows of crude oil transferable to these sites, given the capacity constraints defined by geographical and other considerations on the pipelines between intervening cities. The flow capacities in cubic feet (hundreds) are shown in the figure. Determine the maximum possible flow between all pairs of points.

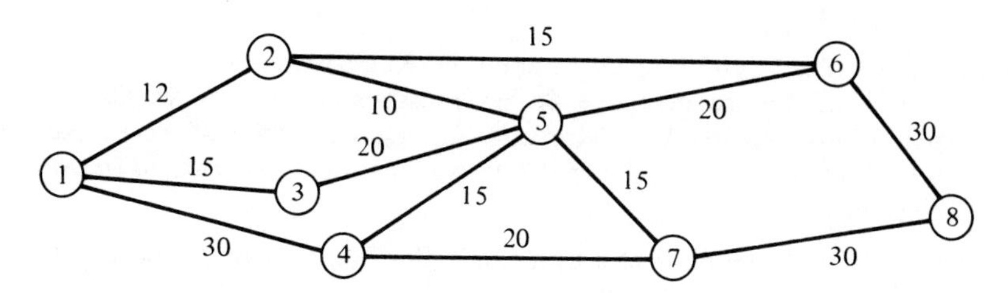

Oilwell-node 1
Cities-nodes 2, 3, 4, 5
Refineries-nodes 6, 7
Distribution center-node 8

63. Seven cities use the same communication system. The figures in the matrix represent the maximum number of calls between any two cities. Find the maximum number of simultaneous calls between all pairs of cities.

		City j						
		1	2	3	4	5	6	7
	1	—	10	—	2	—	—	—
	2	10	—	7	—	3	—	—
	3	—	7	—	—	—	9	—
City i	4	2	—	—	—	8	—	6
	5	—	3	—	8	—	11	6
	6	—	—	9	—	11	—	4
	7	—	—	—	6	6	4	—

REFERENCES

[1] BARR, R. S., F. GLOVER, AND D. KLINGMAN, "An Improved Version of the Out-of-Kilter Method and a Comparative Study of Computer Codes," *Mathematical Programming, 7* (1), 60–87 (1974).

[2] BELLMORE, M., G. BENNINGTON, AND S. LUBORE, "A Multivechicle Tanker Scheduling Problem," *Transportation Science, 5*, 36–47 (1971).

[3] BENNINGTON, G. E., "An Efficient Minimal Cost Flow Algorithm," *Management Science, 19*, 1042–1051 (1973).

[4] BENNINGTON, G. E., "Applying Network Analysis," *Journal of Industrial*

Engineering, *6* (1), 17–25 (January 1974). Portions reproduced by permission of the authors and the American Institute of Industrial Engineers.

[5] BRADLEY, G. H., "Survey of Deterministic Networks," *AIIE Transactions*, *7*, 222–234 (1975).

[6] CHARNES, A., F. GLOVER, D. KARNEY, D. KLINGMAN, AND J. STUTZ, "Past, Present and Future of Large Scale Transshipment Computer Codes and Applications," Center for Cybernetic Studies, University of Texas, Report CS 131, July 1973 (rev. October 1973) (to appear in *Computers and Operations Research*).

[7] CHIEN, R. T., "Synthesis of a Communication Net," *Journal of Research and Development*, *4*, 311–320 (1960).

[8] DANTZIG, G. G., W. BLATTNER, AND M. R. RAO, "Finding a Cycle in a Graph with a Minimum Cost to Time Ratio with Applications to a Ship Routing Problem," *Theory of Graphs International Symposium*, pp. 77–83. Paris: Dunod, and New York: Gordon and Breach, 1966.

[9] DANTZIG, G., "Application of the Simplex Method to a Transportation Problem," in *Activity Analysis of Production and Allocation*, ed. T. C. Koopmans. New York: John Wiley & Sons, Inc., 1951.

[10] DEI ROSSI, J. A., R. S. HEISER, AND N. S. KING, "A Cost Analysis of Minimum Distance TV Networking for Broadcasting Medical Information," RM-6204-NLM RAND Corporation, Santa Monica, Calif., February, 1970.

[11] DIJKSTRA, E. W., "A Note on Two Problems in Connection with Graphs," *Numerishe Mathematik*, *1*, 269–271 (1959).

[12] DREYFUS, S. E., "An Appraisal of Some Shortest Path Algorithms," *Operations Research*, *17*, 395–412 (1969).

[13] DREYFUS, S. E., "A Generalized Equipment Replacement Study," *Journal of the Society for Industrial and Applied Mathematics*, *8*, 425–435 (1960).

[14] EVANS, JAMES R., "Network Modeling in Production Planning," *Proceedings of the 1978 AIIE National Systems Conference*, Montreal, Canada, 1978.

[15] FORD, L. R., AND D. R. FULKERSON, *Flows in Networks*. Princeton, N.J.: Princeton University Press, 1962.

[16] FORD, L. R., AND D. R. FULKERSON, "A Primal–Dual Algorithm for the Capacitated Hitchcock Problem," *Naval Research Logistics Quarterly*, *4* (1), 47–54, 1957.

[17] FOX, B., "Vehicle Scheduling and Driver Run Cutting: RUCUS Package Overview," M71-58, The Mitre Corporation, McClean, Va., September, 1971.

[18] FRANK, H., and I. T. FRISCH, *Communication, Transmission and Transportation Networks*. Reading, Mass.: Addison-Wesley, Publishing Co., Inc., 1971.

[19] GARFINKEL, R. S., AND G. L. NEMHAUSER, *Integer Programming*. New York: John Wiley & Sons, Inc., 1972.

[20] Glover, F., and D. Klingman, "New Advances in the Solution of Large Scale Network and Network-related Problems," *Colloquia Mathematica Societatis Janos Bolyai*, Vol. 12. Amsterdam: North-Holland Publishing Co., 1975.

[21] Glover, F., and D. Klingman, "A Note on Computational Simplification in Solving Generalized Transportation Problems," *Transportation Science*, *7*, 351–361 (1973).

[22] Glover, F., and D. Klingman, "Capsule View of Future Developments on Large Scale Network and Network-related Problems," CCS238, University of Texas at Austin, 1975.

[23] Glover, F., and D. Klingman, "Network Application in Industry and Government," *AIIE Transactions*, *9* (4) (1977).

[24] Glover, F., J. Hultz, and D. Klingman, "Improved Computer Based Planning Techniques—Part I," *Interfaces*, *8* (4) (August 1978).

[25] Gomory, R. E., and T. C. Hu, "Multi-terminal Network Flows," *SIAM Journal of Applied Mathematics*, *9*, 551–571 (1971).

[26] Held, M., and R. Karp, "The Travelling Salesman Problem and Minimum Spanning Trees," *Operations Research*, *18* (6), 1138–1162 (1970).

[27] Held, M., and R. Karp, "The Travelling Salesman Problem and Minimum Spanning Trees: Part II," *Mathematical Programming*, *1*, 6–25 (1971).

[28] Heller, I., and C. B. Tompkins, "An Extension of a Theorem of Dantzigs," in *Linear Inequalities and Related Systems*, ed. H. Kuhn and A. W. Tucker. Princeton, N.J.: Princeton University Press, 1956.

[29] Hitchcock, F. L., "The Distribution of a Product from Several Sources to Numerous Localities," *Journal of Mathematics and Physics*, *20*, 224–230 (1941).

[30] Hoffman, A. J., and S. Winograd, "Finding All Shortest Distances in a Directed Network," *IBM Journal of Research and Development*, *16*, 412–414 (1972).

[31] Hu, T. C., *Integer Programming and Network Flows*. Reading, Mass.: Addison-Wesley Publishing Co., Inc., 1969.

[32] Hu, T. C., "The Maximum Capacity Route Problem," *Operations Research*, *9*, 898–900 (1961).

[33] Jacobs, W. W., "The Caterer Problem," *Naval Research Logistics Quarterly*, *1*, 154–165 (1954).

[34] Klein, M., "A Primal Method Cost Flows," *Management Science*, *14* (3), 205–220 (1967).

[35] Klingman, D., A. Napier, and J. Stutz, "NETGEN—A Program for Generating Large Scale (Un) Capacitated Assignment, Transportation and Minimum Cost Flow Network Problems," *Management Science*, *20* (5), 814–822 (1974).

[36] KLEITMAN, D. H., reported by H. FRANK AND I. FRISCH in "Network Analysis," *Scientific American*, *223*, 94–103 (July 1970).

[37] KLINGMAN, D. H., AND J. HULTZ, "Solving Constrained Generalized Network Problems," Research Report CCS 257, Center For Cybernetic Studies, The University of Texas at Austin, 1976.

[38] KOOPMANS, T. C., "Optimum Utilization of the Transportation System," *Proceedings of the International Statistical Conferences*, Washington, D.C., 1947.

[39] KUHN, H. W., "The Hungarian Method for the Assignment Problem," *Naval Research Logistics Quarterly*, *2*, 83–97 (1955).

[40] LAWLER, E. L., "A Solvable Case of the Traveling Salesman Problem," *Mathematical Programming*, *1*, 267–269 (1971).

[41] LITTLE, J. D. C., et al., "An Algorithm for the Traveling Salesman Problem," *Operations Research*, *11*, (5), 972–989 (1963).

[42] IRI, M., *Network Flow, Transportation and Scheduling*. New York: Academic Press, Inc., 1969.

[43] MINIEKA, E. T., AND D. R. SHIER, "A Note on an Algebra for the *K* Best Routes in a Network," *Journal of the IMA*, *11*, 145–149 (1973).

[44] POLLACK, M., "The Maximum Capacity Route through a Network," *Operations Research*, *8*, 733–736 (1960).

[45] ROSS, G. T., AND R. M. SOLAND, "A Branch and Bound Algorithm for the Generalized Assignment Problem," *Mathematical Programming*, *8*, 91–104 (1975).

[46] SALTMAN, R. G., G. R. BOLOTSKY, AND Z. G. RUTHBERG, "Heuristic Cost Optimization of the Federal Telpak Network," National Bureau of Standards Technical Note 787, 1973.

[47] SHIER, D. A., "Iterative Methods for Determining the *K* Shortest Paths in a Network," *Networks*, *6*, 205–230 (1976).

[48] SHIER, D. A., "Computational Experience with an Algorithm for Finding the *K* Shortest Paths in a Network," *Journal of Research, National Bureau of Standards*, *78B*, 139–165 (July–September 1974).

[49] SMYTHE, W. R., AND L. JOHNSON, *Introduction to Linear Programming with Applications*. Englewood Cliffs, N.J.: Prentice-Hall, Inc., 1966.

[50] WILLIAMS, T. A., AND G. P. WHITE, "A Note on Yen's Algorithm for Finding the Length of All Shortest Paths in *N*-Node Nonnegative Distance Networks," *Journal of the Association of Computing Machinery*, *20* (3), 389–390 (July 1973).

[51] VEINOTT, A. F., JR., AND H. M. WAGNER, "Optimal Capacity Scheduling I," *Operations Research*, *10*, 518–522 (1962).

[52] ZADEH, N., "Theoretical Efficiency and Partial Equivalence of Minimum Cost Flow Algorithms: A Bad Network Problem for the Simplex Method," Opera-

tions Research Center, University of California at Berkeley, ORC 72-7, March 1972.

[53] ZAHN, C. T., "Graph-Theoretical Methods for Detecting and Describing Gestalt Clusters," *IEEE Transactions on Computers*, *C-20* (1), 68–86 (January 1971).

[54] ZANGWILL, W., "A Backlogging Model and a Multi-echelon Model of a Dynamic Economic Lot Size Production System—A Network Approach," *Management Science*, *15*, 506–527 (1969).

THE OUT-OF-KILTER ALGORITHM: GENERALIZED ANALYSIS OF CAPACITATED, DETERMINISTIC NETWORKS

"There must be a better way to do it,
and if we search long enough, I am sure
we will find it."

Quoted from a Contemporary Researcher

Having completed a study of the wide variety of algorithms presented in Chapter 2, the reader might assume that better algorithms cannot be found to solve the problems considered in that chapter. However, as in all areas of systems analysis and optimization, more efficient approaches can be developed if such algorithms are discovered. The out-of-kilter algorithm represents a significant contribution to the theories of network analysis, and is considered to be the most general and widely used algorithm when dealing with capacitated, deterministic network flows. Although sometimes not as efficient as specialized algorithms, its broad applicability and unifying framework warrants the special attention to which we will now address our efforts.

The out-of-kilter algorithm will be discussed in three parts. Part I will present the underlying theory of the algorithm and demonstrate its use by means of a small example. Part II will address the modeling versatility of the

3

algorithm and illustrate network construction procedures to a wide variety of classical network problems. Part III will present formulations and solutions to several real-world applications.

PART I: NETWORK-FLOW OPTIMIZATION WITH THE OUT-OF-KILTER ALGORITHM—THEORETICAL CONCEPTS

The primary objective of Part I is to present the basic underlying theory of the out-of-kilter network-flow algorithm. The out-of-kilter algorithm has been developed to deal with capacitated network-flow problems. The algorithm is carefully developed using the concepts of linear programming duality theory and complementary slackness conditions. The actual mechanics of the solution technique is explained in a general but technically sound framework and a set of decision rules leading to the problem solution is presented. The entire algorithm is then summarized and condensed to five basic steps guided by tabular decision rules. Finally, an example problem is presented and manually solved to illustrate the application of the out-of-kilter solution technique to capacitated network-flow problems.

3.1 Basic Terminology

In order to provide a common ground for communication, the following terminology relevant to network flows will be used in this section. Some definitions have been previously introduced, but will be repeated here for clarity and exposition purposes.

1. A *node* is the basic entity of a network diagram and usually represents a physical origin or termination point, such as a factory, retailer, contract or source of manpower, and so on. A node that generates flow (such as a warehouse in some applications) is called a *source* node. A node that consumes flow (such as a customer) is called a *sink* node. The set of nodes in a network will be designated **N**.

2. *Arcs* are the lines that connect the various nodes in a network and are sometimes called branches. A directed arc is one in which flow is only permitted in a predesignated direction. The set of arcs in the network will be designated **S**.

3. A *network* is a connected set of arcs and nodes, generally representing a physical process in which units move from source(s) to sink(s). A typical network is given in Figure 3.1, in which node 1 is the source node and node 4 is the sink node. A network may have multiple sources and sinks such as that given in Figure 3.2. In this case nodes 2 and 3 are source nodes, and nodes 5 and 6 are sink nodes. Nodes 1 and 7 have been added with the dotted arcs and are called the *super source* and the *super sink* nodes, respectively.

4. If an arc can only tolerate certain magnitudes of flow, such as a

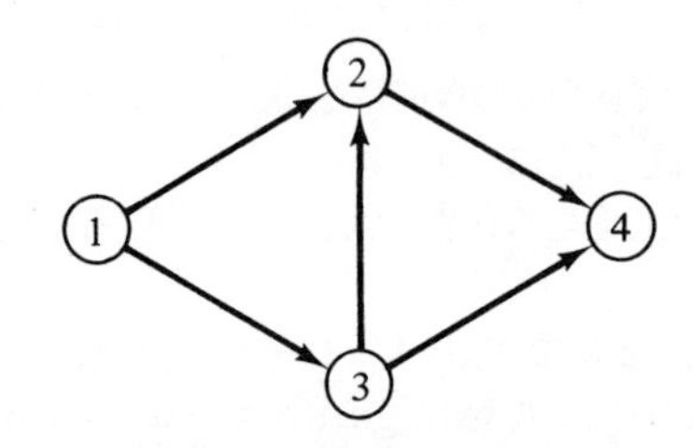

Figure 3.1. Directed network.

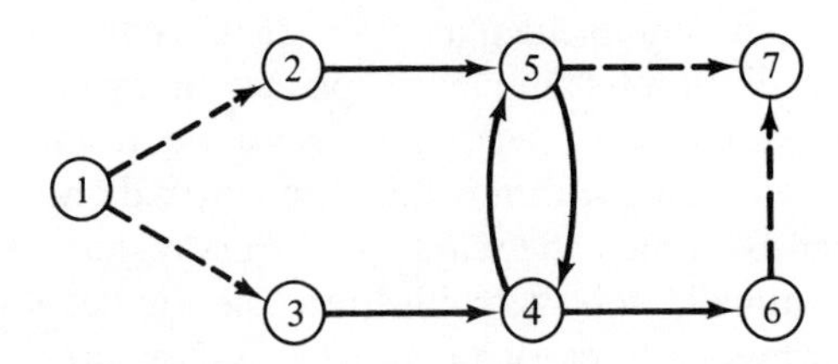

Figure 3.2. Super source and super sink.

designated upper bound and/or a designated lower bound, the arc is said to be *capacitated.*

5. For purposes of discussion it will be convenient to describe an arc as a forward arc or a reverse arc, depending on the node from which it is traversed. If an arc oriented from node i to node j is traversed from node i, it is called a *forward arc.* If it is traversed from node j, it is called a *reverse arc.*

6. A network with capacitated arcs is called a *capacitated network.*

7. A *circulation* is an assignment of flow to arcs such that flow is conserved at each node. That is, the total flow entering the node is equal to the total flow leaving the node. The out-of-kilter algorithm deals with circulations; thus it is often necessary to modify the original networks to provide for circulations. For the networks of Figures 3.1 and 3.2, an additional arc is required to connect the sink with the source. This is called the *return arc* and is illustrated in Figure 3.3. The details of the return arc depend upon the situation described by the network and will be discussed in detail in Part II of this chapter.

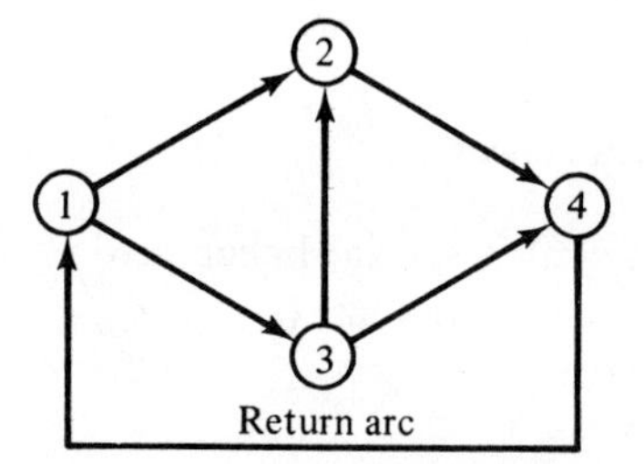

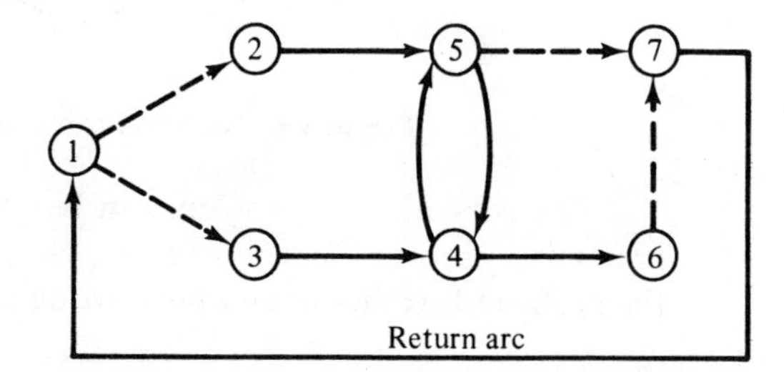

Figure 3.3. Closed networks.

8. In a capacitated network there are always lower and upper bounds on each arc. (These may, in fact, be zero or infinity.) The actual flow through each arc can be any flow between the upper and lower limits, as long as any additional constraints are not violated. Throughout this chapter the following notation will be used.

f_{ij} = flow through arc (i, j)

L_{ij} = lower capacity on arc (i, j)

U_{ij} = upper capacity on arc (i, j)

c_{ij} = cost associated with shipping 1 unit of flow from node i to node j

9. The out-of-kilter algorithm (OKA) is an iterative procedure to find the circulation in a given capacitated network which minimizes the total cost of all flows passing through the arcs of the network.

3.2 Basic Theory

Suppose that we have a directed arc connecting node i to node j. Let the amount of flow passing through this arc be represented by f_{ij}; where we adopt the conventions that if

1. $f_{ij} = 0$, no flow occurs.
2. $f_{ij} > 0$, flow is passing from node i to node j.

Using this notation, a network segment is given in Figure 3.4.

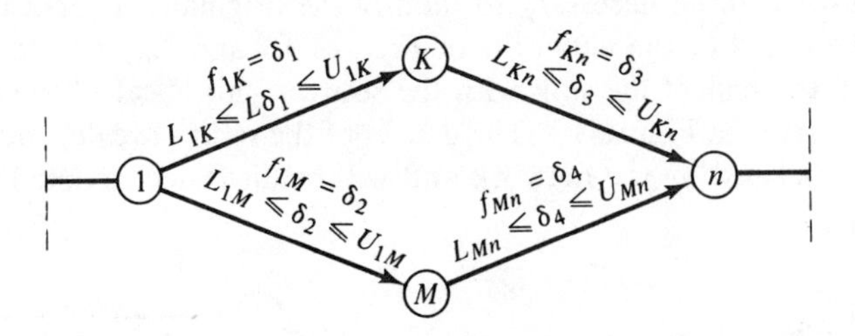

Figure 3.4. Notation for a capacitated network.

The network-flow problem can be represented as a special linear programming problem. Since the cost of shipping 1 unit across arc (i, j) is c_{ij}, the problem becomes one of minimizing total cost:

$$\text{minimize} \sum_{(i,j)\in \mathbf{S}} c_{ij} f_{ij} \tag{3-1}$$

subject to capacity restrictions

$$\begin{aligned} f_{ij} &\leq U_{ij} \qquad (i, j) \in \mathbf{S} \\ f_{ij} &\geq L_{ij} \qquad (i, j) \in \mathbf{S} \end{aligned} \tag{3-2}$$

To ensure that any commodity shipped into a node must also leave, a conservation-of-flow constraint is also imposed, such that

$$\sum_{j\in \mathbf{N}} f_{ji} - \sum_{j\in \mathbf{N}} f_{ij} = 0 \qquad \text{all } i \in \mathbf{N},\ i \neq j \tag{3-3}$$

and nonnegativity is usually required of the flows:

$$f_{ij} \geq 0 \qquad \text{for all arcs } (i, j)$$

Equations (3-1), (3-2), and (3-3) represent the minimum-cost circulation problem as a special linear programming problem. It is from this basic formulation that the out-of-kilter algorithm will be derived, using optimality conditions obtained from the duality theory of linear programming. First, Eqs. (3-1), (3-2), and (3-3) will be rewritten in a more convenient form.

$$\text{maximize} \sum_{(i,j)\in \mathbf{S}} -c_{ij} f_{ij}$$

subject to

$$\sum_{j\in \mathbf{N}} f_{ij} - \sum_{j\in \mathbf{N}} f_{ji} = 0 \qquad \text{all } i \in \mathbf{N} \quad \text{(conservation-of-flow constraints)}$$

$$f_{ij} \leq U_{ij} \qquad \text{(upper-bound constraints)}$$
$$-f_{ij} \leq -L_{ij} \qquad \text{(lower-bound constraints)}$$
$$f_{ij} \geq 0 \qquad \text{(nonnegativity constraints)}$$

where for convenience in future development the objective function has been multiplied through by -1 to change the problem from minimization to maximization. We will call this the *primal problem*. A well-known linear programming result is that for every primal problem there exists a corresponding *dual problem*. In this case the dual is given by

$$\text{minimize} \sum_{(i,j)\in \mathbf{S}} U_{ij}\, \alpha_{ij} - L_{ij}\, \delta_{ij}$$

subject to

$$\pi_i - \pi_j + \alpha_{ij} - \delta_{ij} \geq -c_{ij} \qquad \text{all } (i,j) \in \mathbf{S}$$
$$\pi_i \qquad \text{unrestricted for all } i \in \mathbf{N}$$
$$\alpha_{ij} \geq 0 \qquad \text{for all } (i,j) \in \mathbf{S}$$
$$\delta_{ij} \geq 0 \qquad \text{for all } (i,j) \in \mathbf{S}$$

In this formulation the π variables are associated with the conservation-of-flow constraints of the primal problem. The π's are unrestricted because those constraints are equalities. The α variables of the dual problem are associated with the upper-bound constraints of the primal, and the δ variables are associated with the lower-bound constraints. For each primal variable f_{ij} there is a dual constraint.

As an example, suppose that it is desired to ship 3 units of product from source 1 to sink 4 through the network shown in Figure 3.5, where the triplet on each arc (i, j) represents (U_{ij}, L_{ij}, c_{ij}). To solve this problem, the network must be "closed" by adding the return arc (4, 1). The amount of flow that we wish to ship from source node 1 to sink node 4 will be equal to the amount shipped across arc (4, 1). Thus the shipping quantity can be specified by setting $L_{41} = U_{41} = 3$. No cost is associated with that arc, so that $c_{41} = 0$.

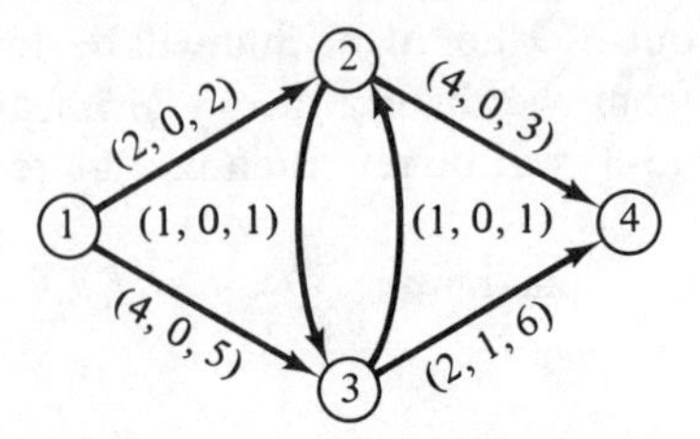

Figure 3.5. Open capacitated network.

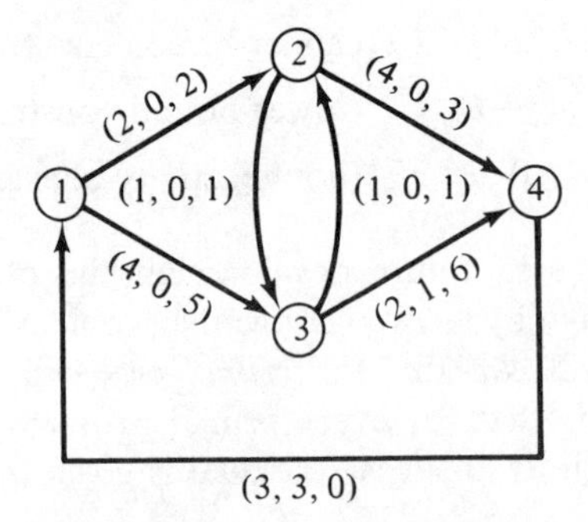

Figure 3.6. Closed capacitated network.

The complete circulation network is represented by Figure 3.6. The primal and dual formulations of this problem are shown in Tables 3.1 and 3.2.

We now address our attentions to the primal and dual formulations in order to find a technique which will:

1. Always yield optimal solutions (when they exist).

Table 3.1. PRIMAL PROBLEM

f_{12}	$+f_{13}$					$-f_{41}$	$=$	0	Nodes
$-f_{12}$		$+f_{23}$	$+f_{24}$	$-f_{32}$			$=$	0	
	$-f_{13}$	$-f_{23}$		$+f_{32}$	$+f_{34}$		$=$	0	
			$-f_{24}$		$-f_{34}$	$+f_{41}$	$=$	0	
f_{12}							$\leq$	2	Upper bounds
	f_{13}						$\leq$	4	
		f_{23}					$\leq$	1	
			f_{24}				$\leq$	4	
				f_{32}			$\leq$	1	
					f_{34}		$\leq$	2	
						f_{41}	$\leq$	3	
					$-f_{34}$		$\leq$	-1	Lower bounds
						$-f_{41}$	$\leq$	-3	
Maximize $-2f_{12} - 5f_{13} - f_{23} - 3f_{24} - f_{32} - 6f_{34}$,								all $f_{ij} \geq 0$	

Table 3.2. DUAL PROBLEM

$$
\begin{array}{lllllllllllll}
\pi_1 & -\pi_2 & & & +\alpha_{12} & & & & & & & & \geq -2 \\
\pi_1 & & -\pi_3 & & & +\alpha_{13} & & & & & & & \geq -5 \\
 & \pi_2 & -\pi_3 & & & & +\alpha_{23} & & & & & & \geq -1 \\
 & \pi_2 & & -\pi_4 & & & & +\alpha_{24} & & & & & \geq -3 \\
 & -\pi_2 & +\pi_3 & & & & & & +\alpha_{32} & & & & \geq -1 \\
 & & \pi_3 & -\pi_4 & & & & & & +\alpha_{34} & & -\delta_{34} & \geq -6 \\
-\pi_1 & & & +\pi_4 & & & & & & & +\alpha_{41} & -\delta_{41} & \geq 0
\end{array}
$$

$$
\text{Minimize} \quad 2\alpha_{12} + 4\alpha_{13} + 1\alpha_{23} + 4\alpha_{24} + 1\alpha_{32} + 2\alpha_{34} + 3\alpha_{41} - \delta_{34} - 3\delta_{41}
$$

$$
\begin{array}{ll}
x_1, x_2, x_3, x_4 & \text{unrestricted} \\
\alpha_{ij} \geq 0 & (i, j) \in \mathbf{S} \\
\delta_{ij} \geq 0 & (i, j) \in \mathbf{S}
\end{array}
$$

2. Yield optimal solutions more efficiently than the general linear programming algorithms.

Conditions for optimality will now be developed using results obtained from primal–dual linear programming theory.

3.3 Fundamental Theorems

Given solutions to the primal and dual problems, these solutions are optimal for their respective problems if and only if

1. Both solutions are feasible.
2. For every positive (nonzero, nonnegative) dual *variable*, the corresponding primal constraint is tight.[1]
3. For every dual *constraint* that is not tight, the corresponding primal *variable* is equal to zero.

The latter two conditions are called the *complementary slackness conditions* and when applied to the minimum-cost circulation problem along with the feasibility constraints, they yield the following necessary and sufficient *conditions for optimality* [7].

Primal feasibility

P_1: $\sum_j f_{ij} - \sum_j f_{ji} = 0$ (conservation of flow) for all $i \in \mathbf{N}$

P_2: $L_{ij} \leq f_{ij} \leq U_{ij}$ (capacity constraints) for all $(i, j) \in \mathbf{S}$

Dual feasibility

D_1: $\pi_i - \pi_j + \alpha_{ij} - \delta_{ij} \geq -c_{ij}$ for all $(i, j) \in \mathbf{S}$

D_2: $\alpha_{ij} \geq 0$ for all $(i, j) \in \mathbf{S}$

D_3: $\delta_{ij} \geq 0$ for all $(i, j) \in \mathbf{S}$

Complementary Slackness

C_1: if $\pi_i - \pi_j + \alpha_{ij} - \delta_{ij} > -c_{ij}$, then $f_{ij} = 0$

C_2: if $\alpha_{ij} > 0$, then $f_{ij} = U_{ij}$

C_3: if $\delta_{ij} > 0$, then $f_{ij} = L_{ij}$

---

[1]A tight constraint is one for which the left-hand side of the constraint equals the right-hand side of the constraint at the given solution.

An equivalent formulation of the conditions for optimality is given by the following relationships [3]:

$$\begin{aligned} &\text{I. If } \pi_j - \pi_i > c_{ij}, && \text{then } \alpha_{ij} > 0 \text{ and } f_{ij} = U_{ij} \\ &\text{II. If } \pi_j - \pi_i < c_{ij}, && \text{then } \delta_{ij} > 0 \text{ and } f_{ij} = L_{ij} \\ &\text{III. If } \pi_j - \pi_i = c_{ij}, && \text{then } U_{ij} \geq f_{ij} \geq L_{ij} \end{aligned}$$

provided that we choose

$$\begin{aligned} &\text{IV. } \alpha_{ij} = \max\,[0; \pi_j - \pi_i - c_{ij}] \\ &\text{V. } \delta_{ij} = \max\,[0; -\pi_j + \pi_i + c_{ij}] \end{aligned}$$

and

$$\text{VI. } \sum_j f_{ij} - \sum_j f_{ji} = 0 \qquad \text{for all } i$$

In terms of seeking the optimum results through successive perturbations of the solution vector, these conditions are very efficient since only conditions I, II, and III need be evaluated, and these do not involve the dual variables α and δ of the relationships IV and V.

Assuming that conditions IV and V are satisfied, and using the definition $\bar{c}_{ij} = c_{ij} + \pi_i - \pi_j$, conditions I, II, III, and VI can be put in a more convenient form:

$$\begin{aligned} &k_1\text{: if } \bar{c}_{ij} < 0, && \text{then } f_{ij} = U_{ij} \\ &k_2\text{: if } \bar{c}_{ij} > 0, && \text{then } f_{ij} = L_{ij} \\ &k_3\text{: if } \bar{c}_{ij} = 0, && \text{then } L_{ij} \leq f_{ij} \leq U_{ij} \\ &k_4\text{: conservation of flow is satisfied} \end{aligned}$$

If two nodes i and j, and their connecting arc, satisfy optimality condition k_1, k_2, or k_3, that arc is said to be *in kilter*. If an arc does not satisfy either k_1, k_2, or k_3 that arc is said to be *out of kilter*. An optimal solution is found when all arcs are in kilter and the conservation-of-flow equations (condition k_4) are satisfied. If no such set of flows exists, the problem has no feasible solution.

3.4 The Out-of-Kilter Algorithm for Solution of Minimal-Cost Circulation Problems

The out-of-kilter algorithm attempts to find values for the π_i's and the f_{ij}'s that satisfy the optimality conditions given by relationships k_1 through k_4. The algorithm can be initiated by assigning to the arcs any set of flows satisfying the conservation of flow constraints and assigning to the nodes an arbitrary set of π_i's. If one looks at individual arcs in terms of the optimality conditions, nine mutually exclusive states of an arc are possible as the algorithm proceeds to optimality (Table 3.3).

Table 3.3. POSSIBLE STATES OF AN ARC

State	$\bar{c}_{ij}$	f_{ij}	*In Kilter?*
α	$\bar{c} > 0$	$f = L$	Yes
β	$\bar{c} = 0$	$L \leq f \leq U$	Yes
δ	$\bar{c} < 0$	$f = U$	Yes
α_1	$\bar{c} > 0$	$f < L$	No
β_1	$\bar{c} = 0$	$f < L$	No
δ_1	$\bar{c} < 0$	$f < U$	No
α_2	$\bar{c} > 0$	$f > L$	No
β_2	$\bar{c} = 0$	$f > U$	No
δ_2	$\bar{c} < 0$	$f > U$	No

By referring to these nine states, systematic changes can be made in arc flows until the optimality conditions are satisfied. The values of $\bar{c}_{ij}$ uniquely indicate whether an arc is in kilter or out of kilter, and also whether or not an increase or decrease in arc flow is needed to bring that particular arc in kilter. Things are not quite as simple as they might seem, however, since the optimality conditions also require conservation of flow. Hence, if flow is increased (or decreased) along a certain arc, say arc (i, j), conservation of flow is violated at the connecting nodes. In order not to violate the conservation-of-flow criterion, *another* path from node j to node i (or from i to j when the flow is reduced) must be found. The arc (i, j) and the path from j to i together form a cycle. Changing the flow on a cycle does not affect the conservation of flow at any node. This alternative path from node j to node i must be chosen such that (i) no in-kilter arcs are thrown out of kilter, and (ii) no out-of-kilter arcs are thrown further out of kilter. Hence, assuming that a path has been found from node j to node i, there should be a functional notation which tells one how to change the flows in all the arcs involved and which path to choose—this is called a *labeling procedure* and proceeds as follows.

3.5 The Labeling Procedure

1. If an arc (i, j) is found in which the flow is to be *increased* to bring that arc in kilter, the arc must be in either state β_1, δ_1, or α_1. Label node j, $[q_j, i^+]$. This says that node j may *receive* q_j additional units *from* node i. If

the arc is in state α_1, then define q_j to be min $[q_j, L_{ij} - f_{ij}]$ and if the arc is in state β_1[2] or δ_1 then define q_j to be min $[q_i, U_{ij} - f_{ij}]$.[3]

2. If the flow along arc (i, j) is to be *decreased*, the arc must be in state β_2, δ_2, or α_2. Label node i, $[q_i, j^-]$. This says that the flow leaving node i and entering node j can be reduced by q_i. If the arc is in state α_2 or β_2,[4] then define q_i as min $[q_j, f_{ij} - L_{ij}]$ and if the arc is in state δ_2 then define q_i as min $[q_j, f_{ij} - U_{ij}]$.

3. If an arc (i, j) is found to be in state α, β, or δ, that arc is in kilter and its flow should not be altered. The exception is state β, in which the flow might be increased or decreased without violating any conditions.

Once an arc (i, j) has been found to be out of kilter, node i or j is labeled by either rule 1 or 2. If the change indicated by the label can be made, the arc can be brought into kilter. However, by previous arguments, an alternative path from node i to j (or j to i) must be found in order that the conservation of flow be maintained. To keep track of the changes needed in each arc along the alternative path, each intermediate node on the connecting path should also be labeled.

Consider an arbitrary arc (x, y) on an alternative path. This intermediate arc will fall in one of the nine mutually exclusive states previously enumerated, and if the arc is out of kilter, a change in flow should never be made which drives it further out of kilter. In like manner, the flow in an arc that is in kilter should never be changed in such a manner as to drive that arc out of kilter. Now, suppose that an observer is standing on an arbitrary labeled node x,[5] and that it is desired to traverse the arc from node x (labeled) to node y (unlabeled).

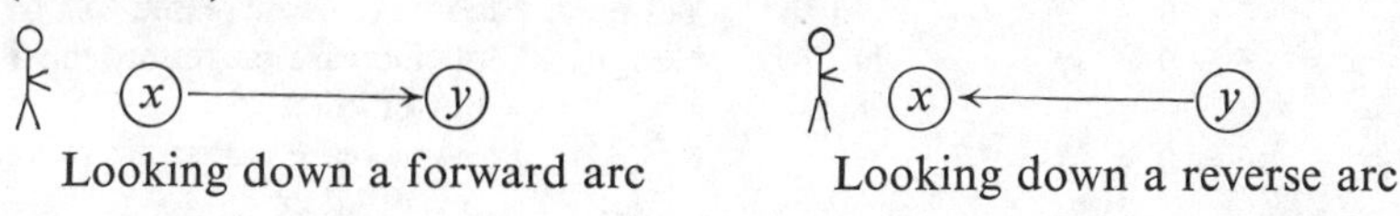

Looking down a forward arc Looking down a reverse arc

From the labeled node, the observer seeks to traverse all forward and reverse arcs incident to that node. If an arc is in the proper state, the node at the other end can be labeled. Several nodes might be labeled from a single node by proceeding in this manner, although only one label is needed to proceed. Once the labeling process has been completed from a given node, that node

[2]Note that even though an increase in flow to the lower bound will bring the arc in kilter, since $\bar{c} = 0$, the flow can actually be increased to the upper bound without driving the arc out of kilter.

[3]These rules will always attempt to put arc (i, j) in kilter while at the same time approaching optimality in a convergent sequence.

[4]Same logic as above.

[5]Since the process always starts at a given node, one node will always be labeled initially.

is marked *scanned*. The observer then moves to a labeled but *unscanned* node. The question that you need to ask as you label *unscanned* nodes on incident arcs is: "Should I *decrease* or *increase* flow along this particular arc?" After this change has been determined, the change in flow along that arc can be indicated. At this point, you may proceed from a scanned node to an unscanned but labeled node. The amount by which you can change the flow in arc (x, y) can be determined from the unique state of that arc, and the decision rules by which the change is made are given in Tables 3.4 and 3.5.

Table 3.4. LABELING PROCESS THROUGH A FORWARD ARC

$$(x, y) \in \mathbf{S} \quad \overset{\pi_x}{\underset{x}{\circ}} \xrightarrow[f_{xy}]{c_{xy}} \overset{\pi_y}{\underset{y}{\circ}}$$

$$\bar{c}_{xy} = c_{xy} + \pi_x - \pi_y$$

Suppose that x is labeled $[q_x, Z^{\pm}]$; can y be labeled? Never label y from x if an *increase* in the flow f_{xy} will make the arc more out of kilter.

State of Arc (x, y)	$\bar{c}_{xy}$	f_{xy}	*In Kilter?*	*Can y Be Labeled?*	*Why?*
α	$\bar{c} > 0$	$f = L$	Yes	No	Flow increase would make the arc out of kilter
β	$\bar{c} = 0$	$f < U$	Yes	Yes	Flow may be increased to U
		$f = U$	Yes	No	Flow cannot be increased
δ	$\bar{c} < 0$	$f = U$	Yes	No	Flow cannot be increased
α_1	$\bar{c} > 0$	$f < L$	No	Yes	Flow may be increased to L
β_1	$\bar{c} = 0$	$f < L$	No	Yes	Flow may be increased to U
δ_1	$\bar{c} < 0$	$f < U$	No	Yes	Flow may be increased to U
α_2	$\bar{c} > 0$	$f > L$	No	No	Flow increase makes arc more out of kilter
β_2	$\bar{c} = 0$	$f > U$	No	No	Flow increase makes arc more out of kilter
δ_2	$\bar{c} < U$	$f > U$	No	No	Flow increase makes arc more out of kilter

Summary: Label y, $[q_y, X^+]$ if $\bar{c}_{xy} < 0$ and $f_{xy} < L_{xy}$;
$q_y = \min [q_x, L_{xy} - f_{xy}]$
$\bar{c}_{xy} \leq 0$ and $f_{xy} < U_{xy}$;
$q_y = \min [q_x, U_{xy} - f_{xy}]$

The problem conceptually now appears to be solved; that is, an out-of-kilter arc (i, j) is chosen, and the flow across that arc is adjusted to bring the arc in kilter. To preserve conservation of flow, an alternative path from node j to node i (or from i to j) is sought, and, once found, the flows along this path are adjusted according to what the final label q_i (or q_j) indicates. Consider a wandering walker that starts on a journey from node j, passing across forward and reverse arcs through scanned nodes approaching node i. Recall

Table 3.5. LABELING PROCESS THROUGH A REVERSE ARC

$(y, x) \in \mathbf{S}$ π_x c_{yx} π_y, arc from y to x with flow f_{yx}

$\bar{c}_{yx} = c_{yx} + \pi_y - \pi_x$

Suppose that x is labeled $[q_x, Z^{\pm}]$ can y be labeled? Never label y from x if a *decrease* in the flow f_{yx} will make the arc more out of kilter.

State of Arc (y, x)	$\bar{c}_{yx}$	f_{yx}	*In Kilter?*	*Can y Be Labeled?*	*Why?*
α	$\bar{c} > 0$	$f = L$	Yes	No	Flow decrease would drive arc out of kilter
β	$\bar{c} = 0$	$L \leq f \leq U$	Yes	Yes	Flow can be decreased by $f - L$
		$L = f$	Yes	No	Flow cannot be decreased
δ	$\bar{c} < 0$	$f = U$	Yes	No	Flow cannot be decreased
α_1	$\bar{c} > 0$	$f < L$	No	No	Flow decrease would make arc more out of kilter
β_1	$\bar{c} = 0$	$f < L$	No	No	Flow decrease would make arc more out of kilter
δ_1	$\bar{c} < 0$	$f < U$	No	No	Flow decrease would make the arc more out of kilter
α_2	$\bar{c} > 0$	$f > L$	No	Yes	Flow may be decreased by $f - L$
β_2	$\bar{c} = 0$	$f > U$	No	Yes	Flow may be decreased by $f - L$
δ_2	$\bar{c} < 0$	$f > U$	No	Yes	Flow may be decreased by $f - U$

Summary: Label y, $[q_y, X^-]$ if $\bar{c}_{yx} \geq 0$ and $f_{yx} > L_{yx}$; $q_y = \min [q_x, f_{yx} - L_{yx}]$
$\bar{c}_{yx} < 0$ and $f_{yx} > U_{yx}$; $q_y = \min [q_x, f_{yx} - U_{yx}]$

that no change is allowed in the flow across an arc if that change either (1) forces an in-kilter arc out of kilter or (2) drives an out-of-kilter arc further out of kilter. Hence, it is possible that the walker reaches a node at which he is unable to continue. It would appear that the "walk" is over and no path from j to i can be found—hence no solution would exist to the given network-flow problem. It would serve no purpose to move to another out-of-kilter arc and start the process anew, since eventually the aforementioned arc must be chosen again before optimality can be achieved. Such an event is called *non-breakthrough*. Fortunately, if non-breakthrough occurs there is one more alternative available in the search for an optimal network flow. Recall that the state of an arc is uniquely determined by checking $\bar{c}_{ij} = c_{ij} + \pi_i - \pi_j$, so that a change in the π values affects the nine possible states of an arc. According to the procedure by which the dual problem was formerly established, each and every node has a π value associated with it, so that there are exactly n dual π variables for a network-flow problem with n nodes. The question that now presents itself, assuming that non-breakthrough has occurred, is: which π variables should be changed to provide a possible path from node j to node i (or from i to j if flow is reduced)? Recall that the walker has started at node j and "walked" through a series of forward and reverse arcs through scanned nodes attempting to reach node i. As he walks, the nodes through which he travels have been labeled according to previous rules.[6] Hence, at non-breakthrough there are two mutually exclusive sets of nodes: (1) scanned labeled nodes and (2) unlabeled nodes. Obviously, the only nodes that we are interested in at non-breakthrough are nodes that will enable us to complete a walk between nodes j and node i. Logically, to continue we must walk from a scanned labeled node (since we can at least get that far) to an unlabeled node. Therefore, the only π numbers that need be considered are those associated with the arcs connecting labeled nodes to unlabeled nodes. *Define the set of all labeled nodes to be* **A**, *and the set of all unlabeled nodes to be* $\bar{\mathbf{A}}$.

If we are standing on an arbitrary scanned labeled node x, looking toward an unlabeled node y; we will either be looking at a forward arc or a reverse arc. If we are looking at a forward arc, flow can pass from **A** to $\bar{\mathbf{A}}$, and if we are looking across a reverse arc, flow can pass from $\bar{\mathbf{A}}$ into **A**.

***Case I*:** Let **B** be the set of all arcs originating at a node in **A** and terminating at a node in $\bar{\mathbf{A}}$ with $\bar{c} > 0$ and flow less than or equal to the upper bound.

***Case II*:** Let $\bar{\mathbf{B}}$ be the set of all arcs originating at a node in $\bar{\mathbf{A}}$ and terminating at a node in **A** with $\bar{c} < 0$ and flow greater than or equal to the lower bound.

Since $\bar{c}$ can be calculated for every arc in sets **B** and $\bar{\mathbf{B}}$, one should proceed as follows:

1. Case I: For any $\bar{c} > 0$.
   Define $\zeta_1 = \min_{\mathbf{B}} [\bar{c}_{xy}]$ if $\mathbf{B} \neq \varnothing$; otherwise, $\zeta_1 = \infty$.
2. Case II: For any $\bar{c} < 0$.
   Define $\zeta_2 = \min_{\bar{\mathbf{B}}} [-\bar{c}_{yx}]$ if $\bar{\mathbf{B}} \neq \varnothing$; otherwise, $\zeta_2 = \infty$.
3. Let $\zeta = \min [\zeta_1, \zeta_2]$.
4. Change all node numbers (π values) in the set $\bar{\mathbf{A}}$ by adding ζ to every π_k, where k is a member of the set $\bar{\mathbf{A}}$.
5. Do not erase any previous labels.

The process then continues by returning to the labeling procedure.

---

[6]Either node i or j is among this set of labeled nodes, since one of these nodes was labeled initially.

If the preceding steps change the state of at least one arc leading from the set of all labeled nodes to the set of all unlabeled nodes, the labeling procedure is continued until (1) breakthrough occurs or (2) non-breakthrough occurs again. If the preceding steps do not change the state of at least one arc leading from $\mathbf{A}$ to $\bar{\mathbf{A}}$, a feasible solution cannot be found. When breakthrough occurs, an alternative path from node j to node i (or from i to j) has been found, and all that is necessary is to retrace this path, changing the flow in each arc on that path as indicated by the final label q_i (or q_j) at each labeled node along that path. At this point all labels are erased, another out-of-kilter arc is chosen, and the procedure begins again. An optimal solution to the network-flow problem has been found when all arcs are in kilter.

In the event that breakthrough does not occur, sets $\mathbf{B}$ and $\bar{\mathbf{B}}$ are reestablished and the node numbers are once again changed according to the previous rules. The labeling process is repeated until either arc (i, j) is in kilter or until a non-breakthrough occurs at which $\zeta = \infty$. In the case where $\zeta = \infty$, there is no optimal solution to the problem as stated and the algorithm terminates.[7] Note that if $\bar{c} = 0$ for each arc, all forward arc flows are at U_{ij}, and all reverse arc flows are at L_{ij}; then no path can be found.

The out-of-kilter algorithm can now be summarized as follows. The network is established and an initial circulation is chosen which satisfies the conservation-of-flow equations. A circulation of zero will always satisfy this condition. Next an arbitrary set of π numbers is assigned to the nodes. By using the labeling procedure, flows are adjusted when breakthrough occurs. Otherwise, new node numbers are found, and the procedure is repeated. The next section will attempt to clarify the algorithmic procedures through a graphical interpretation of the previous discussions.

3.6 A Graphical Interpretation of the Out-of-Kilter Algorithm

The purpose of this section is to develop and discuss a graphical approach that reproduces the corrective actions taken by the out-of-kilter algorithm. The logic of the procedure and its simplicity make this mnemonic device particularly attractive from a computational point of view.

At any stage of the out-of-kilter method, the solutions can be represented by points $(f_{ij}, \bar{c}_{ij})$. The state of any arc (i, j) can be identified by examining the location of the point relative to the corresponding bounds L_{ij} and U_{ij}, as shown in Figure 3.7.

---

[7]Necessary and sufficient conditions for convergence will not be presented here. They are given in references [7] and [3].

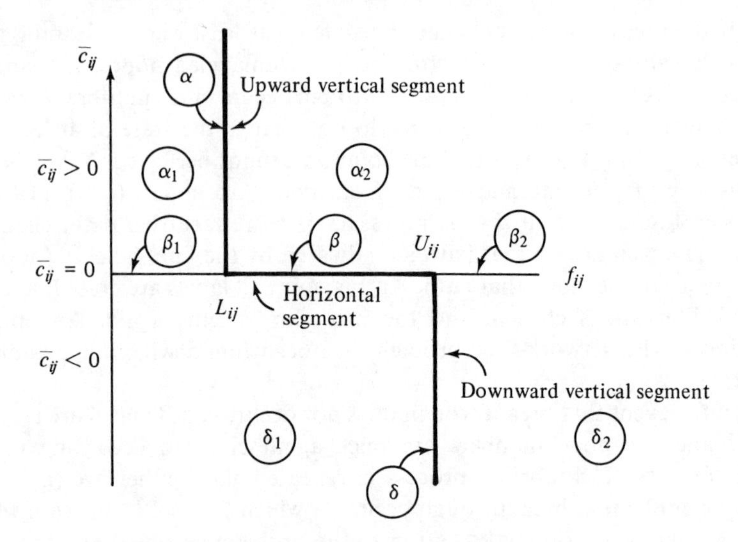

Figure 3.7. Mnemonic device for the out-of-kilter method.

If the arc is in kilter, the point $(f_{ij}, \bar{c}_{ij})$ falls on the dark solid three-segment line shown in Figure 3.7. The upward vertical segment represents the in-kilter state α, the downward vertical segment represents the in-kilter state δ, and the horizontal segment represents the in-kilter state β.

If the arc is out of kilter, the corresponding point falls somewhere else, so we will attempt to take corrective action. There are two categories of possible corrective action:

1. Modifications of flows (f_{ij}'s).
2. Modification of adjusted costs ($\bar{c}_{ij}$'s).

The modification of flows (the labeling procedure) corresponds to horizontal displacements in the graph of Figure 3.7 and the modification of adjusted costs (change of π values) to vertical displacements. The directions of valid horizontal moves are shown in Figure 3.8(a) and (b), for each of the six out-of-kilter states. Valid vertical displacements are shown in Figure 3.8(c) and (d).

The following rules can be used to control the size of the horizontal and vertical displacements. The rules for the horizontal displacements are equivalent to those for the labeling procedure. The rules for the vertical displacements are equivalent to those for changing the dual π values.

3.6.1 Horizontal Displacements

Let us consider any arc (i, j) whose condition is out of kilter. Let this arc be represented by the point $(f_{ij}, \bar{c}_{ij})$. This point can lie to the left or right of either of the three segments of the in-kilter line.

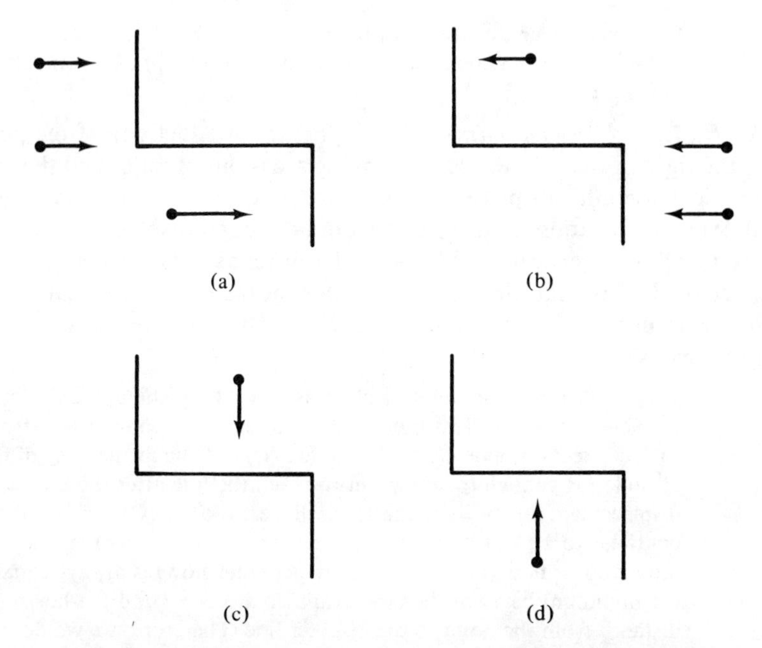

Figure 3.8. Valid directions for horizontal and vertical displacements.

1. *Horizontal Displacements to the Right* (Flow Augmentation). If the point lies to the left of the line, the corresponding arc is out of kilter and the only acceptable horizontal displacement is to the right, as illustrated in Figure 3.8(a). Any modification of flow resulting in horizontal displacements to the left is not allowed, since it would worsen the current status of the arc. In the context of the labeling procedure, this means that node i cannot be labeled from node j, (i, j) being a reverse arc. The following observations apply to forward arcs.

  a. If $\bar{c}_{ij} > 0$, move the point as close as possible to the upward vertical segment of the line. If there is enough flow to reach the line, the arc is now in kilter. If there is not sufficient flow, the arc will continue to be out of kilter but horizontally closer to the line. The increase in flow is always equal to the minimum between the flow available at node i and the size of the step needed to reach the upward vertical segment. If q_i is the amount of flow available at node i, label node j by $[q_j, i^+]$, where $q_j = \min [q_i, L_{ij} - f_{ij}]$.
  b. If $\bar{c}_{ij} = 0$, move the point to the right as much as the available flow permits, without exceeding the upper bound U_{ij}. In this case node j can be labeled by $[q_j, i^+]$, where $q_j = \min [q_i, U_{ij} - f_{ij}]$.
  c. If $\bar{c}_{ij} < 0$, move the point to the right as much as possible according to the available flow, but without exceeding the upper bound U_{ij}. If the downward vertical segment is reached, the arc will be in kilter.

Otherwise, it will continue to be out of kilter but horizontally closer to the line. In the present case, we label node j by $[q_j, i^+]$, where $q_j = \min[q_i, U_{ij} - f_{ij}]$.

2. *Horizontal Displacements to the Left* (Flow Reduction). If the point lies to the right of the line, the corresponding arc is out of kilter and the only acceptable horizontal displacement is to the left, as illustrated in Figure 3.8(b). Any modification of flow resulting in horizontal displacements to the right is not allowed, since it would worsen the current status of the arc. Again, in the context of the labeling procedure, this means that node j cannot be labeled from node i, (i, j) being a forward arc. The following observations apply to reverse arcs.

a. If $\bar{c}_{ij} > 0$, move the point as close as possible to the upward vertical segment of the line. This is equivalent to sending a counter flow from j to i in direction opposite to that of arc (i, j). If the availability of flow at node j is enough for the point to reach the line after the horizontal displacement to the left, the arc will be in kilter. Otherwise, it will continue to be out of kilter but horizontally closer to the line. The reduction of flow (i.e., the size of the counter flow) is always equal to the minimum between the flow available at node j and the horizontal distance from the point to the in-kilter line. Therefore, we would label i from j by $[q_i, j^-]$, where $q_i = \min[q_i, f_{ij} - L_{ij}]$.

b. If $\bar{c}_{ij} = 0$, move the point to the left as much as the available flow permits, without placing the point on the other side of the upward vertical segment of the line. In this case node i can be labeled by $[q_i, j^-]$, where $q_i = \min[q_j, f_{ij} - L_{ij}]$.

c. If $\bar{c}_{ij} < 0$, move the point to the left as much as the available flow permits, without placing the point on the other side of the downward vertical segment. If this segment is reached, the arc will be in kilter. Otherwise, it will continue to be out of kilter but horizontally closer to the line. In this case we label i by $[q_i, j^-]$, where $q_i = \min[q_j, f_{ij} - U_{ij}]$.

3.6.2 Vertical Displacements

As illustrated in Figure 3.8(c) and (d), vertical displacements are successful only when the final position of the point lies in the horizontal segment of the in-kilter line. If we assume that node i is labeled and node j unlabeled, a point with positive ordinate in the graph represents a directed arc from node i to node j. If the ordinate is negative, the point represents a directed arc from node j to node i. Therefore, two cases can be considered:

1. $\bar{c}_{ij} > 0$ and $f_{ij} \leq U_{ij}$.
2. $\bar{c}_{ji} < 0$ and $f_{ji} \geq L_{ji}$.

In case 1 a downward vertical displacement would put the point closer to the horizontal segment of the line, as can be seen in Figure 3.8(c). Recall that by definition $\bar{c}_{ij} = c_{ij} + \pi_i - \pi_j$. Hence, $0 = c_{ij} + \pi_i - (\pi_j + \bar{c}_{ij})$. This implies that for the point to reach the line we must increase the node number π_j by $\bar{c}_{ij}$.

In case 2 an upward vertical displacement would put the point closer to the horizontal segment of the line, as can be seen in Figure 3.8(d). From the definition $\bar{c}_{ji} = c_{ji} + \pi_j - \pi_i$ we obtain $0 = c_{ji} + (\pi_j - \bar{c}_{ji}) - \pi_i$, which implies that the node number π_j must be decreased by $\bar{c}_{ji}$ for the point to reach the line after the vertical displacement.

Let **B** be the set of points representing arcs in the conditions described in case 1 above. We want to displace all these points vertically downward without causing any ordinate to be negative. To accomplish this we can add to each π_j the minimal ordinate of the points under consideration. That is, add $\zeta_1 = \min_{\mathbf{B}} [\bar{c}_{ij}]$ to each π_j.

Similarly, if $\bar{\mathbf{B}}$ is defined as the set of points representing arcs in the conditions given in case 2, we want to displace all these points vertically upward without causing any ordinate to be positive. To accomplish this we can subtract from each node number π_j the maximal ordinate of the points under consideration. That is, add $\zeta_2 = -\max_{\bar{\mathbf{B}}} [\bar{c}_{ji}] = \min_{\bar{\mathbf{B}}} [-\bar{c}_{ji}]$ to each π_j.

If the node numbers π_j's of *all* the unlabeled nodes are to be changed by adding the same amount, we would add $\zeta = \min [\zeta_1, \zeta_2]$ to each π_j. The final conclusion is that all points in **B** and $\bar{\mathbf{B}}$ are brought closer to the line by moving them vertically a step of size equal to the minimal distance from the line. This procedure guarantees that at least one of the out-of-kilter arcs becomes in kilter, provided that **B** and $\bar{\mathbf{B}}$ are not empty.

3.7 Algorithmic Steps

***Step 1*:** Find an out-of-kilter arc, (i, j). If none, stop.

***Step 2*:** Determine if the flow in the arc should be increased or decreased to bring the arc in kilter. If it should be increased, go to step 3. If it should be decreased, go to step 4.

***Step 3*:** Find a path in the network, using the labeling algorithm, from j to i along which the flow can be passed without causing any arcs on the path to become further out of kilter. If a path is found, adjust the flow in the path and increase the flow in (i, j). If (i, j) is now in kilter, go to step 1. If it is still out of kilter, repeat step 3. If no path can be found, go to step 5.

***Step 4*:** Find a path from i to j along which the flow can be passed without causing any arcs to become further out of kilter. If a path is found, adjust the flow in the path and decrease the flow in (i, j). If (i, j) is now in kilter, go to step 1. If (i, j) is still out of kilter, repeat step 4. If no path is found, go to step 5.

Step 5: Change the π values and repeat step 2 for arc (i, j) keeping the same labels on all arcs already labeled. If the node numbers become ∞, stop. There is no feasible flow.

3.8 A Numerical Example

The out-of-kilter algorithm will now be used to completely solve the minimum-cost circulation problem given in Figure 3.6. Only two steps need to be taken in order to apply the algorithm; (1) choose initial values for the dual variables, π_i, $i = 1, 2, 3, 4$; and (2) choose an initial network flow that satisfies the conservation of flow constraint. In performing (1) and (2), it is not necessary that a feasible solution be chosen, only that these variables be explicitly defined. For convenience, the π values can be chosen as $\pi_1 = \pi_2 = \pi_3 = \pi_4 = 0$. A set of network flows satisfying (2) is given by $f_{12} = 0$, $f_{13} = 2$, $f_{23} = 0$, $f_{32} = 2$, $f_{24} = 2$, $f_{34} = 0$, $f_{41} = 2$.[8]

3.8.1 A Numerical Example for the Out-of-Kilter Algorithm

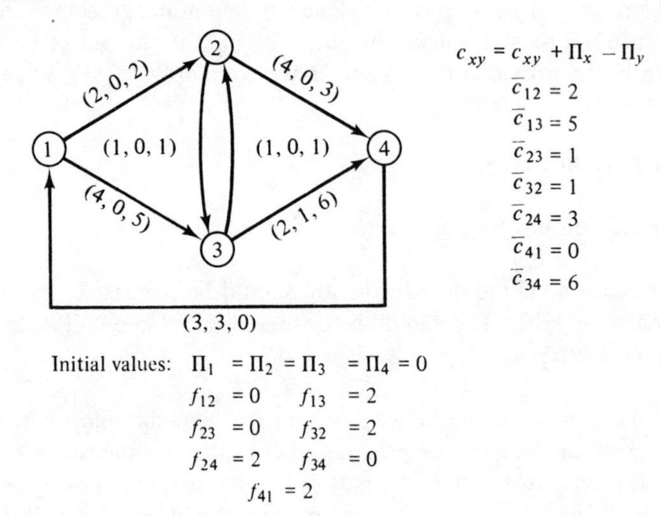

[8]It is intuitively obvious that if near-optimal solutions are chosen initially, the algorithm will terminate more rapidly. These were chosen simply for illustrative purposes. Note also that the solution is not feasible.

Arc (x, y)	$\bar{c}_{xy}$	f_{xy}	*State* (x, y)			*In Kilter?*
(1, 2)	2	0	$\bar{c} > 0$	$f = L$	α	Yes
(1, 3)	5	2	$\bar{c} > 0$	$f > L$	α_2	No
(2, 3)	1	0	$\bar{c} > 0$	$f = L$	α	Yes
(2, 4)	3	2	$\bar{c} > 0$	$f > L$	α_2	No
(3, 2)	1	2	$\bar{c} > 0$	$f > L$	α_2	No
(3, 4)	6	0	$\bar{c} > 0$	$f < L$	α_1	No
(4, 1)	0	2	$\bar{c} = 0$	$f < L$	β_1	No

Iteration 1

1. Pick out-of-kilter arc (4, 1).[9]
2. State of arc (4, 1) is β_1; so increase flow by 1 unit to the lower bound = 3.
3. Find path from 1 to 4 by labeling procedure.

Labeling procedure

Node	*Label*
1	[1, 4⁺] Node 1 is now labeled
2	Cannot be labeled (in kilter)
3	Cannot be labeled (a flow increase would drive the arc further out of kilter)

Non-breakthrough has occurred.

Sets of nodes: $\mathbf{A} = \{1\}$, $\bar{\mathbf{A}} = \{2, 3, 4\}$

Arc subsets: $\mathbf{B} = \{(1, 2), (1, 3)\}$

$$\bar{\mathbf{B}} = \varnothing$$

$$\zeta_1 = \min_{\mathbf{B}} [2, 5] = 2$$

$$\zeta_2 = \min_{\bar{\mathbf{B}}} [\varnothing] = \infty$$

$$\therefore \quad \zeta = 2$$

New dual variables: $\pi_1 = 0; \pi_3 = 2$

$$\pi_2 = 2; \pi_4 = 2$$

[9]Any out-of-kilter arc could have been chosen at this point.

Iteration 2

Arc (x, y)	$\bar{c}_{xy}$	f_{xy}	*State* (x, y)			*In Kilter?*
(1, 2)	0	0	$\bar{c} = 0$	$f = L$	β	Yes
(1, 3)	3	2	$\bar{c} > 0$	$f > L$	α_2	No
(2, 3)	1	0	$\bar{c} > 0$	$f = L$	α	Yes
(2, 4)	3	2	$\bar{c} > 0$	$f > L$	α_2	No
(3, 2)	1	2	$\bar{c} > 0$	$f > L$	α_2	No
(3, 4)	6	0	$\bar{c} > 0$	$f < L$	α_1	No
(4, 1)	2	2	$\bar{c} > 0$	$f < L$	α_1	No

1. Arc (4, 1) is still out of kilter.
2. State of (4, 1) is α_1; so increase flow to lower bound = 3.
3. Find path from 1 to 4 by labeling procedure.

Labeling procedure

Node	*Label*	
1	[1, 4+]	
2	[1, 1+]	
3	[1, 2−]	Flow can be decreased to upper bound without making arc further out of kilter
4	[1, 3+]	

Breakthrough has occurred.
Change flows in network as indicated above.

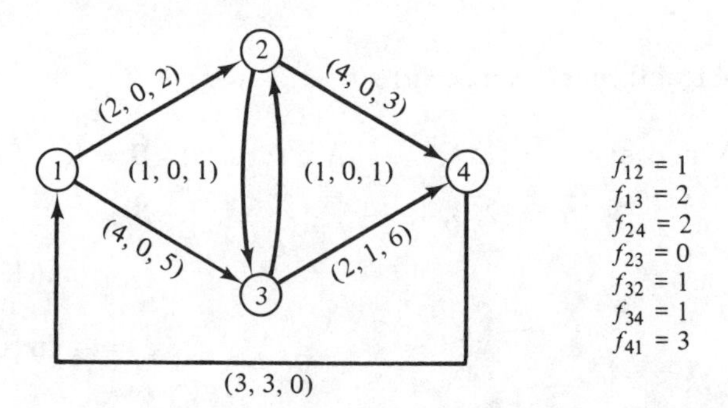

Arc (x, y)	$\bar{c}_{xy}$	f_{xy}	*State* (x, y)			*In Kilter?*
(1, 2)	0	1	$\bar{c} = 0$	$L < f < U$	β	Yes
(1, 3)	3	2	$\bar{c} > 0$	$f > L$	α_2	No
(2, 3)	1	0	$\bar{c} > 0$	$f = L$	α	Yes
(2, 4)	3	2	$\bar{c} > 0$	$f > L$	α_2	No
(3, 2)	1	1	$\bar{c} > 0$	$f = U$	α_2	No
(3, 4)	6	1	$\bar{c} > 0$	$f = L$	α	Yes
(4, 1)	2	3	$\bar{c} > 0$	$f = L$	α	Yes

Iteration 3

1. Pick an out-of-kilter arc, say arc (1, 3).
2. State of arc is α_2; so decrease flow to lower bound; decrease by 2.
3. Find path from 1 to 3 by labeling procedure.

Labeling procedure

Node	*Label*	
1	[2, 3⁻]	
2	[1, 1⁺]	Even though in kilter, flow can be increased in arc (1, 2) by 1 without driving it out of kilter
3	[1, 2⁻]	

Breakthrough accomplished.
Change network flows as indicated by the labeling procedure.

Arc (x, y)	$\bar{c}_{xy}$	f_{xy}	*State (x, y)*			*In Kilter?*
(1, 2)	0	2	$\bar{c} = 0$	$f = U$	β	Yes
(1, 3)	3	1	$\bar{c} > 0$	$f > L$	α_2	No
(2, 3)	1	0	$\bar{c} > 0$	$f = L$	α	Yes
(2, 4)	3	2	$\bar{c} > 0$	$f > L$	α_2	No
(3, 2)	1	0	$\bar{c} > 0$	$f = L$	α	Yes
(3, 4)	6	1	$\bar{c} > 0$	$f = L$	α	Yes
(4, 1)	2	3	$\bar{c} > 0$	$f = L$	α	Yes

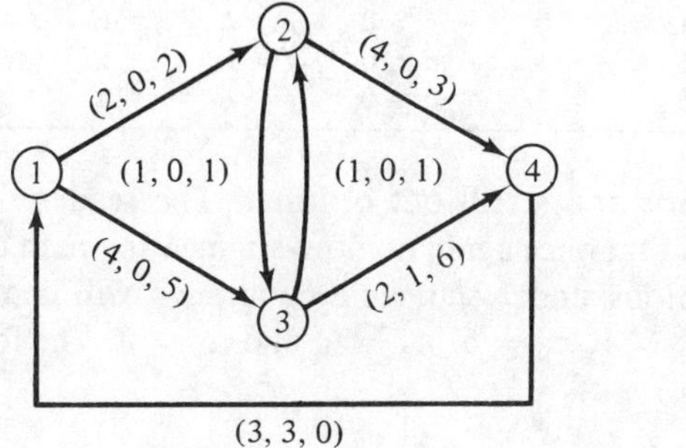

$f_{12} = 2$
$f_{13} = 1$
$f_{24} = 2$
$f_{23} = 0$
$f_{32} = 0$
$f_{34} = 1$
$f_{41} = 3$

We still have two arcs out of kilter.

Iteration 4

1. Pick an out-of-kilter arc, say arc (1, 3).
2. State of arc is α_2; so decrease flow to lower bound: decrease flow by 1 unit.
3. Find path from 1 to 3 by labeling procedure.

Labeling procedure

Node	*Label*
1	$[1, 3^-]$
2	Cannot be labeled
4	Cannot be labeled

Non-breakthrough has occurred.

Sets of nodes: $\mathbf{A} = \{1\}$ Arc subsets: $\mathbf{B} = \{(1, 3)\}$

$\mathbf{A} = \{2, 3, 4\}$ $\bar{\mathbf{B}} = \varnothing$

$$\zeta_1 = \min_{\mathbf{B}} [3] = 3$$

$$\zeta_2 = \min_{\bar{\mathbf{B}}} [\varnothing] = \infty$$

$$\therefore \quad \zeta = 3$$

New dual variables: $\pi_1 = 0; \pi_2 = 5$

$\pi_3 = 5; \pi_4 = 5$

Iteration 5

Arc (x, y)	$\bar{c}_{xy}$	f_{xy}	*State* (x, y)			*In Kilter?*
(1, 2)	−3	2	$\bar{c} < 0$	$f = U$	δ	Yes
(1, 3)	0	1	$\bar{c} = 0$	$U > f > L$	β	Yes
(2, 3)	1	0	$\bar{c} > 0$	$f = L$	α	Yes
(2, 4)	3	2	$\bar{c} > 0$	$f > L$	α_2	No
(3, 2)	1	0	$\bar{c} > 0$	$f = L$	α	Yes
(3, 4)	6	1	$\bar{c} > 0$	$f = L$	α	Yes
(4, 1)	5	3	$\bar{c} > 0$	$f = L$	α	Yes

At this point, only one arc is still out of kilter. The reader should at this point be able to verify that once again no flow-augmenting path can be found. Two successive iterations under the same conditions will change the dual variables to values $\pi_1 = 1$, $\pi_2 = 5$, $\pi_3 = 6$, and $\pi_4 = 8$. The following conditions will now exist.

Arc (x, y)	$\bar{c}_{xy}$	f_{xy}	*State* (s, y)			*In Kilter?*
(1, 2)	−2	2	$\bar{c} < 0$	$f = U$	δ	Yes
(1, 3)	0	1	$\bar{c} = 0$	$U > f > L$	β	Yes
(2, 3)	0	0	$\bar{c} = 0$	$f = L$	β	Yes
(2, 4)	0	2	$\bar{c} = 0$	$U > f > L$	β	Yes
(3, 2)	2	0	$\bar{c} > 0$	$f = L$	α	Yes
(3, 4)	4	1	$\bar{c} > 0$	$f = L$	α	Yes
(4, 1)	7	3	$\bar{c} > 0$	$f = L$	α	Yes

Since each arc is now in kilter, the entire network is in kilter and all conditions for optimality have now been satisfied.

Optimal (minimum-cost) circulation: $f^*_{41} = 3$

$$f^*_{12} = 2 \qquad f^*_{32} = 0$$

$$f^*_{13} = 1 \qquad f^*_{34} = 1$$

$$f^*_{23} = 0 \qquad f^*_{24} = 2$$

Optimal (minimal) cost is 21.

3.9 Summary and Conclusions

The theory of the out-of-kilter algorithm has been discussed in a heuristic setting, using the well-known theory of dual linear programming. All the steps leading to an optimal solution have been examined, and a logical explanation was attempted to justify why those steps were taken. Tables have been presented summarizing the various changes that are possible as the algorithm proceeds, along with step-by-step procedures to be employed when attacking the problem. An example problem was presented and solved using the out-of-kilter algorithm.

The primary purpose of this section was to present in a logical and organized fashion the basic theory and methodology employed in the basic algorithm. Although the solution procedure can become quite tedious for large networks, it is well defined and has been computerized. A computer program for the out-of-kilter algorithm will be given in Part III. Although the out-of-kilter algorithm can be used to solve a wide variety of problems, the solution procedures and the criterion for optimality will not be altered for different problems, only the network configuration. Hence, the power of the out-of-kilter algorithm is revealed in two important concepts:

1. It will effectively solve a wide range of network-flow problems.
2. The algorithm does not require a feasible solution to begin the optimizing procedure, only one in which the conservation-of-flow equations are satisfied. (The null vector will always satisfy the algorithm.)

A residual benefit common to all network-flow procedures is that the problem can be easily visualized, a property not present in linear programming formulations in more than two dimensions.

PART II: NETWORK-FLOW OPTIMIZATION WITH THE OUT-OF-KILTER ALGORITHM—MODELING CONCEPTS

The primary objective of this section is to demonstrate the applicability of the out-of-kilter algorithm to generalized capacitated network-flow problems. The complete network structures for solving the following capacitated network-flow problems are discussed:

1. The transportation problem.
2. The assignment problem.
3. Maximum-flow problems.
4. The shortest-path tree.
5. The transshipment problem.
6. Minimum cost/Maximum flow problems.

In addition, a production planning problem is formulated and solved using the out-of-kilter algorithm. An economic interpretation of the dual solution variables is also attempted relevant to the production planning problem. In Part III, a computer code useful in solving moderate- to medium-sized problems (nodes ≤ 500, arcs ≤ 500) will be presented and used to solve an example.

3.10 Problem Solution Using the Out-of-Kilter Algorithm

To apply the out-of-kilter algorithm, two steps are essential in the problem formulation:

1. Proper representation of the problem as a closed-loop capacitated network-flow problem.
2. Initial values for the dual node variables, π_k, and an initial circulation that satisfies the conservation-of-flow equations.

With regard to step 1, consider an arbitrary network configuration starting with a single sink node and a single source node (which may, in fact, be super source and super sink nodes). The proper configuration is given in Figure 3.9.

In order to form a closed-loop system, an additional arc must always be added connecting node t to node s. Denote this arc as the return arc. This return arc will exhibit the same characteristics as every other arc in the network, namely, that it will have a capacity–cost triplet (U, L, c) and associated

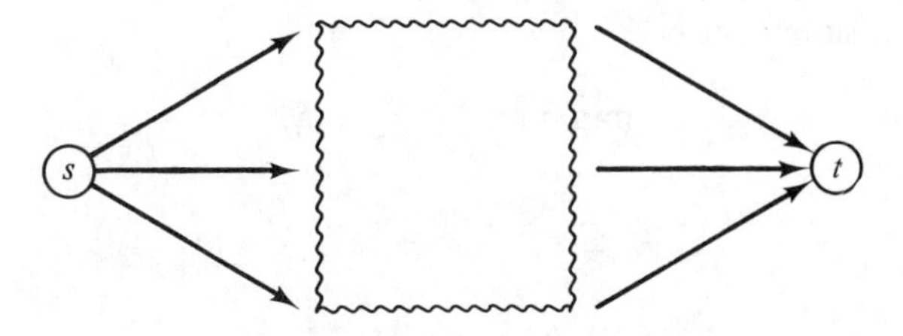

Figure 3.9. Single source and sink.

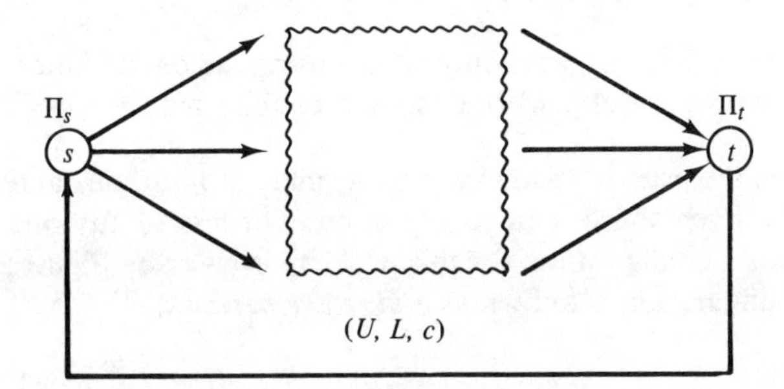

Figure 3.10. Network form and notation.

dual variables π_s and π_t.[10] A complete network configuration would then appear as in Figure 3.10. The values of (U, L, c), π_s, and π_t will subsequently be discussed for specific out-of-kilter examples.

3.11 The Transportation Problem

Suppose that there are m warehouses and n sales outlets. The supply of a particular item is known at each warehouse, and the demand for a particular item is known for each outlet. The cost of shipping 1 unit from each warehouse to each outlet is also known. The object is to find the shipping policy from warehouses to outlets that minimizes the total shipping cost, while at the same time meeting all demands.

Let

f_{ij} = amount shipped from warehouse i to outlet j

a_i = supply at warehouse i

b_j = demand at outlet j

c_{ij} = cost of shipping one item from warehouse i to outlet j

[10] Actually, the variables π_t and π_s are not arc characteristics, but unique node characteristics. They have only been included here for clarity and completeness.

The problem statement is

$$\text{minimize} \sum_{i=1}^{m} \sum_{j=1}^{n} c_{ij} f_{ij}$$

subject to

$$\sum_{j=1}^{n} f_{ij} \leq a_i \qquad i = 1, 2, \ldots, m$$

$$\sum_{i=1}^{m} f_{ij} \geq b_j \qquad j = 1, 2, \ldots, n$$

$$f_{ij} \geq 0 \qquad \text{all } i, \text{ all } j$$

In order to solve a transportation problem using the out-of-kilter algorithm, the following procedural guidelines should be followed.

1. There are exactly m sources and n sinks, or m origins and n destinations. Each source can supply a_i units or less to any one sink. The network configuration for this problem is given by Figure 3-11. Such a configuration is known as a *bipartite network*.

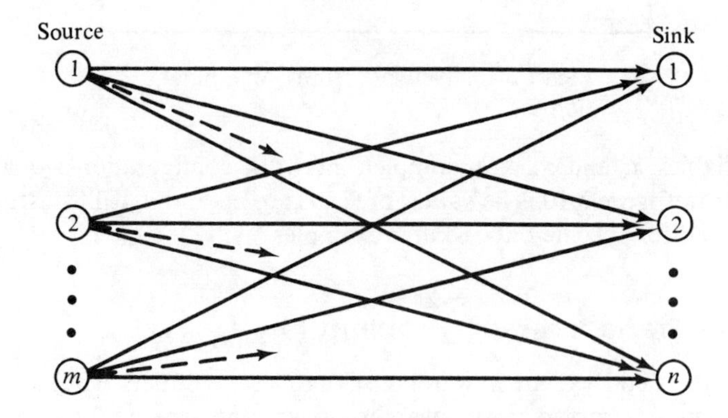

Figure 3.11. Open transportation network.

2. Define the capacity–cost triplet (U, L, c) for each arc to be $(\infty, 0, c_{ij})$.
3. Create a super source node s and a super sink node t and structure the rest of the network according to the following rules:
   a. For each origin, i, create an arc directed from the super source node to that particular origin with capacity–cost triplet defined by $(U, L, c) = (a_i, 0, 0)$.
   b. For each destination, j, create an arc directed from the destination to the super sink with capacity–cost triplet defined by $(U, L, c) = (\infty, b_j, 0)$.
4. Create an arc (t, s) with capacity–cost triplet $(U, L, c) = (\sum_{j=1}^{n} b_j, \sum_{j=1}^{n} b_j, 0)$.

5. Start the algorithm with all flows $f_{ij} = 0$ and all dual variables $\pi_k = 0$, $k = 1, 2, \ldots, (m + n + 2)$.

Note that the arc (t, s) is in state β_1 since $\bar{c}_{ts} = 0$ and the flow is less than the lower bound on the return arc. Hence, the return arc is always initially out-of-kilter.

3.11.1 Example of the Transportation Formulation

Suppose that we are given the following problem matrix:

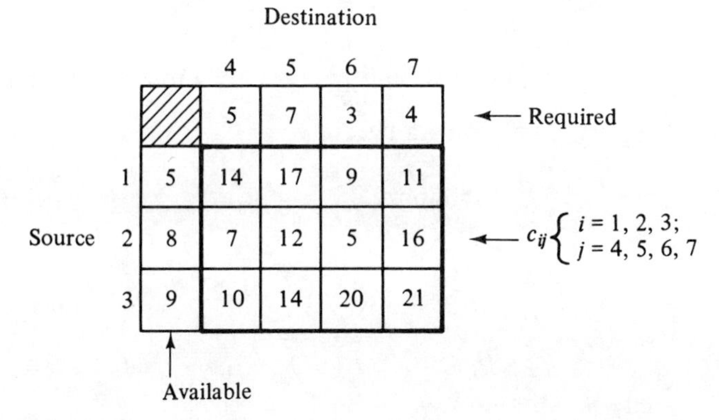

	Available	Destination 4	5	6	7	
Required		5	7	3	4	← Required
Source 1	5	14	17	9	11	
Source 2	8	7	12	5	16	← c_{ij} { $i = 1, 2, 3$; $j = 4, 5, 6, 7$
Source 3	9	10	14	20	21	

The network shown in Figure 3.12 can now be solved for the decision variables f_{ij} using the out-of-kilter algorithm.

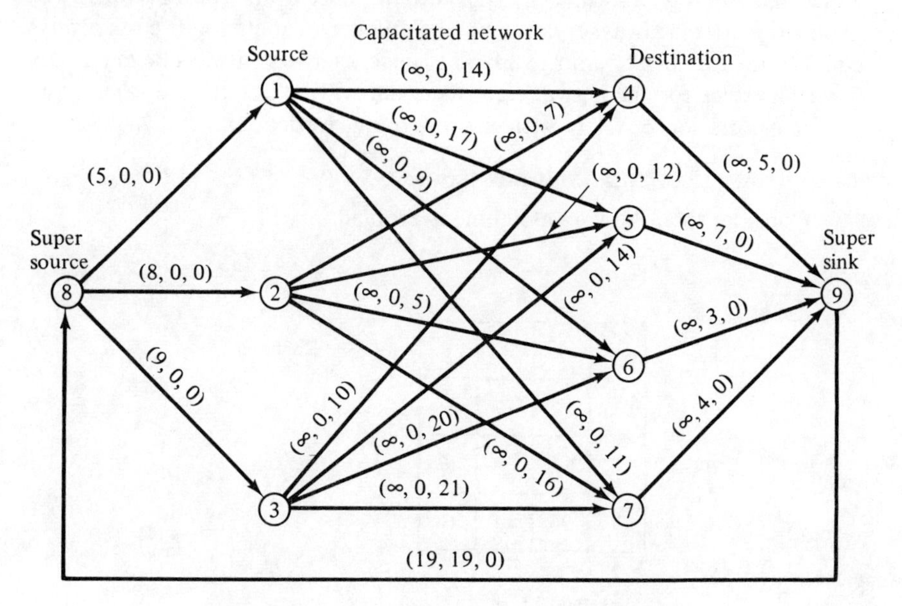

Figure 3.12. Complete transportation network.

3.12 The Assignment Problem

The assignment problem develops when there are exactly q destinations, each desiring only 1 unit from exactly q sources. Each source is capable of supplying only 1 unit to one destination. The cost of shipping 1 unit from each source to each destination is known, and the problem is to minimize the total cost involved.

Let

f_{ij} = amount shipped from source i to destination j

s_i = supply at source i

b_j = demand at destination j

c_{ij} = cost of shipping 1 unit from source i to destination j

The linear programming problem becomes

$$\text{minimize} \sum_{i=1}^{q} \sum_{j=1}^{q} f_{ij}c_{ij}$$

subject to

$$\sum_{j=1}^{q} f_{ij} \leq s_i = 1 \qquad i = 1, 2, \ldots, q$$

$$\sum_{i=1}^{q} f_{ij} \geq b_j = 1 \qquad j = 1, 2, \ldots, q$$

$$f_{ij} \geq 0 \qquad \text{all } i, \text{ all } j$$

The assignment problem is actually a special transportation problem in which $m = n = q$, and $a_i = b_j = 1$ for all sources and destinations. The problem matrix is always square and the number of units required is always equal to the number of units available; hence, there will always be maximum flow. The rules for the network construction will be exactly the same as in the transportation problem with the foregoing restrictions.

3.12.1 Example of the Assignment Formulation

Consider the following assignment problem matrix.

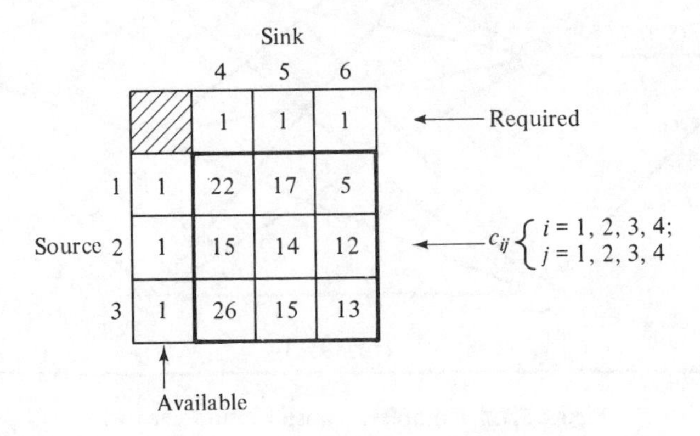

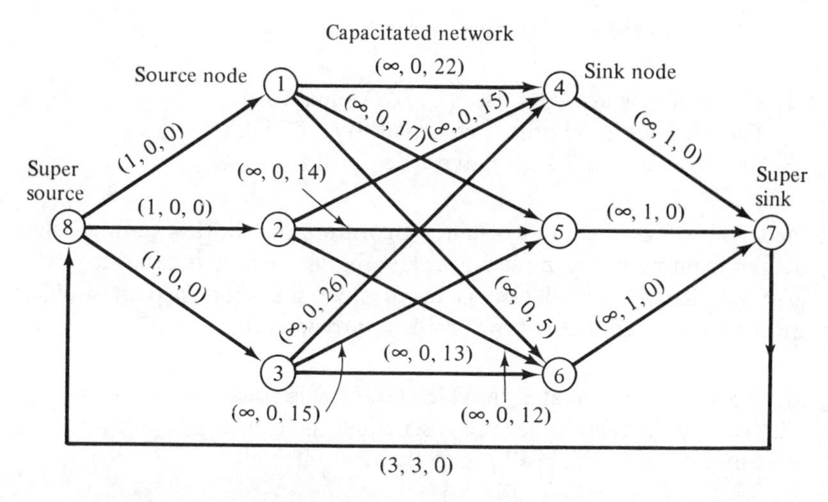

Figure 3.13. Assignment algorithm.

The solution variables f_{ij} can now be obtained using the out-of-kilter algorithm on the network shown in Figure 3.13.

3.13 Maximum Flow Through a Capacitated Network

Suppose that it is desired to ship as much flow as possible from a source s to a destination t across a capacitated network. Assuming that we are given upper and lower bounds for each arc, the following steps should be taken to solve this problem with the out-of-kilter algorithm.

1. Create arc (t, s) and let $(U, L, c) = (\infty, 0, -\infty)$.
2. Set the lower bound on all remaining arcs to zero, and the upper bounds to the capacitated values: $L_{ij} = 0$, all $i \neq t$, all $j \neq s$; $U_{ij} =$ capacitated value, all $j \neq s$.
3. Put zero cost on all remaining arcs: $c_{ij} = 0$, all $i \neq t$, all $j \neq s$.
4. Designate all arc flows as zero: $f_{ij} = 0$ all i, all j, and all dual variables equal to zero, $\pi_k = 0$, all k.

Note that with zero flow in all arcs, every arc is in state β (in kilter) except arc (t, s), which is in state δ_1 (out of kilter).

5. Proceed as usual.

3.14 The Shortest-Path Problem

It is sometimes necessary to find the shortest path from source node s to sink node t, where the arc "costs" are times or distances. This can be accomplished by the following steps.

Let d_{ij} = distance or time from node i to node j.

1. Create a new arc (t, s) with (U, L, c) equal to $(1, 1, 0)$.
2. Set (U, L, c) in all other arcs equal to $(1, 0, d_{ij})$.
3. Set $f_{ij} = 0$ for all i, all j, and let $\pi_k = 0$ for all k.

At the termination of the out-of-kilter algorithm, the shortest path is found by tracing from node s to node t over all arcs whose flow is equal to 1.

An alternative formulation is to consider the shortest-path problem identical to a minimum-cost flow problem, provided that:

1. An arc (t, s) is created in which (U, L, c) is equal to $(1, 0, -\infty)$.
2. (U, L, c) in every other arc is set equal to $(1, 0, d_{ij})$.
3. Set $f_{ij} = 0$ for all i, all j, and set $\pi_k = 0$ for all k.

The algorithm will send only 1 unit of flow through the network, and the shortest path is that path created by arc flows equal to 1.

3.15 The Shortest-Path Tree Problem

In a shortest-path tree problem, the shortest path is sought from an arbitrary node k to every other node in the network, or a selected group of nodes. This can be accomplished by the following steps.

1. Create an arc directed from each node to node k.
2. Set (U, L, c) on these arcs equal to $(1, 1, 0)$.
3. For each arc (i, j) in the network, set $(U, L, c) = (\infty, 0, d_{ij})$. Here d_{ij} is the distance from node i to node j.
4. Set all arc flows $f_{ij} = 0$ for all arcs and let $\pi_k = 0$ at each node.
5. Create a return arc with $(U, L, c) = (\phi, \phi, 0)$, where ϕ = number of arcs added to node k.

This method gives the shortest-path tree in which the path along the tree from node k to any other node is shorter than any other path. The arcs that are in the tree are those with positive flow at termination.

This is not the same as the minimal-spanning-tree problem, in which the sum of the lengths of the arcs in the tree is minimized.

A word of caution might be in order here, namely, that the procedure will always yield the shortest-path tree but not necessarily the minimum total cost.

3.16 The Transshipment Problem

Consider a system in which there are m warehouses and n retail outlets. Each warehouse has a_i, $i = 1, 2, \ldots, m$, items available and each retail outlet requires b_j, $j = 1, 2, \ldots, n$, items. It is possible to ship items from any warehouse to any retail outlet, and in addition items can be supplied through alternative routes involving intermediate warehouses. It is also possible to ship from one warehouse to another, and then to a retail outlet. Conceptually, it would appear that one solution would be to solve the problem using a transportation algorithm. However, if this is done, the amount "available" at each retail outlet and the amount "required" at each outlet would be equal to zero; hence, no transshipment would be possible [6]. It is also conceivable that the cost of shipping 1 unit from source i to outlet j would not equal the cost of shipping 1 unit from outlet j to source i. A typical example of this is shipping routes along one-way streets.

To overcome the first problem, choose an amount θ large enough to cover any possible transshipment and add this to all initial availabilities and all initial requirements. As indicated in Section 2.11, a logical choice for the value of θ would be $\min\left[\sum_{i=1}^{m} a_i, \sum_{j=1}^{n} b_j\right]$. The shipping costs involved must necessarily be determined from existing economic conditions and problem configuration.

Consider a transshipment problem with two sources supplying two retail outlets with the following cost matrix.

	Available	s_1	s_2	o_1	o_2
Required		θ	θ	$3+\theta$	$6+\theta$
s_1	$5+\theta$	1	4	6	2
s_2	$4+\theta$	2	0	8	3
o_1	θ	6	7	0	3
o_2	θ	5	2	9	1

In the above matrix, $\theta = \min\left[\sum a_i, \sum b_j\right] = \min[9, 9] = 9$.

The transshipment problem now appears in exactly the same form as a classical transportation problem, and the out-of-kilter steps to problem solution are exactly the same. It should be noted that in this particular example (a) $m = n$ and (b) $\sum_{i=1}^{m} a_i = \sum_{j=1}^{n} b_j$, but in general this will not be true.

If (a) is not true, this simply means that the cost matrix will not be square. If (b) is not true, $\sum_{i=1}^{m} a_i > \sum_{j=1}^{n} b_j$ must be true for a solution to exist. Note that a dummy destination is not necessary, only that the return arc be labeled $(\sum_j [b_j + \theta], \sum_j [b_j + \theta], 0)$.

3.17 Nonlinear Costs

Certain problems with nonlinear arc costs can be solved with the out-of-kilter algorithm by using a piecewise linear approximation. For instance, Figure 3.14 shows a convex nonlinear cost function for the arc (x, y) with a piecewise linear approximation superimposed.

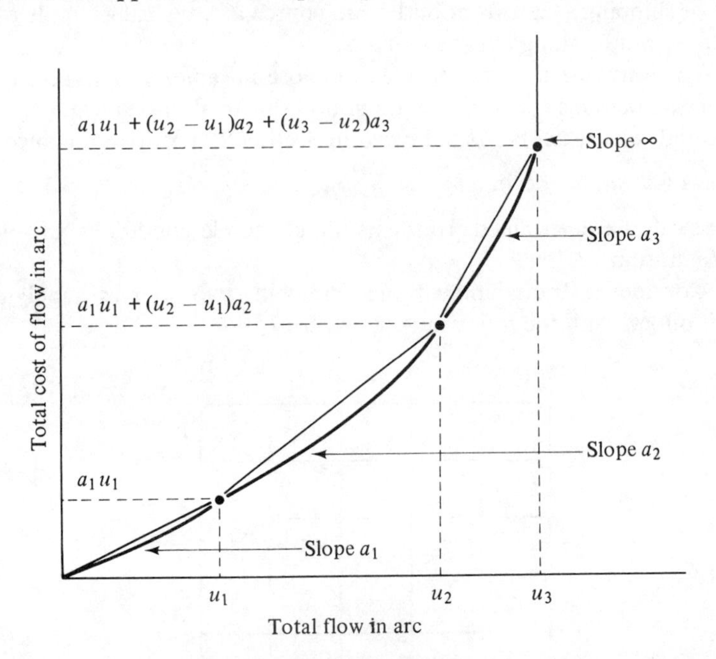

Figure 3.14. Nonlinear costs.

The linearized cost function is represented in the network by three arcs from x to y identified (1), (2), (3), as in Figure 3.15. Because the original cost function is convex, the slopes of the linear segments satisfy the following relationship:

$$a_1 < a_2 < a_3$$

Thus, as flow is increased from x to y by the algorithm, it is first assigned to arc (1). When arc (1) is at full flow, flow is assigned to arc (2). When arc (2)

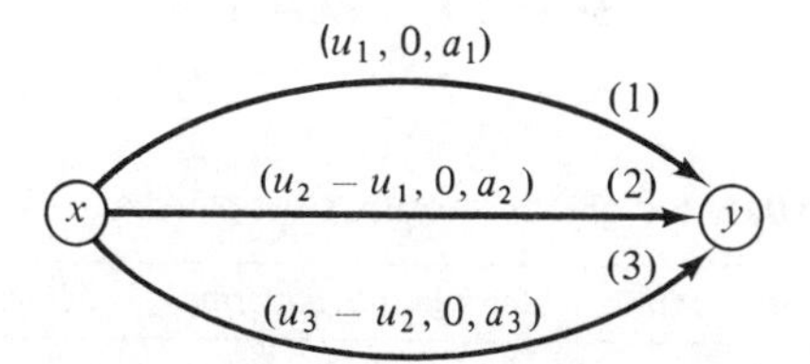

Figure 3.15. Network representation of the piecewise approximation.

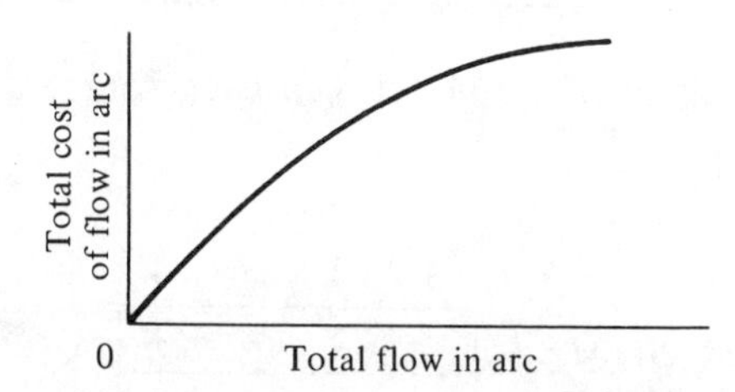

Figure 3.16. Concave nonlinear cost function.

is full, flow is assigned to arc (3). The resulting cost vs. flow curve is the piecewise linear representation in Figure 3.14.

If cost vs. flow is a concave function as in Figure 3.16, it cannot be approximated by a linear network. The out-of-kilter algorithm cannot therefore be used if cost relationships such as this are present. This is a major limitation of the out-of-kilter algorithm. The reader is invited to see Jensen and Barnes for treatment of concave cost flows [5].

As a final example, the use of the out-of-kilter algorithm will be illustrated by solving the following production planning problem [8].

3.18 A Production-Distribution Problem (Phillips and Jensen)

A company manufacturing chairs has four plants located around the country. The cost of manufacture, excluding raw material, per chair and the minimum and maximum monthly production for each plant are shown in Table 3.6. Twenty pounds of wood is required to make each chair. The company obtains wood from two sources. The sources can supply any amount to the company, but contracts specify that the company must buy at least 8 tons of wood from each supplier. The cost of obtaining the wood at the sources is:

Source 1: $0.10/lb
Source 2: 0.075/lb

Table 3.6. MANUFACTURE COSTS AND PRODUCTION LEVELS

Plant	*Cost per Chair*	*Maximum Production*	*Minimum Production*
1	$5	500	0
2	7	750	400
3	3	1000	500
4	4	250	250

Shipping cost per pound of wood between each source and each plant is shown below in cents.

		Plant 1	2	3	4
Wood Source	1	1	2	4	4
	2	4	3	2	2

The chairs are sold in four major cities: New York, Chicago, San Francisco, and Austin. Transportation costs between the plants and the cities are shown in the following table. All costs are in dollars per chair. The maximum and

		City *NY*	*A*	*SF*	*C*
Plant	1	1	1	2	0
	2	3	6	7	3
	3	3	1	5	3
	4	8	2	1	4

minimum demand and the selling price for chairs in each city are shown in Table 3.7.

Using the out-of-kilter algorithm, find:

1. Where each plant should buy its raw materials.
2. How much should be made at each plant.

Table 3.7. PRICE AND DEMAND DATA

City	*Selling Price*	*Maximum Demand*	*Minimum Demand*
NY	$20	2000	500
A	15	400	100
SF	20	1500	500
C	18	1500	500

3. How much should be sold at each city.
4. Where each plant should ship its product.

The out-of-kilter network for this particular problem is given in Figure 3.17. A summary of the network labels is given in Table 3.8. Note that all the costs

Table 3.8. NETWORK LABELS

Arc	*From Node*	*To Node*	*Upper Bound*	*Lower Bound*	*Cost*
1	1	2	2500	800	0
2	1	3	2500	800	0
3	2	4	2500	0	220
4	2	5	2500	0	240
5	2	6	2500	0	280
6	2	7	2500	0	280
7	3	4	2500	0	230
8	3	5	2500	0	210
9	3	6	2500	0	190
10	3	7	2500	0	190
11	4	8	500	0	500
12	5	9	750	400	700
13	6	10	1000	500	300
14	7	11	250	250	400
15	8	12	500	0	100
16	8	13	500	0	100
17	8	14	500	0	200
18	8	15	500	0	0
19	9	12	750	0	300
20	9	13	750	0	600
21	9	14	750	0	700
22	9	15	750	0	300
23	10	12	1000	0	300
24	10	13	1000	0	100
25	10	14	1000	0	500
26	10	15	1000	0	300
27	11	12	250	0	800
28	11	13	250	0	200
29	11	14	250	0	100
30	11	15	250	0	400
31	12	16	2000	500	−2000
32	13	16	400	100	−1500
33	14	16	1500	500	−2000
34	15	16	1500	500	−1800
35	16	1	5400	1600	0

Number of nodes = 16
Number of arcs = 35

In this particular example, there are no physical upper bounds on arcs 1 through 10 and arc 35. The upper bounds on these arcs were set at an arbitrarily large number simply for convenience. In theory, these numbers represent $+\infty$.

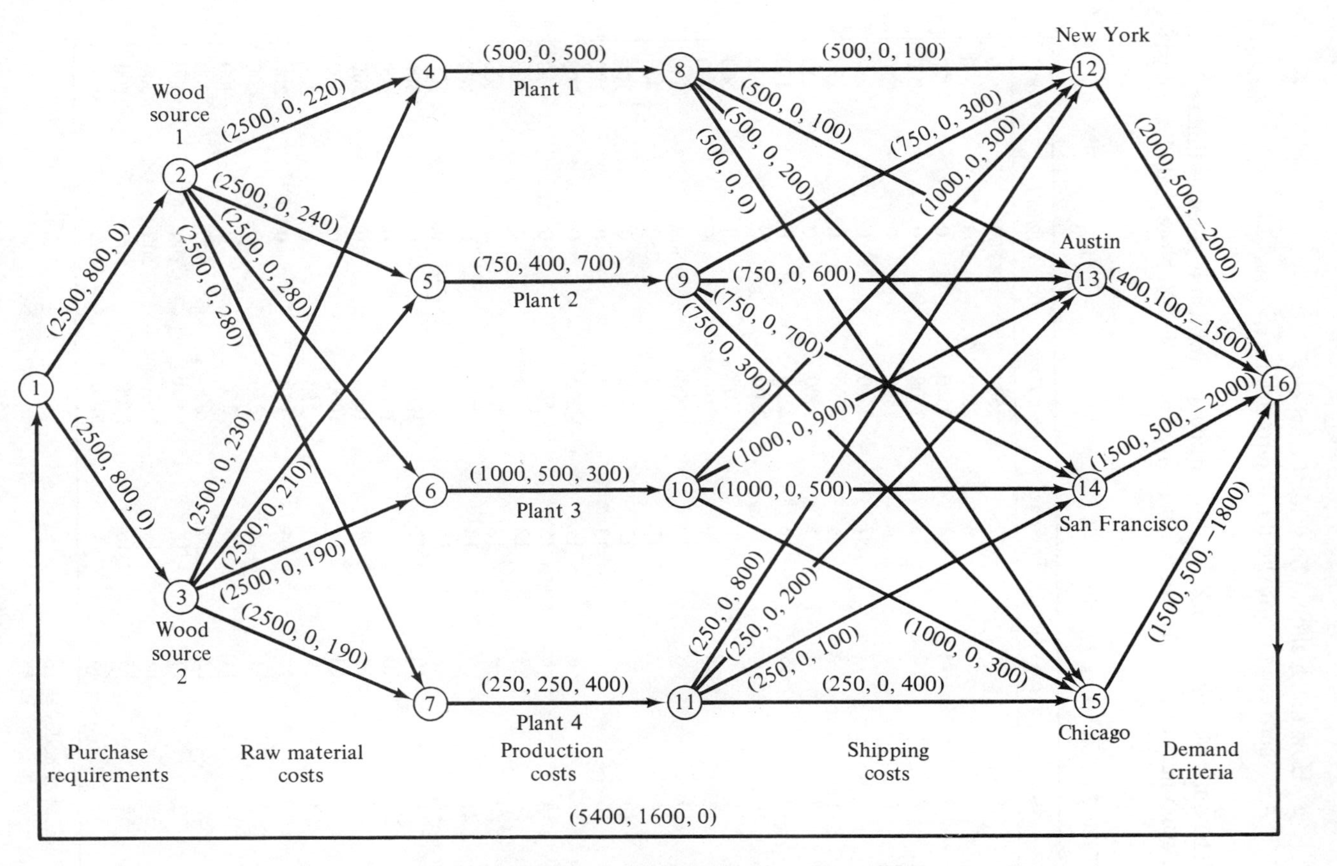

Figure 3.17. Production/distribution network.

are in cents per chair, and all the upper and lower bounds are in chairs. The optimal (minimal-cost) out-of-kilter solution to this problem is given in Table 3.9.

Table 3.9. THE OPTIMAL SOLUTION

Node k	π_k	*Arc* (i, j)	f_{ij}	*Arc* (i, j)	f_{ij}
1	1230	1	800	19	750
2	1200	2	1700	20	0
3	1230	3	500	21	0
4	1420	4	300	22	0
5	1440	5	0	23	650
6	1420	6	0	24	100
7	1420	7	0	25	250
8	3230	8	450	26	0
9	2930	9	1000	27	0
10	2930	10	250	28	0
11	3330	11	500	29	250
12	3230	12	750	30	0
13	3030	13	1000	31	1400
14	3430	14	250	32	100
15	3230	15	0	33	500
16	1230	16	0	34	500
		17	0	35	2500
		18	500		

The original questions can now be answered in terms of the f_{ij} solution variables.

1. Where should each plant buy its raw materials?

   a. Plant 1 should buy 500 chairs or 10,000 lb from wood source 1 and 0 chairs or 0 lb from wood source 2.
   b. Plant 2 should buy 300 chairs or 6000 lb from wood source 1 and 450 chairs or 9000 lb from wood source 2.
   c. Plant 3 should buy 0 chairs or 0 lb from wood source 1 and 1000 chairs or 20,000 lb from wood source 2.
   d. Plant 4 should buy 0 chairs or 0 lb from wood source 1 and 250 chairs or 5000 lb from wood source 2.

2. How many chairs should be made at each plant?

   a. Plant 1 should make 500 chairs.
   b. Plant 2 should make 750 chairs.
   c. Plant 3 should make 1000 chairs.
   d. Plant 4 should make 250 chairs.

3. How many chairs should be sold at each city?
   a. New York should sell 1400 chairs.
   b. Austin should sell 100 chairs.
   c. San Francisco should sell 500 chairs.
   d. Chicago should sell 500 chairs.

4. Where should each plant ship its product?
   a. Plant 1 ships 500 chairs to Chicago.
   b. Plant 2 ships 750 chairs to New York.
   c. Plant 3 ships 650 chairs to New York, 100 to Austin, and 250 to San Francisco.
   d. Plant 4 ships 250 chairs to San Francisco.

The optimal solution to this production planning problem was obtained by solving an equivalent dual problem and at optimality recovering the corresponding primal variables. It is both interesting and constructive to attempt an interpretation of these dual variables. Recall that the out-of-kilter algorithm utilizes the complementary slackness conditions of dual linear programming, and in particular, the dual variables α_{ij} and δ_{ij} are chosen [3, 4] such that:

$$\alpha_{ij} = \max\,[0, \pi_j - \pi_i - c_{ij}]$$
$$\delta_{ij} = \max\,[0, \pi_i - \pi_j + c_{ij}]$$

From the solution variables and Figure 3.17, the following network segment can be constructed involving nodes 1, 2, 3, and 5:

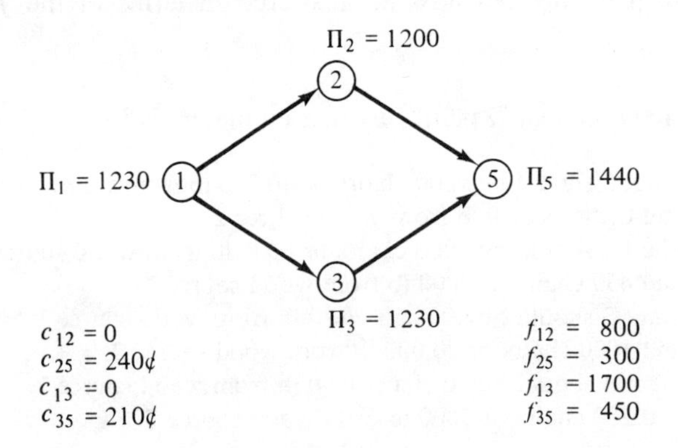

Figure 3.18. A sub-net.

For this network the following results are obtained:

$$\text{arc } (1, 2) \longrightarrow \pi_2 - \pi_1 < c_{12} \Longrightarrow f_{12} = L_{12}$$
$$\text{arc } (2, 5) \longrightarrow \pi_5 - \pi_2 = c_{25} \Longrightarrow L_{25} \leqq f_{25} \leqq U_{25}$$

$$\text{arc } (1, 3) \longrightarrow \pi_3 - \pi_1 = c_{13} \Longrightarrow L_{13} \leqq f_{13} \leqq U_{13}$$
$$\text{arc } (3, 5) \longrightarrow \pi_5 - \pi_3 = c_{35} \Longrightarrow L_{35} \leqq f_{35} \leqq U_{35}$$

From Figure 3.18 the dual variables for arc (1, 2) are

$$\alpha_{12} = \max\,[0, 1200 - 1230 - 0] = 0$$
$$\delta_{12} = \max\,[0, 1230 - 1200 + 0] = 30 \text{ cents}$$

Since the dual α variables are associated with the upper-bound constraints of the primal problem, and the δ variables are associated with the lower-bound constraints, $\delta_{12} = \$0.30$ indicates that if the lower bound on arc (1, 2) is decreased by 1 unit (the flow is at the lower bound), we would save \$0.30 per chair. This is reflected by the fact that a decrease in arc flow of 1 unit from node 1 to node 2 saves \$0.00. A unit decrease in unit flow from node 2 to node 5 saves \$2.40. In order to conserve flow out of node 1 and into node 5, a unit increase from node 1 to node 3 would cost \$0.00, while a unit increase from node 3 to node 5 would cost \$2.10. This results in a net savings of \$0.30, the same savings reflected by the dual variable δ_{12}.

As a second example, consider the following network segment.

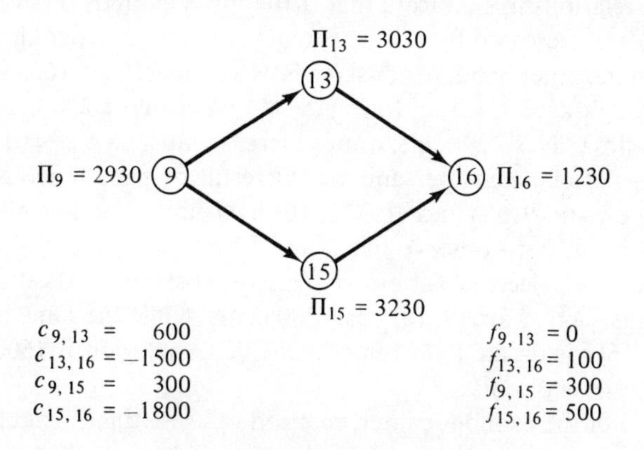

Figure 3.19. Second sub-net.

For this network the following results are obtained:

$$\text{arc } (9, 13) \longrightarrow \pi_{13} - \pi_9 < c_{9,13} \Longrightarrow f_{9,13} = L_{9,13}$$
$$\text{arc } (13, 16) \longrightarrow \pi_{16} - \pi_{13} < c_{13,16} \Longrightarrow f_{13,16} = L_{13,16}$$
$$\text{arc } (9, 15) \longrightarrow \pi_{15} - \pi_9 = c_{9,15} \Longrightarrow L_{9,15} \leq f_{9,15} \leq U_{9,15}$$
$$\text{arc } (15, 16) \longrightarrow \pi_{16} - \pi_{15} < c_{15,16} \Longrightarrow f_{15,16} = L_{15,16}$$

From Figure 3.19 the dual variables for arc (9, 13) are

$$\alpha_{9,13} = \max[0, 3030 - 2930 - 600] = 0$$
$$\delta_{9,13} = \max[0, 2930 - 3030 + 600] = 500 \text{ cents}$$

For arc (9, 15):

$$\alpha_{9,15} = \max[0, 3230 - 2930 - 300] = 0$$
$$\delta_{9,15} = \max[0, 2930 - 3230 + 300] = 0$$

For arc (13, 16):

$$\alpha_{13,16} = \max[0, 1230 - 3030 - (-1500)] = 0$$
$$\delta_{13,16} = \max[0, 3030 - 1230 + (-1500)] = 300 \text{ cents}$$

For arc (15, 16):

$$\alpha_{15,16} = \max[0, 1230 - 3230 - (-1800)] = 0$$
$$\delta_{15,16} = \max[0, 3230 - 1230 + (-1800)] = 200 \text{ cents}$$

These dual relationships indicate that if the lower bounds on arc (9, 13) and arc (13, 16) are decreased by 1, a net savings of 800 cents per chair would be realized. On the other hand, to conserve flow at nodes 9 and 16, a unit increase in flow through arc (9, 15) and arc (15, 16) would cost 200 cents (if a unit decrease in flow saves 200 cents, a unit increase must cost 200 cents). Hence, a net savings of 600 cents per unit would result by simultaneously lowering the bounds on arc (9, 13) and arc (13, 16) by 1 unit. Another way of rationalizing this result is to observe that a unit decrease in arc (9, 13) saves 600 cents, while a unit decrease across arc (13, 16) costs us 1500 cents. Similarly, a unit increase across arc (9, 15) costs 300 cents, while the same unit increase in arc (15, 16) yields an 1800 cents profit. The net gain is (600 − 1500 − 300 + 1800) = 600 cents.

Because of the economic interpretation of these dual variables when the arc costs are in dollars or cents, the dual variables are often referred to as *pricing vectors.*

3.19 Summary and Conclusions

The out-of-kilter algorithm is applicable to any problem that can be described as a minimal-cost circulation problem. This does not require that the physical process itself involve a circulation. Certainly, the transportation problem involves only goods shipped one way, from sources to destinations.

The circulation is provided by adding a "return arc" (t, s). Flow in this are does not necessarily have physical significance.

It is also unnecessary that flow be explicitly present in the problem. The assignment and the shortest-path problems, for instance, are combinatorial problems, not flow problems. Flow introduced by the circulation model is used as a device to find the optimal solution. The potential user should be attentive to possible applications that do not involve circulation or flow.

The key clue to a possible application of the out-of-kilter algorithm is the network. So many situations seem to be easily described by networks that the user with his new tool in hand, rather than having difficulty finding applications, is often driven to misapplication.

The user should remember that his only modeling tool for the out-of-kilter algorithm is the network with its directed arcs. Each arc is described by three numbers: cost per unit flow, lower bound to flow, and upper bound to flow. If there are any side conditions relating the flows on different arcs (except conservation of flow), the out-of-kilter algorithm is ruled out as a solution tool. Often such problems can be modeled as general linear programs and solved using readily available linear programming codes, or transformed into problems amiable to the out-of-kilter structure.

The out-of-kilter algorithm has a wide variety of potential uses. The practicing industrial engineer would do well to add a working knowledge of the basic theory and an efficient computer code embodying this algorithm to his bag of tools.

PART III: APPLICATIONS OF THE OUT-OF-KILTER ALGORITHM[11]

3.20 The "Bottleneck" Assignment Problem

As a first example, let us consider the *bottleneck assignment problem*; operational researchers usually have neat-sounding names for their problems. To start at the beginning, let us describe a commonly encountered assignment problem. Given a group of workers, a group of machines to be operated by these workers, and a known efficiency for each worker operating each machine, the assignment problem is to assign exactly one worker to each machine (or workers to jobs, teachers to classes, etc.) so that the sum of the efficiencies resulting from the assignment is as large as possible. Note that merely assigning each worker to that machine on which he is most efficient

[11] Portions of Part III are taken from material published in *Industrial Engineering* with permission of the American Institute of Industrial Engineers. The examples are due to Swanson, Woolsey, and Hillis [11].

is not necessarily a solution to the problem, as several workers may be most efficient on the same machine, and the problem calls for exactly one worker for each machine. Thus it is likely that some workers will be assigned to machines other than the ones on which they are most efficient in order that the sum of the efficiencies be maximized. This problem can be formulated and solved as a network-flow problem, as indicated in Section 3.12. The optimal flow will indicate an assignment that maximizes the total efficiency. In particular, the network in Figure 3.20 models the assignment situation where 3 men are to be assigned to 3 machines. Nodes 1, 2, and 3 represent the workers, and nodes 4, 5, and 6 represent the machines.

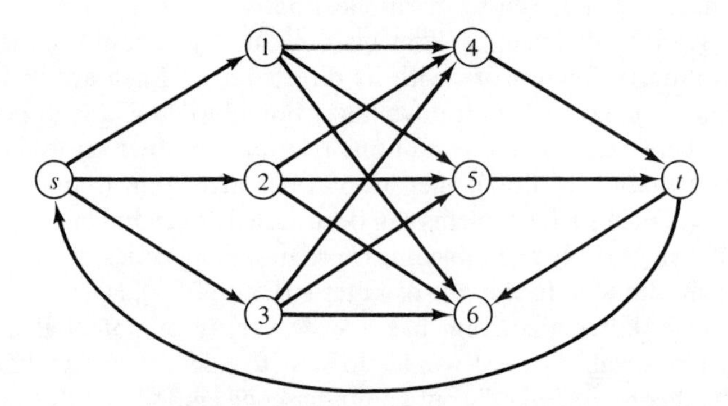

Figure 3.20. Worker-efficiency network.

Now, back to the bottleneck assignment problem, which differs slightly from the assignment problem just described in that the objective is to maximize the minimum efficiency resulting from the assignment. This objective is often realistic, for example, in series-type assembly lines with different work stations. In such a case, the worker with the minimum efficiency (the bottleneck) would determine the rate of output for the whole assembly line. It would be desirable, then, to maximize the minimum efficiency. This action determines an optimal assignment appropriate for this bottleneck situation as opposed to finding that assignment which maximizes the sum of efficiencies. This bottleneck assignment problem has a network solution. In fact, this problem may be solved by sequentially solving related assignment problems. The algorithm is as follows:

1. Solve the (bottleneck assignment) problem as if it were an ordinary assignment problem; this is easily done by using the OKA. As the OKA is a *minimizing* algorithm, minimize the sum of inefficiencies (the negative of efficiency), which is equivalent to maximizing efficiency.
2. Look at the optimal solution from step 1. Determine from this solu-

tion the assignment with the greatest inefficiency (lowest efficiency). If there are ties, pick one arbitrarily.

3. Remove from the network used in step 1 those arcs which have inefficiencies (costs) greater than or equal to that inefficiency determined in step 2 (lowest efficiency–greatest inefficiency from step 1).

4. Go to step 1 with the new network determined by step 3.

5. Continue this process until it is impossible to make the desired number of assignments (as more and more arcs are deleted from the original network). This event is realized when OKA indicates "no feasible solution." The last *feasible* assignment made is then the optimal solution to the bottleneck assignment problem. The justification for this claim is that the OKA is forced to solve a sequence of assignment problems in which the *minimum efficiency* is constantly increasing (as more arcs are being deleted). This algorithm is illustrated on the sample problem defined in Table 3.10.

Table 3.10. AN EFFICIENCY MATRIX

	Job				
Worker	1	3	2	6	0
	4	2	3	8	3
	8	4	1	5	0
	3	5	4	8	8
	2	6	9	5	2

The entry a_{ij} in the matrix represents the efficiency (number of units processed per hour) if worker i is assigned to job j. The zero entries (a_{15}, a_{35}) indicate that workers 1 and 3 are both unqualified for job 5.

Solving the bottleneck problem as if it were an ordinary assignment problem (12 nodes, 36 arcs), the OKA produces the following assignment:

Assignment 1

Worker 1 $\longrightarrow$ job 2	with efficiency of 3
Worker 2 $\longrightarrow$ job 4	with efficiency of 8
Worker 3 $\longrightarrow$ job 1	with efficiency of 8
Worker 4 $\longrightarrow$ job 5	with efficiency of 8
Worker 5 $\longrightarrow$ job 3	with efficiency of 9

The sum of the efficiencies is 36; the lowest efficiency is 3. Thus, delete those entries in the matrix (and corresponding arcs in the network) with efficiencies less than or equal to 3. Now there are 12 nodes and 24 arcs comprising the network. Solving this assignment network with the OKA produces the following solution:

Assignment 2

Worker 1 ⟶ job 4	with efficiency of 6
Worker 2 ⟶ job 1	with efficiency of 4
Worker 3 ⟶ job 2	with efficiency of 4
Worker 4 ⟶ job 5	with efficiency of 8
Worker 5 ⟶ job 3	with efficiency of 9

The sum of the efficiencies has now decreased to 31, but the smallest efficiency has increased to 4. To continue, delete those arcs of the assignment network with efficiencies less than or equal to 4. There are now 12 nodes and 21 arcs. The OKA now indicates that there is no feasible solution to this new network. Therefore, assignment 2 is optimal for the bottleneck assignment problem. There is a $33\frac{1}{3}\%$ increase in output with this solution over the original assignment, even though the *sum* of the efficiencies is smaller (the bottleneck is larger).

This increase in output is related to the number of times one is forced to delete arcs from the network and resolve the problem. Thus, not only does one obtain an optimal solution to the bottleneck assignment problem, but one also reinforces the proverbial work ethic: the more you work, the greater the reward.

3.21 Scheduling Workers to Time-Dependent Tasks

Consider the problem of determining the minimum number of workers necessary to accomplish a fixed schedule of partially overlapping tasks. A different setting for the same problem is to determine the minimum number of machines needed to perform a schedule of tasks with known setup times, or to determine the minimum number of airplanes needed to service a known flight schedule [1]. The OKA has been used to solve this problem to determine the minimum number of busses required to service a regional transportation authority's proposed route and schedule [4]. The network formulation will become evident as we proceed with the following example (due to Paul Jensen, Department of Mechanical Engineering, the University of Texas at Austin):

Ten tasks are to be performed. Their start and end times are shown in Table 3.11. The array shows the time it takes to go from one task to any other (setup time). We are asked to find the minimal number of workers to perform the tasks. From these data it is possible to determine which tasks are

Table 3.11. TASK–TIME MATRIX

Task	*Start*	*End*	**1**	**2**	**3**	**4**	**5**	**6**	**7**	**8**	**9**	**10**
1	1:00 P.M.	1:30 P.M.	—	60	10	230	180	20	15	40	120	30
2	6:00 P.M.	8:00 P.M.	10	—	40	75	40	5	30	60	5	15
3	10:30 P.M.	11:00 P.M.	70	30	—	0	70	30	20	5	120	70
4	4:00 P.M.	5: 00 P.M.	0	50	75	—	20	15	10	20	60	10
5	4:00 P.M.	7:00 P.M.	200	240	150	70	—	15	5	240	90	65
6	12:00 noon	1:00 P.M.	20	15	20	75	120	—	30	30	15	45
7	2:00 P.M.	5:00 P.M.	15	30	60	45	30	15	—	10	5	0
8	11:00 P.M.	12:00 midnite	20	35	15	120	75	30	45	—	20	15
9	8:10 P.M.	9:00 P.M.	25	60	15	10	100	70	80	60	—	120
10	1:45 P.M.	3:00 P.M.	60	60	30	30	120	40	50	60	70	—

"related" to which other tasks; "related" is used here in the sense that a given job can be completed and a new job can be set up before that new job is scheduled to start. For example, job 1 is related to each of jobs 2, 3, 7, 8, and 9, but is not related to jobs 4, 5, 6, or 10 (end of job 1 is at 1:30 P.M.; setup time for job 4 is 230 minutes, equals 3 hours 50 minutes; adding those times it is seen that the 4:00 P.M. start time of job 4 precludes job 1 and job 4 being related). In this manner it is seen that:

—Job 1 is related to jobs 2, 3, 7, 8, 9.
—Job 2 is related to jobs 3, 8, 9.
—Job 3 is not related to any job.
—Job 4 is related to jobs 2, 3, 8, 9.
—Job 5 is related to jobs 3, 8.
—Job 6 is related to jobs 2, 3, 4, 5, 7, 8, 9, 10.
—Job 7 is related to jobs 2, 3, 8, 9.
—Job 8 is not related to any job.
—Job 9 is related to jobs 3, 8.
—Job 10 is related to jobs 2, 3, 4, 8, 9.

Note that the fewest number of individuals required is equal to the maximal number of (unrelated) tasks, no two of which can be performed by the same individual. To solve this problem with the OKA, devise an "assignment network" that assigns certain tasks to certain other tasks (these assigned tasks are grouped so that they may be performed by the same individual). There are to be, in addition to the source and sink nodes, two "columns" of nodes, with as many nodes in each column as there are tasks. The arcs for this network are determined by whether or not two given nodes are related. Thus, the node corresponding to task 1 in the first column of nodes is connected to those nodes in the second column corresponding to tasks 2, 3, 7, 8, and 9. Nodes corresponding to tasks 3 and 8 need not appear in the first column of nodes (but will appear in the second column), since these tasks are not related to any other tasks (but other tasks are related to them). The upper bound on the flow in the "return" or "circulation" (super sink to super source) arc is equal to the number of tasks minus the number of tasks not related to any tasks. In our example, this is 8. The upper bound on flow in each other arc is 1. The lower bound on flow in each arc is 0.

The objective is to find the maximum flow in this capacitated network. This is accomplished by setting the unit cost of flow in the return arc to a negative number (say, −10), and setting the unit cost of flow in each other arc equal to 0. The maximum flow (optimal solution) for this network determines the minimal number of workers necessary to accomplish the schedule of tasks. In particular, arcs that have a positive flow indicate which tasks should be grouped so that they can be performed by a single individual. The number of such groups determines the minimal number of individuals required. The following "assignment arcs" have positive flow at optimality:

Task 1 ⟶ task 7
Task 4 ⟶ task 2
Task 5 ⟶ task 3
Task 6 ⟶ task 10
Task 7 ⟶ task 9
Task 9 ⟶ task 8
Task 10 ⟶ task 4

Recall that arcs exist in the network only if nodes are related. Thus the solution is to be interpreted as follows: Task 1 is related to task 7, which is related to task 9, which is related to task 8.

Task 1 ⟶ task 7 ⟶ task 9 ⟶ task 8. These four tasks are "connected" (each is related to the next) and may be done sequentially by a single individual. This "string" of tasks is broken when task 8 is not related to any other task. There are two additional "strings" of tasks:

Task 6 ⟶ task 10 ⟶ task 4 ⟶ task 2
Task 5 ⟶ task 3

Thus, these three groups or strings of tasks each require a separate individual to perform them; no fewer than three individuals can accomplish these 10 tasks within the given constraints on start, end, and setup times.

3.22 A Wholesale Storage and Marketing Problem

Consider the following inventory problem. The owner of a wholesale outlet bought 50 table radios for $20 each. Because of changing demand, the selling price of these radios fluctuates from month to month. Also, because of the size of the radios and storage conditions, the storage cost fluctuates from month to month. The data are given in Table 3.12.

Table 3.12. DATA FOR THE WHOLESALE STORAGE AND MARKETING PROBLEM

Number of Radios	*0–10*	*11–25*	*26–45*	*46 or More*
	Storage Costs per Radio (dollars)			
Month 1	1.80	1.60	1.40	1.20
Month 2	2.60	2.40	2.20	2.00
Month 3	2.40	2.20	2.00	1.80
	Selling Price per Radio (dollars)			
Month 1	45.00	40.00	35.00	30.00
Month 2	40.00	35.00	30.00	25.00
Month 3	55.00	50.00	45.00	40.00
Month 4	50.00	45.00	40.00	35.00

The decision that needs to be made is how many radios to sell in each of months 1, 2, 3, and 4 to maximize profit. The network that models this decision process is given in Figure 3.21. The arc from node 1 to node 2 represents the time period in which the radios were bought; arcs (2, 3), (3, 4), and (4, 5) allow for holdover storage in months 1, 2, or 3, respectively. Arcs (2, 6), (3, 6), (4, 6), and (5, 6) allow for selling the radios in month 1, 2, 3, or 4, respectively.

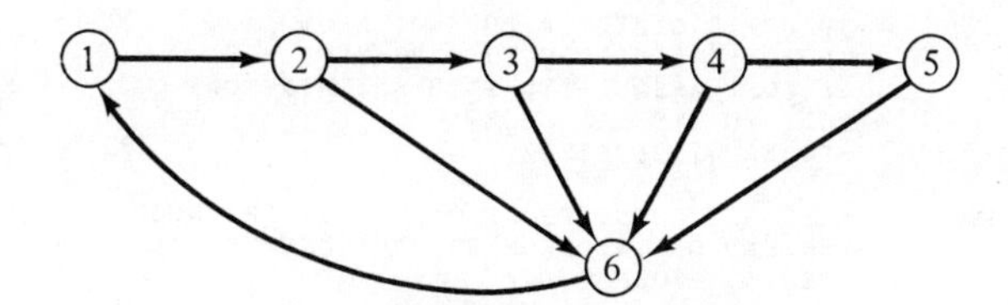

Figure 3.21. Radio sales network.

Actually, since the data show this particular problem to be nonlinear, one might not expect a simple solution. However, this problem and those of identical structure have a fortunate characteristic in that the optimal flow is along a single chain of arcs. That is, the radios will all be sold in the same period; the problem then reduces to finding the proper month in which to sell, and pursuing this policy to the limit of one's inventory. With this prior knowledge, the OKA can be used to find the optimal path ("longest" path) through the network. The optimal policy in this case is to hold all radios in inventory until month 3, and then liquidate the entire stock. The profit is $\$840 = (\$40 - 20 - 1.2 - 2.0)50$.

3.23 Computer Code for Out-of-Kilter Algorithm

```
C***********************************************************************
C*                                                                     *
C*    PURPOSE:                                                         *
C*                                                                     *
C*       TO OBTAIN OPTIMAL SOLUTIONS FOR VARIOUS CAPACITATED           *
C*       NETWORK PROBLEMS. THE FOLLOWING CAPACITATED NETWORK           *
C*       FLOW PROBLEMS CAN BE SOLVED.                                  *
C*                                                                     *
C*       1) THE TRANSPORTATION PROBLEM                                 *
C*       2) THE ASSIGNMENT PROBLEM                                     *
C*       3) MAXIMUN FLOW PROBLEM                                       *
C*       4) THE SHORTEST PATH TREE PROBLEM                             *
C*       5) THE TRANSSHIPMENT PROBLEM                                  *
C*       6) MINIMUN COST-MAXIMUN FLOW PROBLEM                          *
C*                                                                     *
C*    LOCATION:  SUBROUTINE OKALG IN NETWORK OPTIMIZATION              *
C*                                                                     *
C*                                                                     *
C*    USAGE:                                                           *
C*                                                                     *
C*       AT PRESENT THIS PROGRAM HANDLES UP TO 500 ARCS AND 500 NODES. *
C*       THIS NUMBER MAY BE INCREASED BY INCREASING THE DIMENSION      *
C*       STATEMENTS IN SUBROUTINE OKALG AND MAIN PROGRAM.              *
C*                                                                     *
C*    INPUT DATA:                                                      *
C*                                                                     *
C*       SET 1:  SINGLE CARD WITH THE CALL NAME OF THE ALGORITHM:OKAL  *
C*               -FORMAT(A4)                                           *
C*                                                                     *
C*       SET 2:  SINGLE CARD WITH THE NUMBER OF NODES AND ARCS IN THE  *
C*               PROBLEM.                                              *
C*               -FORMAT(2I10)                                         *
C*                                                                     *
C*       SET 3:  THE TOTAL NUMBER OF INPUT CARDS IN THIS SET IS EQUAL  *
C*               TO THE TOTAL NUMBER OF ARCS IN THE NETWORK.           *
C*               THE FOLLOWING QUANTITIES ARE READ FROM EACH CARD:     *
C*               1) START NODE                                         *
C*               2) END NODE                                           *
C*               3) UPPER LIMIT ON THE ARC FLOW.                       *
C*               4) LOWER LIMIT ON THE ARC FLOW.                       *
C*               5) COST ASSOCIATED WITH SHIPPING ONE UNIT FROM START  *
C*               NODE TO THE END NODE                                  *
C*               -FORMAT(2I5,3F10.2)                                   *
C*                                                                     *
C*       SET 4:  THIS SET CONSISTS OF ONLY ONE CARD WHICH INDICATES    *
C*               THE END OF THE PROBLEM. THE USER SHOULD PUNCH THE WORD*
C*               'PEND'  -FORMAT(A4)                                   *
C*                                                                     *
C*       SET 5:  THIS SET CONSISTS OF ONLY ONE CARD WHICH INDICATES    *
C*               THE END OF THE INPUT DECK. THE USER SHOULD PUNCH      *
C*               THE WORD 'EXIT'  -FORMAT(A4)                          *
C*                                                                     *
C*    PROGRAM CONTENTS:                                                *
C*                                                                     *
C*       THE SUBROUTINES WHICH COMPRISE THIS PROGRAM ARE:              *
C*          OKALG: INPUT,OUTPUT AND INITIALIZATION                     *
C*          NETFLOW - LABELING PROCEDURE                               *
C*                                                                     *
C*    DEFINITION OF TERMS USED IN THE PROGRAM                          *
C*                                                                     *
C*       I    = STARTING NODE OF THE ARC                               *
C*       J    = ENDING NODE OF THE ARC                                 *
C*       HI   = UPPER LIMIT ON THE CAPACITY OF THIS ARC                *
C*       LO   = LOWER LIMIT ON THE CAPACITY OF THIS ARC                *
C*       FLOW= ACTUAL FLOW THRU THIS ARC                               *
C*       PI   = VALUES OF THE DUAL VARIABLES                           *
C*       COST= COST ASSOCIATED IN SHIPPING ONE UNIT                    *
C*                                                                     *
```

```
C*      TECHNIQUE USED:                                                  *
C*         THE TECHNIQUE USED IN THIS ALGORITHM IS EXPLAINED IN         *
C*         SECTION                                                       *
C*                                                                       *
C*      REFERENCES                                                       *
C*                                                                       *
C*         D.T.PHILLIPS AND P.A.JENSEN, NETWORK FLOW OPTIMIZATION        *
C*         WITH THE OUT-OF-KILTER ALGORITHM, RESEARCH MEMORANDUM         *
C*         NO.71-2, FEBRUARY 1971, PURDUE UNIVERSITY, W.LAFAYETTE        *
C*                                                                       *
C************************************************************************
```

3.24 Crude Oil Production and Distribution

A series of three oil refineries can obtain crude oil from two sources. One source is from the Gulf of Alaska at a cost of $16.80 per barrel ($0.40/gal) or from the Persian Gulf at a cost of $12.60 per barrel ($0.30/gal). Both sources can furnish the refineries with an unlimited amount and a minimum amount of 300,000 gallons per day. Table 3.13 gives the production capacities and cost of distillation per gallon at each refinery.

Table 3.13. DISTILLATION COSTS AND PRODUCTION RATES

Refinery	*Refining Cost ($/gal)*	*Maximum Production (gal)*	*Minimum Production (gal)*
1	0.03	1,250,000	200,000
2	0.04	1,500,000	300,000
3	0.05	1,350,000	300,000

The refineries have production goals to meet in order to supply four different regions on a daily basis. Table 3.14 gives the associated shipping costs in dollars per gallon. Table 3.15 gives the daily requirement for each region and the selling price per gallon.

Table 3.14. SHIPPING COSTS

Refinery	*New York*	*Atlanta*	*Dallas*	*Los Angeles*
1	0.05	0.06	0.06	0.07
2	0.06	0.07	0.07	0.06
3	0.07	0.06	0.06	0.05

Table 3.15. DAILY REQUIREMENTS

Location	*Maximum Demand*	*Minimum Demand*	*Selling Price ($/gal)*
New York	250,000	200,000	0.85
Atlanta	175,000	90,000	0.70
Dallas	175,000	100,000	0.65
Los Angeles	350,000	200,000	0.75

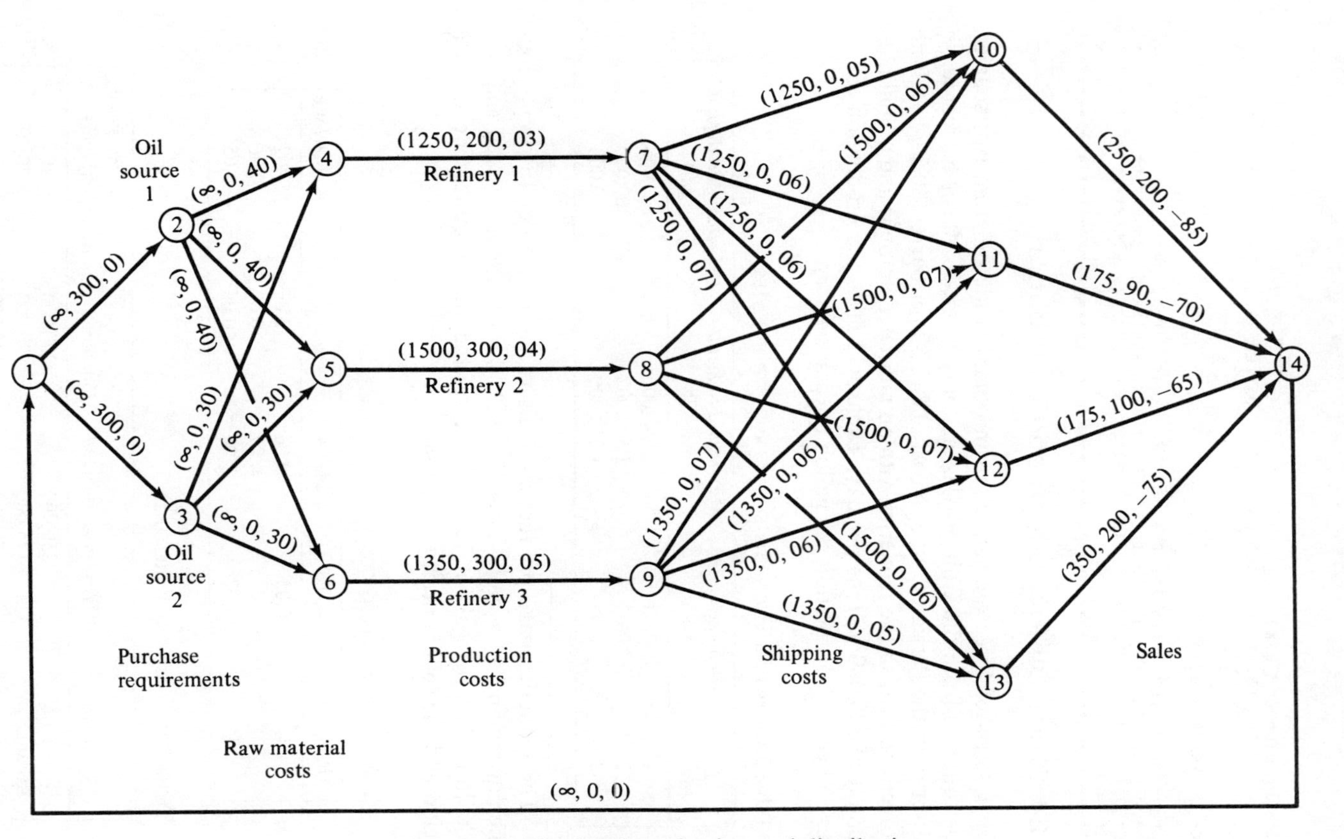

Figure 3.22. Crude oil production and distribution.

For the oil company to maximize its profits, it must supply its regions with the cheapest fuel at the cheapest shipping cost. The network diagram is presented in Figure 3.22 and the solution is shown on the computer printout obtained after running the code described in Section 3.23.

Note that the solution yields a *negative* optimal solution. This indicates that the optimal production/distribution cycle results in *positive* cash flow (profit). This is because the costs were entered as positive numbers and the selling prices as negative numbers. The corresponding optimal production/distribution pattern is easily recovered from the optimal arc flows.

****SOLUTION BY OUT-OF-KILTER ALGORITHM****

FINAL SUMMARY REPORT

NUMBER OF NODES = 14
NUMBER OF ARCS = 28

M	I	J	HI	LO	FLOW	COST
1	1	2	9999999	300	300	0
2	1	3	9999999	300	650	0
3	2	4	9999999	0	0	40
4	2	5	9999999	0	0	40
5	2	6	9999999	0	300	40
6	3	4	9999999	0	350	30
7	3	5	9999999	0	300	30
8	3	6	9999999	0	0	30
9	4	7	1250	200	350	3
10	5	8	1500	300	300	4
11	6	9	1350	300	300	5
12	7	10	1250	0	200	5
13	7	11	1250	0	0	6
14	7	12	1250	0	150	6
15	7	13	1250	0	0	7
16	8	10	1500	0	50	6
17	8	11	1500	0	175	7
18	8	12	1500	0	25	7
19	8	13	1500	0	50	6
20	9	10	1350	0	0	7
21	9	11	1350	0	0	6
22	9	12	1350	0	0	6
23	9	13	1350	0	300	5
24	10	14	250	200	250	−84
25	11	14	175	90	175	−69
26	12	14	175	100	175	−64
27	13	14	350	200	350	−74
28	14	1	9999999	0	950	0

TOTAL PROJECT COST = −29525.00

ARC	FLOW(ARC)
1	300
2	650
3	0
4	0
5	300
6	350
7	300
8	0
9	350
10	300
11	300
12	200
13	0
14	150
15	0
16	50
17	175
18	25
19	50
20	0
21	0
22	0
23	300
24	250
25	175
26	175
27	350
28	950

NODE	PI(NODE)
1	47
2	37
3	47
4	77
5	77
6	77
7	80
8	79
9	80
10	85
11	86
12	86
13	85
14	47

SENSITIVITY ANALYSIS

M	I	J	HI	LO	FLOW	COST
1	1	2	9999999	300	300	0
2	1	3	9999999	300	650	0
5	2	6	9999999	0	300	40
6	3	4	9999999	0	350	30
7	3	5	9999999	0	300	30
9	4	7	1250	200	350	3
10	5	8	1500	300	300	4
11	6	9	1350	300	300	5
12	7	10	1250	0	200	5
14	7	12	1250	0	150	6
16	8	10	1500	0	50	6
17	8	11	1500	0	175	7
18	8	12	1500	0	25	7
19	8	13	1500	0	50	6
23	9	13	1350	0	300	5
24	10	14	250	200	250	−84
25	11	14	175	90	175	−69
26	12	14	175	100	175	−64
27	13	14	350	200	350	−74
28	14	1	9999999	0	950	0

EXERCISES

1. Define and formulate the following problems as flow-network circulation models:
   (a) The transportation problem (five sources and four sinks).
   (b) The assignment problem (three sources and three sinks).
   (c) The maximum-flow problem (general framework).
   (d) The shortest-path problem (general framework).
   (e) The shortest-path tree problem (four nodes in the network with each connected to one another).
   (f) The transshipment problem (four sources and four sinks).

2. A production planning model: consider a manufacturer faced with the problem of meeting the demand for a product over the next three periods. Two production alternatives, regular time and overtime, are available in each period. The following data are provided:

Period	Capacity (units)		Unit Production Cost		Anticipated Demand (units)
	Regular Time	Overtime	Regular Time	Overtime	
1	100	20	$14	$18	60
2	100	10	17	22	80
3	60	20	17	22	140

The cost of carrying 1 unit in inventory from one period to the next is $1. The inventory level at the start of the period 1 is 15 units. Formulate and solve this problem as a circulation network by using the out-of-kilter algorithm. Formulate and solve this problem as a transportation problem using the out-of-kilter algorithm.

3. Find the maximum flow from node 1 to node 5 in the following circulation problem starting with a zero flow for all arcs and zero dual π values for all nodes. Each arc is labeled with a triplet, (L_{ij}, U_{ij}, c_{ij}).

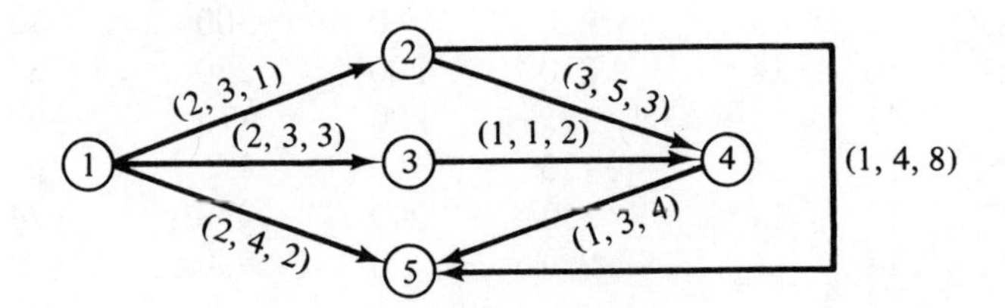

4. Formulate and solve the transportation problem shown as a circulation flow network.

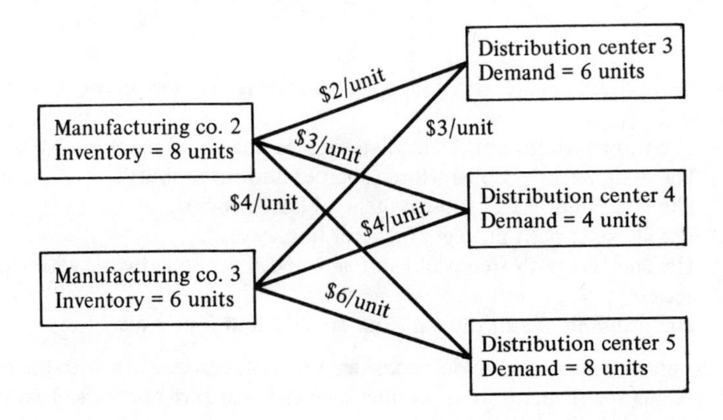

5. Formulate the primal and dual problems for the network shown and write the complementary slackness conditions.

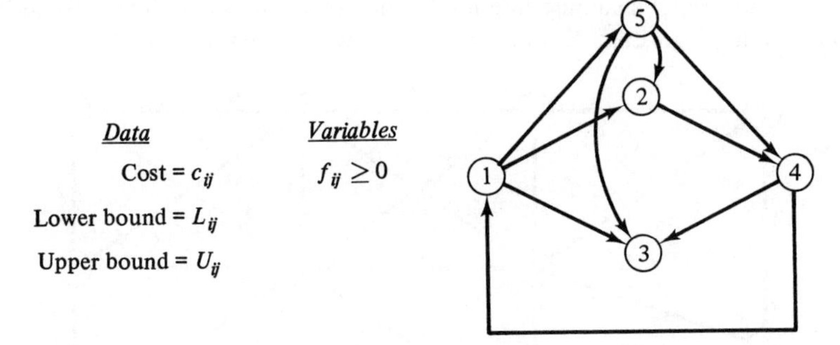

6. Consider the capacitated network shown. Use the out-of-kilter computer code to solve the following problems:

Arc	Upper Bound	Lower Bound	Cost per Unit Flow
(1, 3)	20	0	$12
(1, 4)	15	0	8
(2, 4)	17	0	13
(2, 5)	32	0	9
(3, 4)	41	0	10
(3, 7)	27	0	15
(4, 6)	∞	0	0
(4, 7)	16	0	7
(4, 8)	19	0	9
(5, 4)	23	0	11
(5, 8)	29	0	5
(6, 9)	31	0	7
(6, 10)	14	0	10
(7, 10)	19	0	8
(8, 10)	28	0	14
(9, 11)	34	0	12
(10, 11)	29	0	11

(a) Find the minimum-cost path from node 1 to node 11.
(b) Find the arc flows necessary to supply 50 units of flow to node 11.
(c) Find the maximum flow that can be shipped from nodes 1 and 2 to node 11.

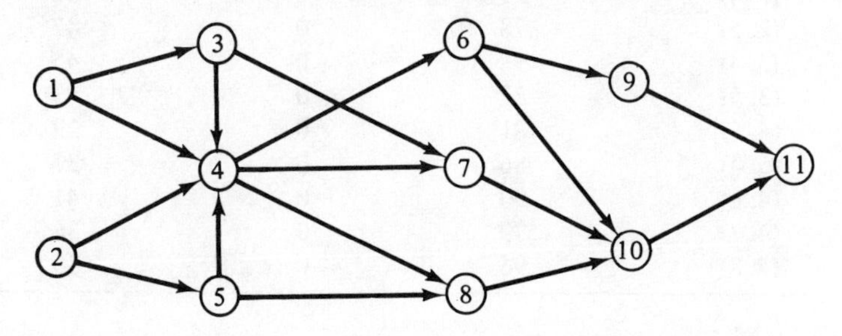

7. Find the minimal spanning tree for the undirected network shown, using the out-of-kilter algorithm. Each arc parameter is a cost per unit flow.

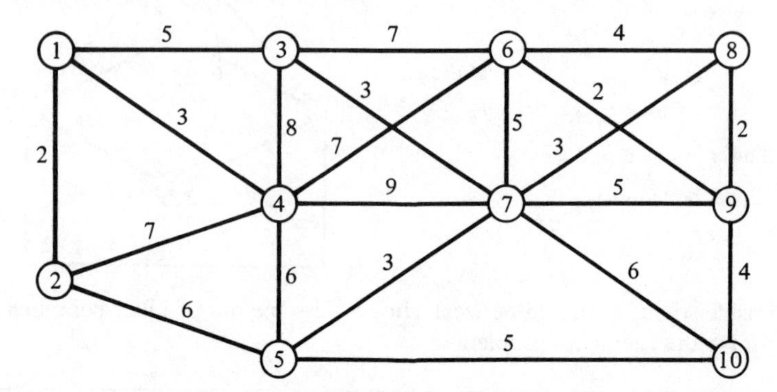

8. Using the definitions presented in Chapter 1, find the minimum arborescent spanning tree rooted at node 1 for the network shown.

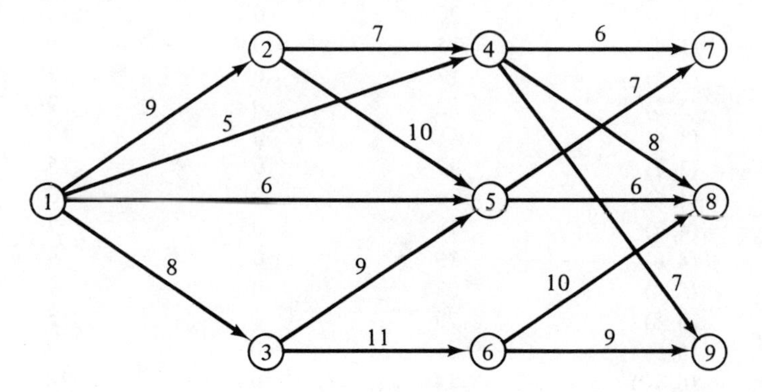

9. Consider the network shown. Use the out-of-kilter code to solve the following problems:

Arc	*Upper Bound*	*Lower Bound*	*Cost per Unit Flow*
(1, 4)	95	0	$46
(1, 5)	86	0	29
(2, 4)	91	0	31
(2, 5)	78	0	47
(3, 4)	98	0	45
(3, 5)	77	0	53
(4, 5)	81	0	29
(5, 4)	86	0	37
(4, 6)	90	0	41
(4, 7)	79	0	36
(4, 8)	96	0	30

Arc	*Upper Bound*	*Lower Bound*	*Cost per Unit Flow*
(5, 6)	77	0	29
(5, 7)	95	0	43
(5, 8)	84	0	27
(6, 7)	82	0	31
(7, 6)	79	0	30
(7, 8)	91	0	29
(8, 7)	85	0	43
(6, 9)	84	29	69
(7, 9)	80	37	68
(8, 9)	93	31	64

(a) Find the minimal-cost circulation from nodes 1 and 3 to node 9.
(b) Find the maximum amount of flow at minimum cost from nodes 1, 2, and 3 to node 9.

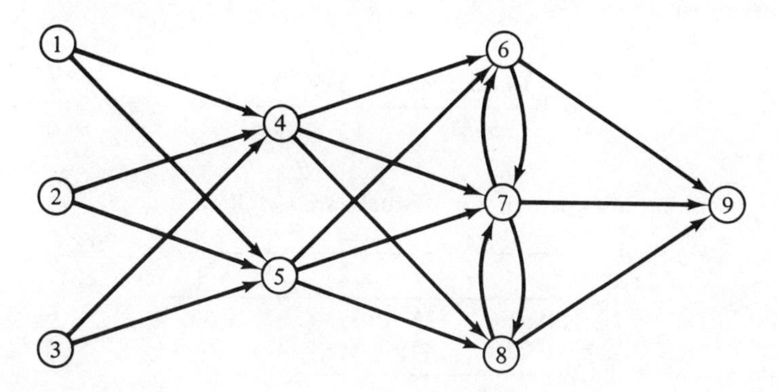

10. What effects do negative arc costs have on the out-of-kilter algorithm?

11. The out-of-kilter algorithm can start with any set of flows that satisfies the conservation-of-flow constraints, feasible or nonfeasible. What is the effect of always starting with a set of flows that is feasible? State the entire set of optimality conditions.

12. What is the difference between a minimal spanning tree and a shortest-path tree? A shortest-path tree and a minimal spanning arborescent tree?

13. Find a shortest-path tree and compare it with the minimal spanning tree in the network shown.

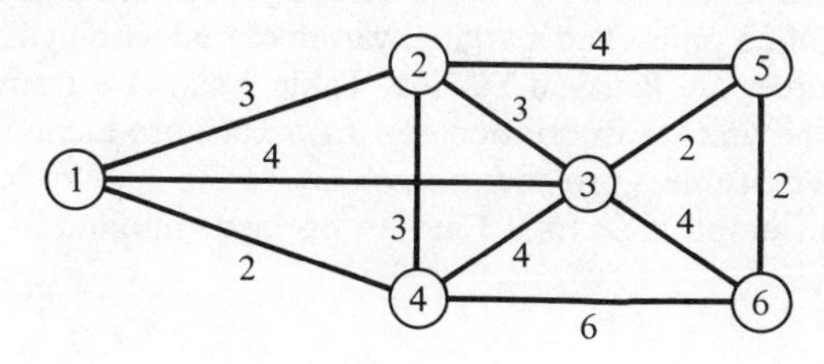

14. Formulate the caterer problem of Section 2.16 as (a) a conventional flow-network problem; (b) a transportation problem using the out-of-kilter algorithm.

15. Solve Exercise 51 of Chapter 2 using the out-of-kilter algorithm.

16. Solve Exercise 52 of Chapter 2 using the out-of-kilter algorithm.

17. The Alamo Brick firm has recently received a request for Mexican brick made at its San Antonio, Texas, plant. The order must be loaded and delivered in the next 24 hours, and the round trip takes 8 hours. Alamo Trucking owns a fleet of four trucks which can carry the following tons of brick per truck:

Alamo Truck	1	2	3	4
Capacity	15	12	13	16

In order to fill customer orders, Alamo Trucking can lease up to three larger trucks from Glider Rent-A-Truck. The available trucks have the following carrying capacity:

Glider Truck	1	2	3
Capacity	7	8	6

The operating costs per trip for all trucks are as follows:

	1	2	3	4
Alamo	25	31	22	30
Glider	44	48	41	—

The total order is for 200 tons of bricks.

(a) Determine a feasible shipping schedule that minimizes total costs.

(b) What is the maximum amount of bricks that could be delivered in 3 days?

(c) The customer just called back and wants 300 tons of brick. The Alamo Brick Plant can purchase brick from the Rio Grande Brick Factory at a cost of $52 per load, and a load contains 10 tons of brick. Determine the new minimal-cost supply strategy.

18. The Tangled Web Production Company currently owns three companies, each of which can ship products to four demand market areas. The costs are not linear, but can be calculated. Each of the three production sources has a maximum capacity of 25 units. The marginal variable production costs for the three production sources are listed in Table 1. Table 2 shows a transportation cost matrix listing the unit transportation cost from each production source to each demand location. Table 3 provides a summary of the market demands in each of the four consumptive sectors. Find an optimal (minimum-cost) operating policy.

Table 1. MARGINAL PRODUCTION COSTS

	Production Source I		*Production Source II*		*Production Source II*	
Production Level	*Marginal Costs*	*Total Variable Costs*	*Marginal Costs*	*Total Variable Costs*	*Marginal Costs*	*Total Variable Costs*
1	$ 50	$ 50	$60	$ 60	$54	$ 54
2	51	101	60	120	54	108
3	51	152	60	180	55	163
4	52	204	60	240	55	218
5	52	256	61	301	56	274
6	53	304	61	362	56	330
7	54	363	62	424	57	387
8	55	418	62	486	57	444
9	57	475	62	548	58	502
10	58	533	63	611	58	560
11	60	593	63	674	58	618
12	61	654	64	738	59	677
13	62	716	64	802	60	737
14	64	780	64	866	61	798
15	65	845	64	930	62	860
16	66	911	65	995	63	923
17	67	978	65	1060	63	986
18	68	1046	66	1126	64	1050
19	69	1115	66	1192	64	1114
20	70	1185	67	1259	65	1179
21	72	1257	67	1326	65	1244
22	75	1352	67	1393	66	1310
23	80	1412	69	1462	66	1376
24	100	1512	70	1532	67	1443
25	120	1632	80	1612	70	1513

Table 2. TRANSPORTATION COST MATRIX

From Production Source	*To Market*			
	A	*B*	*C*	*D*
I	$20	$20	$12	$30
II	45	35	10	20
III	22	10	30	55

Table 3. MARKET DEMANDS

Market	*A*	*B*	*C*	*D*	Total
Demand	10	10	6	9	35

19. The ABC Computer Company has 10 programs available to aid in solving a problem. The output from program 10 is desired. It can take the output from either program 8 or 9 to produce the desired output, but if it works on the output of program 8 it can get the results in 3 seconds, whereas if it works on the results of program 9 it will take 8 seconds to produce the desired output. Similarly, other intermediate output provides the input for programs 8 and 9. This pattern continues for all remaining programs. The problem is which set of programs to use to get the desired results in the least amount of time. The network diagram shown represents the precedent structure and the amount of time it will take to run any program, given the output of any other program. What is the proper sequence of program executions to minimize total execution time?

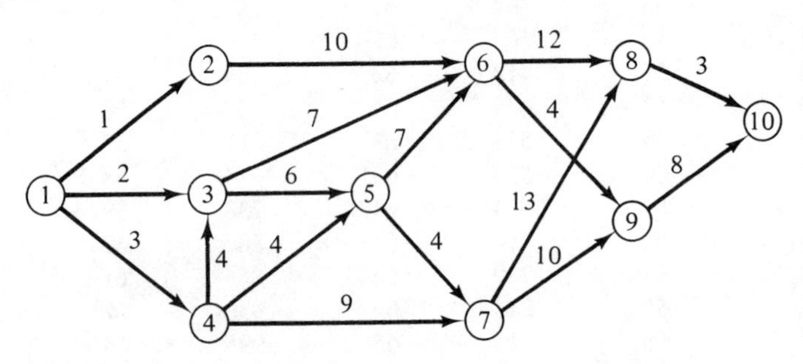

20. Solve Exercise 51 of Chapter 2, assuming that the following transshipment policies are available. (Entries represent the costs in hundreds of dollars.)

From \ To		Mine 1	Mine 2	Mine 3	Mine 4	Mine 5
Mine	1	—	7	4	3	—
	2	6	—	5	7	—
	3	5	5	—	9	—
	4	3	8	10	—	—
Plant	1	—	4	7	2	6
	2	4	—	8	9	2
	3	7	8	—	3	1
	4	3	10	3	—	6
	5	5	2	1	5	—

21. Solve Exercise 35 of Chapter 2 and compare the computer execution time of Dijkstra's method to that of the out-of-kilter method.

22. Solve Exercise 28 of Chapter 2, parts (a) and (b), by the out-of-kilter algorithm.

23. Solve Exercise 29 of Chapter 2, part (a) only, by the out-of-kilter algorithm.

24. Solve Exercise 21 of Chapter 2, part (a), by the out-of-kilter algorithm, and compare your results to the algorithm given in Chapter 2. Can you also solve parts (b) and (c)?

REFERENCES

[1] DURBIN, E. P., "The Out-of-Kilter Algorithm: A Primer," Rand Corporation, Santa Monica, California, December 1967.

[2] FORD, L. R., AND D. R. FULKERSON, "Maximal Flow through a Network," *Canadian Journal of Mathematics* (August 1956).

[3] FORD, L. R., AND D. R. FULKERSON, *Flows in Networks.* Princeton, N.J.: Princeton University Press, 1962.

[4] FULKERSON, D. R., "The Out-of-Kilter Method for Minimal Cost Flow Problems," *Journal of Applied Mathemathics*, **9** (1) (March 1961).

[5] JENSEN, P. A., AND W. BARNES, *Network Flow Programming.* New York: John Wiley & Sons, Inc., 1979.

[6] PHILLIPS, D. T., AND P. A. JENSEN, "Network Flow Optimization with the Out-of-Kilter Algorithm, Part I—Theory," Research Memorandum 71-2, February 1971.

[7] PHILLIPS, D. T., AND P. A. JENSEN, "Network Flow Optimization with the Out-of-Kilter Algorithm, Part II—Applications," Research Memorandum 71-3, February 1971, Purdue University.

[8] PHILLIPS, D. T., AND P. A. JENSEN, "Network Flow Analysis: The Out-of-Kilter Algorithm," *Industrial Engineering* (February 1974). Portions reproduced by permission of the authors and the American Institute of Industrial Engineers.

[9] PHILLIPS, D. T., A. RAVINDRAN, AND J. J. SOLBERG, *Operations Research: Principles and Practice.* New York: John Wiley & Sons, Inc., 1977.

[10] SAHA, J. L., "An Algorithm for Bus Scheduling Problems," *Operational Research Quarterly*, **21** (4) (December 1970).

[11] SWANSON, H. S., R. E. D. WOOLSEY, AND H. HILLIS, "Using the Out-of-Kilter Algorithm," *Industrial Engineering* (March 1974). Portions reproduced by permission of the authors and the American Institute of Industrial Engineers.

[12] SWANSON, H. S., AND R. E. E. WOOLSEY, "An Out-of-Kilter Network Tutorial," *SIGMAP Newsletter* (January 1973).

[13] VAJDA, *Mathematical Programming.* Reading, Mass.: Addison-Wesley Publishing Co., Inc., 1961.

PROJECT MANAGEMENT PROCEDURES

"I'd be glad to improve myself,"
he said, "but I don't know how
to go about it. What shall I do?"

From SAGGY BAGGY ELEPHANT by K. & B. Jackson, © 1947 by Western Publishing Company, Inc. Used by permission.

In adopting the philosophy of the Saggy Baggy Elephant, there are significant computational improvements to be gained when dealing with specific classes of network-flow problems. Such improvements can occur when using networks to represent and control large construction or development projects. Network representations used for CPM/PERT project control and activity scheduling have a very special structure. In particular, such representations contain directed arcs and form circuitless networks. Because of this special structure, very efficient and simple computational procedures have been developed to recover relevant project information. These specialized techniques, along with related problems of cost and resource control, form the basis for this chapter. Chapter 4 concerns itself not only with the "how to go about it" but also with the "what shall I do" once this class of problems has been recognized. CPM/PERT procedures will be dealt with

4

in three distinct parts. Part I concerns itself with the diagrammatical structure, network construction, and computational procedures used in CPM/PERT analysis; Part II addresses resource and cost control in CPM/PERT networks; and Part III summarizes computer software available to solve this class of problems.

PART I: PROJECT MANAGEMENT WITH CPM AND PERT

with contributions by

Warren H. Thomas
State University of New York

Management of complex projects that consist of a large number of interrelated activities poses involved problems in planning, scheduling, and control, especially when the project activities have to be performed in a specified technological sequence. With the help of PERT (program evaluation and review technique) and CPM (critical path method), the project manager can:

1. Plan the project ahead of time and foresee possible sources of complications and delays in completion.
2. Schedule project activities at appropriate times to conform with proper job sequences so that the project is completed as soon as possible.
3. Coordinate and control the project activities so as to stay on schedule in completing the project.

Closely associated with the philosophies of PERT and CPM are the problems of resource balancing/utilization, compression of selected project activities to reduce total project duration, and the analysis of allowable delays known as *slack* or *float* in order to further coordinate the project. Such considerations allow the project manager to perform the following tasks:

1. Sequence and schedule the use of scarce resources throughout the life of the project.
2. Dynamically control the start dates of each activity.
3. Optimally allocate further funds to the project in an effort to reduce the duration of the entire project.
4. Perform cost/time trade-off analysis among various activities using slack and float measures.

This chapter will present the methodologies necessary to execute these objectives, illustrate algorithmic procedures, and discuss computerized solutions.

4.1 Origin and Use of PERT

PERT was developed by the U.S. Navy during the late 1950s to accelerate the development of the Polaris Fleet Ballistic Missile. The development of this weapon involved the coordination of the work of thousands of private contractors and other government agencies. The coordination by PERT was so successful that the entire project was completed 2 years ahead of schedule. This has resulted in further applications of PERT in other weapons development programs in the Navy, Air Force, and Army. At the current time, it is extensively used in industries and other service organizations as well.

The time required to complete the various activities in a research and development project is generally not known a priori. Thus, in its analysis PERT incorporates uncertainties in activity times. It determines the probabilities of completing various stages of the project by specified deadlines, and also calculates the expected time to complete the project. An important and extremely useful by-product of PERT analysis is its identification of various "bottlenecks" in the project. In other words, it identifies the activities that have high potential for causing delays in completing the project on schedule. Thus, even before the project has started, the project manager knows where to expect delays. He or she can take the necessary preventive measures to reduce possible delays so that the project schedule is maintained.

Because of its ability to handle uncertainties in job times, PERT is extensively used in research and development projects.

4.2 Origin and Use of CPM

The critical path method closely resembles PERT in many aspects but was developed independently by E. I. du Pont de Nemours Company. As a matter of fact, the two techniques, PERT and CPM, were developed almost simultaneously. The major difference between them is that CPM does not incorporate uncertainties in job times. Instead, it assumes that activity times are proportional to the amount of resources allocated to them, and that by changing the level of resources the activity times and the project completion time can be varied. Thus, CPM assumes prior experience with similar projects from which relationships between resources and job times are available. CPM then evaluates the trade-off between project costs and project completion time.

CPM is mostly used in construction projects where one has prior experience in handling similar projects. Although PERT and CPM were independently developed and come from different origins, the methods are quite similar. In current practice, the words PERT and CPM are sometimes used interchangeably. The definitions and procedures that follow apply to both PERT and CPM.

4.3 Problem Characteristics

CPM/PERT methods are applicable to projects where a structured sequence of tasks must be completed to achieve a single objective. Illustrations of such projects would be the construction of a building or other structure, a major repair activity, the design and implementation of a complex weapon system, the production of a large piece of equipment, and R&D activities.

These projects have several common characteristics:

1. They are comprised of a well-defined collection of tasks which, when completed, mark the end of the project.

2. The tasks are ordered such that they must be performed in a given sequence.

3. The time to complete each task is known in advance or can be reasonably estimated. In the CPM formulation of the problem, the time to complete a task is assumed to be a single value known in advance (or can be estimated closely). In PERT, greater uncertainty is incorporated by permitting the planner to submit an upper and lower bound on the completion time of each task. The method of task completion time specification represents the only real difference between CPM and PERT.

4. A task once started is allowed to continue without interruption until it is completed.

5. A succeeding task does not need to start immediately upon completion of an immediate predecessor task, although it cannot commence until the prior one has been completed. This characteristic causes some difficulty in the application of CPM or PERT in continuous-processing industries, in which interruption between processing steps is not permitted.

In principle, the duration of a CPM/PERT project network can be determined by methods previously presented in this book. Indeed, the minimum duration of the project is given by the sequence of activities that yields the longest path through the network. This path is called a *critical path* and the activities that comprise the path are called *critical activities*. The activities are called critical because any increase or delay in their completion times will lengthen the entire project. The critical activities will play a major role in the methodologies of CPM/PERT and will not be further explored at this time. However, it is important to note that due to the specialized structure that each CPM/PERT network exhibits, there are easier ways to characterize the project duration than that of conventional maximum-flow algorithms. At this time, note the following structure of a CPM/PERT network:

1. Each arc has an orientation.
2. The project network is constrained to be circuitless.
3. Probabilistic branching is not allowed.
4. An activity cannot be started until *all* preceding activities have been completed.
5. It is always possible to label each activity such that its directed arc always originates at a *lower*-numbered node than the one at which it terminates. Because of this special structure, a very efficient algorithm can be employed to analyze project networks and determine the critical path. Such a procedure will be subsequently presented.

To clarify the discussions that follow, consider the network segment in Figure 4.1, which represents a sequence of six activities, one for each arc.

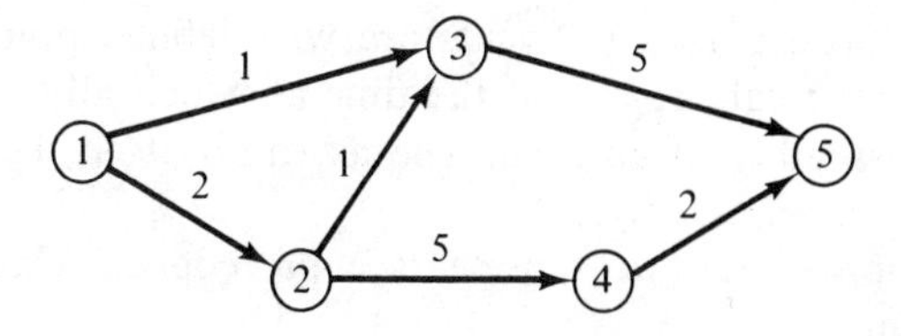

Figure 4.1. An activity network.

The number on each arc is the time required to perform each activity. Note that node 5 will not be realized until activities (3, 5) and (4, 5) are both completed. There are three distinct paths involved in project completion:

I. ① → ③ → ⑤
II. ① → ② → ④ → ⑤
III. ① → ② → ③ → ⑤

Path I has a total duration of 6 units; path II has a duration of 9 units; and path III has a duration of 8 units. As previously indicated, path II is called the *critical path*, since the timely scheduling of those activities on that path is necessary to prevent delay of project completion. Since activities (1, 2), (2, 4), and (4, 5) are on the critical path, no delay in either their start or finish can be tolerated. These are the *critical activities*. In other words, there is no *slack* in either the earliest possible start date or the latest allowable start date for those activities. Each activity *must start on time*. Hence, the earliest possible start time and the latest allowable start time coincide. In like manner, *any* activity could be characterized according to its earliest possible start (ES) and latest allowable start (LS). For example, the ES for activities (1, 3) and (3, 5) are 0 and 3, respectively. Since activity (3, 5) need not be completed until time $T = 9$, the LS for activity (3, 5) is 4. Note that, correspondingly, the latest allowable start (LS) for activity (1, 3) will be time 3. Hence, the *slacks* for activities (1, 3) and (3, 5), defined as LS − ES, are 3 units and 1 unit, respectively. Following this same logic, the reader can verify that activity (2, 3) has slack of 1 unit. For larger project networks, it is easier to calculate activity slack in a tabular fashion. We will now develop the necessary mechanics and mathematical structure for such a procedure.

4.4 Network Construction

The project *network* is a pictorial description of a plan showing the interrelationships among all the tasks required to complete a project. It is composed of directed arcs connecting pairs of nodes. The time-consuming elements of a network (arcs) are known as *activities*. The *nodes* (circles in the network) represent *events*. Events are well-defined points in time. For example, an event might represent the time at which all parts are on hand to permit the assembly of an item. The assembly itself, being a time-consuming element, is an activity.

The direction of the arc represents a precedence relationship. In the network segment

$$(i) \xrightarrow{A} (j)$$

event i must occur before activity A can commence. Similarly, event j cannot occur until activity A has been completed.

The precedence relationship is *transitive* among nodes. If i precedes j and j precedes k, then i precedes k:

$$(i) \longrightarrow (j) \longrightarrow (k)$$

Occasionally, a precedence relationship between activities cannot be represented correctly with conventional activity and event structure. For instance, assume that the network diagram in Figure 4.2 is intended to represent the sequence [activity G follows B and C] and [E follows B (but not C)]. This diagram is incorrect, for it implies that both G and E follow C and B. To permit correct representation, a dummy activity should be used which requires zero time units for its completion. A dummy activity is generally identified by a dashed arrow. The correct representation is given in Figure 4.3, and the dummy activity is labeled X. Dummy activities can be used whenever necessary to show a relationship that cannot otherwise be depicted.

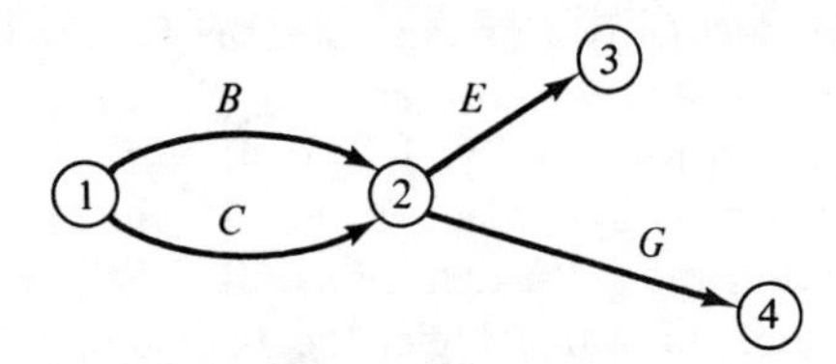

Figure 4.2. Incorrect representation of a precedence relationship.

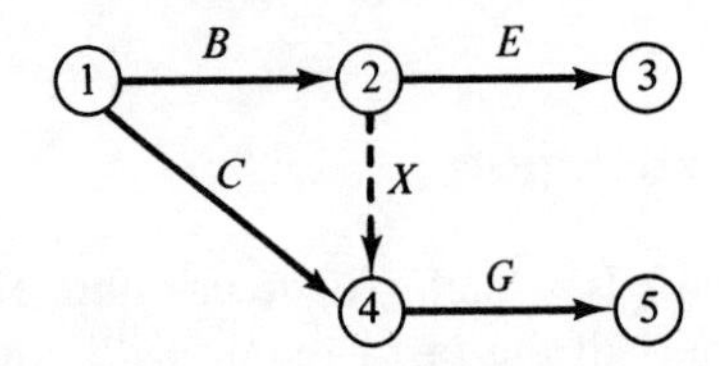

Figure 4.3. Network with a dummy activity.

They are merely a device for forcing a desired relationship without affecting the actual project duration. References [26] and [29] provide further guidance in network formulation. To illustrate the construction of a CPM/PERT diagram, we use the following example.

4.4.1 A Manufacturing Problem

Consider the development of a project network to describe the relationships of the activities involved in the manufacture of a large machine in which subassemblies 1 and 2 are combined into subassembly 4, which is then joined with subassembly 3 to make the final product. Because of the need to match certain items in subassembly 3 with corresponding items in subassembly 2, the third subassembly cannot be built until the parts for subassembly 2 are on hand. Let us assume that the principal activities required to build the machine are those identified in Table 4.1.

The network describing these activities is shown in Figure 4.4. Note that

Table 4.1. EXAMPLE PROBLEM—BUILDING A LARGE MACHINE

Activity		*Duration*	*Immediate*
I.D.	*Description*	*(days)*	*Predecessors*
A	Procure parts for subassembly 1	5	None
B	Procure parts for subassembly 2	3	None
C	Procure parts for subassembly 3	10	None
D	Build subassembly 1	7	*A*
E	Build subassembly 2	10	*B*
F	Build subassembly 4	5	*D* and *E*
G	Build subassembly 3	9	*B* and *C*
H	Final assembly	4	*F* and *G*
I	Final inspection and test	2	*H*

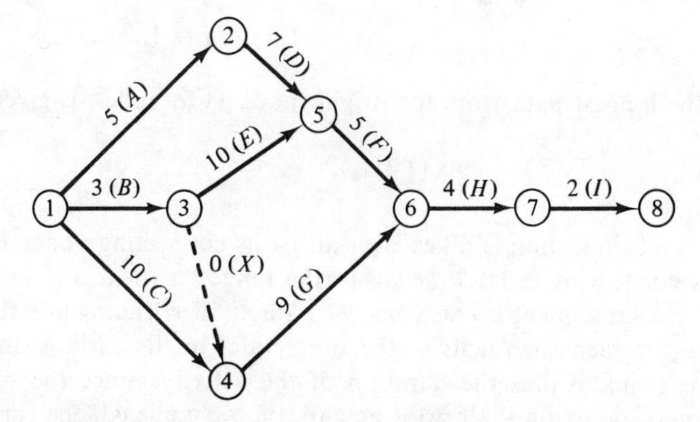

Figure 4.4. Event-on-node network: example problem.

there is a single *origin* event and a single *terminal* event. With the exception of these, every event has at least one activity leading into it and at least one leading from it. Every activity provides a unique connection between two nodes, thus making it possible to identify an activity by the events that it connects. The event nodes have been numbered to provide for their identification. By convention they are numbered so that all activities lead from lower-numbered nodes to higher-numbered nodes.

The ultimate purpose of the network is to provide guidance for the scheduling of individual activities. To do this, one needs first to know the earliest possible and latest permissible times at which each event can occur.

4.5 Earliest Possible Times for Each Event

Recall that an *event* corresponds to a *node* and represents the starting point for one or more activities. Let $T_j(E)$ equal the earliest time event j can occur, $j = 1, 2, \ldots, n$, where n is the number of events (nodes) in the network. Note that $T_j(E)$ is the earliest possible time at which *all* activities incident to node j can be completed. Let d_{ij} represent the duration time of the activity connecting two events (nodes) i and j.

Since an event j cannot occur until all activities leading into node j have been completed, and since an activity cannot commence until its prior event has occurred, we calculate as the earliest time for each event the length of the *longest* path from the origin to that event.

Suppose that there are r paths leading from event 1 to event j. Denote these paths as $\Pi_1, \Pi_2, \ldots, \Pi_r$. Corresponding to each path is a measure equal to the sum of the duration times of all activities on the path. Hence,

$$T_j(\Pi_k) = \sum_m \sum_p d_{mp} \quad m, p \in \Pi_k \qquad k = 1, 2, \ldots, r$$

$$j = 1, 2, \ldots, n$$

Hence, the longest path from the origin (node 1) to node j is given by

$$T_j(E) = \max_k \, [T_j(\Pi_k)] \qquad j = 1, 2, \ldots, n$$

where the maximization is taken over all paths connecting nodes 1 and j.

It is convenient to let $T_1(E) = 0$ (the longest path to the first node is zero). For a subsequent event, consider *all* activities leading into that event. Calculate for each such activity the time equal to the earliest time of the prior event (node) plus the duration of the activity. Since the succeeding event cannot occur until all prior activities are completed, the earliest possible time for the event in question is the maximum of these various times.

In other words, the earliest time for event j is

$$T_j(E) = \begin{cases} 0 & \text{if } j = 1 \quad \text{(origin)} \\ \max\limits_{i<j} [T_i(E) + d_{ij}] & 2 \leq j \leq n \end{cases}$$

where the maximization is taken over all activities terminating at node j from any precedent node i. For a network with n events $j = 1, 2, \ldots, n$, the computation proceeds until the earliest time for terminal event n has been computed.

In the example,

$$T_1(E) = 0$$

$$T_2(E) = T_1(E) + d_{12} = 0 + 5 = 5$$

$$T_3(E) = T_1(E) + d_{13} = 0 + 3 = 3$$

$$T_4(E) = \max \begin{Bmatrix} T_3(E) + d_{34} = 3 + 0 = 3 \\ T_1(E) + d_{14} = 0 + 10 = 10 \end{Bmatrix} = 10$$

$$T_5(E) = 13$$

$$T_6(E) = 19$$

$$T_7(E) = 23$$

$$T_8(E) = 25$$

4.6 Latest Allowable Times for Each Event

Let $T_i(L)$ equal the latest time event i can occur without affecting the completion of the total project (latest allowable finish for the set of all arcs incident to node i). Starting with event n, proceed *backward* through each preceding event. To guarantee that the critical (longest) path will not be exceeded, it is necessary to start the procedure by setting the latest allowable time for the terminal event equal to the earliest possible completion time. Hence,

$$T_n(L) = T_n(E)$$

For any event i that immediately precedes event n, the latest allowable start time is given by

$$T_i(L) = T_n(L) - \min [d_{in}], \qquad i \leq n - 1$$

To compute the latest time for any event i $(i < n)$, consider all activities leading from that event. Compute for each such activity the latest completion time of all subsequent event(s) minus the activity duration. The smallest of these becomes the latest time for event i.

$$T_i(L) = \begin{cases} T_n(E) & i = n \\ \min\limits_{j>i} [T_j(L) - d_{ij}] & 1 \leq i \leq n-1 \end{cases}$$

The minimization is taken over all j connected from i by an activity (i, j). The computation proceeds until the latest time for the origin (event 1) is computed.

Again referring to the previous example, illustrative computations are as follows.

$$T_8(L) = T_8(E) = 25$$

$$T_7(L) = T_8(L) - d_{78} = 25 - 2 = 23$$

$$T_6(L) = T_7(L) - d_{67} = 23 - 4 = 19$$

$$T_5(L) = T_6(L) - d_{56} = 19 - 5 = 14$$

$$T_4(L) = T_6(L) - d_{46} = 19 - 9 = 10$$

$$T_3(L) = \min \begin{Bmatrix} T_5(L) - d_{35} = 14 - 10 = 4 \\ T_4(L) - d_{34} = 10 - 0 = 10 \end{Bmatrix} = 4$$

$$T_2(L) = 7$$

$$T_1(L) = 0$$

4.7 Slack Times and the Critical Path

The maximum time an event can be delayed without causing a corresponding delay in total project completion time is known as the *slack time* for that event. Let

$$S_i = \text{slack time for event } i$$

Then,

$$S_i = T_i(L) - T_i(E)$$

If the latest allowable start time and earliest possible start time for an event are the same ($S_i = 0$), no slippage in the event is permissible. Those events which have zero slack time are events on the *critical path*. Any activities connecting events on the critical path are called *critical activities*. Denote the set of critical activities as SC. In the example, events 1, 4, 6, 7, and 8 are on the critical path. Hence,

$$T_n(L) = T_n(E) = \sum_i \sum_j d_{ij} = 25 \qquad (i, j) \in \text{SC}$$

The critical path is a connected sequence of activities and events extending from the origin to the terminal, any of which if delayed would cause delay of the final project completion. Conversely, those activities and events not on the critical path can be delayed by a nonzero amount of time without affecting total completion time. From Figure 4.4, it can be seen, for example, that the earliest possible start time for event 2 can be delayed $7 - 5 = 2$ units of time without affecting the project's completion.

From a management viewpoint, control of a project should devote primary attention to those activities on the critical path. If reduction in total project completion time is desired, efforts must be applied to the reduction of only those activities on the critical path. Such a reduction of time for any activity on the critical path will be called *crashing*. Optimal crashing procedures will be discussed in Section 4.12.

It should be noted that as the duration of an activity is reduced, it is possible for the critical path to change. For example, the reader should verify that if either activity C or G is reduced more than 1 unit of time, the critical path will change to events 1, 3, 5, 6, 7, 8.

To proceed, it is necessary to distinguish between project *control* and project *time reduction*. Project time reduction is accomplished by allocating more resources to individual activities on the critical path, resulting in shorter activity duration. Project control is the manipulation of activity start/stop dates in order to balance available resources. Time reduction will be discussed in Section 4.12. At this point we discuss project control. As previously stated, control of any project is affected by regulating the starting and stopping of activities, not their durations, since it is the accomplishment of all activities (goals) which comprises the total project. It is therefore desirable to provide specific activity scheduling information in terms of the earliest and latest times that each activity can be started and finished. As we have seen, the critical path passes through those activities for which the earliest and latest start (or latest finish) times are identical, thus permitting no delays.

4.8 Tabular Calculation of the Early Start and Latest Finish Times

The *earliest start* time for an activity is the earliest time the activity can be started assuming that all previous activities are completed as early as possible. Let

$$\text{ES}_{ij} = \text{earliest start for activity } (i, j)$$

Since an *activity* cannot commence until its predecessor *event* has occurred,

$$\text{ES}_{ij} = T_i(E)$$

where the subscript i represents the predecessor event for activity (i, j). It follows that

$$\text{EF}_{ij} = \text{earliest finish for activity } (i, j)$$

$$\text{EF}_{ij} = T_i(E) + d_{ij} = \text{ES}_{ij} + d_{ij}$$

The *latest finish* time for an activity is the latest time it can be completed without delaying the final project completion time. Let

$$\text{LF}_{ij} = \text{latest finish for activity } (i, j)$$

Since an activity can be completed no later than the latest permissible time for successor event j to occur,

$$\text{LF}_{ij} = T_j(L)$$

The *latest start* for activity (i, j) can be computed as follows:

$$\text{LS}_{ij} = \text{LF}_{ij} - d_{ij}$$

Table 4.2 contains the results of these computations applied to each activity of the example. The critical path consists of those activities identified with a star (*). Note that these are activities with zero slack.

Table 4.2. ACTIVITY EARLIEST AND LATEST START AND FINISH TIMES: EVENT-ON-NODE METHOD

				Earliest		*Latest*			
Activity	*Predecessor Event, i*	*Successor Event, j*	*Duration, d_{ij}*	*Start, ES_{ij}*	*Finish, EF_{ij}*	*Start, LS_{ij}*	*Finish, LF_{ij}*	*Slack*	*Critical Path*
A	1	2	5	0	5	2	7	2	
B	1	3	3	0	3	1	4	1	
C	1	4	10	0	10	0	10	0	*
D	2	5	7	5	12	7	14	2	
X	3	4	0	3	3	10	10	7	
E	3	5	10	3	13	4	14	1	
G	4	6	9	10	19	10	19	0	*
F	5	6	5	13	18	14	19	1	
H	6	7	4	19	23	19	23	0	*
I	7	8	2	23	25	23	25	0	*

For the purpose of scheduling the various activities, it is useful to have certain other results, indicating the degree of scheduling freedom under a variety of conditions. These measures, which are extensions to the slack time for an activity, are called *float times*.

4.9 Four Float Measures for Critical Path Scheduling [45]

Float times provide a measure of the scheduling flexibility available in a planning network. They greatly increase the usefulness of early and late start and finish times specified for individual activities. But in many cases, float-time measures are not well understood or used.

Four different float quantities can be defined: *total float*, *free float*, *independent float*, and *safety float*. Float times have several vital uses. The first and most common use is to identify the critical path. A second and more important role of float times is in the actual scheduling of the non-critical path activities. Critical path planning assumes no limitations of resources available to accomplish a project. But, in reality, activities are performed by a labor force with particular skills and specific equipment. It is necessary that activities be allocated to particular time periods so as to balance the demand for various resources. This yields the resource balancing problem which will be discussed in Part II of this chapter.

It may also happen that the organizational unit responsible for a particular activity (department, section, trade grouping, etc.) is also responsible for completing activities that are part of other projects. Float times contain the information needed to accomplish this activity scheduling.

4.9.1 The Computational Procedure

The calculations and meaning of each float measure is best introduced by way of a numerical example. The following situation taken from Thomas [45] might be encountered in a chemical processing plant. Imagine a reactor and storage tank interconnected by a 3-inch insulated process line. Because of erosion, the line needs periodic replacement. Interspersed along the line and at the terminals are valves also in need of replacement. The problem is one of specifing a master timetable for periodic maintenance. Pipe and valves must be ordered. Accurate drawings do not exist and therefore must be made. The line is overhead, so that scaffolding is needed for the replacement operation, but adequate craft labor is available. It is assumed that to minimize downtime on the production unit, work on the project will proceed around the clock. The line cannot be shut down without 30 days' notice. Further, 10 days' lead time is required before the engineering work can be started.

The works engineer has requested that the maintenance and construction superintendent prepare a plan and schedule a review with the operating departments. The plant methods and standards section has furnished the data, shown in Table 4.3, for the various activities involved in the project. Table 4.4 provides a summary of the predecessor and successor relationships for each activity. Using Tables 4.3 and 4.4 together with the results of the

Table 4.3. PIPELINE RENEWAL PROBLEM

	Activity	Immediate Predecessor	Duration (days)
Q	Lead time		10
R	Line available		30
A	Measure and sketch	*Q*	2
B	Develop material list	*A*	1
C	Procure pipe	*B*	30
D	Procure valves	*B*	45
E	Prefabricate sections	*C*	5
F	Deactivate line	*R, B*	1
G	Erect scaffold	*B*	2
H	Remove old pipe and valves	*F, G*	6
I	Place new pipe	*H, E*	6
J	Weld pipe	*I*	2
K	Place valves	*D, F, G*	1
L	Fit up pipe and valves	*K, J*	1
M	Pressure test	*L*	1
N	Insulate	*K, J*	4
O	Remove scaffold	*L, N*	1
P	Cleanup	*M, O*	1

Table 4.4. IMMEDIATE PREDECESSOR AND SUCCESSOR SETS

Activity, i	*Predecessor Set*	*Successor Set*
Start	None	*Q, R*
Q	Start	*A*
R	Start	*F*
A	*Q*	*B*
B	*A*	*C, D, F, G*
C	*B*	*E*
D	*B*	*K*
E	*C*	*I*
F	*B, R*	*H, K*
G	*B*	*H, K*
H	*F, G*	*I*
I	*E, H*	*J*
J	*I*	*L, N*
K	*D, F, G*	*L, N*
L	*J, K*	*M, O*
M	*L*	*P*
N	*J, K*	*O*
O	*L, N*	*P*
P	*M, O*	Finish
Finish	*P*	None

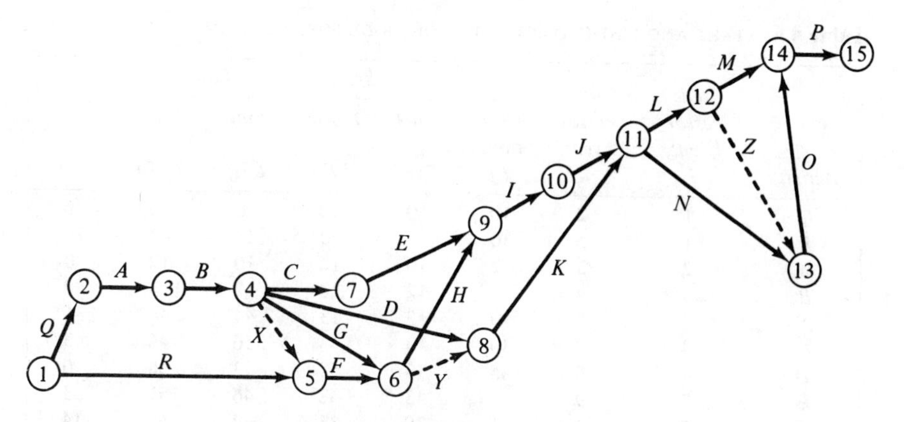

Figure 4.5. Project network.

Table 4.5. EVENT OCCURRENCE TIMES

Event, i	*Earliest,* $T_i(E)$	*Latest,* $T_i(L)$
1	0	0
2	10	10
3	12	12
4	13	13
5	30	44
6	31	45
7	43	46
8	58	58
9	48	51
10	54	57
11	59	59
12	60	63
13	63	63
14	64	64
15	65	65

previous section, Figure 4.5 and Tables 4.5 and 4.6 have been generated, which summarize the project network.

4.9.2 FLOAT CALCULATIONS

The *total float* TF_{ij} of activity (i, j) is the maximum time that activity (i, j) can be delayed without causing delay in the final project completion. It is computed by

$$TF_{ij} = LS_{ij} - ES_{ij}$$

Table 4.6. START AND FINISH TIMES, EVENT-ON-NODE FORMULATION

				Earliest		Latest		
Activity	*Predecessor Event, i*	*Successor Event, j*	*Duration, d_{ij}*	*Start $ES_{ij} = T_i(E)$*	*Finish EF_{ij}*	*Start LS_{ij}*	*Finish $LF_{ij} = T_j(L)$*	*Slack*
Q	1	2	10	0	10	0	10	0
R	1	5	30	0	30	14	44	14
A	2	3	2	10	12	10	12	0
B	3	4	1	12	13	12	13	0
X	4	5	0	13	13	44	44	31
C	4	7	30	13	43	16	46	3
D	4	8	45	13	58	13	58	0
E	7	9	5	43	48	46	51	3
F	5	6	1	30	31	44	45	14
G	4	6	2	13	15	43	45	30
Y	6	8	0	31	31	58	58	27
H	6	9	6	31	37	45	51	14
I	9	10	6	48	54	51	57	3
J	10	11	2	54	56	57	59	3
K	8	11	1	58	59	58	59	0
L	11	12	1	59	60	62	63	3
M	12	14	1	60	61	63	64	3
Z	12	13	0	60	60	63	63	3
N	11	13	4	59	63	59	63	0
O	13	14	1	63	64	63	64	0
P	14	15	1	64	65	64	65	0

or

$$\mathrm{TF}_{ij} = \mathrm{LF}_{ij} - \mathrm{EF}_{ij}$$

An activity with zero total float is on the critical path. Hence, it is clear that the definition of *total float* is identical to that of slack previously introduced. The total float for each of the activities in the example is shown in Table 4.7. The other float measures in Table 4.7 will be discussed shortly. Activities *Q*, *A*, *B*, *D*, *K*, *N*, *O* and *P* are on the critical path.

A fact too often overlooked is the implicit assumption in the total float computation that all predecessor activities (at least all those having any relevance to the activity under consideration) must be completed as early as possible, and that all successor activities are forced to be accomplished as late as possible to realize total float for one activity. Hence, it is generally impossible for each activity actually to realize its total float since the total float for one activity is closely coupled with the total float of others.

Consideration of the example will show that activity *C* can exercise its 3 units of total float only by causing activities *E*, *I*, and *J* to have none.

Hence, total float, although vital for identifying the critical path and

Table 4.7. FLOAT TIMES

Activity, i	*Total Float,* TF_i	*Free Float,* FF_i	*Independent Float,* IF_i	*Safety Float,* SF_i
Q	0	0	0	0
R	14	0	0	14
A	0	0	0	0
B	0	0	0	0
C	3	0	0	3
D	0	0	0	0
E	3	0	0	0
F	14	0	0	0
G	30	1	16	30
H	14	11	0	0
I	3	0	0	0
J	3	3	0	0
K	0	0	0	0
L	3	0	0	3
M	3	3	0	0
N	0	0	0	0
O	0	0	0	0
P	0	0	0	0

useful for overall project scheduling, is of questionable value when scheduling individual activities.

4.9.3 Free Float

Free float is a measure of the maximum time activity (i, j) may be delayed without affecting the start of the successor activities. As above, it is assumed that all prior activities are completed as early as possible. Free float, FF_{ij}, differs from total float in that it measures the time available without delaying succeeding jobs. It is determined by

$$FF_{ij} = T_j(E) - EF_{ij}$$

An activity's free float can, of course, never be larger than its total float.

The free float times for the example are shown in Table 4.7. Although activity *E* has 3 units of total float, it has no free float, since it cannot be delayed without delaying activity *I*. Activity *H*, which has 14 units of total float, has only 11 units of free float. Any delay beyond 11 forces a delay in the earliest possible start of activity *I*.

4.9.4 Independent Float

Planning for the accomplishment of a particular activity could be severely disrupted if prior activities were not completed as early as possible, as is assumed in both total and free float. The *independent float* is a very convenient measure of scheduling freedom. The independent float IF_{ij} of activity (i, j) is the maximum time (i, j) can be delayed without delaying successor activities, if all prior activities are finished as late as possible. It gives a measure of time available if the worst possible conditions prevail in the predecessor activities. It also gives an indication of the possible degree of decoupling the activity from the others in the project, hence the name independent.

The independent float IF_{ij} is found by the expression

$$\mathrm{IF}_{ij} = \max \begin{cases} 0 \\ T_j(E) - [T_i(L) + d_{ij}] \end{cases}$$

The initial maximum term is included to set any negative results to zero.

In the example, only activity G has nonzero independent float. Each of the other activities can be completely deprived of any float time by late predecessor activities.

4.9.5 Safety Float

SF_{ij} is the maximum time an activity can be delayed without affecting the final project completion if predecessor activities are completed as late as possible. This new measure appears to be one of the more useful float measures when scheduling a particular activity. Note that it permits only successor activities to be delayed, not the entire project.

$$\mathrm{SF}_{ij} = \mathrm{LF}_{ij} - [T_i(L) + d_{ij}]$$

or

$$\mathrm{SF}_{ij} = T_j(L) - [T_i(L) + d_{ij}]$$

In the example, only activities R, C, G, and L have nonzero safety float. It can be seen from the network that C can be delayed by delaying E and I while G can be delayed by delaying H. Activities H, E, and I are deprived of any safety float by the possibility of their predecessors being delayed.

Table 4.8 summarizes the four float times according to whether predecessor and successor activities are assumed to be accomplished as early or as late as possible.

One might ask, "Of what value are the various float times?" The first obvious answer is that zero total float identifies an activity on the critical path as one to be closely monitored and controlled. All the floats are useful to determine exactly when an activity is to be accomplished. In many cases, the labor or equipment needed to accomplish an activity must compete with other activities for available supplies of these resources. The various float times provide assistance when actually assigning these resources to each

Table 4.8. CLASSIFICATION OF THE FOUR FLOAT TIMES

		Successors Completed	
		Early	*Late*
Predecessors completed	*Early*	Free float	Total float
	Late	Independent float	Safety float

For Activity (i, j)

Total float: $LS_{ij} - ES_{ij}$ or $LF_{ij} - EF_{ij}$

Free float: $T_j(E) - EF_{ij}$

Independent float: $\max \begin{cases} 0 \\ T_j(E) - [T_i(L) + d_{ij}] \end{cases}$

Safety float: $LF_{ij} - [T_i(L) + d_{ij}]$
or $T_j(L) - [T_i(L) + d_{ij}]$

activity. Float times also permit some freedom in attempting to balance resource demands.

4.10 An Activity-on-Node Formulation

An alternative formulation of the problem, introduced in Section 2.1.2, places all activity times on the nodes of the network. It possesses significant advantages over the older event-on-node method. Because of certain computational and structural advantages over conventional activity on arc procedures, it may become the more commonly used method for computerized project planning.

4.10.1 Network Construction

The primary difference in this method is that activities are represented by *nodes*, with the arcs merely representing precedence relationships. In other words, time is consumed at the nodes and not the arcs (Figure 4.6). Activities *A* and *B* precede activity *C*, which precedes *D* and *E*. Note that the physical process of passing through nodes *A* to *E* is time-consuming. The

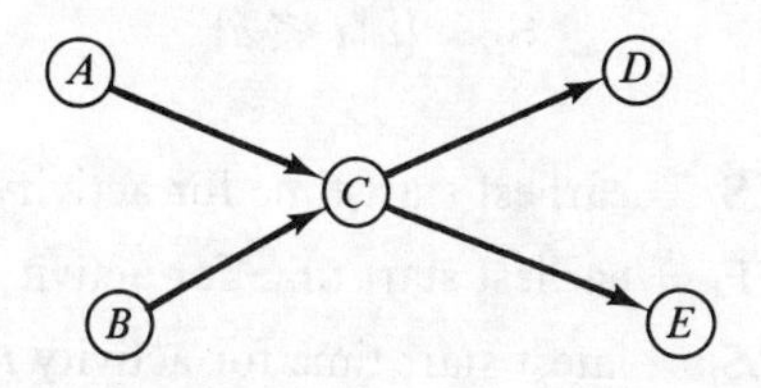

Figure 4.6. Activity-on-node network.

concept of an event is not needed and is not included. All computations directly concern activity start and finish times. Because there is no need to include events, the network construction task is greatly facilitated. Further, as will be seen, there is no need at any time to create dummy activities. Any precedence relationship can be expressed correctly without them.

The construction of the network is facilitated by adding two fictitious activities. The first, which we label "Start," precedes all other activities; the second, referred to as "Finish," follows all activities. Each has zero duration. The example previously considered is represented in the activity-on-node formulation in Figure 4.7. Note that not only is the network very different in appearance, but in addition, the correct precedence relationships are represented without dummy activities. The number in parentheses at each node is the activity time duration.

4.10.2 Computational Procedure

The project is completed when activity P has been completed. The figure shows the network of the project where pseudo-activities START and FINISH, each of duration zero, have been added to provide a single origin and terminal. In this problem, the finish activity could be omitted since there is already a single last activity. It is included as a matter of general procedure. The double-subscript notation necessitated by the previous activity-on-arc formulation is no longer needed. Since all activities are uniquely associated with a single node, a single subscript will suffice. Henceforth, a reference to activity i will implicitly refer to a particular node i. The duration of activity i will now be denoted simply by $d_i, i = 1, 2, \ldots, n$. The duration d_i is usually indicated near each node.

Next it is useful to define a *predecessor* set of activities, $\mathbf{P}_i$, as the set of activities that immediately precede activity i. The immediate *successor* set, $\mathbf{S}_i$, is defined as the set of activities that immediately follow activity i. More precisely, let the symbol $\ll$ denote an immediate precedence relationship. The statement $i \ll j$ implies that i is an immediate predecessor of j or, equivalently, that j is an immediate successor of i. It follows that

$$\mathbf{P}_i = \{a \mid a \ll i\}$$

and

$$\mathbf{S}_i = \{a \mid i \ll a\}$$

Defining

$$\text{ES}_i = \text{earliest start time for activity } i$$

$$\text{EF}_i = \text{earliest start time for activity } i$$

$$\text{LS}_i = \text{latest start time for activity } i$$

$$\text{LF}_i = \text{latest finish time for activity } i$$

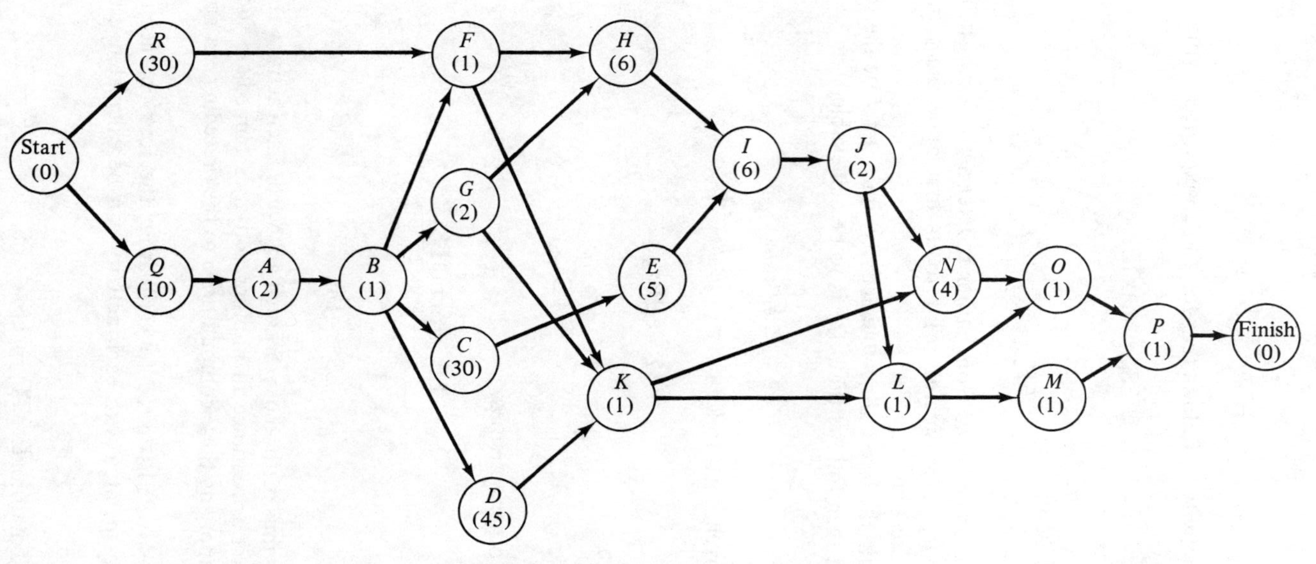

Figure 4.7. Activity-on-node representation.

and letting

$$\mathrm{ES}_{\mathrm{start}} = \mathrm{EF}_{\mathrm{start}} = 0$$

for the initial activity, it follows that for all subsequent activities

$$\mathrm{ES}_i = \max_{X \in \mathbf{P}_i} [\mathrm{EF}_X]$$

and

$$\mathrm{EF}_i = \mathrm{ES}_i + d_i$$

These computations proceed from earlier to later activities in such a manner that each activity is considered prior to its first appearance in another's immediate predecessor set.

To compute the latest allowable activity times, let T be the target completion date. It then follows that, if T is to be achievable,

$$T \geq \mathrm{EF}_{\mathrm{finish}}$$

Again it is conventional to let

$$T = \mathrm{EF}_{\mathrm{finish}}$$

We then obtain

$$\mathrm{LF}_{\mathrm{finish}} = \mathrm{LS}_{\mathrm{finish}} = T$$

Hence, for all other activities,

$$\mathrm{LF}_i = \min_{X \in \mathbf{S}_i} [\mathrm{LS}_X]$$

and

$$\mathrm{LS}_i = \mathrm{LF}_i - d_i$$

This is the previous method of working backward. Each activity is considered prior to its first appearance in another activity's immediate successor set. The result of applying these computations to the preceding example is shown in Table 4.9.

The reader will be asked to verify that the four characterizations of activity float associated with each activity can be calculated using the following results:

$$\text{Total float:} \quad \mathrm{TF}_i = \mathrm{LS}_i - \mathrm{ES}_i \quad \text{or} \quad \mathrm{TF}_i = \mathrm{LF}_i - \mathrm{EF}_i$$

Those activities with zero total float are on the critical path.

Table 4.9. EARLY AND LATE START AND FINISH TIMES

		Earliest		*Latest*	
Activity, i	*Duration, y_i*	*Start, ES_i*	*Finish, EF_i*	*Start, LS_i*	*Finish, LF_i*
Start	0	0	0	0	0
Q	10	0	10	0	10
R	30	0	30	14	44
A	2	10	12	10	12
B	1	12	13	12	13
C	30	13	43	16	46
D	45	13	58	13	58
E	5	43	48	46	51
F	1	30	31	44	45
G	2	13	15	43	45
H	6	31	37	45	57
I	6	48	54	51	57
J	2	54	56	57	59
K	1	58	59	58	59
L	1	59	60	62	63
M	1	60	61	63	64
N	4	59	63	59	63
O	1	63	64	63	64
P	1	64	65	64	65
Finish	0	65	65	65	65

$$\text{Free float:}\quad FF_i = \min_{x \in S_i} [ES_X - EF_i]$$

$$\text{Independent float:}\quad IF_i = \max_{x \in S_i} \begin{cases} 0 \\ \min ES_X - [\max_{Z \in P_i} LF_Z + d_i] \end{cases}$$

$$\text{Safety float:}\quad SF_i = LF_i - \max_{Z \in P_i} [LF_Z + d_i]$$

Using Table 4.7 and applying these results to the previous example, all measures of float can be generated. Of course, these results are identical to those previously obtained in Table 4.6.

4.11 Program Evaluation and Review Technique (PERT)

In previous discussions, the probability considerations in the management of a project were not included. Previous methodologies assumed that the job times are known with certainty. Uncertainty in activity durations is accounted for using PERT.

For each activity in the project network, PERT assumes three estimates for its completion time. They include (1) *a most probable time* denoted by m, (2) *an optimistic time* denoted by a, and (3) *a pessimistic time* denoted by b.

The most probable time is the time required to complete the activity under normal conditions. To include uncertainties, a range of variation in job time is provided by the optimistic and pessimistic times. The optimistic estimate is a good guess on the minimum time required when everything goes according to plans; while the pessimistic estimate is a guess on the maximum time required under adverse conditions such as mechanical breakdowns, minor labor troubles, and shortage of or delays in delivery of material. It should be remarked here that the pessimistic estimate does not take into consideration unusual or prolonged delays or other catastrophes. Because both of these estimates are only qualified guesses, the actual time for an activity could lie outside this range. (From a probabilistic viewpoint we can only say that the probability of a job time falling outside this range is very small.)

Most PERT analysis assumes a Beta distribution for the job times as shown in Figure 4.8, where μ represents the average job duration. The value of μ depends on how close the values of a and b are relative to m.

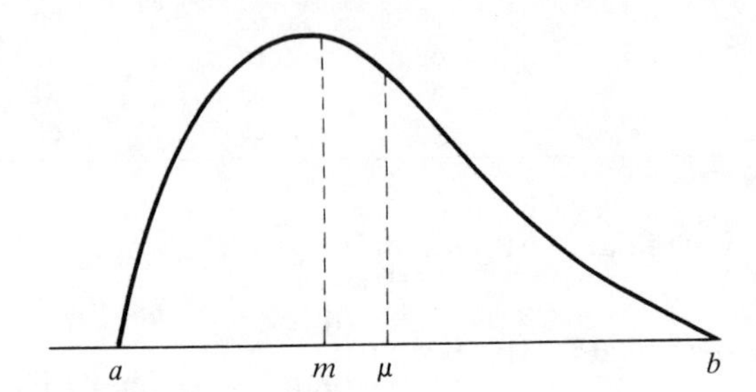

Figure 4.8. Beta distribution for job time.

The expected time to complete an activity is approximated as

$$\mu = \frac{a + 4m + b}{6}$$

Since the actual time may vary from its mean value, we need the variance of the job time. For most unimodal distributions (with single peak values), the end values lie within three standard deviations from the mean value. Thus, the spread of the distribution is equal to six times the standard deviation value (σ).

Thus, $6\sigma = b - a$, or $\sigma = (b - a)/6$. The variance of the job time equals

$$\sigma^2 = \left(\frac{b - a}{6}\right)^2$$

With the three time estimates on all the jobs, PERT calculates the average time and the variance of each job using the foregoing formulas for μ and σ^2. Treating the average times as actual job times, the critical path can be found. The duration of the project (T) is given by the sum of all the job times on the critical path, but the job times are random variables. Hence, the project duration T is also a random variable, and we can talk of the average duration of the project and its variance.

The expected duration of the project is the sum of all the average times of the jobs on the critical path. Similarly, the variance of the project duration is the sum of all the variances of the jobs on the critical path assuming that all the job times are independent.

4.11.1 A PERT Example

Consider a project consisting of nine jobs ($A, B, \ldots, I$) with the precedence relations and time estimates shown in Table 4.10. First, we compute the average time and the variance for each one of the jobs. They are tabulated in Table 4.11.

Table 4.10. PRECEDENCE RELATIONS AND TIME ESTIMATES

Job	*Predecessors*	*Optimistic Time, a*	*Most Probable Time, m*	*Pessimistic Time, b*
A	—	2	5	8
B	*A*	6	9	12
C	*A*	6	7	8
D	*B, C*	1	4	7
E	*A*	8	8	8
F	*D, E*	5	14	17
G	*C*	3	12	21
H	*F, G*	3	6	9
I	*H*	5	8	11

Table 4.11. MEANS AND STANDARD DEVIATIONS FOR JOB TIMES

Job	*Average Time*	*Standard Deviation*	*Variance*
A	5	1	1
B	9	1	1
C	7	$\frac{1}{3}$	$\frac{1}{9}$
D	4	1	1
E	8	0	0
F	13	2	4
G	12	3	9
H	6	1	1
I	8	1	1

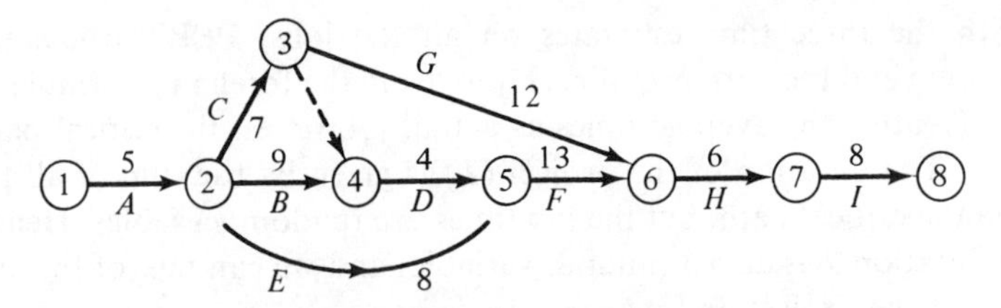

Figure 4.9. Project network of example.

Figure 4.9 shows the project network, using the activity-on-arc representation, where the numbers on the arcs indicate the average job times. Using the average job times, the earliest and latest times of each event are calculated. The critical path is found as ① → ② → ④ → ⑤ → ⑥ → ⑦ → ⑧. Correspondingly, the critical jobs are A, B, D, F, H, and I.

Let T denote the project duration. Then the expected length of the project is

$$E(T) = \text{sum of the expected times of jobs } A, B, D, F, H, \text{ and } I$$
$$= 5 + 9 + 4 + 13 + 6 + 8 = 45 \text{ days}$$

The variance of the project duration is

$$V(T) = \text{sum of the variances of jobs } A, B, D, F, H, \text{ and } I$$
$$= 1 + 1 + 1 + 4 + 1 + 1 = 9$$

The standard deviation of the project duration is

$$\sigma(T) = \sqrt{V(T)} = 3$$

4.11.2 Probabilities of Completing the Project

The project duration T is the sum of all the job times on the critical path. PERT assumes that all the job times are independent and are identically distributed. Hence, by the Central Limit Theorem, T has a normal distribution with mean $E(T)$ and variance $V(T)$. Figure 4.10 exhibits a normal distribution with mean μ and variance σ^2.

In our example T is normally distributed with mean 45 and standard deviation 3. For any normal distribution, the probability that the random variable lies within one standard deviation from the mean is 0.68. Hence, there is a 68% chance that the project duration will be between 42 and 48 days. Similarly, there is a 99.7% chance that T will lie within three standard deviations (between 36 and 54).

We can also calculate the probabilities of meeting specified project deadlines. For example, the management wants to know the probability of

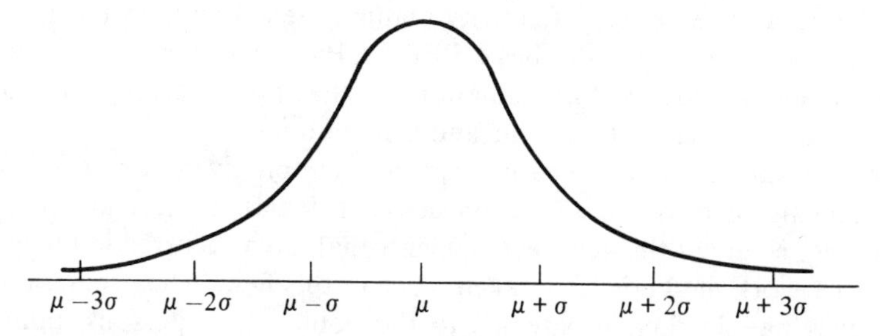

Figure 4.10. Normal distribution with mean μ and standard deviation σ.

completing the project by 50 days. In other words, we have to compute Prob ($T \leq 50$), where $T \sim N$ (45, 9). This can be obtained from a table of the standardized normal distribution whose mean is 0 and standard deviation is 1.

In probability theory the random variable $Z = [T - E(T)]/\sigma(T)$ is proved to be normally distributed with mean 0 and standard deviation 1. Hence,

$$\text{Prob}\,(T \leq 50) = \text{Prob}\left(Z \leq \frac{50 - 45}{3}\right) = \text{Prob}\,(Z \leq 1.67) = 0.95$$

Thus, there is a 95% chance that the project will be completed within 50 days.

Suppose that we want to know the probability of completing the project 4 days sooner than expected. This means that we have to compute

$$\text{Prob}\,(T \leq 41) = \text{Prob}\left(Z \leq \frac{41 - 45}{3}\right) = \text{Prob}\,(Z \leq -1.33) = 0.09$$

Hence, there is only a 9% chance that the project will be completed in 41 days.

PART II: RESOURCE ALLOCATION IN PROJECT NETWORKS

with contributions by

Edward W. Davis
Colgate Darden Graduate School of Business
University of Virginia

The general appearance of program evaluation and review technique (PERT) and critical path method (CPM) techniques in the early 1960s saw an explosion of interest in project network methods. The ensuing popularity of these techniques demonstrated that network models are a useful means of formulating a wide variety of activity planning/scheduling problems. But it was also recognized that the basic PERT/CPM procedures are somewhat naive models of most real-life situations in that they focus on *time* to the exclusion of *resource* requirements and availabilities.

As a result, there has been increasing attention given in recent years to the problems of resource allocation associated with project planning and scheduling. Many of the new developments that have occurred in the general area of network methods have taken place in this field. The proliferation of techniques has, in fact, progressed to the point where persons unfamiliar with these procedures are often confused over exactly what can and cannot be done with regard to resource management. This section is an attempt to allay such confusion by categorizing the types of procedures that have been developed and summarizing the capabilities and limitations of each approach.

4.12 Time vs. Cost: Dollar Allocations

It is often true that the performance of some or all project activities can be accelerated by the allocation of more resources at the expense of the higher activity direct cost. When this is so, there are many different combinations of activity durations that will yield some desired schedule duration. However, each combination may yield a different value of total project cost. Time/cost trade-off procedures are directed at determining the least-cost schedule for any given project duration.

For example, consider the simple eight-activity project shown in Table 4.12. Each activity can be performed at different durations ranging from an upper "normal" value, at some associated "normal" cost, down to a lower,

Table 4.12. ILLUSTRATIVE NETWORK WITH ACTIVITY TIME/COST DATA

	Normal		*"Crash"*		
Activity	*Time (days)*	*Cost*	*Time (days)*	*Cost*	*Cost Slope*
(0, 1)	4	$ 210	3	$ 280	$ 70
(0, 2)	8	400	6	560	80
(1, 2)	6	500	4	600	50
(1, 4)	9	540	7	600	30
(2, 3)	4	500	1	1100	200
(2, 4)	5	150	4	240	90
(3, 5)	3	150	3	150	—[a]
(4, 5)	7	600	6	750	150
		$3050		$4280	

[a]This activity cannot be expedited.

"crash" value, with an associated higher cost. Note that if time/cost trade-off values for each activity are assumed to be linear, the cost of intermediate activity durations between the normal and crash durations is easily determined from the single cost "slope" value for each activity [e.g., the cost of performing activity (0, 2) in 7 days instead of 8 equals $400 + $80, or $480].

If all activity durations are set at "normal" values, the project duration is 22 days, as determined by the critical path consisting of activities (0, 1), (1, 2), (2, 4), (4, 5) in Figure 4.11. The calculations are given in Table 4.13. The associated cost of project performance is $3050, as indicated in Figure 4.12. Note that this cost could be increased to $3870 through unintelligent decision-making by "crashing" all activities not on the critical path, with no decrease in project duration. Between these upper and lower cost values for a project duration of 22 days there are several other possible values, depending upon the number of noncritical activities crashed.

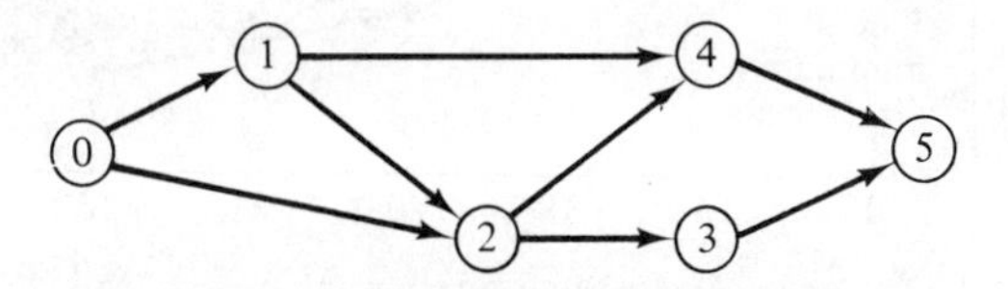

Figure 4.11. Activity network for cost crashing problem.

Table 4.13. CRITICAL PATH ANALYSIS

		Earliest		*Latest*			
Activity	*Duration*	*ES*	*EF*	*LS*	*LF*	*Slack*	*CP*
(0, 1)	4	0	4	0	4	0	*
(0, 2)	8	0	8	2	10	2	
(1, 2)	6	4	10	4	10	0	*
(1, 4)	9	4	13	6	15	2	
(2, 3)	4	10	14	15	19	5	
(2, 4)	5	10	15	10	15	0	*
(3, 5)	3	14	17	19	22	5	
(4, 5)	7	15	22	15	22	0	*
					Total duration:		22

If all activity durations are set at "crash" values, the project duration can be decreased to 17 days, with a total cost of $4280, as shown by the extreme upper left point of Figure 4.12. However, a duration of 17 days can also be achieved at lower cost by not "crashing" activities unnecessarily. Thus, activity (0, 2) can be set at 7 instead of 6, activity (1, 4) at 8 instead of 7, and activity (2, 3) at 4 instead of 1. With all other activities set at crash value, the associated cost of performance for 17-day project duration is

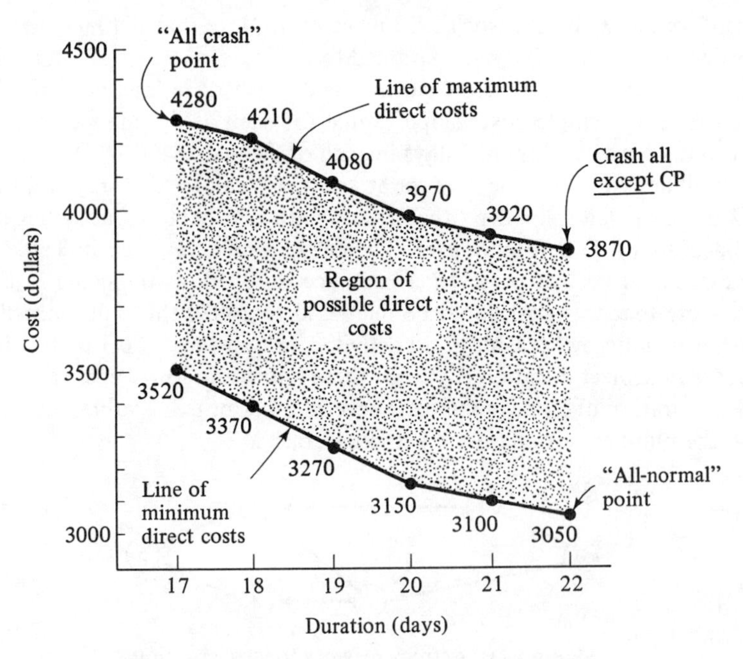

Figure 4.12. Project duration versus direct cost.

reduced to \$3520. This value is the lowest possible value for 17-day project duration, as may easily be determined by experimentation.

Between 22 and 17 days there are several possible values of project duration, as shown in Figure 4.12. For each such duration there is a range of possible cost values, depending upon the duration of individual activities and whether activities are crashed unnecessarily or not. Figure 4.12 shows the curve of both maximum and minimum costs and the region of possible costs for each duration between these curves.

In this simple example the minimum-direct-cost curve is easily determined by trial and error. But in more realistic cases consisting of dozens or even hundreds of activities, such trial-and-error determination becomes extremely tedious, if not impossible. Thus various systematic computation schemes, including mathematical programming, have been developed to aid in the quick determination of the minimum-cost curve for every possible value of project duration. Some of these routines are designed to handle cases in which the activity time/cost trade-off relationships are nonlinear; many will also produce the lowest total cost curve (i.e., sum of direct plus indirect costs), as illustrated in Figure 4.13.

To illustrate the fundamentals of total cost crashing, consider once again the previous example with a fixed indirect cost of \$130 per day. In this case, the curve of Figure 4.13 becomes that of Figure 4.14. The calculations are summarized in Table 4.14.

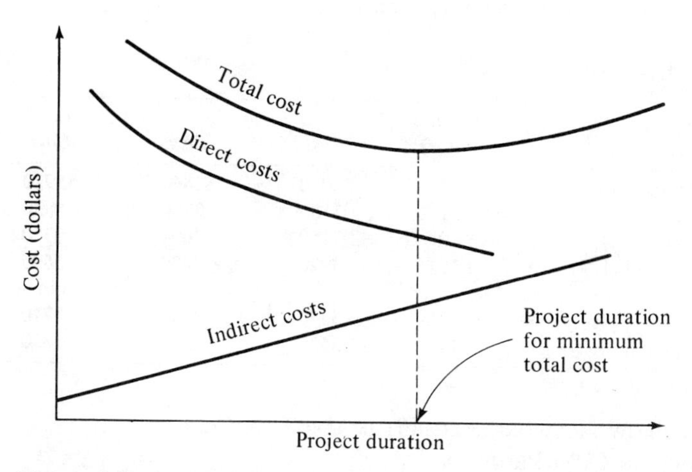

Figure 4.13. Determining project schedule for minimum total cost.

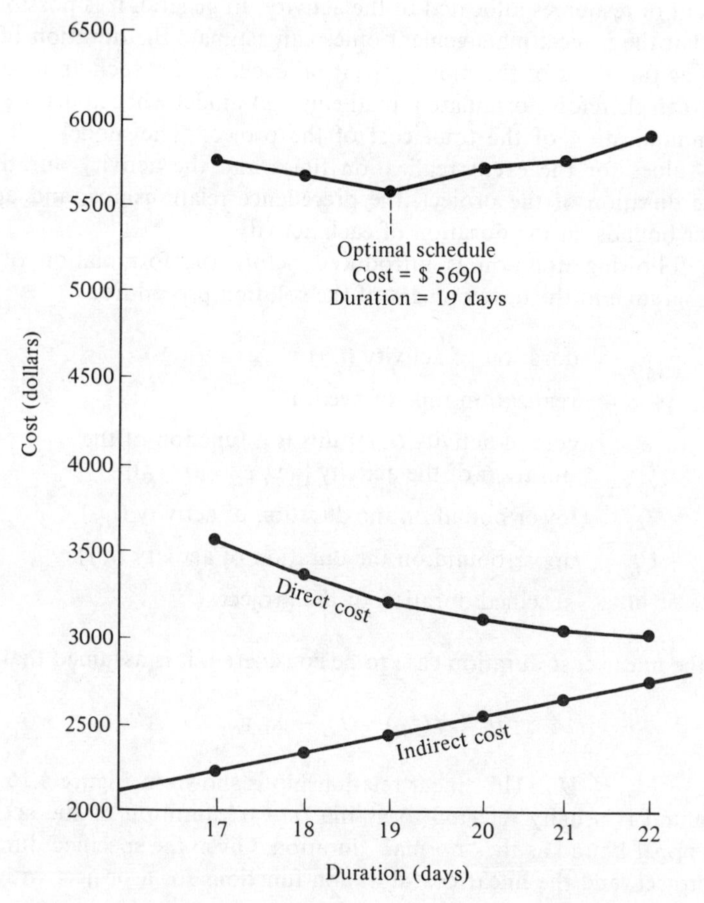

Figure 4.14. Optimal crashing with both direct and indirect costs.

Table 4.14. COST CRASHING

Crash Activity	*Duration (days)*	*Direct Cost*	*Indirect Cost*	*Total Cost*
(Critical path)	22	\$3050	\$2860	\$5910
(1, 2)	21	3100	2730	5830
(1, 2)	20	3150	2600	5750
(2, 4) and (1, 4)	19	3220	2470	5690 (optimal)
(4, 5)	18	3370	2340	5710
(0, 1) and (0, 2)	17	3520	2210	5730

4.12.1 A Flow-Network Algorithm for Time/Cost Trade-offs in CPM Projects

The duration of any activity of a project can be controlled by limiting the amount of resources allocated to the activity. In general, it is possible to assume that the project management office can estimate the duration of the activities as functions of the money spent on each. Under such an assumption, one can therefore formulate a mathematical model whose purpose will be the minimization of the total cost of the project. The model will find optimal values for the event realization times and the activity durations, given the duration of the project, the precedence relationships, and upper and lower bounds on the duration of each activity.

The following notation is introduced before the formulation of the linear program and the development of the solution procedure:

y_{ij} duration of activity (i, j)

t_i realization time of event i

c_{ij} cost of activity (i, j); this is a function of the duration of the activity [i.e., $c_{ij} = f(y_{ij})$]

L_{ij} lower bound on the duration of activity (i, j)

U_{ij} upper bound on the duration of activity (i, j)

T specified duration of the project

In the linear cost-duration case to be considered, it is assumed that

$$c_{ij} = f(y_{ij}) = b_{ij} - a_{ij}y_{ij}$$

where $L_{ij} \leq y_{ij} \leq U_{ij}$. This linear relationship is shown in Figure 4.15. The lower bound is usually referred to as the "crash" duration of the activity, and the upper bound as the "normal" duration. Given the specified duration of the project, and the linear cost-duration functions for a project with net-

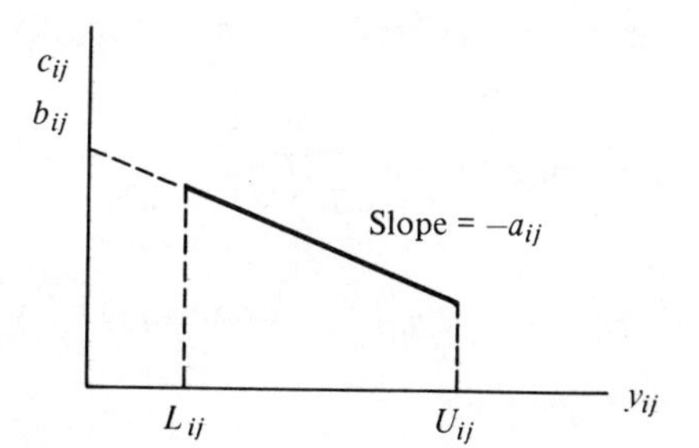

Figure 4.15. Linear cost-duration function.

work representation $G = (\mathbf{N}, \mathbf{A})$, we want to know which activities in the set $\mathbf{A}$ must be crashed and which must be left at their normal durations.

The problem under consideration can be formulated in terms of a linear program, as shown below. It is assumed that the set of nodes $\mathbf{N}$ can be defined as $\mathbf{N} = \{1, 2, \ldots, n\}$, where node 1 represents the start of the project and node n the end of it.

$$\text{minimize} \sum_{(i,j) \in \mathbf{A}} (b_{ij} - a_{ij}y_{ij})$$

subject to

$$\begin{aligned} t_i - t_j + y_{ij} &\leq 0, && \text{for all } (i, j) \text{ in } \mathbf{A} \\ -t_1 + t_n &= T \\ y_{ij} &\leq U_{ij}, && \text{for all } (i, j) \text{ in } \mathbf{A} \\ y_{ij} &\geq L_{ij}, && \text{for all } (i, j) \text{ in } \mathbf{A} \end{aligned}$$

This model is equivalent to the following linear program with the objective function being maximized. The nonnegativity constraints for t_i and y_{ij} are not explicitly stated in the model, which implies that all the constraints of the corresponding dual program must be of the equality type.

$$\text{maximize} \sum_{(i,j) \in \mathbf{A}} a_{ij}y_{ij}$$

subject to

$$\begin{aligned} t_i - t_j + y_{ij} &\leq 0, && \text{for all } (i, j) \text{ in } \mathbf{A} \\ -t_1 + t_n &= T \\ y_{ij} &\leq U_{ij}, && \text{for all } (i, j) \text{ in } \mathbf{A} \\ -y_{ij} &\leq -L_{ij}, && \text{for all } (i, j) \text{ in } \mathbf{A} \end{aligned}$$

Let f_{ij}, v, γ_{ij}, and δ_{ij} be the dual variables corresponding to the foregoing linear constraints. The dual variables have been listed in the same order in which the constraints were incorporated in that model. The dual program can be formulated as follows:

$$\text{minimize } \{Tv + \sum_{i,j} U_{ij}\gamma_{ij} - \sum_{i,j} L_{ij}\delta_{ij}\}$$

subject to

$$f_{ij} + \gamma_{ij} - \delta_{ij} = a_{ij}, \qquad \text{for all } (i, j) \text{ in } \mathbf{A}$$

$$\sum_j f_{1j} - v = 0$$

$$\sum_j [f_{ij} - f_{ji}] = 0, \qquad \text{for } i = 2, \ldots, n - 1$$

$$-\sum_j f_{jn} + v = 0$$

$$f_{ij}, \gamma_{ij}, \delta_{ij} \geq 0, \qquad \text{for all } (i, j) \text{ in } \mathbf{A}$$

From the mathematical structure of the dual program, it can be concluded that the dual variables f_{ij} may be viewed as flows in a capacitated network. The second, third, and fourth groups of constraints correspond to flow constraints for the source, the intermediate, and the terminal nodes, respectively. In particular, the third group of constraints corresponds to the well-known flow-conservation constraints.

The complementary slackness conditions for linear programs can be used to derive the following results, which must be true for a solution to be optimal:

$$\begin{aligned} t_i - t_j + y_{ij} < 0 &\qquad f_{ij} = 0 \\ y_{ij} < U_{ij} &\qquad \gamma_{ij} = 0 \\ y_{ij} > L_{ij} &\qquad \delta_{ij} = 0 \\ f_{ij} > 0 &\qquad t_i - t_j + y_{ij} = 0 \\ \gamma_{ij} > 0 &\qquad y_{ij} = U_{ij} \\ \delta_{ij} > 0 &\qquad y_{ij} = L_{ij} \end{aligned}$$

Note that the last two conditions jointly imply that γ_{ij} and δ_{ij} cannot be simultaneously positive, under the assumption that $L_{ij} \neq U_{ij}$. Since $f_{ij} + \gamma_{ij} - \delta_{ij} = a_{ij}$, we can easily conclude that the nonnegative values of γ_{ij} and δ_{ij} are given by

1. $\gamma_{ij} = a_{ij} - f_{ij}$, when $\delta_{ij} = 0$.
2. $\delta_{ij} = f_{ij} - a_{ij}$, when $\gamma_{ij} = 0$.

Therefore, we can write

1. $\gamma_{ij} = \max [0; a_{ij} - f_{ij}]$, when $\delta_{ij} = 0$.
2. $\delta_{ij} = \max [0; f_{ij} - a_{ij}]$, when $\gamma_{ij} = 0$.

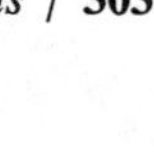

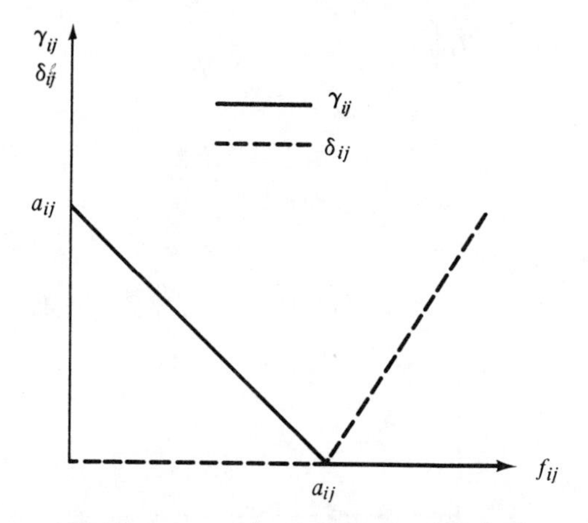

Figure 4.16. γ_{ij} and δ_{ij} as functions of f_{ij}.

The graphical representation of γ_{ij} and δ_{ij} as linear functions of f_{ij} is given in Figure 4.16.

Three cases can be identified if all possible values for γ_{ij}, δ_{ij}, and f_{ij} are investigated:

Case 1: $\gamma_{ij} > 0$ and $\delta_{ij} = 0$; $0 \leq f_{ij} < a_{ij}$ and $y_{ij} = U_{ij}$.
Case 2: $\gamma_{ij} = 0$ and $\delta_{ij} = 0$; $f_{ij} = a_{ij}$ and $L_{ij} \leq y_{ij} \leq U_{ij}$.
Case 3: $\gamma_{ij} = 0$ and $\delta_{ij} > 0$; $f_{ij} > a_{ij}$ and $y_{ij} = L_{ij}$.

Based on the complementary slackness conditions, the following optimality conditions are obtained for each case:

Case 1

$$0 < f_{ij} < a_{ij} \quad \text{and} \quad t_i - t_j + U_{ij} = 0$$

or

$$f_{ij} = 0 \quad \text{and} \quad t_i - t_j + U_{ij} < 0$$

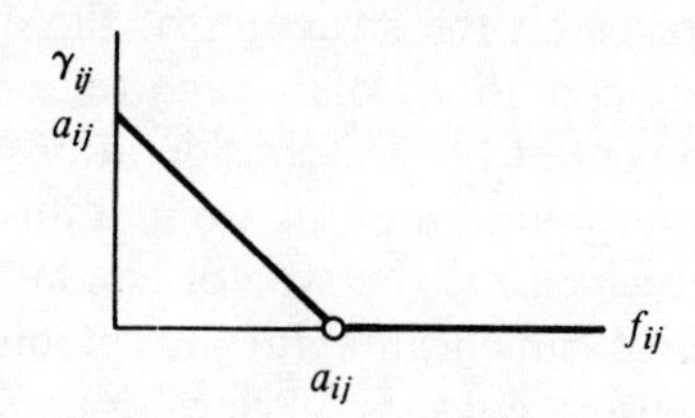

Case 2

$$f_{ij} = a_{ij} \quad \text{and} \quad t_i - t_j + y_{ij} = 0, \qquad L_{ij} \leq y_{ij} \leq U_{ij}$$

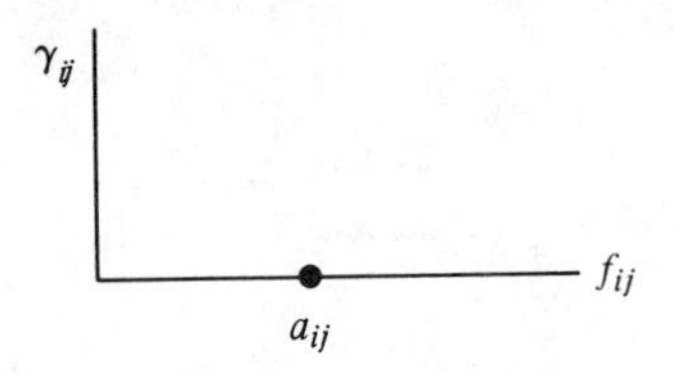

Case 3

$$a_{ij} < f_{ij} < \infty \quad \text{and} \quad t_i - t_j + L_{ij} = 0$$

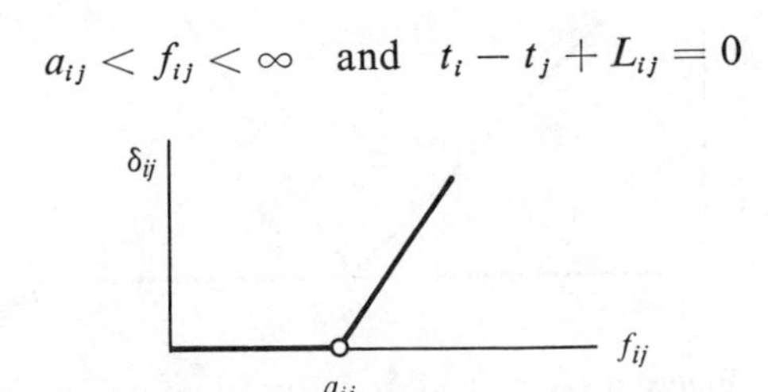

Let us define the following additional symbols:

$$\bar{U}_{ij} = t_i - t_j + U_{ij}$$
$$\bar{L}_{ij} = t_i - t_j + L_{ij}$$
$$\bar{y}_{ij} = t_i - t_j + y_{ij}$$

Therefore, the optimality conditions for each case can be rewritten more compactly as follows:

Case 1: $0 < f_{ij} < a_{ij}$ and $\bar{U}_{ij} = 0$, or $f_{ij} = 0$ and $\bar{U}_{ij} < 0$.
Case 2: $f_{ij} = a_{ij}$ and $\bar{y}_{ij} = 0$.
Case 3: $a_{ij} < f_{ij} < \infty$ and $\bar{L}_{ij} = 0$.

These conditions are essential in the development of the network-flow algorithm, which will be discussed next.

The algorithm. The procedure we are going to discuss in this section was developed by Fulkerson [15] and can be viewed as a dual approach because its derivation relies on the structure of the dual program and the complementary slackness conditions of linear programs. Before describing the algorithm, some simple but helpful considerations are presented.

The most typical interpretation of an arc in a flow network is that of a conduit or pipe through which a given type of commodity is transported. A slightly more sophisticated but meaningful interpretation in the context of the algorithm to be discussed consists of viewing an arc as a two-compartment conduit, such as the one shown in Figure 4.17.

The upper compartment of the representation given in Figure 4.17 has a limited flow capacity while the lower compartment has unlimited capacity.

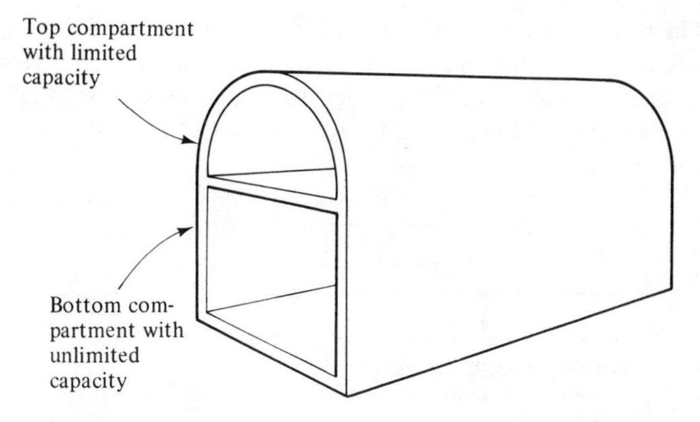

Figure 4.17. Interpretation of an arc as a two-compartment conduit.

Under the current interpretation, it is assumed that flow goes through the bottom compartment only when the capacity of the top compartment is exceeded. Otherwise, the bottom compartment will remain unused.

From the graphical representation of the functions given in Figure 4.16 it can be verified that case 1 occurs when the top compartment is not yet saturated, case 2 occurs when the top compartment is exactly full, and case 3 when the flow is forced through the bottom compartment, whose capacity, as mentioned before, is assumed unlimited. Therefore, we can interpret the absolute value of the slope of the linear cost-duration function of a given activity as the capacity of the top compartment of the arc representing that activity in the flow network model.

The algorithm can be used to construct a project cost-duration curve. In this case, the algorithm is started with a maximal project duration and additional budgets are estimated at each step of the method where a reduction in project duration is achieved. An illustration of such a curve is given in Figure 4.18.

The algorithm consists of three fundamental steps. The first step is a feasibility test to verify if it is possible to reduce a specified duration of the

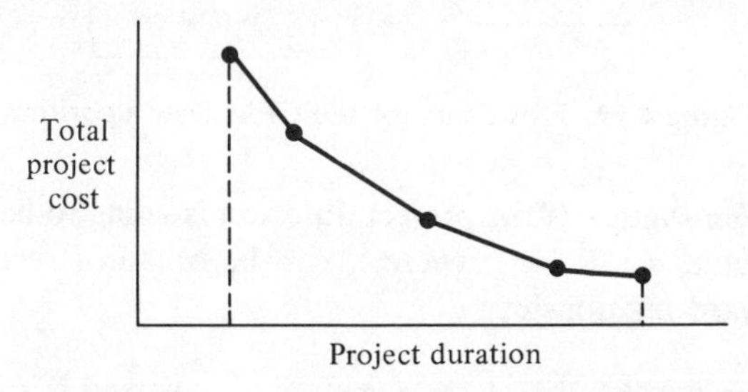

Figure 4.18. Project cost-duration curve.

project. The second step is a labeling procedure to modify the flows through the network corresponding to the dual program. The third step performs the reductions of the duration of the project, when a non-breakthrough results from the second step of the algorithm. A flowchart of the algorithm is given in Figure 4.19.

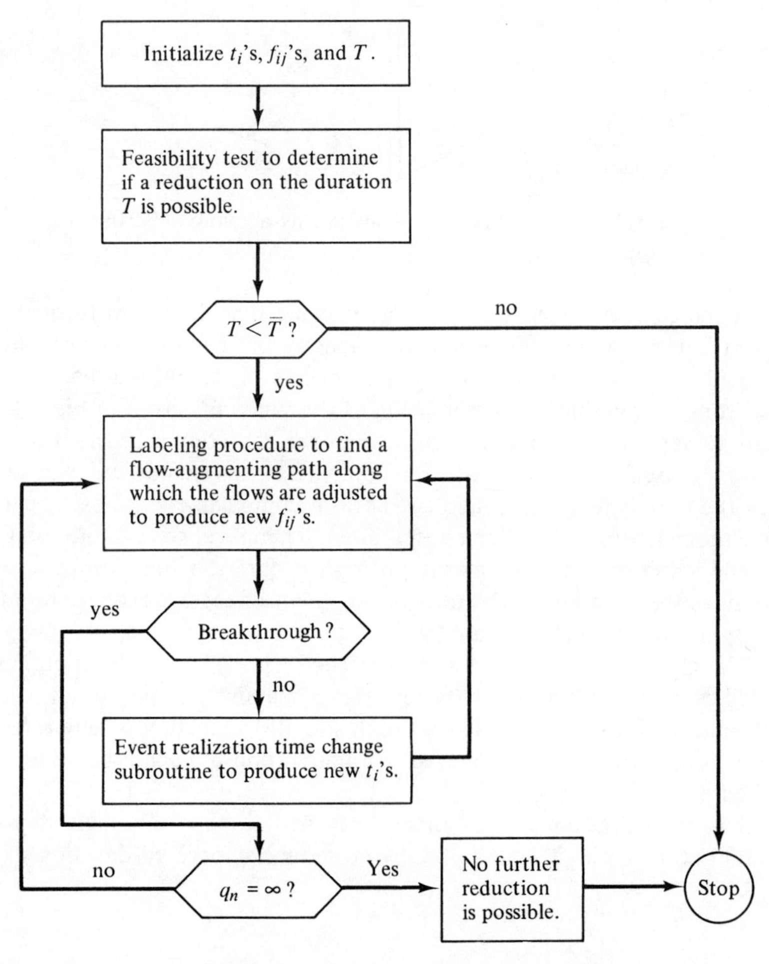

Figure 4.19. Flowchart for the CPM flow algorithm.

Initialization stage. If the project duration is going to be reduced from its maximal value, set $\bar{T} = t_n$, where t_n can be obtained recursively by the following forward relationship:

$$t_j = \max\,[t_i + U_{ij}], \qquad j = 2, 3, \ldots, n; \text{ and } t_1 = 0.$$

Also, all flows can be set equal to zero or to any other values that satisfy the flow-conservation constraints.

Step 1*: *Feasibility test. Set all $y_{ij} = U_{ij}$. If there is a path P from node 1 to node n such that $\bar{L}_{ij} = 0$ for each arc (i, j) in P, this would mean that $y_{ij} = L_{ij}$ for each arc (i, j) in the path. But $U_{ij} \neq L_{ij}$ in nontrivial cases, and therefore $T < \bar{T}$ is not feasible. If no path P exists, the project duration $T < \bar{T}$ is feasible. In order to identify the nodes in the path P, we can label them according to the following rule: label node j from node i by $[\infty, i]$ if arc (i, j) is in P. In this procedure node 1 has a permanent label $[\infty, 0]$.

Step 2*: *Labeling procedure. We divide the discussion of the labeling procedure into two parts: (a) increases in flow along forward arcs, and (b) reductions in flow along reverse arcs.

(a) Forward arcs: If arc (i, j) belongs to case 1, the condition $\bar{U}_{ij} = 0$ is necessary for optimality. Therefore, the flow f_{ij} can be increased by the minimum between the availability of flow at node i and the amount needed to reach the value a_{ij}, which represents the flow condition for case 2. Thus, we can label node j by $[q_j, i]$, where

$$q_j = \min [q_i; a_{ij} - f_{ij}]$$

If arc (i, j) is in case 3, the flow f_{ij} can be increased by any amount, since in this case the arc capacity becomes unlimited. Therefore, we can move to node j all the flow available at node i. That is, we can label node j by $[q_j, i]$, where

$$q_j = q_i.$$

(b) Reverse arcs: If arc (i, j) is in case 1, $\bar{U}_{ij} = 0$ would continue to be a necessary condition for optimality after any reduction of the flow f_{ij}. This implies that we can send a counter flow from j to i whose value is limited by the availability of flow at node j. The purpose of the counter flow is to reduce the value of the flow f_{ij}. Therefore, we can label i from j by $[q_i, j]$, where

$$q_i = \min [q_j, f_{ij}]$$

If arc (i, j) is in case 3, $\bar{L}_{ij} = 0$ would continue to be a necessary condition for optimality if the flow f_{ij} is reduced without causing the arc (i, j) to be in case 1. Thus, the arc must continue to be in case 3 or move to case 2. Then, we can label node i from node j by $[q_i, j]$, where

$$q_i = \min [q_j, f_{ij} - a_{ij}]$$

Step 3*: *Event time change procedure. This step is performed only when a non-breakthrough has occurred, that is, when it has been impossible to assign a label to node n. Let $\mathbf{E}$ be the set of labeled nodes and $\bar{\mathbf{E}}$ the set of unlabeled nodes. The purpose of this step of the algorithm is to reduce the realization time of each node in the set $\bar{\mathbf{E}}$.

Let $i \in \mathbf{E}$ and $j \in \bar{\mathbf{E}}$. If arc (i, j) satisfies at least one of the conditions $\bar{L}_{ij} < 0$ and $\bar{U}_{ij} < 0$, the realization time of node j can be decreased to satisfy the condition on y_{ij} which is necessary for optimality. To see this, let us introduce the following definitions:

$$\mathbf{B} = \{(i, j) \mid i \in \mathbf{E}, j \in \bar{\mathbf{E}}, \bar{L}_{ij} < 0 \text{ or } \bar{U}_{ij} < 0\}$$

$$\zeta_1 = \min_{\mathbf{B}} [\bar{y}_{ij} = y_{ij} + t_i - t_j < 0]$$

The condition $\bar{y}_{ij} = y_{ij} + t_i - t_j$ can be rewritten as indicated below:

$$0 = y_{ij} + t_i - (t_j + \bar{y}_{ij})$$

which implies that t_j can be reduced by $-\bar{y}_{ij}$ (note that $\bar{y}_{ij} < 0$) in order to satisfy the condition that the new value $\bar{y}_{ij}$ computed with the new t_j is equal to zero, as required for optimality.

Similarly, if arc (j, i) meets at least one of the conditions $\bar{L}_{ji} > 0$ and $\bar{U}_{ji} > 0$, the realization time of node j can be decreased to satisfy the condition on y_{ji} which is necessary for optimality. As before, introduce the following definitions:

$$\bar{\mathbf{B}} = \{(j, i) \mid i \in \mathbf{E}, j \in \bar{\mathbf{E}}, \bar{L}_{ji} > 0 \text{ or } \bar{U}_{ji} > 0\}$$

$$\zeta_2 = \min_{\bar{\mathbf{B}}} [\bar{y}_{ji} = y_{ji} + t_j - t_i > 0]$$

The condition $\bar{y}_{ji} = y_{ji} + t_j - t_i$ can be rewritten as follows:

$$0 = y_{ji} + (t_j - \bar{y}_{ji}) - t_i$$

which implies that t_j can be reduced by $\bar{y}_{ji}$ (note that $\bar{y}_{ji} > 0$) in order to satisfy the condition that the new value $\bar{y}_{ji}$ computed with the new t_j is equal to zero, as required for optimality.

If the same reduction is performed on the realization time of each node j in set $\bar{\mathbf{E}}$, the following procedure can be used:

1. Find ζ_1.
2. Find ζ_2.
3. Find $\zeta = \min [\zeta_1, \zeta_2]$.
4. Reduce t_j by ζ for each node $j \in \bar{\mathbf{E}}$.

Once all t_j values have been changed, we begin the labeling procedure (step 2) keeping all labels $[\infty, i]$ to save computational work.

Example of optimal time/cost trade-off. Find the cost-duration curve for the project shown in Figure 4.20. The additional data needed to solve this problem are given in Table 4.15.

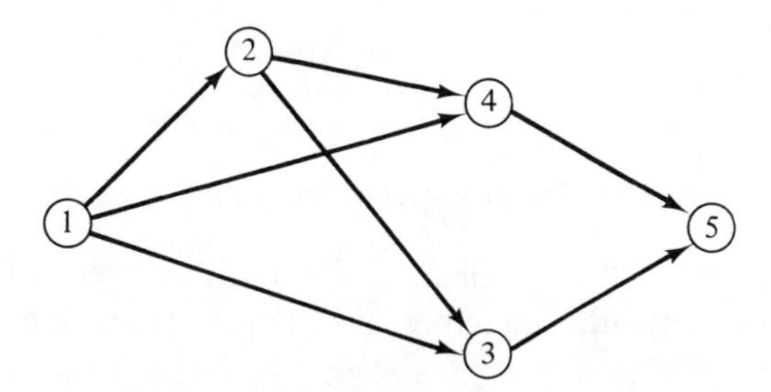

Figure 4.20. Network for sample problem.

Table 4.15. DURATION BOUNDS AND COST DATA

Arc (i, j)	L_{ij}	U_{ij}	a_{ij}
(1, 2)	2	5	3
(1, 3)	1	3	2
(1, 4)	4	6	2
(2, 3)	2	4	5
(2, 4)	3	7	1
(3, 5)	0	3	4
(4, 5)	3	3	—

Each arc is labeled with four numbers $[L_{ij}, U_{ij}, a_{ij}, f_{ij}]$. The current realization time of each event is represented by $[t_i]$, written next to node i. All the initial flow values are set equal to zero. The initial values for the t_i's can be obtained as indicated in Table 4.16. The results for the initial solutions are shown in Figure 4.21.

Table 4.16. INITIAL EVENT REALIZATION TIMES

Node	*Incident Arcs*	t_i
1	—	0
2	(1, 2)	max [0 + 5] = 5
3	(1, 3), (2, 3)	max [0 + 3, 5 + 4] = 9
4	(1, 4), (2, 4)	max [0 + 6, 5 + 7] = 12
5	(3, 5), (4, 5)	max [9 + 3, 12 + 3] = 15

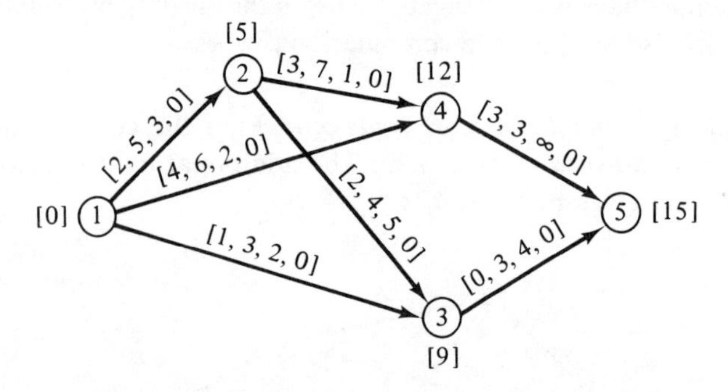

Figure 4.21. Initial solutions for sample problem.

Feasibility test: No path with all $\bar{L}_{ij} = 0$ is possible after setting $y_{ij} = U_{ij}$ for all arcs of the network. Therefore, $T < \bar{T} = 15$ is feasible.

Iteration 1

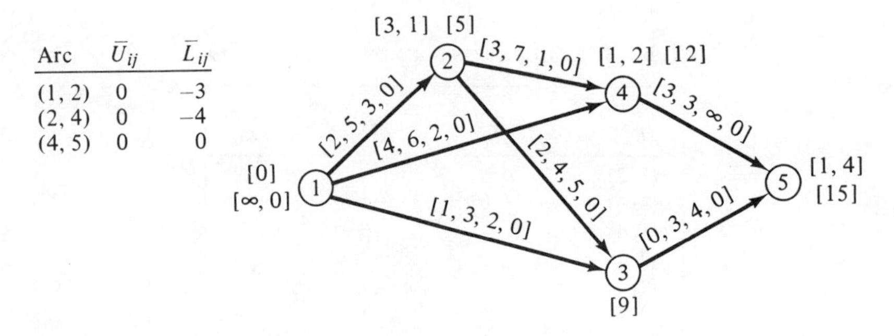

Arc	$\bar{U}_{ij}$	$\bar{L}_{ij}$
(1, 2)	0	−3
(2, 4)	0	−4
(4, 5)	0	0

The result of this iteration is a breakthrough with $q_5 = 1$. The flow on each arc of the flow-augmenting path (1, 2)–(2, 4)–(4, 5) is increased by 1.

Iteration 2

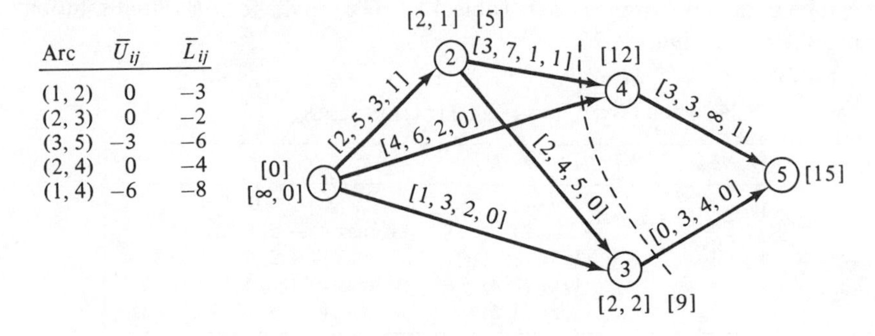

Arc	$\bar{U}_{ij}$	$\bar{L}_{ij}$
(1, 2)	0	−3
(2, 3)	0	−2
(3, 5)	−3	−6
(2, 4)	0	−4
(1, 4)	−6	−8

Since it has not been possible to label node 5, we obtain a non-breakthrough:

$$\mathbf{B} = \{(3, 5), (2, 4), (1, 4)\}, \qquad \zeta_1 = 3$$
$$\bar{\mathbf{B}} = \varnothing, \qquad \zeta_2 = \infty$$

Therefore, $\zeta = 3$. Both t_4 and t_5 must be decreased by the amount $\zeta = 3$. That is,

$$t_4 = 9, t_5 = 12.$$

Iteration 3

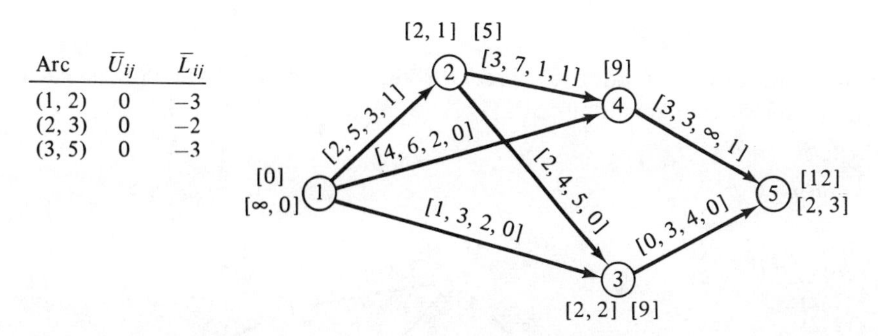

Arc	$\bar{U}_{ij}$	$\bar{L}_{ij}$
(1, 2)	0	−3
(2, 3)	0	−2
(3, 5)	0	−3

We have a breakthrough with $q_5 = 2$. The flow on each arc of the flow-augmenting path (1, 2)–(2, 3)–(3, 5) is increased by 2.

Iteration 4

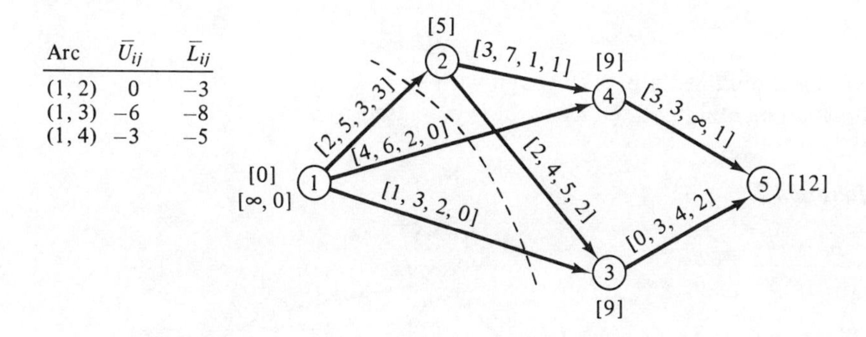

Arc	$\bar{U}_{ij}$	$\bar{L}_{ij}$
(1, 2)	0	−3
(1, 3)	−6	−8
(1, 4)	−3	−5

Since we could not label node 5, the result is a non-breakthrough:

$$\mathbf{B} = \{(1, 2), (1, 3), (1, 4)\}, \qquad \zeta_1 = 3$$
$$\bar{\mathbf{B}} = \varnothing, \qquad \zeta_2 = \infty$$

Therefore, $\zeta = 3$, and $t_2 = 2$, $t_3 = 6$, $t_4 = 6$, and $t_5 = 9$.

Iteration 5

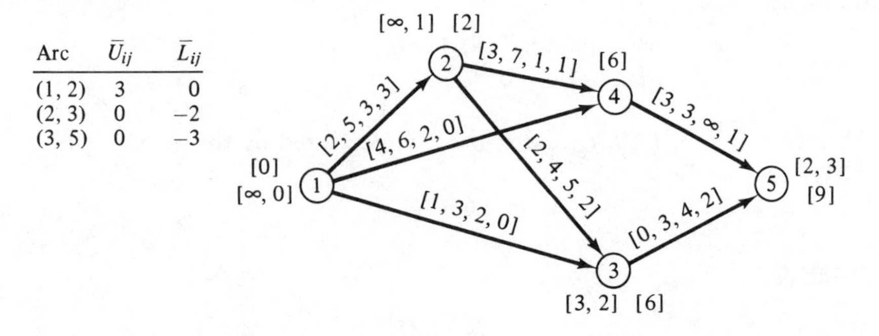

Arc	$\bar{U}_{ij}$	$\bar{L}_{ij}$
(1, 2)	3	0
(2, 3)	0	−2
(3, 5)	0	−3

The result of this iteration is a breakthrough with $q_5 = 2$. The flow on each arc of the flow-augmenting path (1, 2)–(2, 3)–(3, 5) is increased by 2.

Iteration 6

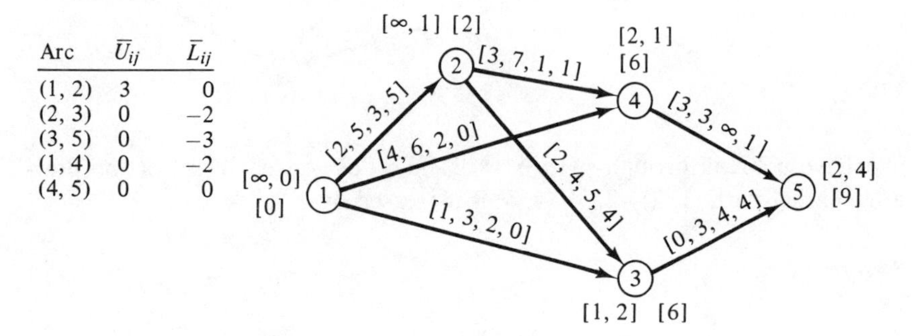

Arc	$\bar{U}_{ij}$	$\bar{L}_{ij}$
(1, 2)	3	0
(2, 3)	0	−2
(3, 5)	0	−3
(1, 4)	0	−2
(4, 5)	0	0

We have obtained a breakthrough with $q_5 = 2$. The flow on each arc of the flow-augmenting path (1, 4)–(4, 5) is increased by 2.

Iteration 7

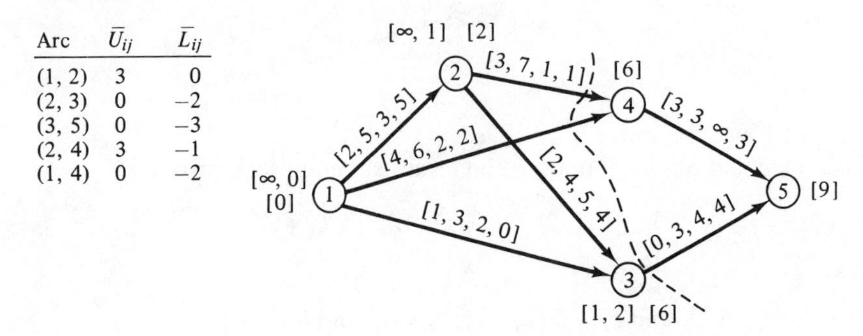

Arc	$\bar{U}_{ij}$	$\bar{L}_{ij}$
(1, 2)	3	0
(2, 3)	0	−2
(3, 5)	0	−3
(2, 4)	3	−1
(1, 4)	0	−2

Since it has not been possible to label node 5, we obtain a non-breakthrough:

$$\mathbf{B} = \{(1, 4), (2, 4), (3, 5)\}, \qquad \zeta_1 = 1$$
$$\bar{\mathbf{B}} = \varnothing, \qquad \zeta_2 = \infty$$

Therefore, $\zeta = 1$, and $t_4 = 5$, $t_5 = 8$.

Iteration 8

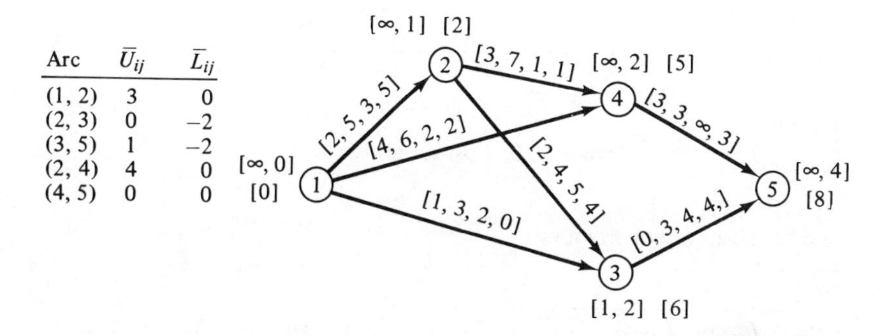

Arc	$\bar{U}_{ij}$	$\bar{L}_{ij}$
(1, 2)	3	0
(2, 3)	0	−2
(3, 5)	1	−2
(2, 4)	4	0
(4, 5)	0	0

We have obtained a breakthrough with $q_5 = \infty$. Therefore, this is the optimal solution. The results are summarized in Table 4.17. The corresponding project cost-duration curve is given in Figure 4.22.

Table 4.17. OPTIMAL SOLUTION

Iteration	t_5	$F = \sum f_{i5}$	*Reduction in Duration, R*	$R \cdot F$	*Cumulative Increase*
2	15	1	3	3	3
4	12	3	3	9	12
7	9	7	1	7	19
8	8	—	(Terminate)	—	—

The activity durations corresponding to each of the break points of the project cost-duration curve given in Figure 4.22 can be obtained using the complementary slackness conditions. Since the flow values f_{ij} and the parameters a_{ij} are known at each iteration of the algorithm, we can compute the values of γ_{ij} and δ_{ij} from the results

$$\gamma_{ij} = \max\,[0;\, a_{ij} - f_{ij}]$$
$$\delta_{ij} = \max\,[0;\, f_{ij} - a_{ij}]$$

By the complementary slackness conditions of linear programs, if $\gamma_{ij} > 0$,

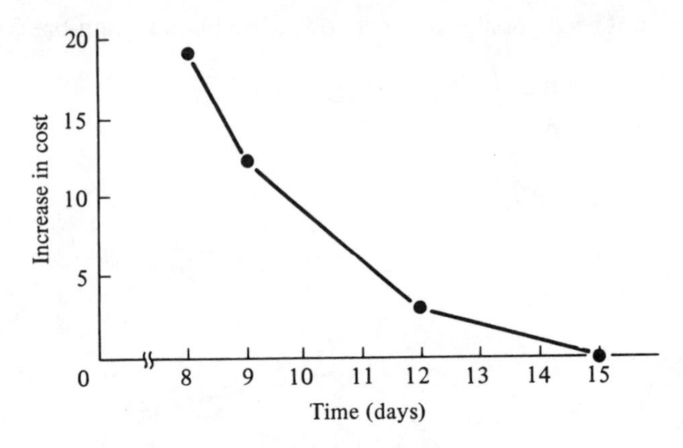

Figure 4.22. Cost-duration curve.

Table 4.18. SUMMARY OF RESULTS

Arc	$T=8$				$T=9$				$T=12$				$T=15$			
(i,j)	f_{ij}	γ_{ij}	δ_{ij}	y_{ij}	f_{ij}	γ_{ij}	δ_{ij}	y_{ij}	f_{ij}	γ_{ij}	δ_{ij}	y_{ij}	f_{ij}	γ_{ij}	δ_{ij}	y_{ij}
(1, 2)	5	0	2	2	5	0	2	2	3	0	0	5[a]	1	2	0	5
(1, 3)	0	2	0	3	0	2	0	3	0	2	0	3	0	2	0	3
(1, 4)	2	0	0	5[a]	0	2	0	6	0	2	0	6	0	2	0	6
(2, 3)	4	1	0	4	4	1	0	4	2	3	0	4	0	5	0	4
(2, 4)	1	0	0	3[a]	1	0	0	4[a]	1	0	0	4[a]	1	0	0	7[a]
(3, 5)	4	0	0	2[a]	4	0	0	3[a]	2	2	0	3	0	4	0	3
(4, 5)	3	∞	0	3	1	∞	0	3	1	∞	0	3	1	∞	0	3

[a] Since both γ_{ij} and δ_{ij} are equal to zero, the complementary conditions only indicate that the duration lies in a range. In this problem the activity is set at its maximal feasible duration since the model seeks to minimize the total cost of the project.

then $y_{ij} = U_{ij}$, and if $\delta_{ij} > 0$, then $y_{ij} = L_{ij}$. Otherwise, $L_{ij} \leq y_{ij} \leq U_{ij}$. The results for the numerical problem under consideration are given in Table 4.18.

4.12.2 Applications of Time/Cost Procedures

Examples of the basic time/cost trade-off procedure are given in the original papers by Kelley and Walker [21, 23] and Fulkerson [15]. The procedure described in these papers is based on linear programming and is implemented through a network-flow algorithm programmed for computers which has long been available through the IBM SHARE library [11]. Such a program will automatically produce the minimum-cost curve of project duration and detailed activity start–finish times associated with every point on the curve. One advantage of this network-flow routine is that if integer-activity-duration data are used as input, the "breakpoints" (possible project durations) of the resulting project cost curve are given as integer values. Assumptions of linear time/cost relationships are required, however.

A different approach to the problem, for other assumptions of activity time/cost relationships, is given in the original *NASA/DOD Guide to PERT Cost* [32]. Another useful variation of the basic procedures was developed by Fondahl [14]. A hand-computational scheme capable of handling a variety of activity time/cost assumptions, it involves alternate shortening and lengthening of project duration to "wiggle in" on the minimum cost for a desired project duration. Most of the approaches to time/cost trade-off analysis developed through 1966 are described more fully in [11].

Some of the relatively few new developments in time/cost trade-off procedures since 1966 include that of Siemens [42], who describes a simple approach which may be performed manually or programmed for computer; it handles linear relationships optimally and approximates the optimal in cases of nonlinearity. Benson and Sewall [2] describe a commercially available computer program for time/cost analysis which will handle networks of up to 40,000 activities, and Hollander [17] describes another large-scale program that will handle various assumptions about activity–cost relationships. Moder and Phillips [29] give a good clear discussion and comparison of time/cost procedures and list several other computer programs having time/cost analysis features.

An important point to note about time/cost procedures is that resource limitations are usually not considered explicitly. None of the procedures mentioned here will automatically produce a minimum-cost schedule that will satisfy stated constraints on the availability of resources such as cost, manpower, or equipment. The Benson and Sewall program, for example, while unusually sophisticated in many respects, does not attempt to automatically schedule production. The program produces reports showing how individual cost or time constraints have been exceeded, but leaves to the human scheduler the task of making detailed changes necessary to meet overall constraints.

Another point to note about time/cost trade-off procedures is that they have not been widely implemented in practice. The major reasons for this lie in the amount and complexity of input data required for analysis, coupled with the typical uncertainty of these data. The collection of detailed time/cost information for projects consisting of hundreds of activities, for example, is usually not an easy task. Thus, although the information available from time/cost procedures is potentially quite valuable under certain conditions, such procedures today enjoy a relatively unimportant position in the overall hierarchy of network-based resource allocation techniques.

4.13 Resource Loading

One of the major advantages of the network format of project representation is the ability to easily generate information on the time-phased requirements of such project resources as manpower, equipment, and money. This information is a by-product of the usual critical-path method scheduling computations. The only requirement is that the resource demands associated with each project activity represented in the network diagram be identified separately.

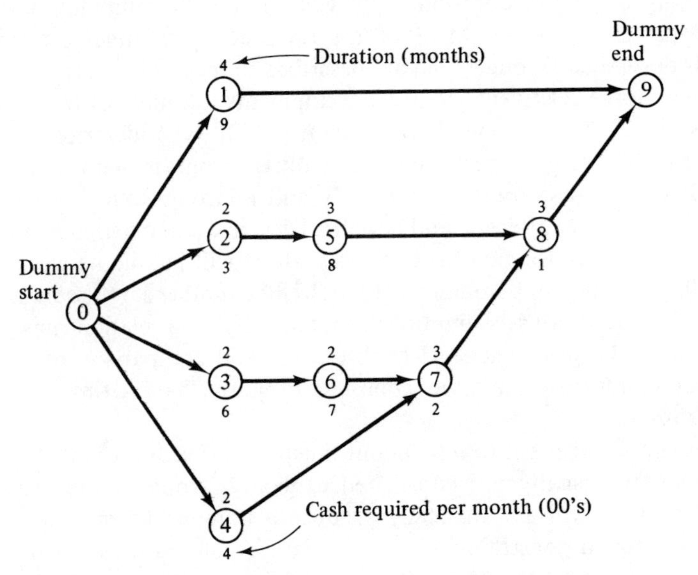

Figure 4.23. Illustrative project network with durations and resource requirements.

Figure 4.23, for example, shows the activity-on-node network for a simple hypothetical project. Beside each network node representing a project activity is shown the duration in months and the cash requirements in hundreds of dollars per month. By application of the usual CPM calculations, an "early start time" (ES) and "late finish time" (LF), among other data, will be produced for each activity. From this information a bar-chart diagram, such as shown in Figure 4.24(a), can be drawn, with all activities beginning at their respective ES times (EST). The bar-chart diagram, in turn, is used to produce the profile of cash flow requirements over time, as shown in Figure 4.24(b). This type of analysis can be performed manually or automatically by the computer program which does the CPM calculations.

These examples use cash requirements for illustration. Obviously, data on other resources, such as manpower, could be used instead; in such cases

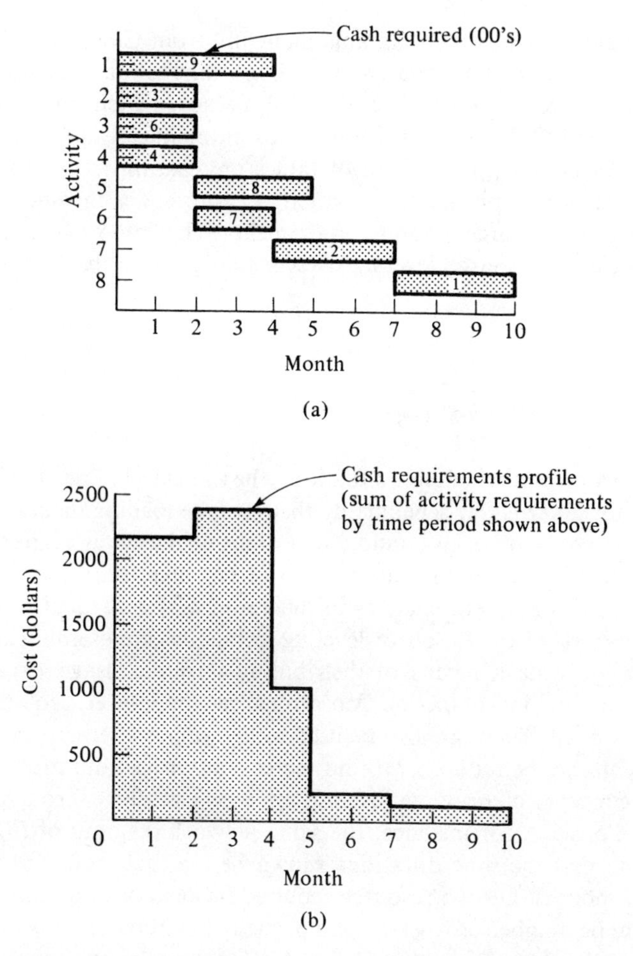

Figure 4.24. Development of resource loading diagram for illustrative network: (a) bar chart diagram generated from network (all jobs at est.); (b) resource loading diagram.

the profile of resource usage over time is commonly referred to as a *resource-loading diagram*. In cases of multiple resource types, a separate loading diagram for each resource type is obtained. Such loading diagrams are extremely useful in project management; they highlight the resource implications of a particular project schedule and provide the basis for more rational project planning. One major pharmaceutical firm, for example, regularly obtains resource-loading information from all CPM schedules for R&D projects to provide resource feasibility checks on the amounts of highly skilled medical and research personnel required at any time. Such procedures are also widely used in the construction industry, with some smaller firms often plotting manpower-loading diagrams manually from CPM output. In most cases the resource-loading data are more important as rough planning guides rather than as precise goals to be obtained, because of schedule uncertainty.

Most good CPM programs today will automatically produce resource loading information (in the form of tables or diagrams) even though they may lack the more sophisticated capability of automatic schedule adjustment to meet desired resource-loading requirements. This latter capability falls in the categories of resource leveling or scheduling under fixed resource constraints.

4.14 Resource Leveling

In many project scheduling situations, the total level of resource demands projected for a particular schedule by the resource loading diagram may not be of major concern because ample quantities of the required resources are available. But it may be that the pattern of resource usage has undesirable features, such as frequent changes in the amount of a particular manpower skill category required. Resource-leveling techniques are useful in such situations; they provide a means of distributing resource usage over time to minimize the period-by-period variations in manpower, equipment, or money expended. They can also be used to determine whether peak resource requirements can be reduced with no increase in project duration.

The concepts of resource leveling are easily grasped through a simple example. Consider, for instance, the same network diagram of Figure 4.23, but assume that the time durations shown beside each node are now days instead of months. Let the resource required, instead of hundreds of dollars per month, be number of workers of a particular skill required per day. Then the resource-loading diagram of Figure 4.25(a) shows an undesirable pattern of manpower requirements over time because of the extreme variation from 24 workers on days 3 and 4 down to 1 worker on days 8, 9 and 10. Resource leveling will produce a more even pattern of manpower utilization without increasing project duration beyond 10 days.

As the bar chart of Figure 4.25(a) shows, jobs 1, 2, 4, and 5 have slack and can be delayed within the limits of their slack without delaying other jobs. Job 1, for example, can obviously be delayed until the last 4 days of project performance, as shown in Figure 4.25(b). This reduces peak manpower requirements from 24 to 14, and the resulting manpower profile is more even than before. However, it still contains considerable fluctuations, with an undesirable "valley" on day 6. By subsequently rescheduling jobs 2 and 5, a more even pattern is obtained, as shown in Figure 4.25(c). This final profile still contains what might be considered undesirable peaks on days 2 and 4, but these cannot be eliminated without either increasing project dura-

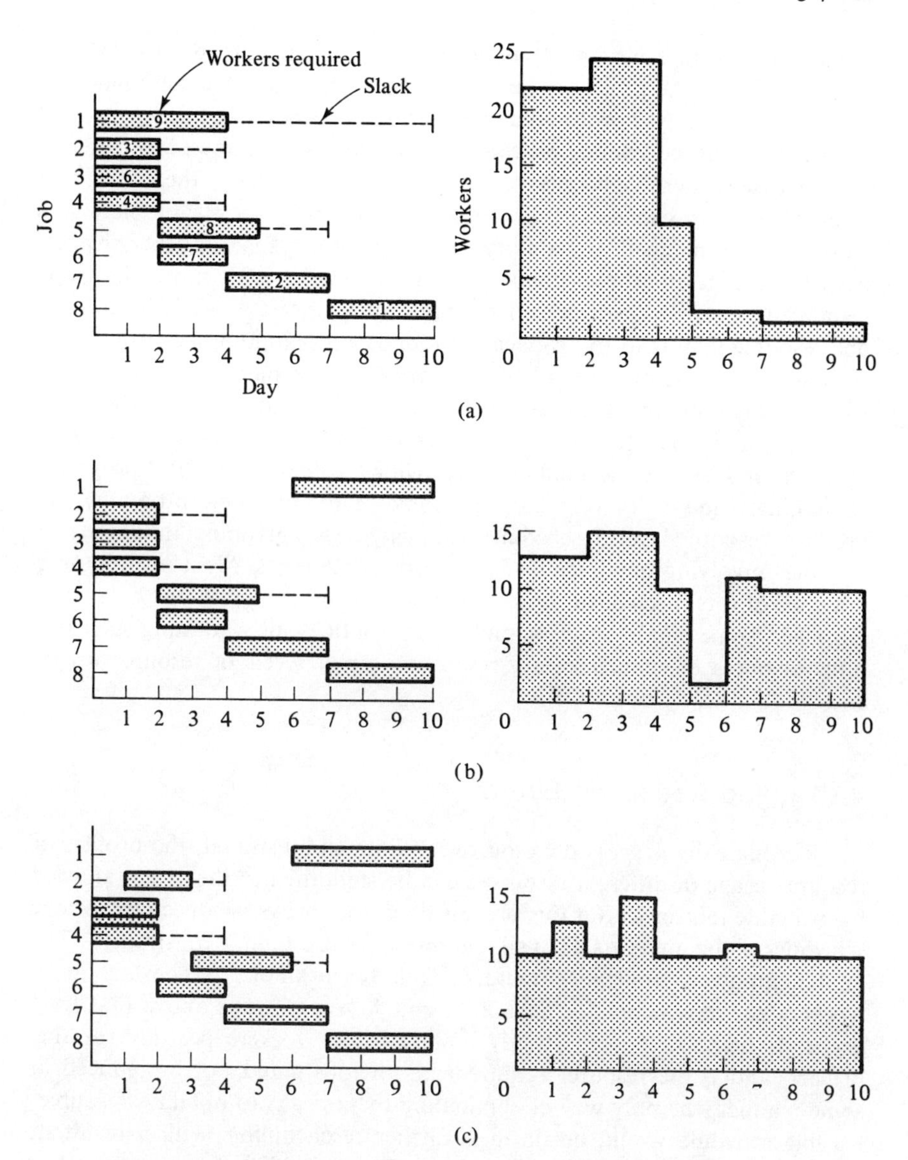

Figure 4.25. Resource leveling: (a) bar chart and manpower profile with all jobs at est.; (b) with job 1 rescheduled; (c) with jobs 1, 2, and 5 rescheduled.

tion or changing the characteristics of some activities so that the associated manpower requirements are changed.

As this simple example shows, the essential idea of resource-leveling centers about the "juggling" or rescheduling of jobs within the limits of available slack (float) to achieve better distribution of resource usage. The slack available in each activity is determined from the standard CPM calculations. Thus, the project duration (i.e., the critical path) is determined by a "time only" analysis, without consideration of resource requirements, and during the resource-leveling process this duration is not allowed to increase.

In cases where only one resource type is concerned, the process of resource leveling can readily be conducted manually for sizable networks with the aid of magnetic scheduling boards, strips of paper, or other physical devices. But in situations involving large networks with more than one resource type, the process becomes complicated, since actions taken to level one resource may tend to produce more imbalances in another resource. In such situations the process is best conducted by computer. CPM programs with resource-leveling features have been available since the early 1960s [11]; Levy and Wiest [26] give a good description of how one such program operates, and Moder and Phillips [29] provide a recent list of CPM programs containing resource-leveling features. Some of the available programs will produce resource-leveled schedules for networks containing thousands of activities involving hundreds of different resource types. Most operate on a principle of rote mechanistic application of some juggling rule or "heuristic." However, some employ quite sophisticated routines incorporating such features as pools of supplementary resources, target levels of resource usage, and "look-ahead" routines.

4.15 Fixed Resource Limits

Through the process of resource leveling, as illustrated, the profiles of resource usage of different resources can be smoothed to the extent allowed by available job slack. But this process does not always produce satisfactory schedules if the amounts of available resources are tightly constrained. The final manpower profile of Figure 4.25(c), for example, is constant at 10 workers per day except for days 2, 4, and 7, where peaks above that level occur. If manpower were strictly limited to 10 workers per day for this project, and if the resource requirements of jobs could not be reduced in some manner, the only way of eliminating these peaks to obtain a resource-feasible schedule would be through further rescheduling, with a resultant increase in project duration. Looking at Figure 4.25(c), for example, it is readily apparent that the manpower peaks can be eliminated by delaying the start of job 1 from day 7 until day 8. This permits job 5 to be delayed 1 day, making it concurrent with job 7, and also permits delaying job 2 for 1 day to make it concurrent with job 6. Since the resulting manpower requirements in every period total exactly 10 or less, the schedule is resource-feasible. However, the project duration has increased by 1 day, as the resulting schedule of Figure 4.26(a) shows.

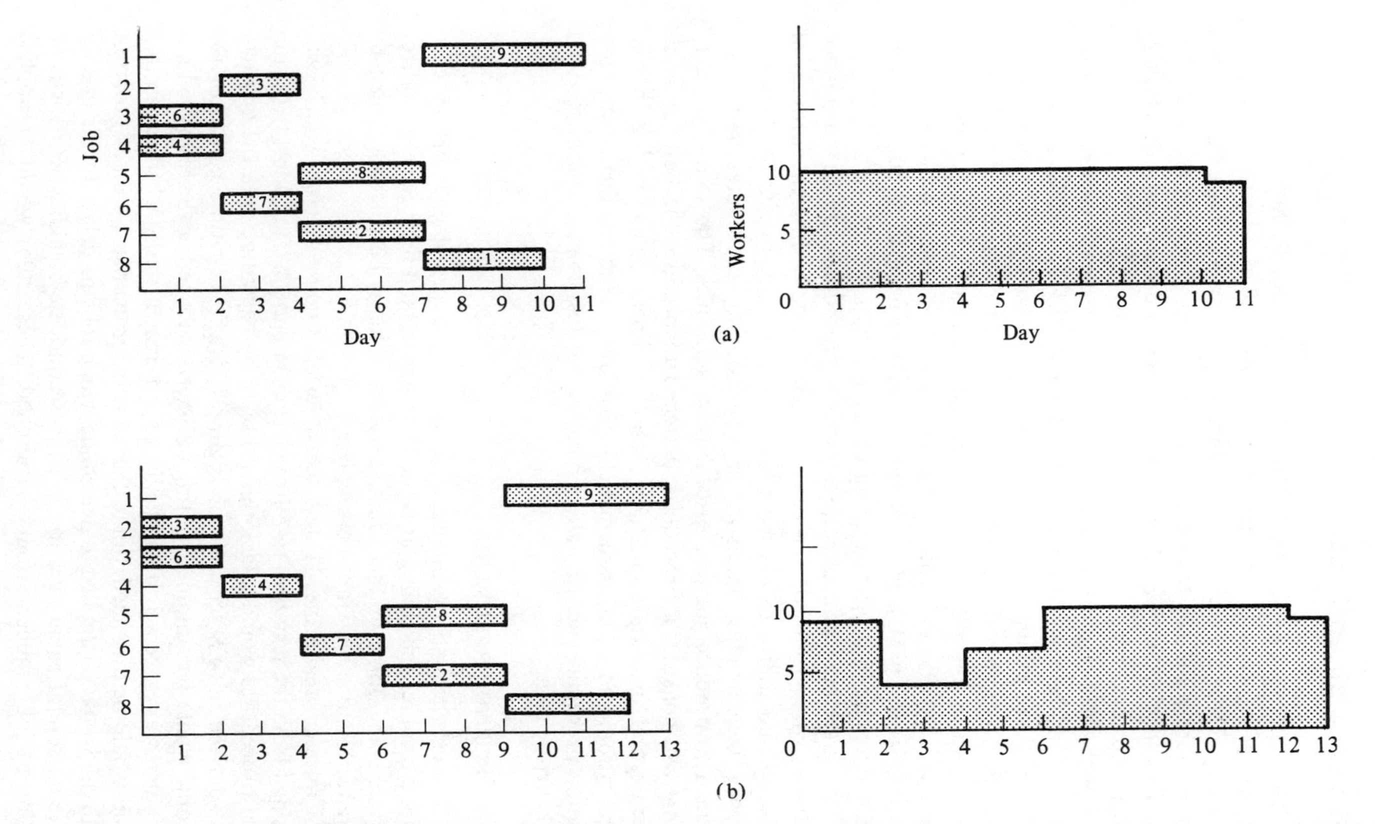

Figure 4.26. Constrained-resource scheduling with fixed limit of 10 workers available: (a) 11-day schedule obtained by further rescheduling of resource leveling results; (b) 13-day schedule obtained with "shortest-job-first" rule.

4.16 Constrained Resources

The type of rescheduling illustrated by this example falls into another category of procedures for network analysis, called *constrained-resource scheduling*. These are techniques designed to produce schedules that will not require more resources than are available in any given period, and have durations that are increased beyond the original critical path duration as little as possible.

Constrained-resource network scheduling procedures were first proposed shortly after the general appearance of CPM/PERT models. Kelley and Walker [22], for example, described in an early paper the development of an approach programmed for the IBM 650 computer which was capable of handling up to four different manpower types per job and up to nine types per project. By 1961 the procedure had been used on several projects, achieving reductions in peak manpower requirements of from 35 to 50%, with an average increase in project duration of around 5%.

The great bulk of constrained-resource procedures can be categorized into two major groups on the basis of their distinctly different methods of approach and utility to management. The first—and by far the largest—category includes the heuristic, or approximate, procedures which are designed to produce good resource-feasible schedules. The second category in contrast, consists of procedures designed to produce the best (optimal) schedule and includes approaches based on linear programming, partial enumeration, and other mathematical techniques. These mathematical optimization procedures are capable of handling far less complex problems than the heuristic category.

4.16.1 Heuristic Approaches

Because heuristic procedures enjoy such a prominent role in constrained-resource scheduling, it is useful to understand their general mode of operation and some of their characteristics. This is best accomplished by reference to some illustrative examples.

Consider again the 11-day schedule of Figure 4.26(a), which was obtained from the results of resource leveling, Figure 4.25(c). Most heuristic-based procedures would attack this problem by beginning with the original, CPM-derived, all-ES schedule of Figure 4.25(a). Each time period of the schedule would be examined, testing to determine whether the resource limit of 10 men was exceeded. If so, the list of jobs concurrently scheduled in that period would be examined. Some chosen rule, or heuristic, such as "shortest job first," would be applied to determine which jobs to delay. In the case of ties (e.g., several equal-duration jobs), a second, tie-breaking heuristic such as "lowest job number" would be applied. These rules would be applied mechanistically, working sequentially by time period (and resource type, if more than one were involved) through the schedule, delaying selected jobs and their followers until all jobs are scheduled and a resource-feasible schedule is obtained. As an illustration of this process, application of the "shortest job first" heuristic as described here produces the 13-day schedule shown in Figure 4.26(b). This schedule is indeed resource-feasible, but it also exhibits considerable manpower fluctuations and is obviously not as satisfactory as the previous 11-day schedule.

It is easy to see here why application of the "shortest job first" rule produced such results. In time period 1, for example, mechanistic application of the heuristics resulted in job 4 being delayed, with jobs 2 and 3 scheduled concurrently. But it is obvious that a better choice would have been to delay job 2 and schedule jobs 3 and 4 concurrently. This would have permitted different arrangements of jobs in succeeding periods, resulting in a shorter schedule with generally higher levels of manpower usage.

Other considerably more sophisticated heuristics could have been used here for illustration, and would have produced the "best possible" 11-day schedule obtained earlier. But in practice, even with more sophisticated rules, it is not possible to tell in advance which particular heuristic or combination of heuristics will produce best results for a given problem. Heuristics that perform poorly on one problem perform well on others. Figure 4.27, for example, shows another, slightly more complex three-resource network. The original critical path duration of 18 days is not feasible because of the indicated resource constraints. Figure 4.28 shows the schedules obtained with two different heuristics compared with the optimum-duration schedule determined with an algorithm given in [10]. In this case the "shortest job first" rule produced better results than the resource scheduling method (RSM) [41] and was fairly close to the optimum-duration schedule.

Heuristic procedures for constrained-resource scheduling are in wide use. One reason for this is because they offer the only available means of obtaining resource-feasible schedules for the types of large, complex problems that often occur in practice. Another reason is that the schedules produced, even though possibly not the best obtainable, are often good enough to use for planning purposes, in view of the typical uncertainties associated with actual activity durations and resource requirements. Furthermore, because of the speed with which such schedules can be produced by the computer, it is often possible to try several different heuristics and to select the best one obtained; in this way optimal or near-optimal results can often be obtained (although this is rarely known for sure).

One of the outstanding characteristics of developments in the area of network methods has been the appearance of a tremendous number of elaborate, computer-based heuristic routines for constrained-resource scheduling. Many of these have been developed by organizations for the own or outside use on a proprietary basis, with the consequence that operat-

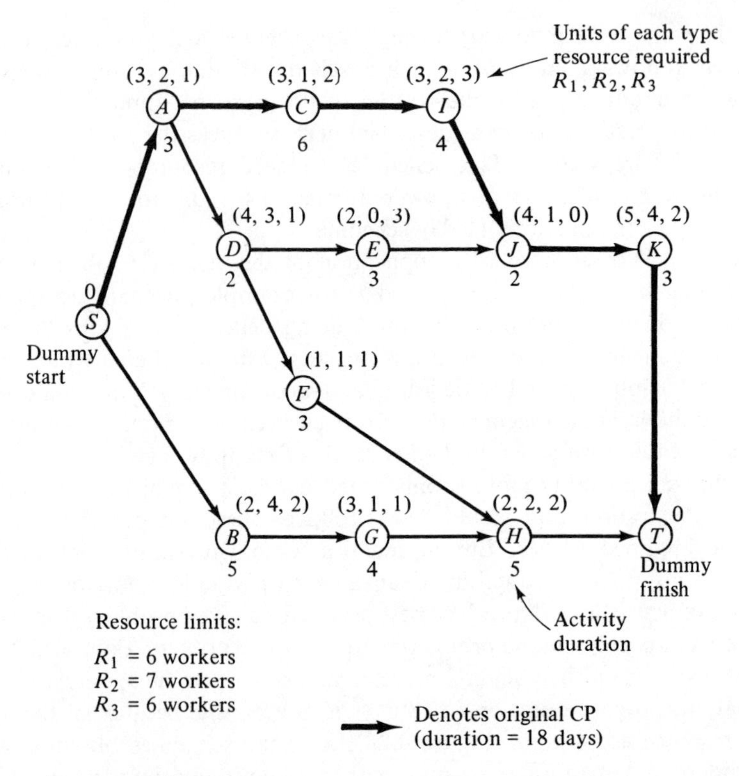

Figure 4.27. Constrained-resource problem with three resource types.

ing details of very few are available in the open literature. However, some are available on the open market with details available to users. For example, Table 4.19 presents a sampling of the more than 25 such programs known to be available on a commercial basis in the USA, with a summary of their major features. These programs are capable of producing a wide variety of resource-oriented reports for management in tabular or graphic form.

4.16.2 Optimal Approaches

The second major category of constrained-resource scheduling procedures—those for producing optimal solutions—has been characterized by relatively less practical progress than the heuristic methods. The optimization procedures that have been developed can be divided into two subcategories:

1. Procedures based on linear programming (LP).
2. Procedures based on enumerative and other mathematical techniques.

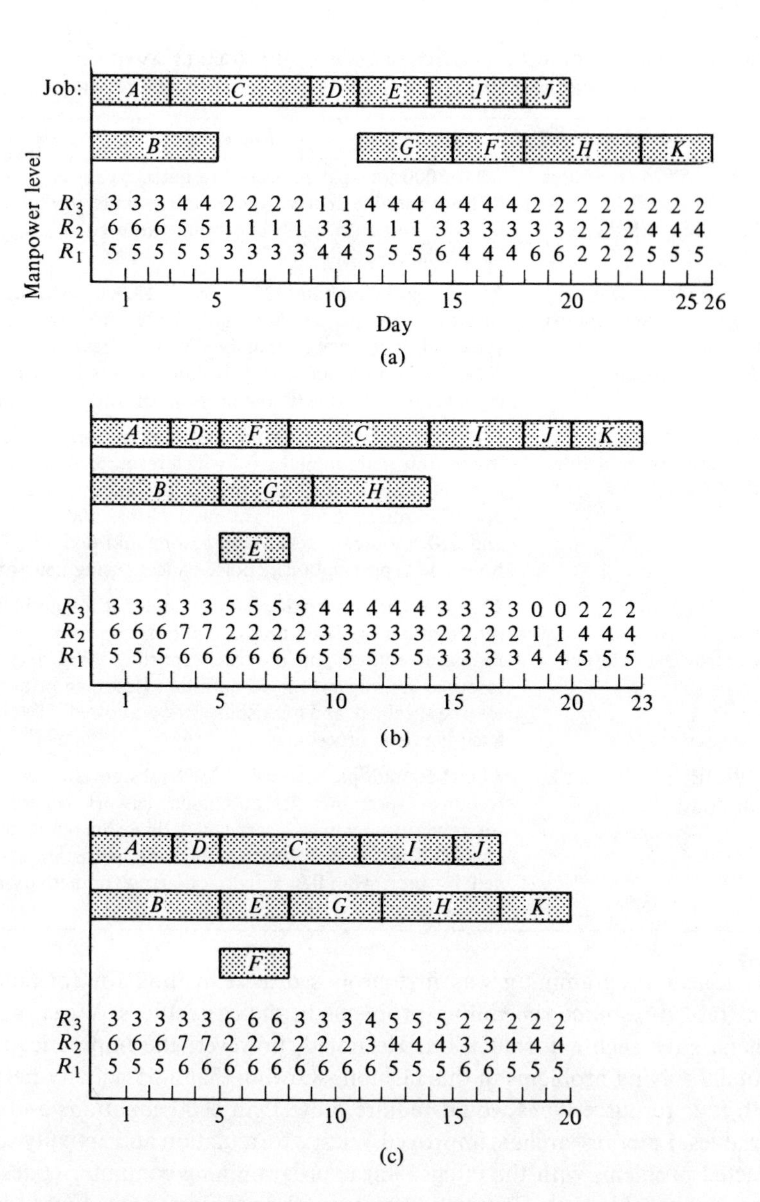

Figure 4.28. Comparison of heuristic-based and optimum schedules for three-resource problem: (a) 26-day resource-feasible schedule obtained with "RSM" heuristic; (b) 23-day resource-feasible schedule obtained with "shortest-job first" heuristic; (c) 20-day (optimum duration) resource-feasible schedule.

Table 4.19. SAMPLE CHARACTERISTICS OF SOME COMMERCIALLY AVAILABLE COMPUTER PROGRAMS WITH CONSTRAINED-RESOURCE NETWORK SCHEDULING CAPABILITIES

Program Name	*Characteristics*
CPM-RPSM (Resource Planning and Scheduling Method), CEIR, Inc.	2000–8000 jobs per project, 4 resource types per project, 26 total variable or constraint resource limits, job splitting allowed, job start/finish constraints allowed. Uses fixed scheduling heuristic.
MSCS (Management Scheduling and Control System), McDonnell Automation Co.	Multiproject capability (25 projects), 18,000 activities, 12 resource types per activity, many flexible assumptions of job conditions, easy updating. Allows project costing and includes report generation. Scheduling heuristics based on complex priority function approach, controllable by user.
PMS/360 (Project Management System), IBM Corp.	A large complex management information system consisting of four main modules (of which resource allocation is one). Handles activity-on-arrow or precedence diagrams; up to 225 multiple projects allowed with 32,000 activities, and 250 resource types. Many costing and updating features and report options, choice of sequencing heuristics.
PPS IV (Project Planning System), Control Data Corp.	2000 jobs per project, 20 resource types per job and project, multiple or single projects, allows overlapping jobs, resource costing, and progress reporting. Will also do resource leveling with fixed duration. Resource priorities may be specified, and multishift work is allowed. Uses one fixed heuristic procedure.
PROJECT/2, Project Software, Inc.	Allows 50 multiple networks, 32,000 jobs, several hundred resource types. Includes automatic network generation for repetitive sequences, easy updating and many cost-analysis features. Choice of sequencing heuristics, specified by user. Handles activity-on-arrow or activity-on-node input.

Linear programming was first proposed as a method for formulating constrained-resource scheduling problems in the early 1960s. Wiest, among others, gave such a formulation. He noted, however, the impracticality of actually solving problems in this fashion, showing that a 55-activity network with four resource types would require more than 5000 equations and 1600 variables. Later researchers improved Wiest's formulation and actually solved selected problems with the integer linear programming computer codes that became available [16]. Generally, however, the problems solved were small, illustrative examples consisting for the most part of fewer than 15 activities and three resource types. In recent years there have appeared a number of new and interesting LP formulations of constrained-resource scheduling [4, 13, 38], and some of these have been applied to slightly more complex problems. But linear programming procedures have still not developed to the point where they are capable of solving the type of problems easily handled by heuristic procedures; they remain today primarily an interesting research topic for academicians.

The second subcategory of optimization procedures consists primarily of techniques that have appeared on the scene fairly recently. These techniques are based on implicit enumeration of all possible activity-sequencing combinations and include the "branch-and-bound" optimization procedures. These methods were first applied to constrained-resource network scheduling in 1967 by Mueller-Mehrbach and by Johnson [20]. These researchers demonstrated that such procedures were capable of optimally solving some single-resource networks with as many as 300 activities. In 1968 a somewhat different implicit enumeration scheme was proposed by Davis [8], which was later developed to handle networks of up to about 50 activities and five resource types [10].

Since the mid-1960s there have been numerous new developments reported, primarily in the areas of constrained-resource scheduling and in procedures for stochastic network analysis. Heuristic scheduling techniques, computer-programmed to give "good" schedules in mechanistic fashion, have been the basis for all practical systems developed to date. There are programs available today capable of scheduling the largest projects imaginable under almost any desired conditions of resource usage and availabilities. They have been applied in such diverse fields as the control of production in repetitive manufacturing operations, the scheduling of construction operations, and the scheduling of radio-TV programs.

PART III: A COMPARISON OF COMMERCIALLY AVAILABLE CPM/PERT COMPUTER PROGRAMS[1]

with contributions by

Larry A. Smith
Florida International University

Peter Mahler
Concordia University
Montreal, Canada

In other sections of this book, computer programs have been provided to execute the majority of the techniques described. We will not attempt to produce a CPM/PERT code since there currently exists a wide variety of commercial software. This section presents a comparison of available programs.

---

[1]Part III is reproduced from *Industrial Engineering* [43], with the permission of the authors and the American Institute of Industrial Engineering.

Over the past 15 years, computer applications in network analysis have undergone major technological changes. The software programs in the 1960s were primarily developed by the mainframe manufacturers. This was due to the knowledge and experience of the hardware manufacturers and the lack of software houses. The number of users was also limited since most packages were very restrictive and costly.

The 1970s marked a considerable change in the software field. The use of programs became less complicated. There has been an increasing number of software brokerage houses over the last few years, designing network packages. This increase in brokerage houses has come about primarily because of the expansion in the number of users in industry. It is not just the large corporations or governments that can feasibly use these packages; medium-sized and small businesses are finding them economical and beneficial in planning and controlling their operations.

A published survey of CPM/PERT computer programs, "Project management with CPM and PERT," was conducted by Moder & Phillips [29] in 1968. An unpublished survey was conducted at Stanford University in the early part of 1976 by Woolpert, "Computer Software for Project Planning and Control" [47]. In this section a comparison is made of 20 major packages. The better known programs have been selected. To facilitate the comparison, a characteristic matrix has been developed, listing the most relevant features.

The characteristic matrix should provide a handy reference and comparison of major CPM/PERT software programs. The summary of characteristics was obtained from the latest available information and the author's interpretation of the documentation provided by the companies listed. There is no attempt to compare the efficiency or effectiveness of the characteristics between programs. Therefore, one program might have a better, more flexible, and efficient sorting routine than another. This cannot be determined from the matrix.

The matrix of characteristics that appears in Table 4.20 is self-explanatory. An X indicates that the particular characteristic is available in that program in some form and an L indicates limited availability; an empty cell in the matrix indicates that the characteristic is not available in the program.

The network schemes considered in Table 4.20 are classified as (a) activity-on-arc representations, (b) activity-on-node representations, and (c) activity-on-node representations with precedence constraints. The first two models, introduced in Section 2.1.2, were fully discussed in Sections 4.4 through 4.10. The third model is an extension of the second one to allow specifications on lags and delays among activity start and finish times. For more details on this model, the reader is advised to consult the book by Moder and Phillips [29].

Table 4.20. A PERT/CPM SOFTWARE SURVEY

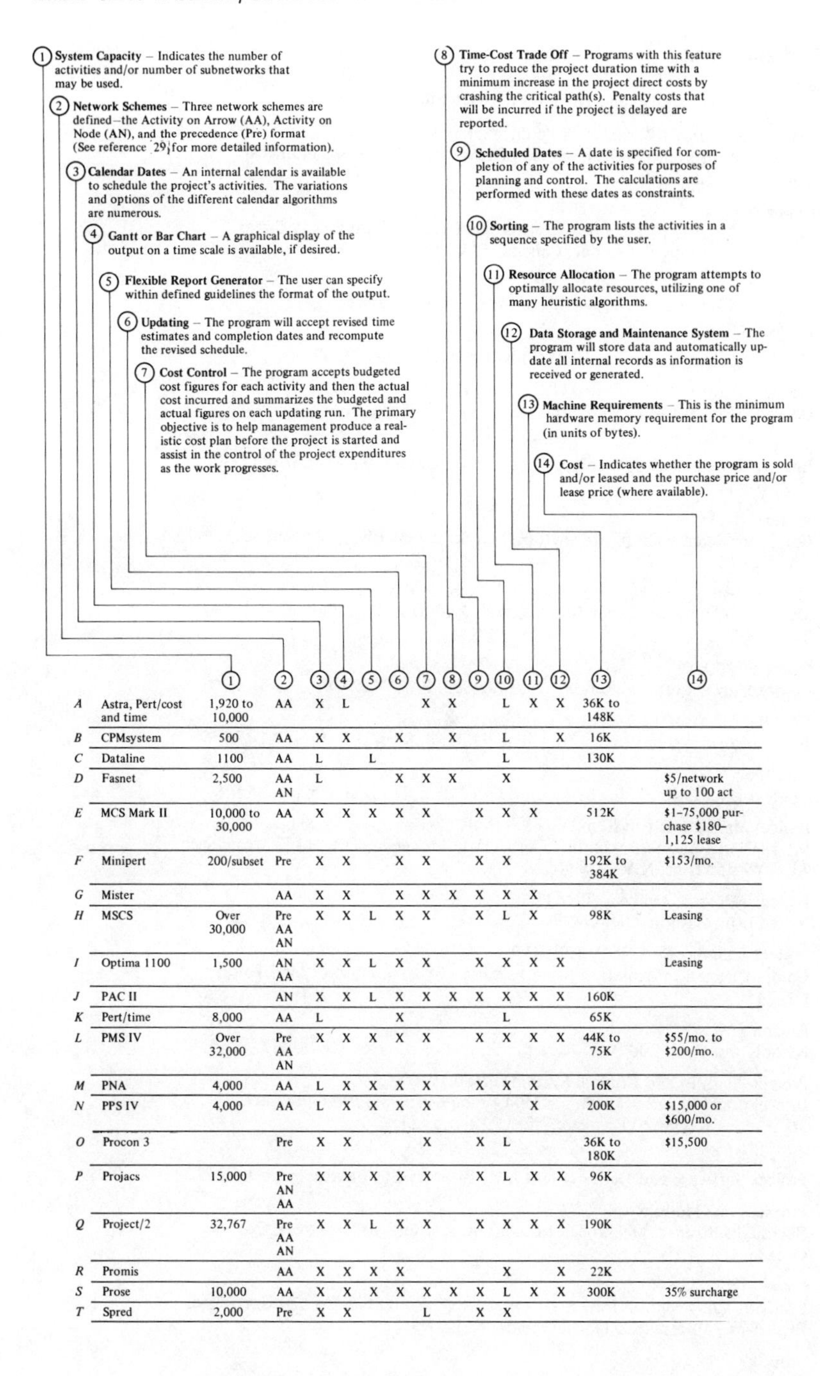

(1) **System Capacity** – Indicates the number of activities and/or number of subnetworks that may be used.

(2) **Network Schemes** – Three network schemes are defined–the Activity on Arrow (AA), Activity on Node (AN), and the precedence (Pre) format (See reference 29 for more detailed information).

(3) **Calendar Dates** – An internal calendar is available to schedule the project's activities. The variations and options of the different calendar algorithms are numerous.

(4) **Gantt or Bar Chart** – A graphical display of the output on a time scale is available, if desired.

(5) **Flexible Report Generator** – The user can specify within defined guidelines the format of the output.

(6) **Updating** – The program will accept revised time estimates and completion dates and recompute the revised schedule.

(7) **Cost Control** – The program accepts budgeted cost figures for each activity and then the actual cost incurred and summarizes the budgeted and actual figures on each updating run. The primary objective is to help management produce a realistic cost plan before the project is started and assist in the control of the project expenditures as the work progresses.

(8) **Time-Cost Trade Off** – Programs with this feature try to reduce the project duration time with a minimum increase in the project direct costs by crashing the critical path(s). Penalty costs that will be incurred if the project is delayed are reported.

(9) **Scheduled Dates** – A date is specified for completion of any of the activities for purposes of planning and control. The calculations are performed with these dates as constraints.

(10) **Sorting** – The program lists the activities in a sequence specified by the user.

(11) **Resource Allocation** – The program attempts to optimally allocate resources, utilizing one of many heuristic algorithms.

(12) **Data Storage and Maintenance System** – The program will store data and automatically update all internal records as information is received or generated.

(13) **Machine Requirements** – This is the minimum hardware memory requirement for the program (in units of bytes).

(14) **Cost** – Indicates whether the program is sold and/or leased and the purchase price and/or lease price (where available).

		(1)	(2)	(3)	(4)	(5)	(6)	(7)	(8)	(9)	(10)	(11)	(12)	(13)	(14)
A	Astra, Pert/cost and time	1,920 to 10,000	AA	X	L			X	X		L	X	X	36K to 148K	
B	CPMsystem	500	AA	X	X		X		X		L		X	16K	
C	Dataline	1100	AA	L		L					L			130K	
D	Fasnet	2,500	AA AN	L			X	X	X		X				$5/network up to 100 act
E	MCS Mark II	10,000 to 30,000	AA	X	X	X	X	X		X	X	X		512K	$1–75,000 purchase $180–1,125 lease
F	Minipert	200/subset	Pre	X	X		X	X		X	X			192K to 384K	$153/mo.
G	Mister		AA	X	X		X	X	X	X	X	X			
H	MSCS	Over 30,000	Pre AA AN	X	X	L	X	X		X	L	X		98K	Leasing
I	Optima 1100	1,500	AN AA	X	X	L	X	X		X	X	X	X		Leasing
J	PAC II		AN	X	X	L	X	X	X	X	X	X	X	160K	
K	Pert/time	8,000	AA	L			X				L			65K	
L	PMS IV	Over 32,000	Pre AA AN	X	X	X	X	X		X	X	X	X	44K to 75K	$55/mo. to $200/mo.
M	PNA	4,000	AA	L	X	X	X	X		X	X	X		16K	
N	PPS IV	4,000	AA	L	X	X	X	X				X		200K	$15,000 or $600/mo.
O	Procon 3		Pre	X	X			X		X	L			36K to 180K	$15,500
P	Projacs	15,000	Pre AN AA	X	X	X	X	X		X	L	X	X	96K	
Q	Project/2	32,767	Pre AA AN	X	X	L	X	X		X	X	X	X	190K	
R	Promis		AA	X	X	X	X				X		X	22K	
S	Prose	10,000	AA	X	X	X	X	X	X	X	L	X	X	300K	35% surcharge
T	Spred	2,000	Pre	X	X			L		X	X				

Table 4.20. (cont.)

Legend for the Software Survey

A Astra II Resource Allocation Program, Pert/time, Pertcost
Honeywell Information Systems, 800 Dorchester Blvd. West, Montreal, Quebec, Canada H3B 1X9

B CPM System
Canadian General Electric Information Services, 2001 University Street, Suite 1120, Montreal, Quebec, Canada H3A 2A6

C Dataline
Dataline, 40 St. Clair Ave. West, Toronto, Ontario, Canada

D Fasnet
University Computing Co., 1930 Hiline Dr., Dallas, TX 75207

E Management Control System (MCS) Mark II
Construction Management System, P.O. Box 90, Haddonfield, NJ 08033

F Minipert
International Business Machines Corp., 1133 Westchester Ave., White Plains, N.Y. 10604

G Mister
Computer Science Corp., Infonet, 650 N. Sepulveda Blvd., El Segundo, CA 90245

H Management Scheduling and Control System (MSCS)
McDonnell Douglas Automation Co., 500 Jefferson Bldg., Suite 400, Houston, TX 77002

I Optima 1100
Sperry Univac Corp., P.O. Box 500, Blue Bell, PA 19422

J PAC II
International Systems Inc., 150 Allendale Road, King of Prussia, PA 19406

K Pert/time
Control Data Corp., 2200 Berkshire Land North, Minneapolis, MN 55441

L Project Management System IV (PMS IV)
International Business Machines Corp., Data Processing Div., 1133 Westchester Ave., White Plains, N.Y. 10604

M Project Network Analysis (PNA)
NCR Corp., Dayton, OH 45479

N Project Planning System IV (PPS IV)
Computing and Information Sciences Corp., 810 Thompson Bldg., Tulsa, OK 74103

O Procon 3
Nichols and Co., 1900 Avenue of the Stars, Los Angeles, CA 90067

P Project Analysis and Control System (PROJACS)
International Business Machines, IBM France, Program Product Center, 36, Avenue Raymond Poincare, 75116 Paris, France

Q Project/2
Project Software and Development Inc., 14 Story St., Cambridge, MA 02138

R Promis-Time Module
Burroughs Business Machines Ltd., 980 St. Antoine St., Montreal, Quebec, Canada, H3C 1AB

S Prose
Concom, Construction Industry Computer Consultants Ltd., 620 Dorchester Blvd. West, Montreal, Quebec, Canada, H3B 1N8

T Spred
Computer Science Corp., Infonet, 605 N. Sepulveda Blvd., El Segundo, CA 90245

EXERCISES

1. A building contractor is trying to plan construction activities for a custom home. The sequence of activities, precedence relationships, and time durations are given in the table. Construct the appropriate CPM/PERT diagram for the sequence of activities using both (a) activity-on-arc and (b) activity-on-node methods.

Activity	*Description*	*Immediate Predecessors*	*Time Duration (days)*
a	Start		0
b	Excavate and pour footings	*a*	4
c	Pour concrete foundation	*b*	2
d	Erect wooden frame, including rough roof	*c*	4
e	Lay brickwork	*d*	6
f	Install basement drains and plumbing	*c*	1
g	Pour basement floor	*f*	2
h	Install rough plumbing	*f*	3
i	Install rough wiring	*d*	2
j	Install heating and ventilating	*d, g*	4
k	Fasten plaster board and plaster (including drying)	*i, j, h*	10
l	Lay finish flooring	*k*	3
m	Install kitchen fixtures	*l*	1
n	Install finish plumbing	*l*	2
o	Finish carpentry	*l*	3
p	Finish roofing and flashing	*e*	2
q	Fasten gutters and downspouts	*p*	1
r	Lay storm drains for rain water	*c*	1
s	Sand and varnish flooring	*o, t*	2
t	Paint	*m, n*	3
u	Finish electrical work	*t*	1
v	Finish grading	*q, r*	2
w	Pour walks and complete landscaping	*v*	5
x	Finish	*s, u, w*	0

2. Consider a project to promote a new product. The activity durations to complete the project are given in the table. Find the minimum total time to complete the project.

No.	*Activity*	*Time Duration (weeks)*	*Precedence Activities*
0	Lead-time planning	3	—
1	Develop training plan	6	0
2	Select trainees	4	0
3	Draft brochure	3	0
4	Conduct training course	1	1, 2, 3
5	Deliver sample products	4	0
6	Print brochure	5	3
7	Prepare advertising	5	0
8	Release advertising	1	7
9	Distribute brochure	2	6

3. A project consists of the activities shown in the table. The durations represent the optimistic (a), most likely (m), and pessimistic (b) times (days) for each activity. Find the critical path for this PERT Network.

	Duration (days)		
Activity	a	m	b
(1, 2)	5	8	10
(1, 3)	18	20	22
(1, 4)	26	33	40
(2, 5)	16	18	20
(2, 6)	15	20	25
(3, 6)	6	9	12
(4, 7)	7	10	12
(5, 7)	5	7	8
(6, 7)	3	4	5

4. The possible activity durations of a certain project are given in the table. These figures represent optimistic (a), most likely (m), and pessimistic (b) times, respectively. The scheduled completion time is 17.5 days. Find the probability of completing the project in the scheduled time.

	Duration (days)		
Activity	a	m	b
(1, 2)	6	8	10
(1, 3)	4	6	7
(1, 4)	4	8	12
(2, 5)	5	6	8
(3, 5)	7	8	9
(4, 6)	7	10	14
(5, 6)	3	4	5

5. What are the three uses of dummy activities and dummy events in the activity-on-arc representation? Illustrate use by examples.

6. A hospital building foundation consists of four consecutive sections. The activities for each section need excavation, reinforcement, and filling concrete. The excavation of one section cannot start until the preceding one is completed. The same applies to filling concrete. After excavating all sections, the plumbing job can be started, but only 15% of the job can be completed before any concrete is poured. After completion of each section of the foundation, an additional 10% of the plumbing can be started provided that the preceding 10% portion is complete. Develop an activity network for this project.

7. Consider the activity network shown. Assume that the latest allowable project completion date is set at 49 days. [$T_8(L) \neq T_8(E)$.] Find the critical path for this network.

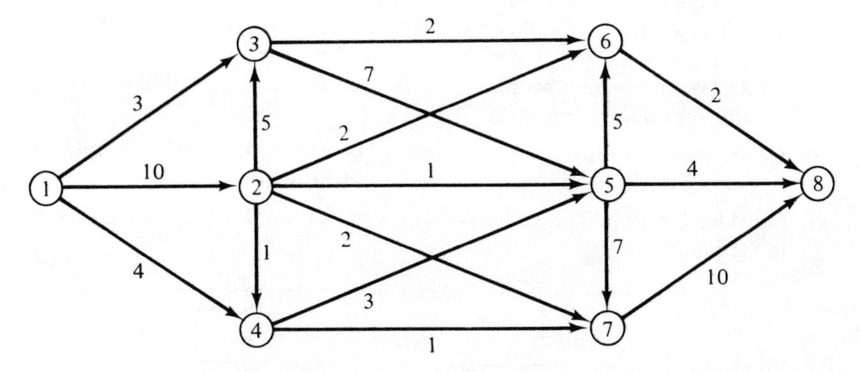

8. Calculate the total float, free float, safety float, and independent float for each of the activities in Exercise 2.

9. Consider a project to promote a new product. The probabilistic time durations to complete the activities are given in the table.

			Time Duration (weeks)		
Precedence Activities	*No.*	*Activities*	*Optimistic, a*	*Most Likely, m*	*Pessimistic, b*
None	0	Lead-time planning	2	3	5
0	1	Develop training plan	2	6	10
1	2	Select trainees	3	4	5
0	3	Draft brochure	1	3	4
9, 10, 5, and 8	4	Field test material	1	1	1
13	5	Deliver sample products	3	4	4
3	6	Print brochure	4	5	6
3	7	Prepare advertising	2	5	7
7	8	Release advertising	1	1	1

(*to be continued*)

Precedence Activities	No.	Activities	Time Duration (weeks) Optimistic, a	Most Likely, m	Pessimistic, b
6	9	Distribute brochure	2	2	3
6 and 2	10	Train sales force	3	5	6
4	11	Review market survey	2	4	5
0	12	Develop prototype product	5	7	8
12	13	Manufacture sample products	2	3	4

(a) Compute the mean and variance for each activity.
(b) Compute the critical path using the activity-on-node representation.
(c) Compute the critical path using the activity-on-arc representation.
(d) What is the probability that the network will be realized in less than 30 days? 40 days? 50 days?
(e) Determine the number of project days for which the probability of exceeding this duration is only 10%.

10. Determine the following for Exercise 1: (a) the critical path; (b) total float; (c) free float; (d) safety float; (e) independent float.

11. Consider the simplified activity-on-arc description of a project network given in the table.

Task	Activity	Duration (days)
A	(1, 2)	8
B	(2, 3)	10
C	(2, 4)	2
D	(3, 4)	16
E	(3, 5)	4
F	(4, 5)	8
G	(3, 6)	7
H	(4, 6)	12
I	(5, 7)	3
J	(6, 7)	8
K	(7, 8)	2

(a) Draw the CPM network for this task structure.
(b) Determine the project duration.
(c) Assume that each task requires exactly one person to perform the associated activity. Draw a bar chart and a resource-loading diagram that profiles the resource usage through time (assume that all jobs start at the earliest possible time).

(d) Use the resource-leveling procedures given in Section 4.14 to minimize the total manpower usage per day over the entire project duration.

12. Data have been established for the following set of activities.

	Time (days)		*Cost*		
Job	*Normal*	*Crash*	*Normal*	*Crash*	*Precedence Activities*
A	7	4	\$ 95	\$100	None
B	6	3	90	97	None
C	5	4	86	104	None
D	7	7	92	98	*A* and *C*
E	6	5	87	93	*A*, *B*, and *C*
F	7	5	112	120	*B*
G	8	5	101	113	*F*
H	9	6	97	109	*K*, *E*, and *D*
I	12	10	95	100	*A* and *C*
J	10	7	100	110	None
K	9	8	105	114	*F*

(a) Draw the CPM network using both the activity-on-arc and activity-on-node methods.
(b) Find the critical path using the activity-on-node representation.
(c) Find the critical path using the activity-on-arc representation.
(d) Crash the network to minimum-time duration along a minimum-cost schedule. Plot the results.

13. Consider the following CPM network (activity-on-node) with the activity durations (weeks) above the node and the crew sizes required to perform the activity below the node.

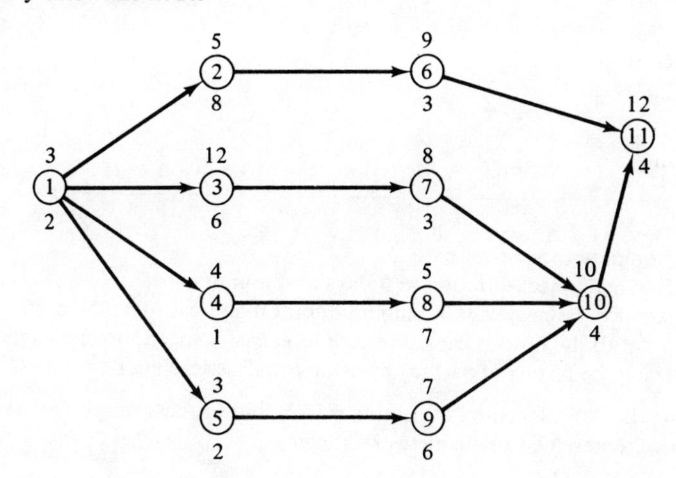

(a) Find the critical path for this project network.
(b) Compute the total float, free float, safety float, and independent float for all activities.
(c) Draw a bar chart and a resource-loading diagram showing the resource requirements profile for this project over the total time period (assume all jobs at earliest possible start).
(d) Use resource-leveling techniques to "smooth out" the manpower requirements.

14. Consider the following activity-on-arc CPM network with associated cost data. Resources required are dollars per day to run the project.

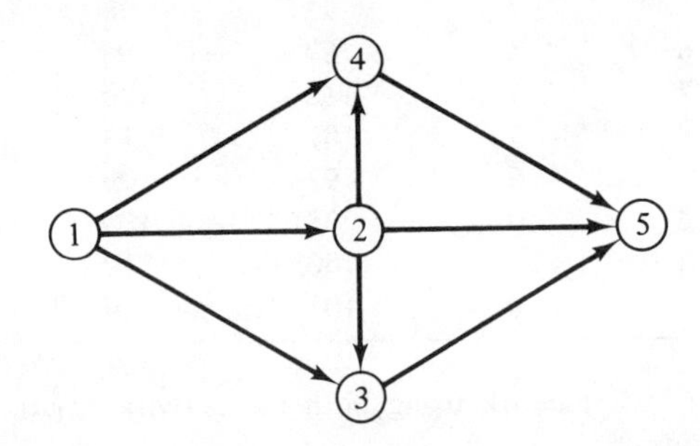

Activity	Normal Duration (days)	Normal Cost	Crash Duration (days)	Crash Cost	Resources Required
(1, 4)	2	$ 80	1	$130	400
(1, 2)	3	70	1	190	370
(1, 3)	6	110	5	135	510
(2, 4)	4	60	3	100	200
(2, 5)	7	85	6	115	150
(2, 3)	2	90	1	100	300
(2, 5)	7	85	6	115	450
(3, 5)	3	50	2	70	400
(4, 5)	4	105	3	175	270

(a) Compute the critical path.
(b) Reduce project duration to 9 days at minimum cost.
(c) Reduce the project to minimum duration and calculate the cost.
(d) Smooth the flow of capital (create level cash expenditures) into the project using the results of part (a) by using resource-leveling techniques.

15. Using the flow algorithm of Section 4.12.1, find the cost-duration curve for the project represented by the network shown.

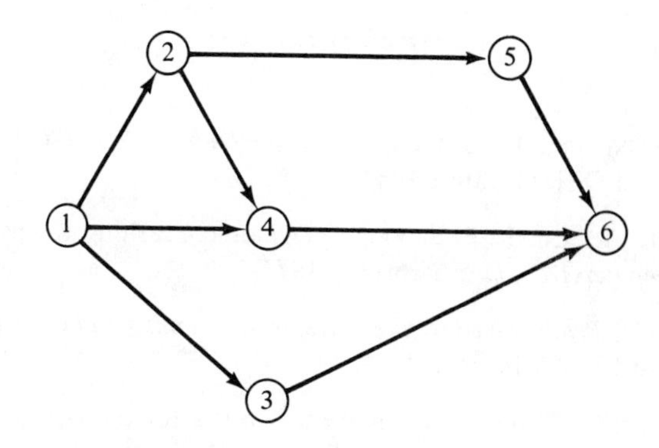

Arc	L_{ij}	U_{ij}	a_{ij}
(1, 2)	4	6	8
(1, 3)	4	8	9
(1, 4)	3	5	3
(2, 4)	3	3	—
(2, 5)	3	5	4
(3, 6)	8	12	20
(4, 6)	5	8	5
(5, 6)	6	6	—

16. Given the following network with activities having normally distributed durations, determine the numbers w_i such that the probability of realization of each node by time w_i is at least .95. The numbers assigned to the arcs are the means, and all variances are equal to 4. Formulate the chance-constrained event realization model.

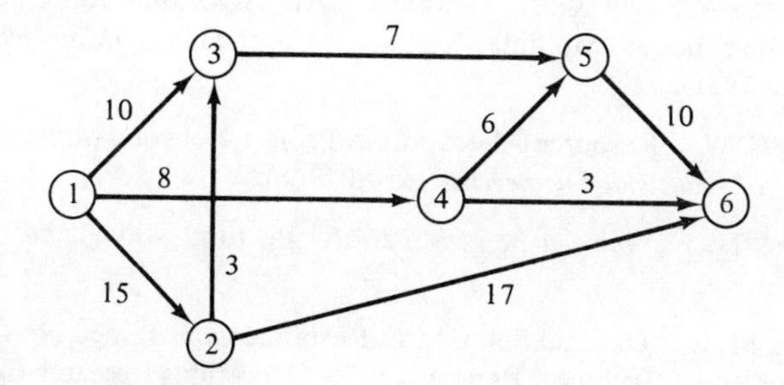

17. Consider a PERT network with p independent paths exponentially distributed with mean duration (path length) equal to $1/b$. Write an expression in terms of b and p for the probability that the project is finished between 10 and 12 time units.

REFERENCES

[1] Baker, B. N., and R. L. Eris, *An Introduction to PERT-CPM*. Homewood, Ill.: Richard D. Irwin, Inc., 1964.

[2] Benson, L. A., and R. F. Sewall, "Dynamic Crashing Keeps Projects Moving," *Computer Decisions* (February 1972).

[3] Buffa, E. S., *Production-Inventory Systems: Planning and Control*. Homewood, Ill.: Richard D. Irwin, Inc., 1968.

[4] Burton, M. R., "Some Mathematical Models for the Allocation of Limited Resources to Critical Path Type Scheduling Problems." Unpublished dissertation, University of Illinois, Urbana, October 1967.

[5] Crowston, W. B. S., "Decision CPM: Network Reduction and Solution," *Operational Research Quarterly* (1970).

[6] Crowston, W. B. S., "Decision Network Planning Models," Management Sciences Research Report No. 138, Carnegie-Mellon University, Pittsburgh, Pa., 1968.

[7] Davis, E. W., "Networks: Resource Allocation," *Journal of Industrial Engineering* (April 1974). Portions reproduced by permission of the authors and the American Institute of Industrial Engineers.

[8] Davis, E. W., "An Exact Algorithm for the Multiple Constrained Resource Project Scheduling Problem." Unpublished dissertation, Yale University, New Haven, Conn., 1968.

[9] Davis, E. W., Private communication, Fall 1979.

[10] Davis, E. W., and G. E. Heidorn, "An Algorithm for Optimal Project Scheduling under Multiple Resource Constraints," *Management Science* (August 1971).

[11] Davis, E. W., "Resource Allocation in Project Network Models—A Survey," *Journal of Industrial Engineering* (April 1966).

[12] Evarts, H. F., *Introduction to PERT*. Boston: Allyn and Bacon, Inc., 1967.

[13] Fisher, M. L., "Optimal Solution of Resource Constrained Network Scheduling Problems," Technical Report No. 56, Operations Research Center, Massachusetts Institute of Technology, 1970.

[14] Fondahl, J. W., "A Noncomputer Approach to the Critical Path Method for the Construction Industry," Construction Institute, Technical Report No. 9, Construction Institute, Stanford University, 1962.

[15] Fulkerson, D. R., "A Network Flow Computation for Project Cost Curves," *Management Science* (January 1961).

[16] Gomory, R. E., and C. W. Wade, "Write-up for Integer Programming Z, 7090, PK IPO2 and PK IPM2," IBM Corporation, Yorktown Heights, N.Y., 1961.

[17] Hollander, G., "Integrated Project Control," *Project Management Quarterly* (April 1973).

[18] Horowitz, J., *Critical Path Scheduling*. New York: The Ronald Press Co., 1967.

[19] *ICP Software Directory*. 2506 Willowbrook Parkway, Indianapolis, Ind.

[20] Johnson, T. J. R., "An Algorithm for the Resource-constrained Project Scheduling Problem." Unpublished Ph.D. thesis, School of Management, Massachusetts Institute of Technology, 1967.

[21] Kelley, J. E., and M. R. Walker, "Scheduling Activities to Satisfy Resource Constraints," in *Industrial Scheduling*, by J. Muth and G. Thompson. Englewood Cliffs, N.J.: Prentice-Hall, Inc., 1963.

[22] Kelley, J. E., and M. R. Walker, "Critical Path Planning and Scheduling," *Proceedings of the Eastern Joint Computer Conference*, 1959.

[23] Kelley, J. E., and M. R. Walker, "Critical Path Planning and Scheduling: Mathematical Basis," *Operations Research* (May–June 1961).

[24] Levin, R. I., and C. A. Kirkpatrick, *Planning and Control with PERT/CPM*. New York: McGraw-Hill Book Company, 1966.

[25] Levy, F. K., G. L. Thompson, and J. D. Wiest, "The ABC's of the Critical Path Method," *Harvard Business Review, 41* (5) (September–October 1963).

[26] Levy, F. K., and J. D. Wiest, *A Management Guide to PERT/CPM*, Englewood Cliffs, N.J.: Prentice-Hall, Inc., 1969.

[27] Lockyer, K. G., *An Introduction to Critical Path Analysis*. London: Sir Isaac Pitman & Sons Ltd., 1967.

[28] Miller, R. W., *Schedule, Cost and Profit Control with PERT*. New York: McGraw-Hill Book Company, 1963.

[29] Moder, J. J., and C. R. Phillips, *Project Management with CPM and PERT*. New York: Van Nostrand Reinhold Company, 1964; 2nd ed., 1970.

[30] Morris, L. N., *Critical Path Construction and Analysis*. Elmsford, N.Y.: Pergamon Press, Inc., 1967.

[31] Muth, J. F., and G. L. Thompson, eds., *Industrial Scheduling*. Englewood Cliffs, N.J.: Prentice-Hall, Inc., 1963.

[32] *NASA/DOD Guide, PERT/Cost*, Office of the Secretary of Defense and National Aeronautics and Space Administration, Washington D.C., 1964.

[33] PERT-Summary Report Phase 1, Special Projects Office, Bureau of Ordinance, Dept. of the Navy, Washington, D.C., July 1958.

[34] PHILLIPS, D. T., G. L. HOGG, AND M. J. MAGGARD, "GERTS III or: A GERTS Simulator for Labor-limited Queueing Systems," Research Memo No. 72-2, Purdue University, 1972.

[35] PRITSKER, A. A. B., L. J. WATTERS, P. M. WOLFE, AND W. HAPP, "GERT: Graphical Evaluation and Review Technique, Part I," *Journal of Industrial Engineering* (1966).

[36] PRITSKER, A. A. B., L. J. WATTERS, P. M. WOLFE, AND G. E. WHITEHOUSE, "GERT: Graphical Evaluation and Review Technique, Part II," *Journal of Industrial Engineering* (1966).

[37] PRITSKER, A. A. B., L. J. WATTERS, P. M. WOLFE, AND R. BURGESS, "The GERT Simulation Programs: GERTS III, GERTS IIIQ, GERTS IIIC, GERTS IIIR," Department of Industrial Engineering, Virginia Polytechnic Institute, Blacksburg, Va.

[38] PRITSKER, A. A. B., L. J. WATTERS, AND P. M. WOLFE, "Multi-project Scheduling with Limited Resources: A Zero–One Programming Approach," *Management Science* (1969).

[39] RIGGS, J. L., AND C. O. HEATH, *Guide to Cost Reduction through Critical Path Scheduling*. Englewood Cliffs, N.J.: Prentice-Hall, Inc., 1966.

[40] ROPER, D. E., "Critical Path Scheduling," *Journal of Industrial Engineering*, *15*(2) (March–April 1964).

[41] SHAFFER, L. R., J. B. RITTER, AND W. L. MEYER, *The Critical Path Method*. New York: McGraw-Hill Book Campany, 1964.

[42] SIEMENS, N., "A Simple CPM Time–Cost Tradeoff Algorithm," *Management Science* (February 1971).

[43] SMITH, L. A., AND P. MAHLER, "Comparing Commercially Available CPM/PERT Computer Programs," *Industrial Engineering*, Vol. 10, No. 4, 1978. Portions reproduced by permission of the author and the American Institute of Industrial Engineers.

[44] SMITH, K. M., *Critical Path Planning*. London: Management Publications Ltd., 1971.

[45] THOMAS, W. H., "Four Float Measures for Critical Path Scheduling," *Industrial Engineering* (October 1969).

[46] VIARISIO, G., "Management Decision System Based on Decision CPM," Presented at INTERNET 72 Congress, Stockholm, 1972.

[47] WOOLPERT, B., "Computer Software for Project Planning and Control," a paper for James R. Freeland and Charles A. Holloway, Stanford University, Graduate School of Business, May 1966.

ADVANCED TOPICS

"I can try," said Peter
"You might need some help," said Sasha
"Then come along," said Peter

"Peter and the Wolf," © Walt Disney Productions

As Peter soon discovered, pursuit of the wolf proved an arduous task, and before he was subdued, the help of Sasha was sorely needed. Peter's experience is best noted by the serious student of network-flow analysis when advanced topics are discussed, for the journey often requires "some help" from other sources. This chapter provides the reader with the fundamental elements required for an introductory discussion of three advanced topics: (1) generalized networks with gains and losses, (2) stochastic networks, and (3) multicommodity network flows. Each of these topics expands the fundamental structure and applicability of the models that were presented in Chapters 1 through 4.

5

PART I: GENERALIZED NETWORKS—NETWORKS WITH GAINS AND LOSSES

In all previous algorithms considered in this book, whenever a unit of flow entered an intermediate node the same unit had to exit the node. This implies that fractional units of flow were neither created nor destroyed in flow adjustments across paths connecting the source and terminal nodes. Clearly, there is a wide variety of possible network formulations requiring modifications of flow resulting in gains or losses of the flow through a given network. Applications include water reservoir models (evaporation, rainfall), structural design problems (modification of stress/strength characteristics), investment models (gains and losses of capital), production and distribution models (quality control rejects, order cancellations), crew scheduling problems (absenteeism), aircraft scheduling formulations (flight maintenance problems), and many more. A network that allows for flow generation or reduction across its arcs will be referred to as a *generalized network*.

More specifically, the flows that are transmitted across the arcs of a generalized network are modified by gain/loss factors so that the amount of flow entering an arc is not necessarily equal to the amount of flow leaving the arc. The value of the incoming flow is actually multiplied by the gain/loss factor to produce the value of the outgoing flow. To facilitate further discussion, consider a directed network consisting of arcs (i, j) such that a flow f_{ij} is passed from node i to node j at a unit cost c_{ij}. The allowable amount of flow along (i, j) must be at least equal to L_{ij} and at most equal to U_{ij}. Additionally, the arc multiplier will be represented by the symbol A_{ij}. These arc parameters will be denoted graphically as shown in the following diagram.

$$\textcircled{i} \xrightarrow[A_{ij}]{(L_{ij},\, U_{ij},\, c_{ij})} \textcircled{j}$$

Note that if $A_{ij} = 1$, a pure or conventional network formulation exists; if $A_{ij} > 1$, the flow is augmented (gains); and if $A_{ij} < 1$, the flow is decreased (losses). Since the flow is adjusted across arcs, conservation of flow is still maintained at all nodes. However, it should be apparent that flow values need not be integer and might now assume any positive real value. Despite the wide range of applications and the generality of structure, generalized network problems (GNP) have not previously received the attention that they deserve. Only recently have attempts been made to construct efficient solution procedures and develop commercial computer codes.

5.1 Application of Generalized Networks

As already indicated, generalized networks can successfully model many problems that have no pure network equivalent. This is made possible by the ability to interpret arc multipliers in two ways. First, multipliers can be viewed as modifying the amount of flow of some particular item. By means of flow modification, generalized networks can model such situations as evaporation, seepage, deterioration, breeding, interest rates, sewage treatment, purification processes with varying efficiencies, machine efficiency, and structural strength design. Second, it is possible to interpret the multipliers as transforming one type of item into another. This interpretation provides a way to model such processes as manufacturing, production, fuel to energy conversions, blending, crew scheduling, manpower requirements, and currency exchanges. (See [4, 8, 9, 13] for more detailed discussions.) The following applications illustrate some practical uses of generalized networks.

A complete water distribution system with losses has been modeled by Bhaumik [1] as a generalized network problem. The model is primarily concerned with the movement of water through canals to various reservoirs. However, the model also considers the retention of water over several time periods. The multipliers in this case represent the loss effect due to both evaporation and seepage.

Kim [19] has utilized generalized networks to represent copper refining processes. The electrolytic refining procedure, in this case, is modeled by a large dc electrical network. The arcs are current paths, with the multipliers representing the appropriate resistances. In this way, it is possible to investigate the effect of short circuits in the refining process.

Charnes and Cooper [4] identify applications of generalized networks for both plastic-limit analysis and warehouse funds-flow models. In plastic-limit analysis, the network is generated by the equations for horizontal and vertical equilibrium and by employing a coupling technique. The warehouse funds-flow model is actually a multi-period model. The arcs are used to represent sales, production, and holding of both products and cash. The multipliers are introduced to facilitate the conversions between cash and products.

A cash management problem has been modeled as a generalized network by Crum [5]. This model for the multinational firm incorporates transfer pricing, receivables and payables, collections, dividend payments, interest payments, royalties, and management fees. The arcs represent possible cash flow patterns and the multipliers represent costs, savings, liquidity changes, and exchange rates. Other applications include machine loading problems [4, 6, 28], blending problems [4, 28], the caterer problem [6, 28], and scheduling problems such as production and distribution problems, crew scheduling, aircraft scheduling, and manpower training [4, 6, 28].

5.2 GNP as a Linear Programming Problem

The generalized network problem can be viewed as a minimum-cost flow problem with gains and losses. The objective is to supply F units of flow to the sink node at minimum cost, or to dispatch a flow of $\bar{F}$ units from the source to the sink node at minimum cost. We will derive an algorithm that can solve both problems, but for formulation purposes we will prefer to use the former objective. The minimum-cost flow problem with gains and losses can be formulated as in Eqs. (5-1) to (5-5).

$$\text{Minimize} \sum_i \sum_j c_{ij} f_{ij} \tag{5-1}$$

Subject to

$$\sum_j f_{sj} - \sum_j A_{js} f_{js} \leq \bar{F} \tag{5-2}$$

$$\sum_j f_{ij} - \sum_j A_{ji} f_{ji} = 0 \qquad i \neq s, t \tag{5-3}$$

$$\sum_j f_{tj} - \sum_j A_{jt} f_{jt} = -F \tag{5-4}$$

$$0 \leq f_{ij} \leq U_{ij} \text{ for all } (i, j) \tag{5-5}$$

There are several important assumptions inherent in this particular formulation.

1. All arc multipliers are real positive numbers.
2. The lower bounds on all arc flows are zero. This can be assumed without loss of generality, since any positive lower bounds can be considered through the simple transformation $\bar{f}_{ij} = f_{ij} - L_{ij}$.
3. Multiple sources and sinks have been combined to form a single source node s and a single sink node t. This can be done by creating a super source and a super sink as in the out-of-kilter problem.
4. Unlike pure network formulations, the input flow $\bar{F}$ need not equal the output flow F due to flow adjustments caused by the arc multipliers. Hence, the required flow F will be defined with respect to the sink node. The input flows are unknown until the final solution is obtained, but can be constrained to be less than or equal to a value $\bar{F}$. Note that under this assumption, it may not be possible to deliver a flow of F to the sink from an input $\bar{F}$.

The solution to this problem will be presented through the development of two distinct algorithms. The first will deal with a special case of GNP, while the second will provide a general solution framework. Before proceeding, we will need to establish some basic properties of generalized networks.

5.3 Network Characteristics

Recall that a set of directed arcs connecting the source node to the sink node is called a *chain* or a *directed chain*. The product of all arc multipliers in a chain **E** will be called the *chain change* and will be denoted by

$$C_E = \prod_{(i,j)\in E} A_{ij} \tag{5-6}$$

Recall that if a chain starts and ends at the same node, it is called a *cycle*; correspondingly, the associated product of arc multipliers considered in Eq. (5-6) will be called the *cycle change*.

Consider the network of Figure 5.1 in isolation from a larger GNP. A chain connecting nodes 1 and 4 is given by the set of arcs $\mathbf{E} = \{(1, 2), (2, 3), (3, 4)\}$ with chain change of $C_E = (1)(2)(6) = 12$. Note that 1 unit of flow entering arc (1, 2) would reach node 2 as 1 unit; 1 unit of flow entering arc (2, 3) would exit that arc as 2 units at node 3; and 2 units entering arc (3, 4) at node 3 would result in 12 units at node 4. Such a chain change is called a *chain gain*, since flow is increased when passed through that chain. On the other hand, the chain $\mathbf{E} = \{(1, 2), (2, 4)\}$ from node 1 to node 4 changes a unit of flow at node 1 to $\frac{1}{4}$ unit at node 4. The chain change is hence $C_E = (1)(\frac{1}{4}) = \frac{1}{4}$ and is known as a *chain loss*. In summary:

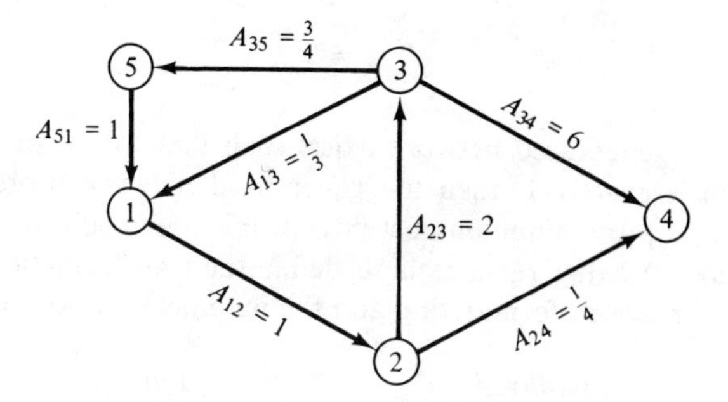

Figure 5.1. Network with gains and losses.

Chain change

$$C_E = \prod_{(i,j)\in E} A_{ij} \begin{cases} = 1 & \text{no change} \\ > 1 & \text{chain gain} \\ < 1 & \text{chain loss} \end{cases}$$

Note further that the sequence of arcs $\mathbf{E} = \{(1, 2), (2, 3), (3, 1)\}$ forms a cycle with $C_E = (1)(2)(\frac{1}{3}) = \frac{2}{3}$. Such a cycle is called a *flow absorbing cycle.* If $\mathbf{E} = \{(1, 2), (2, 3), (3, 5), (5, 1)\}$, then $C_E = (1)(2)(\frac{3}{4})(1) = 1.50$. Such a cycle would be a *flow-generating cycle.* Clearly, any flow that enters the cycle at node 1 is amplified to a flow of 1.50 if passed through the flow-generating cycle. Node 1 will be called a *tapping node* since flow can be tapped from the cycle through this node. Indeed, any flow F sent into such a cycle is amplified by the cycle gain. In a later section, these concepts will be used to structure a general solution algorithm.

5.4 Case I: Generalized Networks with No Flow-Generating or Flow-Absorbing Cycles

If the arc multipliers in a network are not restricted to be less than 1, then in general the minimum-cost flows from source to sink may not form a simple directed chain. Truemper [26] has proved the following theorem.

***Theorem*:** Any generalized network formulation with arbitrary arc multipliers, such that for every possible cycle $\mathbf{E}$ $C_E = \prod_{(i,j)\in E} A_{ij} = 1$, can be converted into a pure network formulation with all arc multipliers $A_{ij} = 1$.

***Corollary*:** A necessary and sufficient condition for this result to hold is that there exist a node number M_i, $i = 1, 2, \ldots, n$, for each node i in the network, such that for each arc (i, j) Eq. (5-7) will be satisfied:

$$\frac{M_i A_{ij}}{M_j} = 1 \tag{5-7}$$

Hence, if a generalized network exists such that for all possible cycles the cycle gain is exactly 1, then the generalized network problem can be transformed to a pure minimum-cost flow problem and solved via any convenient method. All that remains is to define the transformation. Consider the generalized network formulation and the pure network formulation.

Generalized	*Pure*
Minimize: $\sum_i \sum_j c_{ij} f_{ij}$	Minimize: $\sum_i \sum_j c_{ij} f_{ij}$
subject to	subject to
$\sum_j f_{sj} - \sum_j A_{js} f_{js} \leq \bar{F}$	$\sum_j f_{sj} - \sum_j f_{js} \leq \bar{F}$
$\sum_j f_{ij} - \sum_j A_{ji} f_{ji} = 0$	$\sum_j f_{ij} - \sum_j f_{ji} = 0$
$\sum_j f_{tj} - \sum_j A_{jt} f_{jt} = -F$	$\sum_j f_{tj} - \sum_j f_{jt} = -F$
$L_{ij} \leq f_{ij} \leq U_{ij}$	$L_{ij} \leq f_{ij} \leq U_{ij}$

If the results of the previous theorem and corollary hold, then multiply each conservation-of-flow constraint in the generalized formulation by $1/M_i$, $i = 1, 2, \ldots, n$, to obtain the pure formulation, where M_i is determined as follows:

***Step 1*:** Set $M_1 = 1$.

***Step 2*:** For each arc (i, j) successively calculate

$$M_j = M_i A_{ij} \qquad i = 1, 2, \ldots, n;$$
$$j = \text{index of all arcs connected to node } i$$

5.4.1 Example Formulation

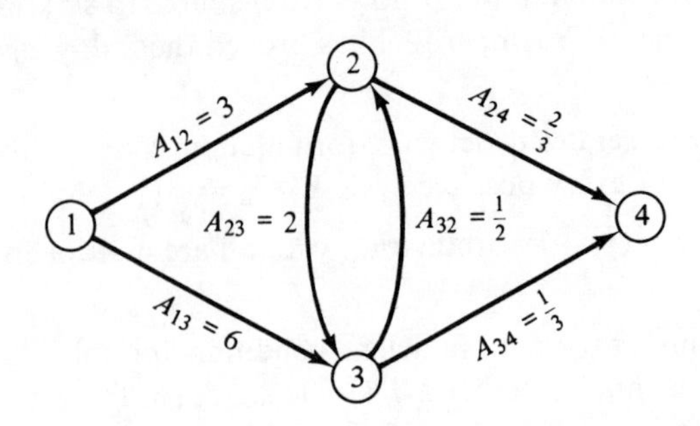

Arc (i, j)	M_i	A_{ij}	M_j
(1, 2)	1	3	3
(1, 3)	1	6	6
(3, 4)	6	$\frac{1}{3}$	2

Hence,

$$M_1 = 1 \qquad M_3 = 6$$
$$M_2 = 3 \qquad M_4 = 2$$

Linear programming formulation (generalized)

$$\text{Minimize: } \sum_i \sum_j c_{ij} f_{ij}$$

S.T.	(1, 2)	(1, 3)	(2, 3)	(3, 2)	(2, 4)	(3, 4)	
	f_{12}	$+ f_{13}$					$\leq \bar{F}$
	$-3f_{12}$		$+ f_{23}$	$- \frac{1}{2}f_{32}$	$+ f_{24}$		$= 0$
		$- 6f_{13}$	$- 2f_{23}$	$+ f_{32}$		$+ f_{34}$	$= 0$
					$- \frac{2}{3}f_{24}$	$- \frac{1}{3}f_{34}$	$= F$

Note that $1/M_1 = 1$, $1/M_2 = \frac{1}{3}$, $1/M_3 = \frac{1}{6}$, $1/M_4 = \frac{1}{2}$. Multiplying the ith row by $1/M_i$, $i = 1, 2, 3, 4$, we obtain the following model.

Linear programming formulation (pure)

$$\text{Minimize: } \sum_i \sum_j c_{ij} f_{ij}$$

S.T.	(1, 2)	(1, 3)	(2, 3)	(3, 2)	(2, 4)	(3, 4)	
	f_{12}	$+ f_{13}$					$\leq \bar{F}$
	$-f_{12}$		$+ \frac{1}{3}f_{23}$	$- \frac{1}{6}f_{32}$	$+ \frac{1}{3}f_{24}$		$= 0$
		$- f_{13}$	$- \frac{1}{3}f_{23}$	$+ \frac{1}{6}f_{32}$		$+ \frac{1}{6}f_{34}$	$= 0$
					$- \frac{1}{3}f_{24}$	$- \frac{1}{6}f_{34}$	$= F$

Now let

$$\bar{f}_{23} = \tfrac{1}{3} f_{23}$$
$$\bar{f}_{32} = \tfrac{1}{6} f_{32}$$
$$\bar{f}_{34} = \tfrac{1}{6} f_{34}$$

Making this change of variables and adjusting the upper and lower bound appropriately, we obtain the desired results.

Finally, the reader should verify that for each possible cycle **E** in the network,

$$\prod_{(i,j)\in \mathbf{E}} \frac{M_i A_{ij}}{M_j} = 1$$

Note that arc multipliers when reversing flow through arcs are the reciprocal of those in the original direction, as illustrated below:

5.5 Case II: Generalized Networks with Flow-Generating and/or Flow-Absorbing Cycles

The algorithm to solve the general problem of networks with gains and losses is far more complex than the algorithms previously discussed in this book. Solution algorithms have been reported by Minieka [22], Oettli and Prager [24], Grinold [14], Jensen and Barnes [17], Glover and Klingman [13], Jarvis and Jezior [15], and Jewell [18]. The algorithm which will be presented in this section is attributed to Jensen and Bhaumik [16]. The only restriction imposed by this algorithm is that no cycle in the network may have a negative cost. The algorithm is based upon a simple labeling procedure which extends the basic labeling strategies employed in pure shortest-route algorithms. Initially, we will briefly describe the three distict phases of the algorithm, and then follow this description with a technical discussion.

5.5.1 Phase I: Initial Flow

a. A minimum-cost chain is identified to connect the source and the sink nodes. If such a chain cannot be found, the problem has no solution.
b. Flow is increased from source to sink along the minimum-cost chain until at least one arc is saturated.

5.5.2 Phase II: The Marginal Network

Once flow has been routed from source to sink as indicated in phase I, the original network is modified to reflect the current flow status, and to investigate the possibility of further flow adjustments in the network. The modified network will be referred to as the *marginal network*. This marginal network consists of all the unsaturated arcs in the original network plus one "mirror arc" for each original arc carrying a strictly positive flow.

5.5.3 PHASE III: THE FLOW-AUGMENTATION PROCESS

Using the marginal network created in phase II, a new flow-augmenting path from source to sink is found and maximum flow at minimum cost allocated to this path. At the conclusion of phase III, either the desired flow has been obtained or phases I and II are repeated. The three phases of the algorithm will now be described in detail following the presentation by Jensen and Bhaumik [16].

5.6 Phase I: The Minimum-Cost Flow-Augmenting Chain

The minimum-cost chain for a generalized network is defined as the chain that can deliver flow at the lowest possible cost from the source to the sink. As shown by Jewell [18], networks with gains (arc multipliers greater than 1) may have flow-generating cycles. Although a network may initially have no arc multipliers greater than 1, during successive iterations of the algorithm, mirror arcs are created with gain factors greater than 1. A mirror arc is defined as an arc that provides for reverse flow with multiplier being equal to the reciprocal of the multiplier of the corresponding "parent" arc. Therefore, in general, the minimum-cost chain may be either a simple directed chain from the source to the sink, or may be a composite chain containing a flow-generating cycle.

To discuss the minimum-cost chain it is necessary to define a new variable, V_i, for each node, i, in the network. The new variable will be referred to as a *node potential*. The potential of a node can be economically interpreted as the cost of supplying 1 unit of flow from that node. In a directed chain, this is the lowest cost of transporting 1 unit of flow from the source node to the node under consideration. In a cycle, it would be the cost incurred in generating 1 unit of flow from that node. The potential of a node is controlled by the costs and arc multipliers of the chain that delivers the flow to that node. If the cost per unit of flow to reach a node i is V_i, the cost per unit of flow to reach another node j, through arc (i, j) is given by

$$\bar{V}_{ij} = \frac{V_i + c_{ij}}{A_{ij}} \tag{5-8a}$$

where c_{ij} is the cost of transmitting 1 unit of flow from node i to node j, and A_{ij} is the multiplier of arc (i, j).

After applying Eq. (5-8a) to all arcs terminating at node j, the potential of this node can be obtained by

$$V_j = \min_{i \in \boldsymbol{\beta}_j} [\bar{V}_{ij}] = \min_{i \in \boldsymbol{\beta}_j} \left[\frac{V_i + c_{ij}}{V_{ij}} \right] \tag{5-8b}$$

In Eq. (5-8b), $\boldsymbol{\beta}_j$ is the set of nodes directly connected to node j by arcs with orientation from i to j. An alternative statement of relationship (5-8b) is

$$V_j \leq \frac{V_i + c_{ij}}{A_{ij}}, \text{ for all } i \in \boldsymbol{\beta}_j \tag{5-8c}$$

with equality obtained for at least one arc entering node j. If there is exactly one arc connecting node i to node j, the potential of node j can be obtained directly from Eq. (5-8a), since in that case $V_j = \bar{V}_{ij}$, that is,

$$V_j = \frac{V_i + c_{ij}}{A_{ij}} \tag{5-8d}$$

Starting with the potential of the source node set equal to zero ($V_s = 0$) and the potential of all other nodes set equal to infinity ($V_i = \infty, i \neq s$), it is possible to iteratively reduce the value of the node potentials until Eq. (5-8c) is satisfied for each arc. The process is accomplished through an arc scanning procedure. For each arc (i, j) Eq. (5-8c) is tested; if it is satisfied, no change is made. Otherwise, the node potential V_j is reset equal to $(V_i + c_{ij})/A_{ij}$. This process is continued until all arcs are scanned. Arcs might be scanned more than once as node potentials are changed. When there are no node potential changes in a complete iteration, the scanning procedure is terminated. Although the arcs may be scanned in any order, it is convenient to start with the source node and proceed scanning along forward arcs recursively, using Eq. (5-8c) to adjust the node potentials.

Any backtracking procedure can be used to trace the minimum-cost chain. In our discussion, a backward pointer p_j will be used to designate the node i for which V_j is a minimum in Eq. (5-8b). Initially, all the backward pointers are set to zero. If at any time $V_j > (V_i + c_{ij})/A_{ij}$ for node j, then p_j is reset equal to i, and, as already indicated, V_j is changed to $(V_i + c_{ij})/A_{ij}$. Therefore, at the conclusion of the algorithm each node i carries a label (p_i, V_i), where V_i is the minimum cost of delivering 1 unit of flow from the node, and p_i is the node immediately preceding the node in the minimum-cost path. If the sink node receives a label $(0, \infty)$, this would imply that no flow-augmenting chain ends at the sink of the network and the algorithm is terminated.

5.7 Phase II: Constructing the Marginal Network

Once flow is introduced into the network through phase I, the residual capacities of unsaturated arcs are obtained by subtracting the current flows from the original arc capacities. These residual capacities can be further adjusted (either increased or decreased) by modifying the flows in the original network. To perform such adjustments the original network is converted into a new network referred to as the *marginal network*. This marginal network, as previously indicated, consists of all unsaturated arcs of the original network plus one mirror arc for each original arc carrying positive flow. For arc (i, j) the corresponding mirror arc will have an orientation from j to i, and, therefore, its flow would reduce that of the parent arc if the two arcs were superimposed. The arc multipliers of the mirror arcs are the reciprocals of the arc multipliers of the parent arcs, and the flows arriving at the ending nodes of the mirror arcs, through these arcs, are equal to the flow reductions for the corresponding parent arcs.

If f_{ij} is the flow along an arc (i, j), then $U_{ij} - f_{ij}$ is the residual capacity of the arc in the marginal network. If A_{ij} is the gain factor of this arc, then $f_{ij}A_{ij}$ is the maximum flow that can be removed from (i, j). Since this reduction would be performed by using the mirror arc (j, i), the capacity of this arc in the marginal network must be defined as being equal to $f_{ij}A_{ij}$. Note that $1/A_{ij}$ is the gain factor of the mirror arc whose flow would reduce $f_{ij}A_{ij}$ units of flow at node j to f_{ij} units at node i. Since a cost of c_{ij} is already incurred per unit of flow along the arc (i, j), the recovery of this expense from the elimination of flow along this arc is reflected by a unit cost of $-c_{ij}/A_{ij}$ on the mirror arc (j, i). After any iteration, mirror arcs are created only for those parent arcs that transmit flow at that iteration, and the original saturated arcs are not included in the marginal network. (Note that a mirror arc is still added for a saturated arc.)

In summary, for each arc $(i, j)^*$ of the marginal network, the following parameters are defined, given the updated values of the flows f_{ij} through the original network:

$$\begin{aligned}
c_{ij}^* &= c_{ij} && \text{if } f_{ij} < U_{ij} \\
c_{ji}^* &= -c_{ij}/A_{ij} && \text{if } f_{ij} > 0 \\
U_{ij}^* &= U_{ij} - f_{ij} && \text{if } f_{ij} \geq 0 \\
U_{ji}^* &= f_{ij}A_{ij} && \text{if } f_{ij} > 0 \\
A_{ij}^* &= A_{ij} && \text{if } f_{ij} < U_{ij} \\
A_{ji}^* &= 1/A_{ij} && \text{if } f_{ij} > 0
\end{aligned}$$

5.8 Phase III: Executing the Flow-Augmentation Process

Once the minimum-cost chain through a generalized network has been identified, it is necessary to route the maximum flow through this chain. This can be performed in a number of different ways. One way, for example, is to simply increase flow in a separate pass through the minimum-cost chain once it has been determined. Flow values are easily computed by back-tracking from sink to source and assigning appropriate flows utilizing the arc bounds.

The minimum-cost chain is closely related to the basis in the linear programming context [32]. Additionally, the interpretation of node potentials is very similar to that of the dual variables in the out-of-kilter algorithm [16].

As previously noted, if the arc gain factors in a network are not restricted to be equal to 1, then the minimum-cost chain may not be a simple directed chain from source to sink. It may contain a flow-generating cycle that does not include the source. Flow-generating cycles are easily recognized in the algorithmic process. If node i receives a label (p_i, V_i) in the arc scanning procedure and later, in the labeling methodology, it is assigned a new label $(p_i, \bar{V}_{ij})$, $\bar{V}_{ij} < V_i$, then a cycle is detected. In this case, the arc scanning process will continue and all node potentials in the cycle are again reduced. Node potentials in the cycle continue to be reduced by a factor equal to the reciprocal of the cycle gain factor. Jensen and Bhaumik [16] have shown that the node potentials in the cycle converge toward a final value if the process continues.

To continue with this discussion the following additional notation will be used. Let $\mathbf{N}_C$ be the set of nodes in a cycle, and let $\mathbf{A}_C$ be the set of arcs in the same cycle. Let d_r and a_r be the cost per unit of flow and the arc multiplier, respectively, for arc $r \in \mathbf{A}_C$. When a cycle is detected, the recursive relationships (5-8a) and (5-8b) cannot be immediately used to determine the node potentials for the nodes in the cycle. Let h be the element of $\mathbf{N}_C$ directly linked to a simple chain connecting the flow-generating cycle $L = (\mathbf{N}_C, \mathbf{A}_C)$ to the sink node t. If $t \in \mathbf{N}_C$, $h = t$. Defining node h as the *tapping node*, its potential is computed by

$$V_h = \frac{g}{g-1} \sum_{j=1}^{n_c} \left[d_j \middle/ \prod_{i=j}^{n_c} a_i \right] \tag{5-9}$$

In Eq. (5-9), g is the cycle gain factor and j is an index for the elements of $\mathbf{A}_C$. It is assumed that the elements of $\mathbf{A}_C$ have been indexed $1, 2, \ldots, n_c$, starting with the arc whose initial node is h, and proceeding along connected arcs. The potential of any other node j in the cycle can be computed from the potential of its preceding node i by using Eq. (5-8d) recursively, once V_h has been computed. If the tapping node h is different from node t, labels for those nodes connecting node h to node t can be similarly recovered using Eq. (5-8d).

As an illustration of the procedure under discussion, the flow-generating cycle of Figure 5.2 will be considered. Node 2 will be arbitrarily chosen as the tapping node.

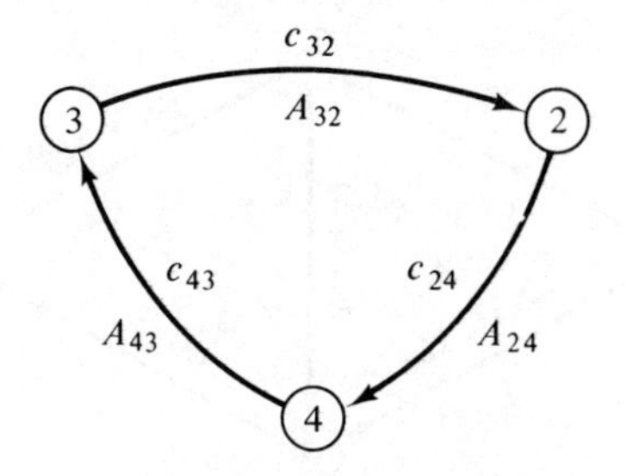

Figure 5.2. Flow-generating cycle.

The sets of nodes and arcs for the sample cycle of Figure 5.2 are defined as follows:

$$\mathbf{N}_C = \{2, 4, 3\}$$
$$\mathbf{A}_C = \{(2, 4), (4, 3), (3, 2)\}$$

According to the interpretation given for index j in Eq. (5-9), we will refer to arc (2, 4) as the first arc, (4, 3) as the second arc, and (3, 2) as the third arc in the cycle. The corresponding parameters for these arcs are $d_1 = c_{24}$, $d_2 = c_{43}$, $d_3 = c_{32}$, $a_1 = A_{24}$, $a_2 = A_{43}$, and $a_3 = A_{32}$. The gain factor of the cycle is equal to $g = a_1 a_2 a_3$ and is assumed to be strictly larger than 1. From Eq. (5-9), the potential of node 2 is given by

$$V_2 = \left[\frac{d_1}{a_1 a_2 a_3} + \frac{d_1}{a_2 a_3} + \frac{d_3}{a_3}\right]\left[\frac{a_1 a_2 a_3}{a_1 a_2 a_3 - 1}\right]$$

Similarly, node 4 or node 3 could be considered as the tapping node and Eq. (5-9) would give the values of V_4 and V_3. Alternatively, once the potential of the tapping node is obtained from Eq. (5-9), the other node potentials can be determined by using Eq. (5-8d).

5.9 A Numerical Example with Gains and Losses

The labeling algorithm for a flow network with gains and losses will now be applied to a hypothetical example taken from the work by Jensen and Bhaumik [16]. The problem will be solved in four iterations.

***Iteration 1*:** In further iterations of the algorithm, the network given below will be referred to as the original network.

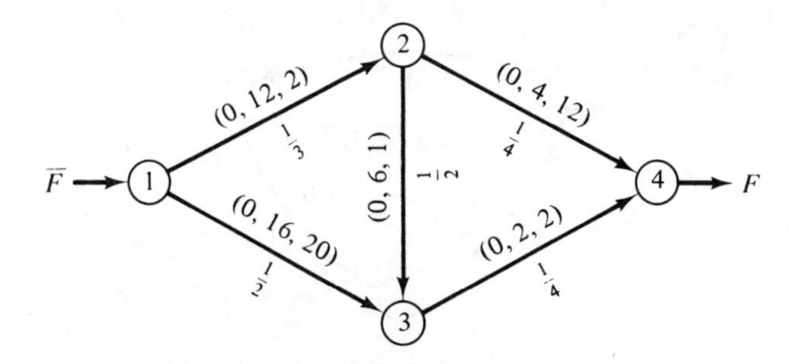

The following table summarizes the results for the application of Eq. (5-8b) to each arc of the original network.

Node j	Incoming Arc	c_{ij}	A_{ij}	V_i	$\bar{V}_{ij}$	V_j	Label
1	—	—	—	0	0	0	(0, 0)
2	(1, 2)	2	1/3	0	6	6	(1, 6)
3	(1, 3)	20	1/2	0	40		
	(2, 3)	1	1/2	6	14	14	(2, 14)
4	(2, 4)	12	1/4	6	72		
	(3, 4)	2	1/4	14	64	64	(3, 64)

The minimum-cost chain is identified as the sequence of arcs (1, 2), (2, 3), and (3, 4).

The maximum flow that can be transported from the source node to the sink node through the minimum-cost chain is a function of both the arc multipliers and the arc upper bounds associated with the chain. It can be proved that the value of the maximum flow at the sink node t is given by

$$F_t = \min_{1 \le k \le m} [U_k \prod_{r=k}^{m} a_r] \tag{5-10a}$$

where m is the number of arcs in the chain, U_k the upper bound on the flow across the kth arc in the chain, and a_k is the multiplier for the kth arc. Eq. (5-10a) can be simplified by defining

$$\bar{F}_k = U_k \prod_{r=k}^{m} a_r \tag{5-10b}$$

In this case,

$$F_t = \min_{1 \le k \le m} [\bar{F}_k] \tag{5-10c}$$

Once the value of F_t is obtained, the corresponding arc flows in the minimum-cost chain are identified by using a backward recursive relationship starting with the sink node. The following table summarizes the results at the completion of iteration 1.

$m = 3$

Arc (i, j)	k	U_k	$\prod_{r=k}^{m} a_r$	$\bar{F}_k$
(1, 2)	1	12	1/24	1/2
(2, 3)	2	6	1/8	3/4
(3, 4)	3	2	1/4	1/2

From Eq. (5-10c), $F_t = \min[1/2, 3/4, 1/2] = 1/2$. Defining the flow on arc (i, j) at iteration I as $f_{ij}^{(I)}$, and backtracking through the minimum-cost chain, the following results are obtained for $I = 1$.

$$f_{34}^{(1)} = \frac{1/2}{1/4} = 2$$

$$f_{23}^{(1)} = \frac{2}{1/2} = 4$$

$$f_{12}^{(1)} = \frac{4}{1/3} = 12$$

$$f_{13}^{(1)} = f_{24}^{(1)} = 0$$

Hence, $F = 1/2$ and $\bar{F} = 12$, implying that a flow of 1/2 unit is generated at a cost of (1/2)(64) = 32. Arcs (1, 2) and (3, 4) are saturated, and are therefore removed from the original network. Mirror arcs (2, 1), (3, 2), and (4, 3) are included in the marginal network to investigate the possibility of flow reductions in the corresponding parent arcs.

Iteration 2:

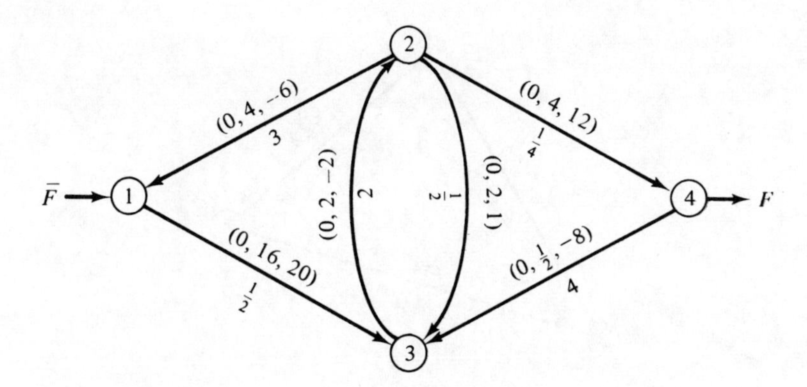

Node 1 is assigned the label (0, 0) at iteration 1. With $V_1 = 0$ and the parameters of the arcs in the marginal network, Eq. (5-8b) can be used recursively to yield the results that are summarized in the following table.

Node j	*Incoming Arc*	c_{ij}	A_{ij}	V_i	$\bar{V}_{ij}$	V_j	*Label*
3	(1, 3)	20	1/2	0	40	—	
2	(3, 2)	−2	2	40	19	19	(3, 19)
1	(2, 1)	−6	3	19	13/3	0	(0, 0)
3	(2, 3)	1	1/2	19	40	—	
4	(2, 4)	12	1/4	19	124	124	(2, 124)
3	(4, 3)	−8	4	124	29	29	(4, 29)

According to the results just obtained, a flow-generating cycle is detected. The cycle consists of the arcs (3, 2), (2, 4), and (4, 3), as shown in Figure 5.3. The gain factor of this cycle is $g = 2$, and therefore $g/(g - 1) = 2$. Using Eq. (5-9), and considering node 4 as the tapping node, V_4 is computed as

$$V_4 = 2\left[\frac{-8}{(4)(2)(1/4)} + \frac{-2}{(2)(1/4)} + \frac{12}{(1/4)}\right] = 80$$

V_3 and V_2 can now be computed from Eq. (5-8d) as indicated below:

$$V_3 = \frac{V_4 + c_{43}}{A_{43}} = \frac{80 - 8}{4} = 18$$

$$V_2 = \frac{V_3 + c_{32}}{A_{32}} = \frac{18 - 2}{2} = 8$$

The nodes of the cycle shown in Figure 5.3 receive new labels (4, 18), (3, 8), and (2, 80), corresponding to nodes 3, 2, and 4, respectively.

Since node 4 is the sink node, an increase in flow can be produced at minimum cost by tapping the cycle at node 4. Additional flow can be obtained

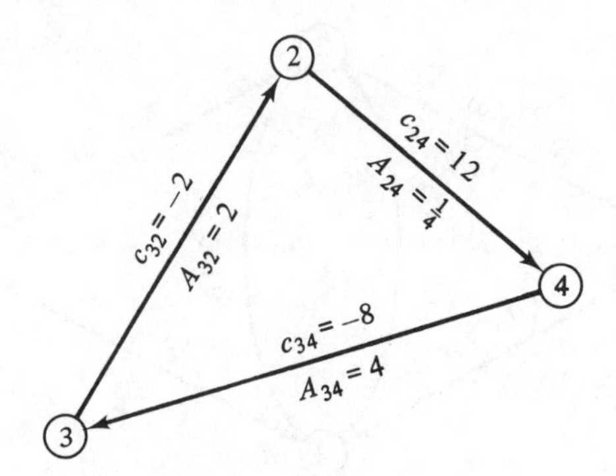

Figure 5.3. Subnetwork.

at a cost of 80 per unit. As in iteration 1, the maximum flow that can be generated at the tapping node depends upon the arc multipliers and the arc upper bounds. However, the cycle maximum flow must be adjusted by the cycle gain factor. The cycle maximum flow generation at the tapping node is given by the following relationship [16]:

$$F_h = \frac{g-1}{g}\left\{\min_{1\leq k\leq m}\left[U_k \prod_{r=k}^{m} a_r\right]\right\} \tag{5-10d}$$

where h is the index of the tapping node, g the cycle gain factor, m the number of arcs in the cycle, U_k the upper bound on the flow allowed on the kth arc of the cycle, and a_k the arc multiplier for the kth arc. It is assumed that the arcs in the cycle have been indexed as described for Eq. (5-9). Note that Eq. (5-10d) is simply Eq. (5-10a) adjusted by a function of the cycle gain factor. The computations for iteration 2 are summarized in the following table.

$m = 3$

Arc (i, j)	k	U_k	$\prod_{r=k}^{m} a_r$	$\bar{F}_k$
(4, 3)	1	1/2	2	1
(3, 2)	2	2	1/2	1
(2, 4)	3	4	1/4	1

From Eq. (5-10d), $F_h = (1/2) \min [1, 1, 1] = 1/2$. Hence, the maximum flow generated at the tapping node is 1/2.

The determination of the appropriate adjustments in arc flows using simple serial backtracking procedures is not possible when flow-generating cycles are detected. However, the general result of Eq. (5-10e) can be used in such cases [16] to predict the value of the flow.

$$\bar{f}_k = \frac{F_h\left[\dfrac{g}{g-1}\right]}{\prod_{r=k}^{m} a_r} \tag{5-10e}$$

Applying Eq. (5-10e) to the cycle shown in Figure 5.3, for which $m = 3$, $F_h = 1/2$, and $g/(g-1) = 2$, the following flow adjustments are obtained:

Arc (i, j)	k	$\prod_{r=k}^{m} a_r$	$\bar{f}_k$
(4, 3)	1	2	1/2
(3, 2)	2	1/2	2
(2, 4)	3	1/4	4

By construction, all adjustments not shown in the above table are equal to zero. The updated arc flow values are equal to the values computed at iteration 1 plus the adjustments corresponding to the arc flows through the marginal network just considered. That is,

$$f_{12}^{(2)} = f_{12}^{(1)} + 0 = 12 + 0 = 12$$

$$f_{13}^{(2)} = f_{13}^{(1)} + 0 = 0 + 0 = 0$$

$$f_{23}^{(2)} = f_{23}^{(1)} - \frac{2}{1/2} = 4 - 4 = 0$$

$$f_{24}^{(2)} = f_{24}^{(1)} + 4 = 0 + 4 = 4$$

$$f_{34}^{(2)} = f_{34}^{(1)} - \frac{1/2}{1/4} = 2 - 2 = 0$$

The incremental flow gain is equal to 1/2, and, after iteration 2, $F = 1$ and $\bar{F} = 12$. The cost of generating 1 unit of flow is equal to 32 (iteration 1) plus $80(1/2) = 72$.

The generalization of the above results is stated as follows. Let $f_{ij}^{(I-1)}$ be the flow on arc (i, j) of the original network at iteration $I - 1$. Let $\bar{f}_{ij}^{(I)}$ be the value of the flow on arc (i, j) of the marginal network constructed in the Ith iteration. Finally, let $\bar{f}_{ji}^{(I)}$ be the value of the flow on the mirror arc (j, i) of the marginal network constructed in the Ith iteration. If arc (i, j) is not saturated at iteration $I - 1$, its incoming flow can be increased by the same amount of flow leaving node i in the marginal network of iteration I. That is,

$$f_{ij}^{(I)} = f_{ij}^{(I-1)} + \bar{f}_{ij}^{(I)} \tag{5-11a}$$

On the other hand, if the flow on arc (i, j) in the original network is strictly positive, it can be reduced by the same amount of flow arriving at node i through the mirror arc (j, i) in the marginal network of iteration I. That is,

$$f_{ij}^{(I)} = f_{ij}^{(I-1)} - \bar{f}_{ji}^{(I)}/A_{ij} \tag{5-11b}$$

It can be verified, by tracing the $f_{ij}^{(2)}$ arc flows, that the updated value of the flow at the sink node is 1 unit, at a cost of 72. Arcs (1, 2) and (2, 4), in the original network, are now saturated and can be removed. Mirror arcs (2, 1) and (4, 2) are included to investigate possible flow reductions.

Iteration 3:

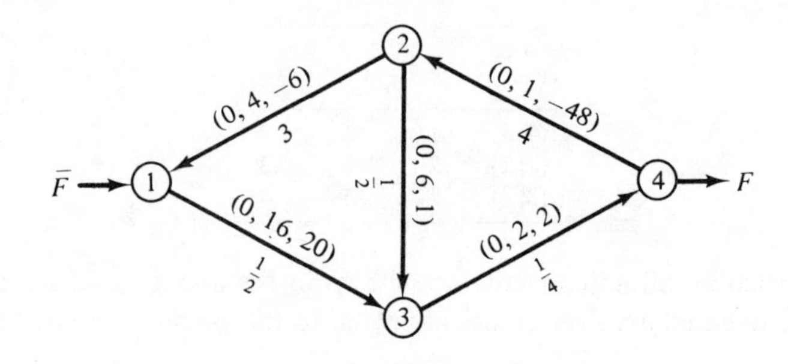

The label of node 1 remains (0, 0), that is, $V_1 = 0$. The following table summarizes the results from the labeling procedure.

Node j	Incoming Arc	c_{ij}	A_{ij}	V_i	$\bar{V}_{ij}$	V_j	Label
3	(1, 3)	20	1/2	0	40	40	(1, 40)
4	(3, 4)	2	1/4	40	168	168	(3, 148)
2	(4, 2)	−48	4	168	30	30	(4, 30)
1	(2, 1)	−6	3	30	8	0	(0, 0)
3	(2, 3)	1	1/2	30	64	—	

No cycles are detected, and all nodes now have labels. The minimum-cost chain is given by the sequence of arcs (1, 3) and (3, 4). The maximum flow at the sink node can be obtained by Eq. (5-10a). A summary of the results needed for the application of this equation is given in the following table.

$m = 2$

Arc (i, j)	k	U_k	$\prod_{r=k}^{m} a_r$	$\bar{F}_k$
(1, 3)	1	16	1/8	2
(3, 4)	2	2	1/4	1/2

From Eq. (5-10a), $F_t = \min [2, 1/2] = 1/2$. By performing backtracking as indicated in iteration 1, it is possible to obtain the flow adjustments given below:

$$\bar{f}_{34}^{(3)} = \frac{1/2}{1/4} = 2$$

$$\bar{f}_{13}^{(3)} = \frac{2}{1/2} = 4$$

$$\bar{f}_{21}^{(3)} = \bar{f}_{23}^{(3)} = 0$$

The updated flows across the arcs of the original network can be computed by using Eqs. (5-11a) and (5-11b), as shown below:

$$f_{12}^{(3)} = f_{12}^{(2)} - \bar{f}_{21}^{(3)}/A_{12} = 12 - 0 = 12$$

$$f_{13}^{(3)} = f_{13}^{(2)} + \bar{f}_{13}^{(3)} = 0 + 4 = 4$$

$$f_{34}^{(3)} = f_{34}^{(2)} + \bar{f}_{34}^{(3)} = 0 + 2 = 2$$

Additionally, since $\bar{f}_{23}^{(3)} = \bar{f}_{24}^{(3)} = 0$,

$$f_{23}^{(3)} = f_{23}^{(2)} = 0$$

$$f_{24}^{(3)} = f_{24}^{(2)} = 4$$

Hence, the incremental gain in flow is again 1/2 unit, which yields a total flow of $F = 1\frac{1}{2}$ and $\bar{F} = 16$. The cost of generating this flow is 32 (iteration 1) plus 40 (iteration 2) plus 1/2 (168) = 156. Arcs (1, 2), (2, 4), and (3, 4) are removed, and mirror arcs (4, 3) and (3, 1) are included to investigate possible flow reductions in the corresponding parent arcs.

Iteration 4:

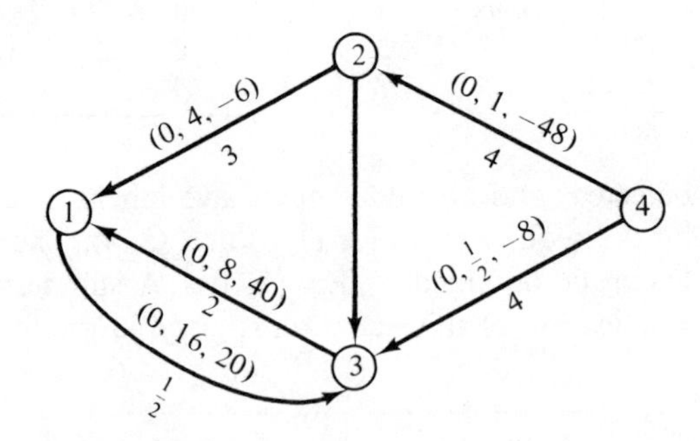

Node 4 must be given the label (0, ∞). Since no flow-augmenting chain can be constructed, the solution currently available is optimal. As can be observed in the following figure, 16 units of flow enter node 1 and $1\frac{1}{2}$ units exit at node 4. The corresponding minimum cost is equal to 12(2) + 4(20) + 4(12) + 2(2) = 156, as predicted by the results obtained at iteration 3.

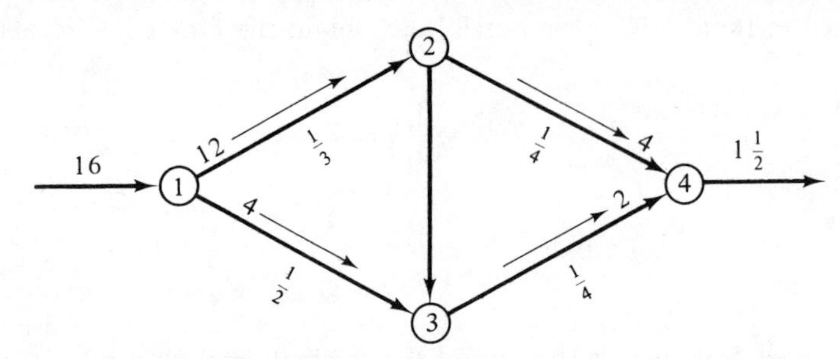

5.10 Summary and Conclusions

The algorithms presented in Part I were intended to illustrate the complexity of generalized networks, and to provide the reader with a basic framework through which other classes of network problems might be solved. Although the labeling algorithm is general in nature, when computerized it is both core-consuming and relatively inefficient for large-scale models. In their book, Jensen and Barnes [17] extend the fundamental concepts discussed in Sections 5.5 to 5.9, and propose a computationally attractive approach. In addition, a highly efficient algorithm based upon linear programming duality theory is developed to solve large-scale models for either linear, strictly convex, or strictly concave objective functions. The serious student of generalized networks is encouraged to examine this reference.

Although not specifically illustrated in this section, flow-reducing cycles can be investigated in the same fashion as flow-generating cycles. In maximum-flow problems, however, flow-reducing cycles are of no interest and can be ignored if encountered in the algorithmic process.

PART II: STOCHASTIC NETWORKS—GRAPHICAL EVALUATION AND REVIEW TECHNIQUE (GERT)

In all previous discussions we have considered only systems with deterministic network representations. In typical project networks, all arcs of the network must be realized for a realization of the overall network. This condition implies that feedback operations cannot be included in the model, since they are represented by branches that close loops, and the existence of these loops in turn implies that the final node of an activity must be realized before the starting node. Two situations have been widely studied in the field of deterministic networks. When there is exactly one realization time for each arc we have the critical path model, and when there are several possible realization times for each arc we have a PERT model.

Often, in modeling industrial systems, a more flexible and powerful representation is provided by a network with a stochastic structure. A stochastic network can be defined as one that can be realized with only a subset of the arcs, with each arc (activity) duration chosen according to a probability distribution. The nature of a stochastic network is such that for the realization of a node, it is not necessary to realize all the arcs incident to the node. For this reason, cycles and self-loops are also allowed in the representation of the system.

5.11 Network Representation

The nodes of a stochastic network can be interpreted as the states of the system. The arcs represent transitions from one state to another. Such transitions can be viewed as generalized activities, characterized by a unique probability density or mass function and a probability of realization.

Each intermediate node in a stochastic network performs two functions, one on the receiving side and another on the emitting side. Usually, these two functions are classified as input and output functions:

1. *Input function:* It indicates the condition under which the node can be realized.
2. *Output function:* It indicates the branching conditions following the node realization. In other words, the output function defines if only one or all the activities emanating from the node are undertaken.

Notice that the initial node of the network performs only an output function, while the ending node performs an input function. There are three types of input functions and two types of output functions, as defined below.

5.11.1 Input Functions

***Type 1*:** The node is realized when all arcs leading into it are realized.

***Type 2*:** The node is realized when any arc leading into it is realized.

***Type 3*:** The node is realized when any arc leading into it is realized, under the condition that only one arc can be realized at a given time.

5.11.2 Output Functions

***Type 1*:** All arcs emanating from the node are undertaken if the node is realized. This function is referred to as a deterministic output function.

***Type 2*:** Exactly one arc emanating from the node is undertaken if the node is realized. The selection of the arc can be described by means of a unique probability. Hence, this output is called a probabilistic function.

In this section we consider only two types of nodes: (a) nodes with type 3 input function and deterministic output function, and (b) nodes with type 3 input function and probabilistic output function. A network consisting of only these two types of nodes is called a GERT network. The two symbols shown in Figure 5.4 are used to represent GERT nodes [34, 35].

To illustrate the use of these nodes, let us consider a simple quality control system, where after an inspection operation it is decided that parts should be sold for scrap, reworked, or sent to the assembly line. After a part is reworked, it can be sold to a secondary retail outlet, scrapped, or sent directly to assembly for internal use (see Figure 5.5).

Note that $p_1 + p_2 + p_3 = 1$ and $p_4 + p_5 + p_6 = 1$. Also, arcs (4, 2), (1, 4), (1, 2), (1, 3), (4, 3), and (4, 5) represent physical processes which might be described by probability density functions. For example, activities (1, 4)

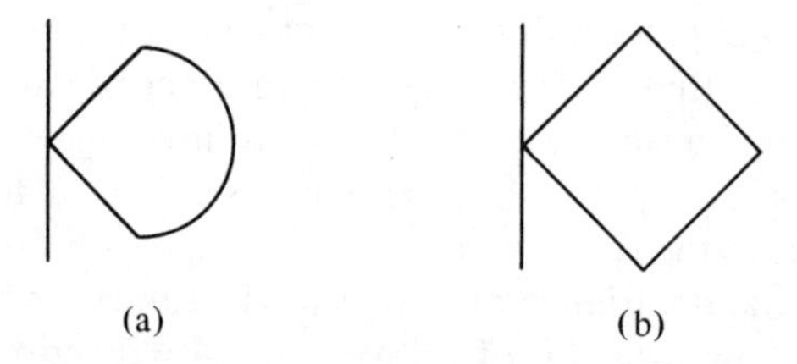

Figure 5.4. Type of GERT nodes: (a) deterministic output; (b) probabilistic output.

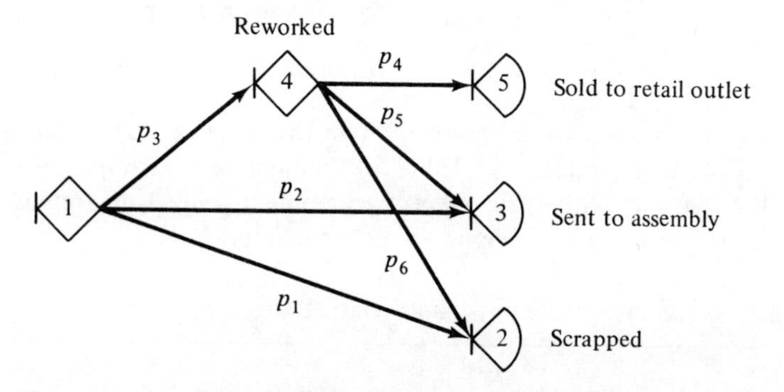

Figure 5.5. Simple GERT network.

and (4, 5) might be normally distributed, activities (1, 2) and (4, 3) exponentially distributed, and activities (1, 3) and (4, 2) uniformly distributed. The GERT methodologies are directed toward answering the following typical questions:

1. What is the probability that a part will be scrapped?
2. What is the probability that a part will be used in assembly?
3. What is the probability that a part will be sold at retail?
4. What is the mean and variance of the time needed to produce a part for assembly?
5. How much time will be lost if a part is scrapped?

In the following sections, we will develop the analytical machinery necessary to answer these questions.

5.12 GERT Basic Procedures

Consider a network $G = (\mathbf{N}, \mathbf{A})$ with only GERT nodes in the set $\mathbf{N}$. Let the random variable Y_{ij} be the duration of activity (i, j). By definition, activity (i, j) can be undertaken only if node i is realized. Therefore, we must know the conditional probability (discrete case) or density (continuous case) function of Y_{ij} given that node i is realized, in order to study the realization of the activity. This, in turn, would allow us to investigate the realization of the overall network. In particular, we would be able to find the moments of the distribution of the realization time of the network, from which we can estimate the mean and variance of the network realization time.

Let f_{ij} be the conditional probability or density function of the duration of activity (i, j). The conditional moment generating function of the random variable Y_{ij} is defined as $M_{ij}(s) = E[e^{sY_{ij}}]$. That is,

$$M_{ij}(s) = \begin{cases} \int e^{sy_{ij}} f(y_{ij})\, dy_{ij} & \text{(continuous random variables)} \\ \sum e^{sy_{ij}} f(y_{ij}) & \text{(discrete random variables)} \end{cases}$$

A particular case is $y_{ij} = a =$ constant. In this case, $M_{ij}(s) = E[e^{sa}] = e^{sa}$. When $a = 0$, then $M_{ij}(s) = 1$. Table 5.1 contains several important probability distributions with their respective moment-generating functions and first (mean) and second moments about the origin.

Table 5.1. MOMENTS AND MOMENT-GENERATING FUNCTIONS

Type of Distribution	$M_E(s)$	*Mean*	*Second Moment*
Binomial (B)	$(pe^s + 1 - p)^n$	np	$np(np + 1 - p)$
Discrete (D)	$\dfrac{p_1e^{sT_1} + p_2e^{sT_2} + \cdots}{p_1 + p_2 + \cdots}$	$\dfrac{p_1T_1 + p_2T_2 + \cdots}{p_1 + p_2 + \cdots}$	$\dfrac{p_1T_1^2 + p_2T_2^2 + \cdots}{p_1 + p_2 + \cdots}$
Exponential (E)	$\left(1 - \dfrac{s}{a}\right)^{-1}$	$\dfrac{1}{a}$	$\dfrac{2}{a^2}$
Gamma (GA)	$\left(1 - \dfrac{s}{a}\right)^{-b}$	$\dfrac{b}{a}$	$\dfrac{b(b+1)}{a^2}$
Geometric (GE)	$\dfrac{pe^s}{1 - e^s + pe^s}$	$\dfrac{1}{p}$	$\dfrac{2 - p}{p^2}$
Negative binomial (NB)	$\left(\dfrac{p}{1 - e^s + pe^s}\right)^r$	$\dfrac{r(1-p)}{p}$	$\dfrac{r(1-p)(1+r-rp)}{p^2}$
Normal (NO)	$e^{sm+(1/2)s^2\sigma^2}$	m	$m^2 + \sigma^2$
Poisson (P)	$e^{\lambda(e^s-1)}$	λ	$\lambda(1 + \lambda)$
Uniform (U)	$\dfrac{e^{sa} - e^{sb}}{(a-b)s}$	$\dfrac{a+b}{2}$	$\dfrac{a^2 + ab + b^2}{3}$

Let p_{ij} be the conditional probability that activity (i, j) will be undertaken given that node i is realized. Define a *W-function* for the random variable Y_{ij} as [34, 35]

$$W_{ij}(s) = p_{ij}M_{ij}(s) \tag{5-12}$$

By using the transformation of Eq. (5-12) we can always define a network G' with identical structure to that of G, but with a single arc parameter W_{ij}

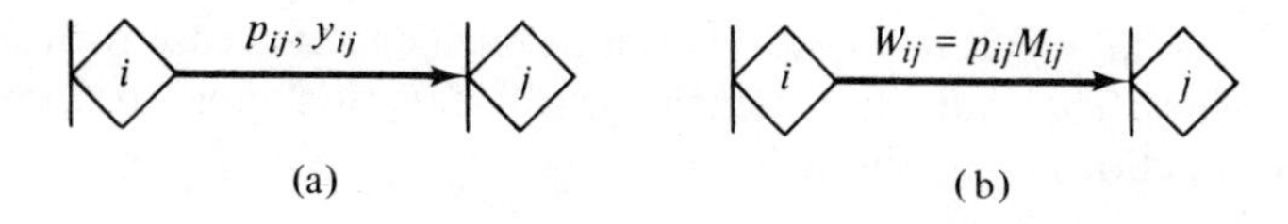

Figure 5.6. Networks G and G': (a) element of network G; (b) element of network G'.

instead of the two parameters p_{ij} and y_{ij}. Figure 5.6 shows a generic element of the type of arc used in G and G'.

In our discussion of GERT we have considered as an arc parameter the duration of the corresponding activity. Actually, it is possible to consider any generic arc parameter that can be added along the arcs of any path.

G' has several attractive computational properties under the assumption that the durations of the activities of network G are independent random variables. To show these properties, we will consider three special cases: (1) G' consists of two arcs in series, (2) G' consists of two branches in parallel, and (3) G' consists of one branch and one self-loop.

5.12.1 Arcs in Series

Consider the simple network given in Figure 5.7. This network consists of two branches in series. These two branches can be substituted by an equivalent branch, as indicated. The original branches have W-transformations

$$W_{ij}(s) = p_{ij}M_{ij}(s)$$
$$W_{jk}(s) = p_{jk}M_{jk}(s)$$

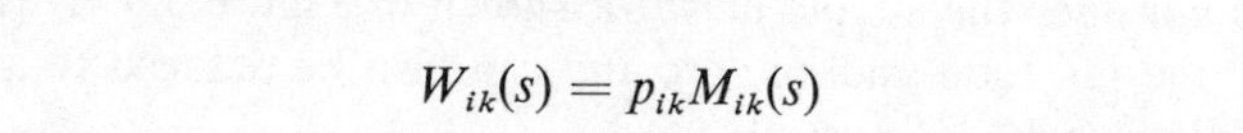

Figure 5.7. Branches in series and their equivalent representation.

The equivalent branch (i, k) has W-function

$$W_{ik}(s) = p_{ik}M_{ik}(s)$$

Recall that the moment-generating function of the sum of two independent random variables is equal to the product of the individual moment-generating functions. Hence, since $p_{ik} = (p_{ij})(p_{jk})$ and $M_{ik}(s) = [M_{ij}(s)][M_{jk}(s)]$, we conclude that

$$\begin{aligned} W_{ik}(s) &= [p_{ij}M_{ij}(s)][p_{jk}M_{jk}(s)] \\ &= W_{ij}(s)W_{jk}(s) \end{aligned} \tag{5-13}$$

The fundamental result of Eq. (5-13) can be extended to a case with three or more branches. The equivalent branch has a W-function equal to the product of the W-functions of the branches in series.

5.12.2 Branches in Parallel

Consider the simple network shown in Figure 5.8. This network consists of two branches in parallel. It will be proved that these branches can be sub-

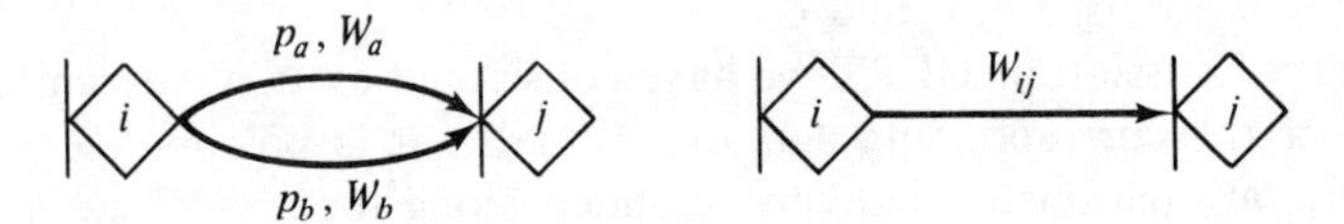

Figure 5.8. Branches in parallel and their equivalent representation.

stituted by an equivalent branch. Let (i, j) be the equivalent branch. By definition, $W_{ij}(s) = p_{ij}M_{ij}(s)$. In this case,

$$p_{ij} = p_a + p_b$$

and

$$M_{ij}(s) = \frac{p_a M_a(s) + p_b M_b(s)}{p_a + p_b}$$

Therefore,

$$\begin{aligned} W_{ij}(s) &= [p_a + p_b]\left[\frac{p_a M_a(s) + p_b M_b(s)}{p_a + p_b}\right] \\ &= W_a(s) + W_b(s) \end{aligned} \tag{5-14}$$

Again, the result given in Eq. (5-14) can be generalized for networks consisting of three or more parallel branches. The equivalent branch has its W-function equal to the sum of the W-functions of the arcs in parallel.

5.12.3 Self-Loops

Let us consider the simple network shown in Figure 5.9. This network consists of one self-loop and one arc and can also be reduced to an equivalent one-arc network.

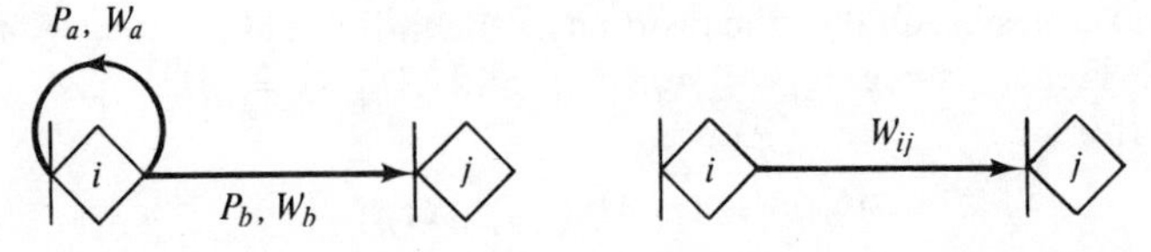

Figure 5.9. Network with a self-loop and its equivalent representation.

Note that the network under consideration can be transformed into that given in Figure 5.10. This new network consists of an infinite sequence of parallel chains, each chain being a sequence of branches in series. Therefore, we can first reduce the chains to equivalent single arcs, and then these arcs can be reduced to the one-branch network equivalent to the original system.

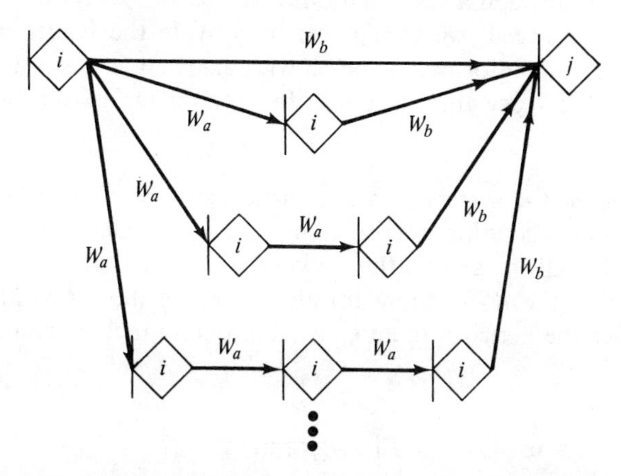

Figure 5.10. Parallel-series network representation of self-loops.

Let (i, j) be the equivalent branch for the network of Figure 5.10. Using Eqs. (5-13) and (5-14), its weight is therefore given by

$$W_{ij} = W_b + W_a W_b + W_a^2 W_b + \ldots = W_b[1 + \sum_{m=1}^{\infty} W_a^m]$$

where we have temporarily omitted the argument s of the W-functions. The expression above can be further simplified after noting that the binomial series $(1 - W_a)^{-1}$ can be expanded as

$$(1 - W_a)^{-1} = 1 + W_a + W_a^2 + W_a^3 + \ldots = 1 + \sum_{m=1}^{\infty} W_a^m$$

Therefore, we can finally write

$$\begin{aligned} W_{ij}(s) &= W_b(s)[1 - W_a(s)]^{-1} \\ &= \frac{W_b(s)}{1 - W_a(s)} \end{aligned} \tag{5-15}$$

Hence, the network of Figure 5.10 reduces to a single equivalent branch with W-function equal to $W_{ij}(s) = W_b(s)/[1 - W_a(s)]$. Note that this procedure can be used for general loops, since any loop involving multiple nodes can be reduced to a self-loop using the series results of Eq. (5-13), and can then be analyzed as above.

Summarizing, if a GERT network consists of parallel chains, series chains, and/or loops, it is always possible to reduce the network to an equivalent single branch network. Actually, this result can be extended to any GERT network through a combination of the basic reductions.

Before we proceed, we would like to provide the reader with fundamental background information on flowgraphs that is needed for further development. We close this section with a summary of the basic GERT steps:

1. Represent the system as a stochastic network with GERT nodes.
2. Determine conditional probabilities and moment-generating functions for all the arcs of the network.
3. Compute the W-functions for all the arcs of the network.
4. Reduce the network to an equivalent one-branch network.

5.13 Basic Concepts of Flowgraphs

A system can be defined as a collection of active and interactive elements that perform a function. In our present discussion we will only consider systems whose elements and relationships among elements can be represented by a set of linear equations. One of the most popular diagrammatic representations of such systems is a flowgraph. In a flowgraph, the elements of the system are represented by nodes and the relationships or transfer functions by arcs.

The fundamental element of a flowgraph is a directed branch from node i to node j with arc parameter t_{ij}. The direction of the branch indicates the input/output relationship between the two variables represented by the nodes of the branch. Node i corresponds to an independent variable x_i and node j to a dependent variable x_j. The parameter t_{ij} is usually referred to as the *arc transmittance*, and indicates the factor used to transform the value of x_i before it is considered as part of the value of x_j. The basic property of flowgraphs establishes that the value of a node is equal to the sum of the transformed values of the nodes incident to the node under consideration. As an illustration, let us consider the elementary flowgraph shown in Figure 5.11.

The fundamental properties of flowgraphs can be used to develop methods that permit direct manipulation of the elements of a graph to transform it into an equivalent graph with simpler structure, and hence easier to solve than the original graph. The topological equation developed by Mason [34] is the best known contribution in this area. It can be used to solve

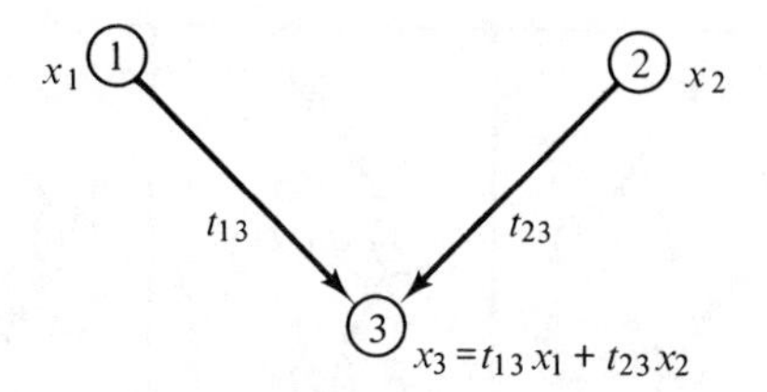

Figure 5.11. Basic property of flowgraphs.

flowgraphs of arbitrarily complex structure. Before discussing Mason's equation, some definitions are necessary.

5.14 Definitions

Loop: a connected sequence of directed branches with every node being common to exactly two branches. A loop is usually referred to as a first-order loop to indicate that it does not contain another loop, and that each node can be reached from every other node. A self-loop can be viewed as a degenerate first-order loop.

Loop of order n: a set of n disjoint first-order loops.

Closed flowgraph: a graph in which each branch belongs to at least one loop.

As an illustration, the flowgraph shown in Figure 5.12 will be examined to identify all the loops and classify them according to their order.

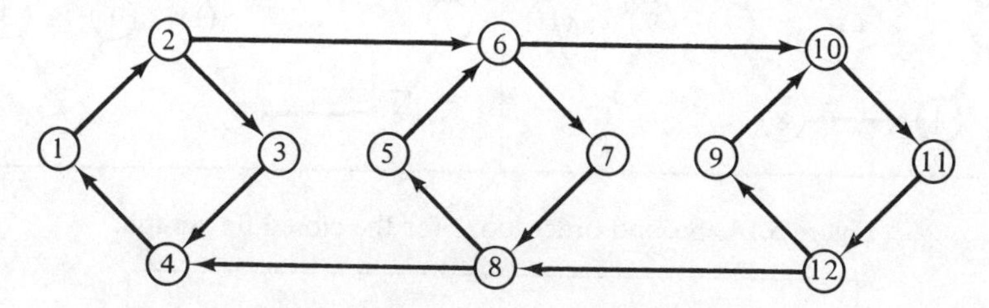

Figure 5.12. Flowgraph with loops.

It can be seen that the flowgraph in Figure 5.12 is closed since it is composed entirely of loops. Figures 5.13 to 5.15 contain the first-, second-, and third-order loops of the graph given in Figure 5.12. As has been defined, a second (third)-order loop is a set of two (three) disjoint first-order loops. For example, L_1 and L_2 form a second-order loop, while L_1, L_2, and L_3 form a third-order loop.

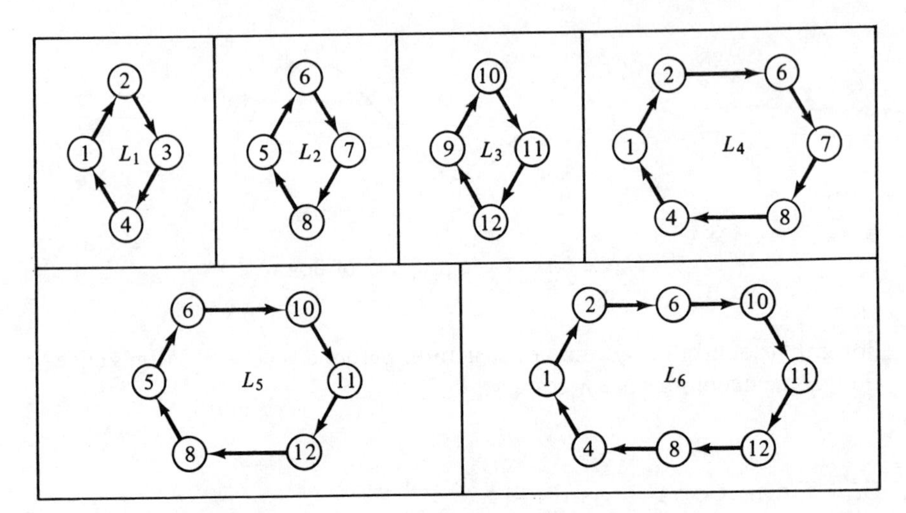

Figure 5.13. First-order loops for the closed flowgraph.

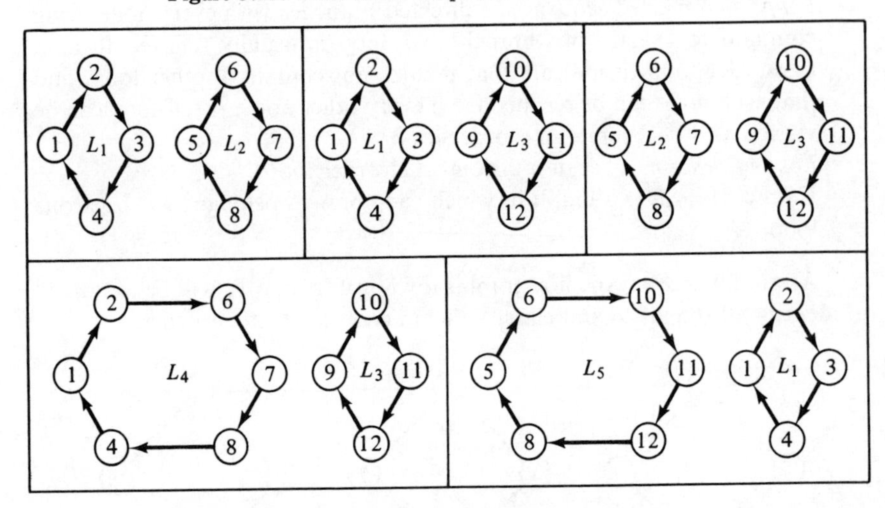

Figure 5.14. Second-order loops for the closed flowgraph.

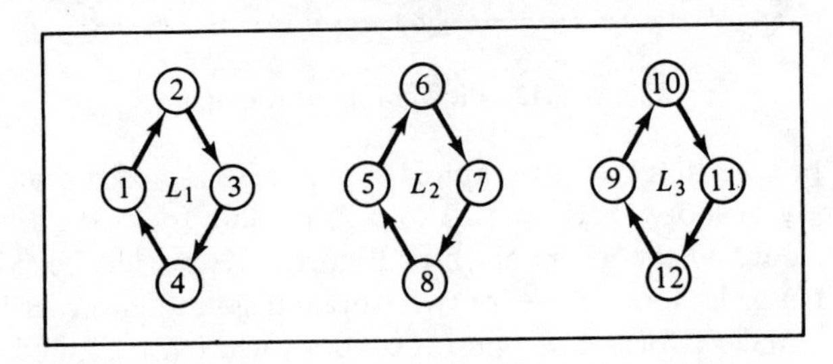

Figure 5.15. Third-order loops for the closed flowgraph.

A first-order loop can be viewed as a chain of directed branches in series with both end nodes being coincident. Therefore, the transmittance of the equivalent flowgraph is equal to the product of the transmittances of the individual branches.

By definition, a loop of order n consists of n disjoint first-order loops. If each of these first-order loops is reduced to an equivalent one-branch flowgraph, the original loop of order n can be viewed as a connected sequence of such one-branch flowgraphs. The immediate result is that the transmittance of the loop of order n is equal to the product of the transmittances of its n first-order loops.

In general, let $L_{11}, L_{21}, L_{31}, \ldots, L_{n1}$ be the n disjoint first-order loops of a given loop of order n. It was previously stated that the equivalent transmittance, T_k, of any disjoint first-order loop L_{k1} equals the product of the transmittances of the branches belonging to the loop. This idea can be expressed by

$$T_k = \prod_{(i,j) \in L_{k1}} t_{ij} \tag{5-16}$$

Hence, the equivalent transmittance, $T(L_n)$, for a loop of order n is given by

$$T(L_n) = \prod_{k=1}^{n} T_k = \prod_{k=1}^{n} \left[\prod_{(i,j) \in L_{k1}} t_{ij} \right] \tag{5-17}$$

The fundamental result of Eq. (5-17) will be utilized to characterize the behavior of systems that can be represented as GERT networks.

5.15 Mason's Rule for Closed Flowgraphs

The purpose of using GERT in stochastic network analysis is to obtain estimates for the mean and the variance of network realization time (here considered to be a generic network parameter) and probabilities of sink node(s) realization(s). It should now be evident that the transmittance of an arc in a GERT network is the corresponding W-function. Recall that the W-function of a given arc is defined as the product of the probability of undertaking the arc and the moment-generating function (MGF) of the duration of the activity represented by the arc.

In the previous section we described how to determine all loops for a closed flowgraph. In order to apply these results to an open network, an additional arc with W-function $W_A(s)$ must be added to the network to connect the terminal node t to the source node s. Then we must obtain *all* the loops (up to the highest possible order) associated with the modified network. The purpose of using $W_A(s)$ is to obtain an expression for the equivalent W-function of the original network. In Figure 5.16, the original network is represented by a black box with W-function $W_E(s)$.

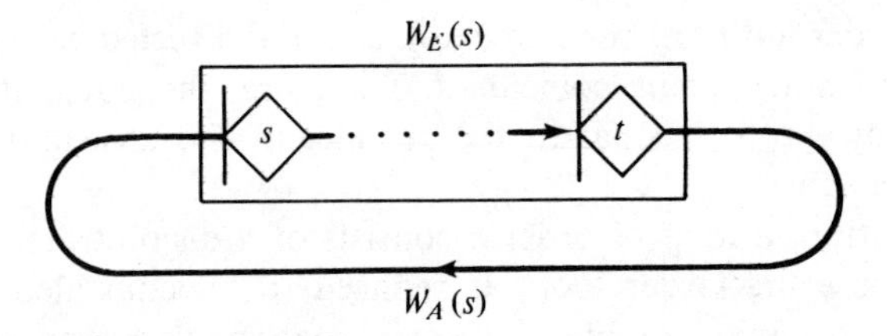

Figure 5.16. Closed stochastic network.

Recall that the equivalent transmittance for a loop of order n is equal to the product of the transmittances of n disjoint first-order loops; that is,

$$T(L_n) = \prod_{k=1}^{n} T_k \tag{5-18}$$

The topology equation for closed graphs, also known as Mason's rule, is defined as in (5-19),

$$\begin{aligned} H = 1 - \sum T(L_1) + \sum T(L_2) \\ - \sum T(L_3) + \ldots + (-1)^m \sum T(L_m) + \ldots = 0 \end{aligned} \tag{5-19}$$

where $\sum T(L_i)$ represents the sum of the equivalent transmittances for all the possible ith-order loops. In words, the following steps must be perfomed when using the topology equation:

1. Obtain the equivalent transmittance for every possible loop of order m.
2. Add the expressions for all the loops of the same order m and multiply by $(-1)^m$. Notice that for loops with an even order $(-1)^m$ will be positive, while for loops with an odd order $(-1)^m$ will be negative. In the topology Eq. (5-19), notice the plus or minus sign in front of each summation.
3. Add 1 to the expression obtained in (2) and equate to zero.

As an example, consider Figure 5.16. The closed flowgraph has one first-order loop, with equivalent transmittance equal to $W_A(s)W_E(s)$. By using Mason's rule we obtain that

$$1 - W_A(s)W_E(s) = 0 \quad \text{or} \quad W_A(s) = \frac{1}{W_E(s)}$$

Note that $W_A(s)$ is contained in the topology equation since it is an element of at least one first-order loop. The important result of this example is that if we substitute $W_A(s)$ with $1/W_E(s)$ in the topology equation and solve for $W_E(s)$, an equivalent W-function for the original stochastic network is obtained.

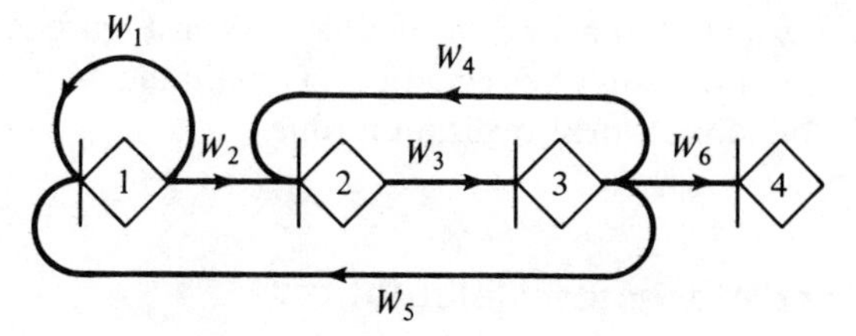

Figure 5.17. Open stochastic network.

As another example, consider the open network given in Figure 5.17. To obtain an equivalent W-function for the network, we must perform the following steps:

1. Close the network with an arc going from node 4 to node 1.
2. Obtain all the possible loops of order n.
3. Use the topology Eq. (5-19) to obtain an expression for $W_E(s)$.

Figure 5.18 contains the modified network, in which for convenience we have omitted the argument s of the W-functions.

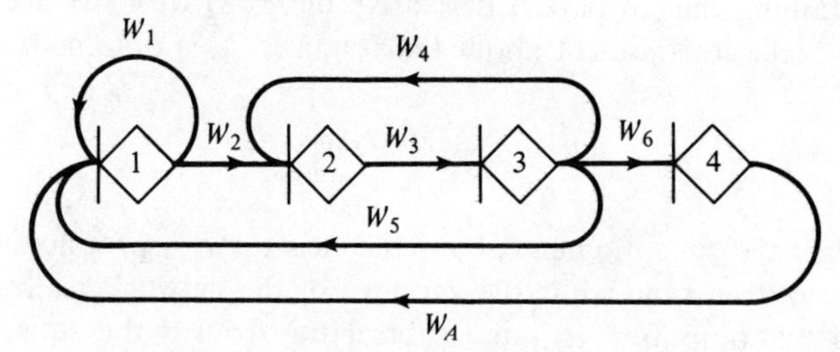

Figure 5.18. Closed stochastic network.

If $W_A(s)$ is substituted with $1/W_E(s)$, the following loop transmittances are obtained using Eq. (5-17):

$$\text{Loops of order 1:} \quad W_1,\ W_3W_4,\ W_2W_3W_5,\ W_2W_3W_6\frac{1}{W_E}$$

$$\text{Loop of order 2:} \quad W_1W_3W_4$$

By using the topology Eq. (5-19) we obtain

$$H = 1 - W_1 - W_3W_4 - W_2W_3W_5 - W_2W_3W_6\frac{1}{W_E} + W_1W_3W_4 = 0$$

Therefore

$$W_E(s) = \frac{W_2W_3W_6}{1 - W_1 - W_3W_4 - W_2W_3W_5 + W_1W_3W_4}$$

which is the equivalent W-function for the network in Figure 5.17. The next section will explain how to use this equation to obtain the mean and variance of the network (or subnetwork) realization time.

5.16 Mean and Variance Calculation

By using the topology Eq. (5-19) we have been able to obtain an expression for the equivalent W-function of the network, $W_E(s)$. Recall that when $s = 0$, $M_E(s) = 1$. Since $W_E(s) = p_E M_E(s)$ we obtain that $p_E = W_E(0)$, from which

$$M_E(s) = \frac{W_E(s)}{p_E} = \frac{W_E(s)}{W_E(0)} \tag{5-20}$$

Notice that at this point we have an expression for $W_E(s)$ in terms of all or some of the W-functions of the branches in the original network. It is easy to evaluate $W_E(0)$; we just need to set $s = 0$ in the expression for $W_E(s)$ given by Eq. (5-19).

By obtaining the jth partial derivative of $M_E(s)$ with respect to s, and setting $s = 0$ the jth moment about the origin, μ_{jE}, is obtained. That is,

$$\mu_{jE} = \frac{\partial^j}{\partial s^j}[M_E(s)]\bigg|_{s=0} \tag{5-21}$$

In particular, the first moment about the origin, μ_{1E}, produces the mean network realization time while the variance of the network realization time is obtained by computing μ_{2E} and subtracting from it the square of μ_{1E}; that is,

$$\sigma^2 = \mu_{2E} - (\mu_{1E})^2 \tag{5-22}$$

5.17 GERT Applications

5.17.1 Production of a High-Risk Item

A government contractor has agreed to produce a high-risk item. Because of stringent inspections, there is only a 20% probability that a successful item will be produced from raw material. What is the expected number of trials necessary to achieve two good items?

In this problem it is assumed that the time needed to produce an item is constant; recall that for the case in which the random variable Y_{ij} is equal to a constant a, then

$$M_{ij}(s) = E[e^{sa}] = e^{sa}$$

By letting $a = 1$, $M_{ij}(s) = e^s$, which will be the moment-generating function for the duration of every arc in the original network. The GERT network for this problem is shown in Figure 5.19. Since there is a 20% probability of success,

$$W_1 = W_3 = 0.80e^s$$
$$W_2 = W_4 = 0.20e^s$$

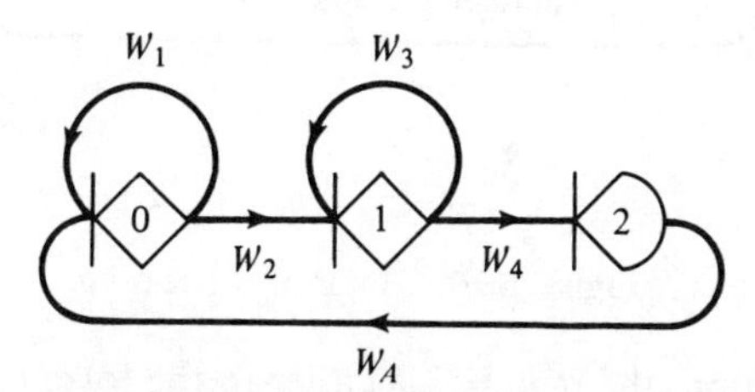

Figure 5.19. GERT network for a high-risk item.

Using the topology equation, we obtain

$$H = 1 - W_1 - W_3 - W_2W_4W_A + W_1W_3 = 0$$
$$1 - W_1 - W_3 - W_2W_4\frac{1}{W_E} + W_1W_3 = 0$$
$$W_E(s) = \frac{W_2W_4}{1 - W_1 - W_3 + W_1W_3}$$

Using the expressions for W_1, W_2, W_3, and W_4, we obtain

$$W_E(s) = \frac{0.04e^{2s}}{1 - 1.6e^s + 0.64e^{2s}}$$

Since $p_E = 1$, $W_E(s) = M_E(S)$. Thus, the expected number of trials necessary to produce two good items is:

$$\mu_{1E} = \left.\frac{\partial M_E(s)}{\partial s}\right|_{s=0} = 10$$

Similarly, the variance of the total number of trials is given by $\sigma^2 = 40$.

5.17.2 Material Processing (Pritsker)

We will now consider the processing of semiconductor material. As shown in Figure 5.20, the material is first inserted into the furnace to alter its impurities. The output of the furnace is either retreated in the furnace, declared inferior, or declared acceptable and cut (sliced) into wafers. After

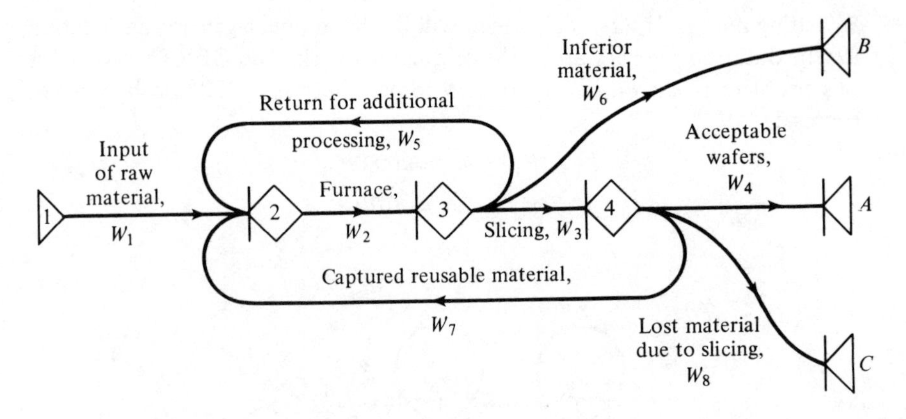

Figure 5.20. Material processing.

the slicing operation, the material is either in the form of acceptable wafers, lost due to the slicing process, or captured and returned as raw material. Let us calculate the average time and variance for obtaining an acceptable wafer (i.e., for arriving at state A). Table 5.2 contains the information related to each branch in the network.

Table 5.2. INFORMATION FOR MATERIAL-PROCESSING ACTIVITIES

Branch	p_i	*Type of Distribution*	*Parameters*	*MGF*
(1, 2)	1	Constant	$a = 1$ hr	$\exp(1s)$
(2, 3)	1	Normal	$m = 0.5$ hr $\sigma = 0.1$ hr	$\exp[0.5s + \frac{1}{2}(0.01)s^2]$
(3, 2)	0.12	Normal	$m = 0.1$ hr $\sigma = 0.1$ hr	$\exp[0.1s + \frac{1}{2}(0.01)s^2]$
(3, B)	0.03	Constant	$a = 0.25$ hr	$\exp(0.25s)$
(3, 4)	0.85	Normal	$m = 0.25$ hr $\sigma = 0.20$ hr	$\exp[0.25s + \frac{1}{2}(0.04)s^2]$
(4, A)	0.75	Constant	$a = 0.20$ hr	$\exp(0.20s)$
(4, C)	0.05	Constant	$a = 0.05$ hr	$\exp(0.05s)$
(4, 2)	0.20	Constant	$a = 0.10$ hr	$\exp(0.10s)$

Since we are interested in the mean and variance for the realization time of an acceptable wafer, we must introduce arc (A, 1), as in Figure 5.21. Using $W_A(s) = 1/W_E(s)$, the topology equation becomes

$$H = 1 - W_1 W_2 W_3 W_4 \frac{1}{W_E} - W_2 W_5 - W_2 W_3 W_7 = 0$$

Note that arcs (3, B) and (4, C) are not considered in the analysis, since they cannot be taken if node A is to be realized.

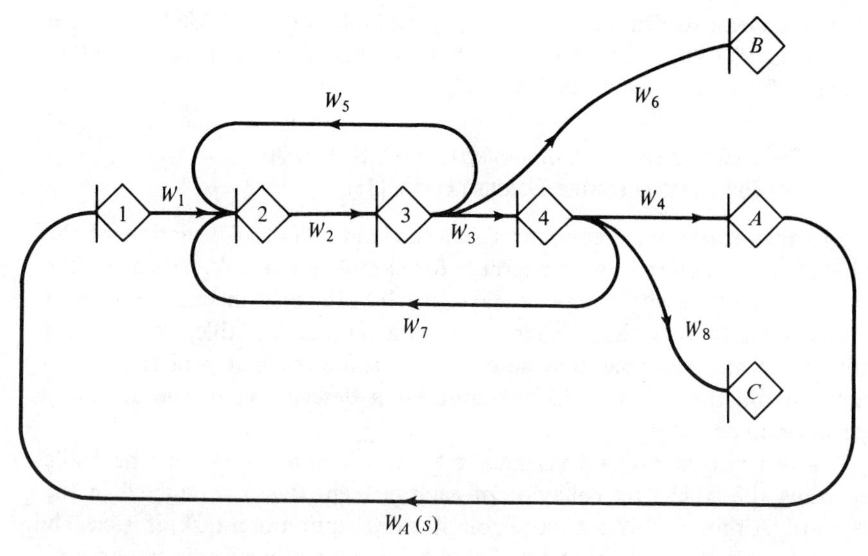

Figure 5.21. Network for the analysis of the realization time of an acceptable wafer.

Solving the topology equation for $W_E(s)$, we can write

$$W_E(s) = \frac{W_1W_2W_3W_4}{1 - W_2W_5 - W_2W_3W_7}$$

After substituting each W-function with its respective probability and MGF, we obtain

$$W_E(s) = \frac{0.6375 \exp(1.95s + 0.025s^2)}{1 - 0.12 \exp(0.6s + 0.01s^2) - 0.17 \exp(0.85s + 0.025s^2)}$$

It can be verified that for this problem, $W_E(0) = 0.8979$. This is interpreted as the probability that an acceptable wafer will be produced.

$$M_E(s) = \frac{W_E(s)}{W_E(0)} = \frac{0.71 \exp(1.95s + 0.025s^2)}{1 - 0.12 \exp(0.6s + 0.01s^2) - 0.17 \exp(0.85s + 0.025s^2)}$$

Now, by obtaining the first and second partial derivatives of $M_E(s)$ with respect to s and setting $s = 0$, we obtain

$$\mu_{1E} = \left.\frac{\partial M_E(s)}{\partial s}\right|_{s=0} = 2.255 \text{ hr}$$

$$\mu_{2E} = \left.\frac{\partial^2 M_E(s)}{\partial s^2}\right|_{s=0} = 5.477 \text{ hr}^2$$

$$\sigma^2 = \mu_{2E} - [\mu_{1E}]^2 = 0.392 \text{ hr}^2$$

Thus, the mean realization time for an acceptable wafer is 2.255 hours, with a variance of 0.392 hour^2. The same analysis can be conducted to obtain similar information for states B and C.

5.17.3 Determination of Probabilistic Time Standards for Tasks Performed under Uncertainty [31]

In this example we compute the mean and standard deviation of the standard time required by an operator for completing a task. Whenever the worker is involved with composite tasks, such as the ones in this problem, it is reasonable to view the standard time as a random variable, possessing a finite mean and variance, and described by some relevant probability distribution function. Figure 5.22(a) contains a description of the assembly operation under study.

In order to investigate variance estimates, assumptions must be made regarding the stochastic behavior of each element (task) conducted in the standard. Although this is a more complex description of a task, it is clearly more accurate than a single time. Table 5.3 contains the mean and variance for each element of the standard. For this example, the stochastic behavior of each element can be described by a normal density function.

The GERT network presented in Figure 5.22(b) is formed by nodes that represent the beginning and ending points of all individual tasks, and arcs that represent the actual time duration of each task. Table 5.4 contains the W-functions associated with the arcs of the GERT network under study. Before we start defining the loops for Figure 5.22(b) it is very useful to perform three simplifications in the network. These changes are presented in Figure 5.23. Thus, we can construct the equivalent network in Figure 5.24, for which it is much easier to identify the loops.

Using Figure 5.24, the equivalent transmittances for loops of first and second order are the following:

Loops of order 1: W_{24}, $W_8W_{10}(W_{11}+W_{12}+W_{13})$,

$$W_1W_2W_8W_9W_{14}W_{20}W_{22}[W_3+W_4(W_5+W_6+W_7)][W_{15}+W_{16}(W_{17}+W_{18}+W_{19})],$$

$$W_1W_2W_8W_9W_{14}W_{20}W_{21}\frac{1}{W_E}[W_3+W_4(W_5+W_6+W_7)][W_{15}+W_{16}(W_{17}+W_{18}+W_{19})]$$

Loop of order 2: $W_8W_{10}W_{24}(W_{11}+W_{12}+W_{13})$

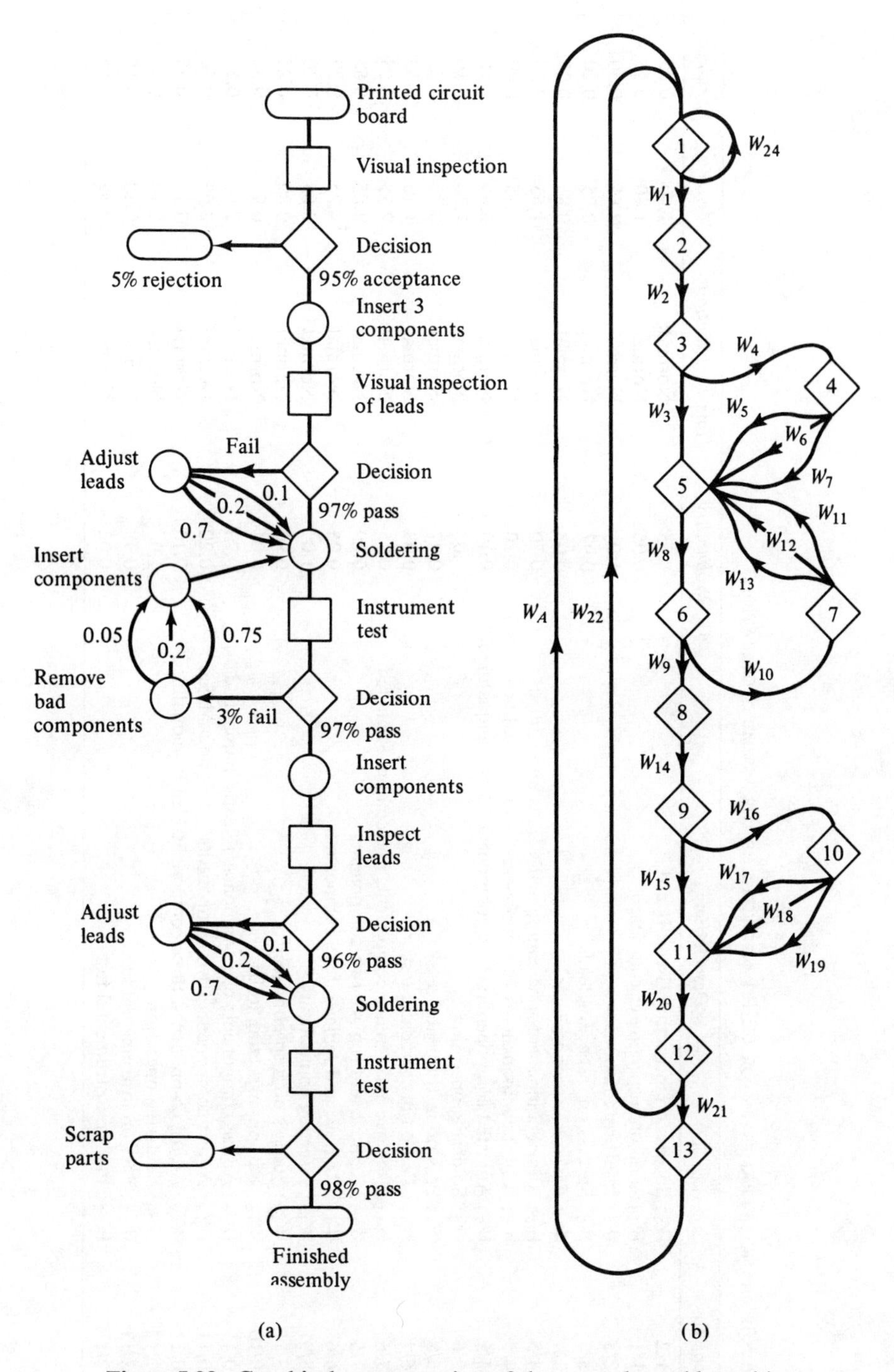

Figure 5.22. Graphical representation of the example problem: (a) stochastic assembly chart; (b) GERT network.

Table 5.3. BRANCH PARAMETERS FOR GERT NETWORK OF THE ASSEMBLY OPERATION

From/To	*Description*	*Probability*	*Time Distribution*	*Mean*	*Variance*
1/2	W_1; Success on inspecting circuit board	0.95	Normal	1.40	0.35
1/1	W_{24}; Fail on inspecting circuit board	0.05	Normal	1.40	0.35
2/3	W_2; Inserting components	1.0	Normal	2.65	0.30
3/5	W_3; Success on inspecting leads	0.97	Normal	0.75	0.20
3/4	W_4; Fail on inspecting leads	0.03	Normal	0.85	0.25
4/5	W_5; Adjust component; one component's leads unqualified	0.70	Normal	1.65	0.35
4/5	W_6; Adjust components; two components' leads unqualified	0.20	Normal	2.15	0.40
4/5	W_7; Adjusted components; three components' leads unqualified	0.10	Normal	2.65	0.55
5/6	W_8; Soldering operation	1.0	Normal	1.30	0.10
6/8	W_9; Pass on instrumental test	0.97	Normal	0.90	0.25
6/7	W_{10}; Fail on instrumental test	0.03	Normal	1.10	0.30
7/6	W_{11}; Remove and replace component; one component failed	0.75	Normal	2.75	0.55
7/6	W_{12}; Remove and replace two components	0.20	Normal	3.65	0.68
6/6	W_{13}; Remove and replace three components	0.05	Normal	4.25	0.79
8/9	W_{14}; Inserting components	1.0	Normal	2.65	0.30
9/11	W_{15}; Success on inspecting leads	0.96	Normal	0.80	0.25
9/10	W_{16}; Fail on inspecting leads	0.04	Normal	0.95	0.30
10/11	W_{17}; Adjust component; one components' leads unqualified	0.70	Normal	2.0	0.5
10/11	W_{18}; Adjust components; two components' leads unqualified	0.20	Normal	2.65	0.60
10/11	W_{19}; Adjust components, three components' leads unqualified	0.10	Normal	3.15	0.75
11/12	W_{20}; Soldering operation	1.0	Normal	2.10	0.40
12/13	W_{21}; Pass on instrumental test	0.98	Normal	0.80	0.15
12/1	W_{22}; Fail on instrumental test	0.02	Normal	0.85	0.15

Table 5.4. GERT *W*-FUNCTIONS FOR THE EXAMPLE NETWORK

		Branch	*Probability*	*W-Function*
Start	*End*	*(i,j)*	p_{ij}	$W = p \exp(\mu t + t^2\sigma^2/2)$
1	2	1	0.95	$0.95 \exp(1.4t + 0.175t^2)$
2	3	2	1.0	$1.0 \exp(2.65t + 0.15t^2)$
3	5	3	0.97	$0.97 \exp(0.75t + 0.10t^2)$
3	4	4	0.03	$0.03 \exp(0.85t + 0.125t^2)$
4	5	5	0.70	$0.70 \exp(1.65t + 0.175t^2)$
4	5	6	0.20	$0.20 \exp(2.15t + 0.20t^2)$
4	5	7	0.10	$0.10 \exp(2.6t + 0.275t^2)$
5	6	8	1.0	$1.0 \exp(1.3t + 0.05t^2)$
6	8	9	0.97	$0.97 \exp(0.9t + 0.125t^2)$
6	7	10	0.03	$0.03 \exp(1.1t + 0.15t^2)$
7	8	11	0.75	$0.75 \exp(2.75t + 0.275t^2)$
7	8	12	0.20	$0.20 \exp(3.65t + 0.34t^2)$
7	8	13	0.05	$0.05 \exp(4.25t + 0.395t^2)$
8	9	14	1.0	$1.0 \exp(2.65t + 0.15t^2)$
9	11	15	0.96	$0.96 \exp(0.8t + 0.125t^2)$
9	10	16	0.04	$0.04 \exp(0.95t + 0.15t^2)$
10	11	17	0.70	$0.70 \exp(2t + 0.25t^2)$
10	11	18	0.20	$0.20 \exp(2.65t + 0.325t^2)$
10	11	19	0.10	$0.10 \exp(3.15t + 0.378t^2)$
11	12	20	1.0	$1.0 \exp(2.1t + 0.20t^2)$
12	13	21	0.98	$0.98 \exp(0.8t + 0.075t^2)$
12	1	22	0.02	$0.02 \exp(0.85t + 0.075t^2)$
13	1	23	1.0	$1.0/W_E$
1	1	24	0.05	$0.05 \exp(1.4t + 0.175t^2)$

Using the topology equation, the resultant equivalent *W*-function is:

$$W_E(s) = \frac{W_1W_2W_8W_9W_{14}W_{20}W_{21}[W_3 + W_4(W_5 + W_6 + W_7)] \cdot [W_{15} + W_{16}(W_{17} + W_{18} + W_{19})]}{1 - W_{24} - W_8W_{10}(W_{11} + W_{12} + W_{13}) - W_1W_2W_8W_9W_{14}W_{20}W_{22} \cdot [W_3 + W_4(W_5 + W_6 + W_7)][W_{15} + W_{16}(W_{17} + W_{18} + W_{19})]}$$

The calculations necessary to obtain the mean and the variance will not be shown here and are left for the reader as an exercise.

However, note that $W_E(s)$ is a moment-generating function $W_E(s) = \text{MGF}_{13,1}$. Since this moment-generating function is expressed in terms of only the transformation variable s, the first two central moments μ_1 and μ_2 about the origin can be obtained by differentiating $W_E(s)$ with respect to s and evaluating the first and second derivatives at $s = 0$. Since μ_1 is actually the expected value of the standard and $\mu_2 - (\mu_1)^2$ is by definition the variance of that standard, the result we seek is achieved. After considerable

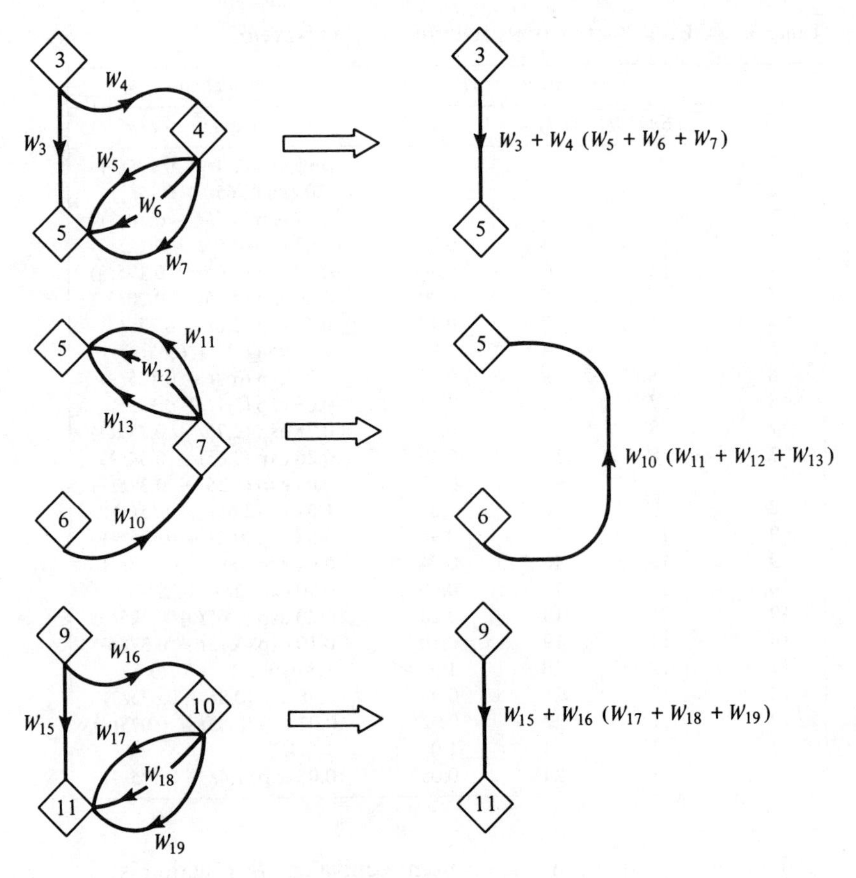

Figure 5.23. Simplifications for the GERT network.

numerical calculations, one can verify that $\mu_1 = 14.032$ minutes and $\sigma^2 = 7.15$ minutes^2.

Note that this result provides a great deal of useful information about the standard. Using Chebyshev's inequality, we can assert that the actual processing time per part will vary between 6.104 minutes and 21.96 minutes over 99% of the time. In this particular example, much stronger statements can be made. Since all component times are normally distributed, the standard time is also normally distributed. With this knowledge, exact probability statements concerning the time to completion can be made. In addition, if a standard is set against this result, accurate hypothesis tests governing changes in standards can be run to maintain standards in the future. It is also instructive to note that the GERT approach presents a unique alternative to the traditional processes of setting time standards. Using conventional methods, the individual elements of time are treated as *constants*, and after summation,

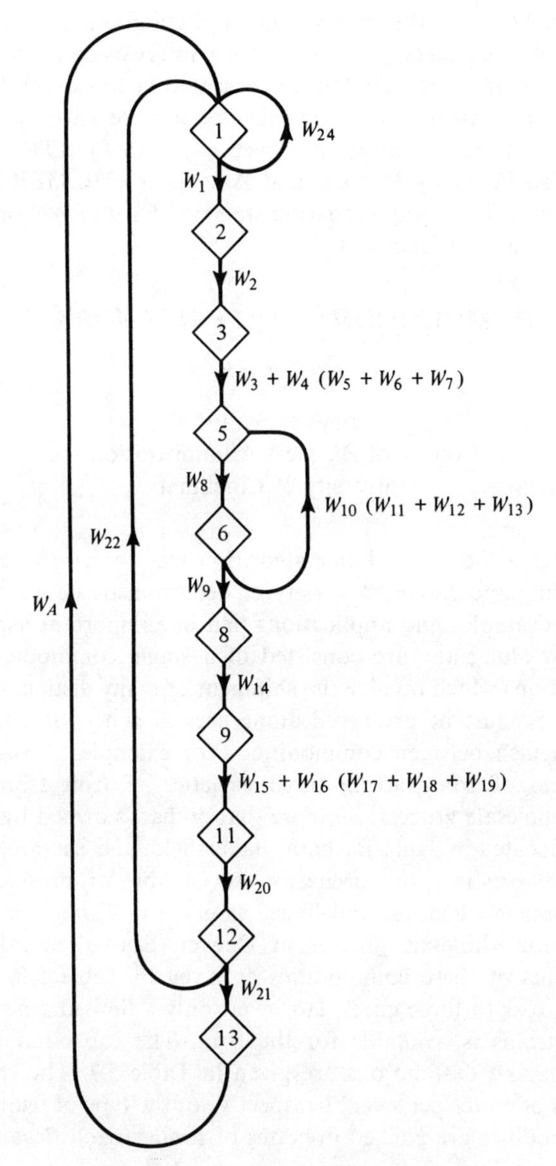

Figure 5.24. Simplified GERT network.

an adjustment is made to simulate random fluctuations or absorb imbalances in the actual processing times. On the other hand, the GERT approach incorporates random variation and uncertainties *directly* into each time element. Hence, the resultant standard *already* includes all random imbalances and needs no further adjustment except for the appropriate personal and fatigue allowances. A bonus is the by-product of the amount of variance in the standard, enabling construction of confidence intervals on the standard time.

Finally, although the calculations presented in this example were complicated and time-consuming, hand calculations are entirely unnecessary. The GERT numerical procedure has been coded in FORTRAN-IV and is maintained/distributed by Pritsker and Associates [33]. GERT analysis is executed routinely by simply preparing standard FORTRAN data cards for *any* continuous density functions.

PART III: MULTICOMMODITY NETWORK FLOWS

contributed by

James R. Evans
College of Business Administration
University of Cincinnati

In Chapter 3 the out-of-kilter algorithm was shown to be a powerful tool in modeling and solving a variety of deterministic network-flow problems. All the examples and applications had one important aspect in common: the flow along any arc consisted of a single commodity. There are many applications which involve the shipment of many distinct commodities whose identities must be preserved along arcs of a network. One must be able to distinguish between commodities. For example, consider the simplified problem of transporting three varieties of fruit from California orchards to wholesale grocers. Suppose that orchards owned by a particular company are located in Santa Barbara, Bakersfield, and Sacramento. During the peak of harvesting, the orchards are capable of producing various amounts of oranges, lemons, and limes, as given in Table 5.5. Contractual agreements with wholesale grocers in Denver, Seattle, and Kansas City require amounts of these commodities as given in Table 5.6. The fruit is shipped by boxcar to these cities. However, only a limited amount of space on outgoing trains is available for the fruit. The capacities between the various origins and destinations are given in Table 5.7. The capacities are given in terms of crates per week, irrespective of the type of fruit (we assume that all commodities are packed in crates of similar size). Because of differ-

Table 5.5. PRODUCTION RATE (crates/week)

	Oranges	*Lemons*	*Limes*
Santa Barbara	800	700	700
Bakersfield	1000	800	500
Sacramento	500	1000	500

Table 5.6. DEMAND RATE (crates/week)

	Oranges	*Lemons*	*Limes*
Denver	900	900	700
Seattle	700	1000	500
Kansas City	700	600	500

Table 5.7. TRAIN CAPACITIES (crates/week)

	To		
From	*Denver*	*Seattle*	*Kansas City*
Santa Barbara	1200	900	1000
Bakersfield	1200	1000	1100
Sacramento	950	1300	1000

Table 5.8. UNIT TRANSPORTATION COSTS (cents/crate)

	To		
From	*Denver*	*Seattle*	*Kansas City*
Santa Barbara			
Oranges	3.2	2.5	5.6
Lemons	2.4	1.9	4.3
Limes	2.3	1.7	4.0
Bakersfield			
Oranges	3.0	2.1	5.5
Lemons	2.3	1.7	4.4
Limes	2.0	1.4	4.3
Sacramento			
Oranges	3.1	2.0	5.7
Lemons	2.2	1.6	4.8
Limes	2.0	1.5	4.3

ences in packaging, the weight of the crates vary by commodity; so also then do the transportation costs. Unit shipping costs are given in Table 5.8.

The problem of determining a minimal-cost shipping policy is a multicommodity network-flow problem. Notice that this example is similar in structure to the transportation problem developed in Section 2.10 in that shipments can only occur between origins and destinations. The major distinguishing features of multicommodity flow problems are that several nonhomogeneous commodities share common arcs, and that the flow of all commodities on an arc is constrained by the arc capacity. Were this not so, one would be able to solve a minimal-cost transportation problem for each commodity independently of the others, using an algorithm such as the out-of-kilter. The capacity constraints, however, make these problems much more difficult to solve.

5.18 Linear Programming Formulations

As in the single-commodity case, multicommodity network-flow problems can be formulated as linear programs. The fruit distribution problem is an example of the *multicommodity transportation problem*, hereafter denoted MCTP. Define x_{ij}^k to be the flow of commodity k from origin i to destination j having unit transportation cost c_{ij}^k. Also, let a_i^k and b_j^k be the supply at node i and demand at node j, respectively, of commodity k, and let u_{ij} represent the capacity of arc (i, j). The mathematical model for the MCTP is given in Eqs. (5-23) to (5-27).

$$\text{minimize} \sum_{k=1}^{r} \sum_{i=1}^{m} \sum_{j=1}^{n} c_{ij}^k x_{ij}^k \tag{5-23}$$

subject to

$$\sum_i x_{ij}^k = b_j^k \qquad \text{all } j, k \tag{5-24}$$

$$\sum_j x_{ij}^k = a_i^k \qquad \text{all } i, k \tag{5-25}$$

$$\sum_k x_{ij}^k \leq u_{ij} \qquad \text{all } i, j \tag{5-26}$$

$$x_{ij}^k \geq 0 \qquad \text{all } i, j, k \tag{5-27}$$

It is assumed, as in the single-commodity transportation problem, that total supply equals total demand for each commodity, that is,

$$\sum_i a_i^k = \sum_j b_j^k \qquad \text{for all } k \tag{5-28}$$

Many multicommodity models include transshipment nodes. At transshipment nodes there is neither a supply nor a demand; the only requirement is that conservation of flow be satisfied. An example of a multicommodity transshipment network is given in Figure 5.25. Nodes 1 and 2 are sources

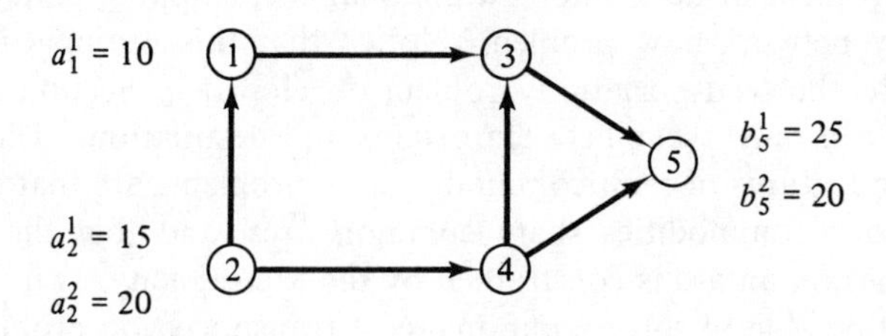

Figure 5.25. Multicommodity transshipment network.

for commodity 1, while node 2 is the only source for commodity 2. Node 5 is the destination for both commodities; nodes 3 and 4 are transshipment nodes. The general linear programming formulation for the multicommodity transshipment problem is given in Eqs. (5-29) to (5-34).

$$\text{minimize} \sum_{k=1}^{r} \sum_{(i,j)\in \mathbf{A}} c_{ij}^k x_{ij}^k \tag{5-29}$$

subject to

$$\sum_j x_{ij}^k - \sum_j x_{ji}^k = a_i^k \quad \text{if node } i \text{ is a source for commodity } k \tag{5-30}$$

$$\sum_j x_{ij}^k - \sum_j x_{ji}^k = 0 \quad \text{if node } i \text{ is a transshipment node} \tag{5-31}$$

$$\sum_i x_{ij}^k - \sum_i x_{ji}^k = -b_j^k \quad \text{if node } j \text{ is a sink for commodity } k \tag{5-32}$$

$$\sum_k x_{ij}^k \leq u_{ij} \quad \text{for } (i, j) \in \mathbf{A} \tag{5-33}$$

$$x_{ij}^k \geq 0 \quad \text{for all } k \text{ and } (i, j) \in \mathbf{A} \tag{5-34}$$

In this formulation, **A** is the set of arcs in the network.

One of the difficulties encountered in solving multicommodity network-flow problems is caused by the fact that optimal solutions may be noninteger. For example, consider the two-commodity transportation problem shown in Figure 5.26. Arc labels represent $(c_{ij}^1, c_{ij}^2, u_{ij})$. The optimal linear programming solution is given in Table 5.9.

Noninteger solutions arise because linear programming formulations do not have unimodular constraint matrices in general. Recall that unimodularity was introduced in a preceding chapter and provides a sufficient

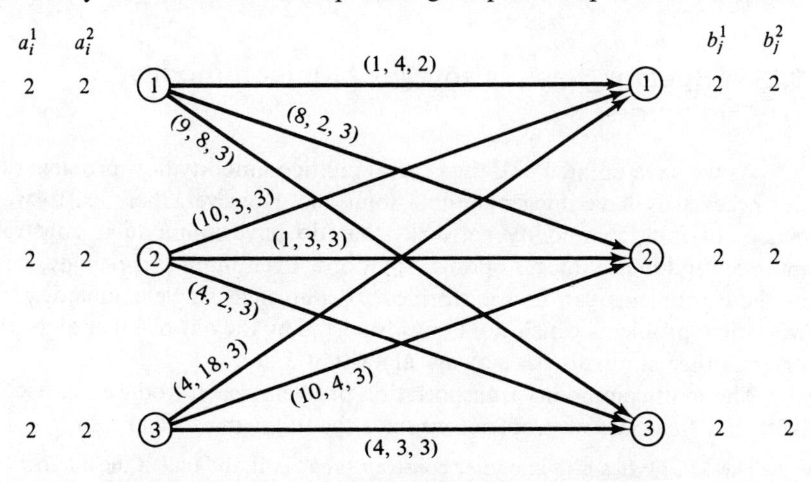

Figure 5.26. Two-commodity transportation problem.

Table 5.9

Arc	*Commodity 1*	*Commodity 2*
(1, 1)	$\frac{3}{2}$	$\frac{1}{2}$
(1, 2)	0	$\frac{3}{2}$
(1, 3)	$\frac{1}{2}$	0
(2, 1)	0	$\frac{3}{2}$
(2, 2)	2	0
(2, 3)	0	$\frac{1}{2}$
(3, 1)	$\frac{1}{2}$	0
(3, 2)	0	$\frac{1}{2}$
(3, 3)	$\frac{3}{2}$	$\frac{3}{2}$

condition for integer optimal solutions. Since single-commodity problems have this property, it is possible to devise highly efficient algorithms whose computations are essentially composed of only additions and subtractions. Division and matrix inversion, for instance, are not required. From a computer standpoint, arithmetic operations performed in integers are much faster than using floating-point (decimal) numbers. This leads to computer implementations that can solve extremely large problems very quickly.

Special algorithms for multicommodity flow problems have been developed which exploit the structure and properties of this class of problems. They have been shown to be faster than general-purpose linear programming procedures, but are not nearly as fast as corresponding single-commodity algorithms. A full discussion of these methods is beyond the scope of this book. However, in the remainder of this chapter we discuss a variety of special topics related to multicommodity flow problems and solution techniques.

5.19 A Special Class of Integer Multicommodity Networks

As we have pointed out, the general multicommodity flow problem does not necessarily have integer optimal solutions. However, there exist several classes of multicommodity networks that do have unimodular constraint matrices and hence integer optimal solutions. Even more surprisingly, many of these problems can be transformed to equivalent single-commodity network flow problems which can be easily solved by the out-of-kilter algorithm or any other minimal-cost network algorithm.

The multicommodity transportation problem was introduced in Section 5.18. For this class of problems we have the following special result:

> The MCTP has a unimodular constraint matrix if and only if the number of sources (m) or the number of destinations (n) is less than or equal to 2.

This result states that an optimal integer solution is guaranteed to exist whenever $m \leq 2$ or $n \leq 2$ *regardless* of the number of commodities. Clearly, if m or n equals one, the solution is trivial. However, if both are greater than 1, the problem is a more general linear program. We shall show that there is a much easier solution method.

Let us reformulate the MCTP defined in Eqs. (5-23) to (5-27) by adding slack variables to the capacity constraints. The new results are given in Eqs. (5-35) to (5-39).

$$\text{minimize} \sum_{k=1}^{r} \sum_{i=1}^{m} \sum_{j=1}^{n} c_{ij}^k x_{ij}^k \tag{5-35}$$

subject to

$$\sum_i x_{ij}^k = b_j^k \qquad \text{all } j, k \tag{5-36}$$

$$\sum_j x_{ij}^k = a_i^k \qquad \text{all } i, k \tag{5-37}$$

$$\sum_k x_{ij}^k + s_{ij} = u_{ij} \qquad \text{all } i, j \tag{5-38}$$

$$x_{ij}^k, s_{ij} \geq 0 \qquad \text{all } i, j, k \tag{5-39}$$

For $m = 2$ the model in (5-35) to (5-39) reduces to the linear program defined by (5-40) to (5-46).

$$\text{minimize} \sum_{k=1}^{r} \sum_{j=1}^{n} [(c_{2j}^k - c_{1j}^k)x_{2j}^k + c_{1j}^k b_j^k] \tag{5-40}$$

$$\sum_k x_{2j}^k + s_{2j} = u_{2j} \qquad \text{all } j \tag{5-41}$$

$$-\sum_j x_{2j}^k = -a_2^k \qquad \text{all } k \tag{5-42}$$

$$-\sum_j s_{2j} = -\sum_j u_{2j} + \sum_k a_2^k \tag{5-43}$$

$$x_{2j}^k + x_{1j}^k = b_j^k \qquad \text{all } j, k \tag{5-44}$$

$$s_{2j} + s_{1j} = u_{2j} + u_{1j} - \sum_k b_j^k \qquad \text{all } j \tag{5-45}$$

$$x_{ij}^k, s_{ij} \geq 0 \qquad \text{all } i, j, k \tag{5-46}$$

By closely examining this linear program, we see that every variable x_{2j}^k and s_{2j} appears exactly twice and with opposite signs in Eqs. (5-41) to (5-43). Therefore, the coefficient matrix of these three constraints has the form of a node–arc incidence matrix. Notice also that x_{1j}^k and s_{1j} appear only in Eqs. (5-44) and (5-45) and are nonnegative. Thus, we may think of these as slack variables and eliminate them by writing Eqs. (5-44) and (5-45) as

$$x_{2j} \leq b_j^k \qquad \text{all } j, k$$

$$s_{2j} \leq u_{2j} + u_{1j} - \sum_k b_j^k \qquad \text{all } j$$

The right-hand-sides of Eqs. (5-41) to (5-43) represent supplies (if positive) and demands (if negative). The structure of the network characterized by Eqs. (5-41) to (5-43) is a transportation model as shown in Figure 5.27. The right-hand-sides of Eqs. (5-44) and (5-45) are capacities of the arcs represented by x_{2j}^k and s_{2j}. To solve this, one may use the out-of-kilter algorithm to find optimal flows x_{2j}^k and s_{2j}. Eqs. (5-44) and (5-45) could then be used to solve for x_{1j}^k and s_{1j}.

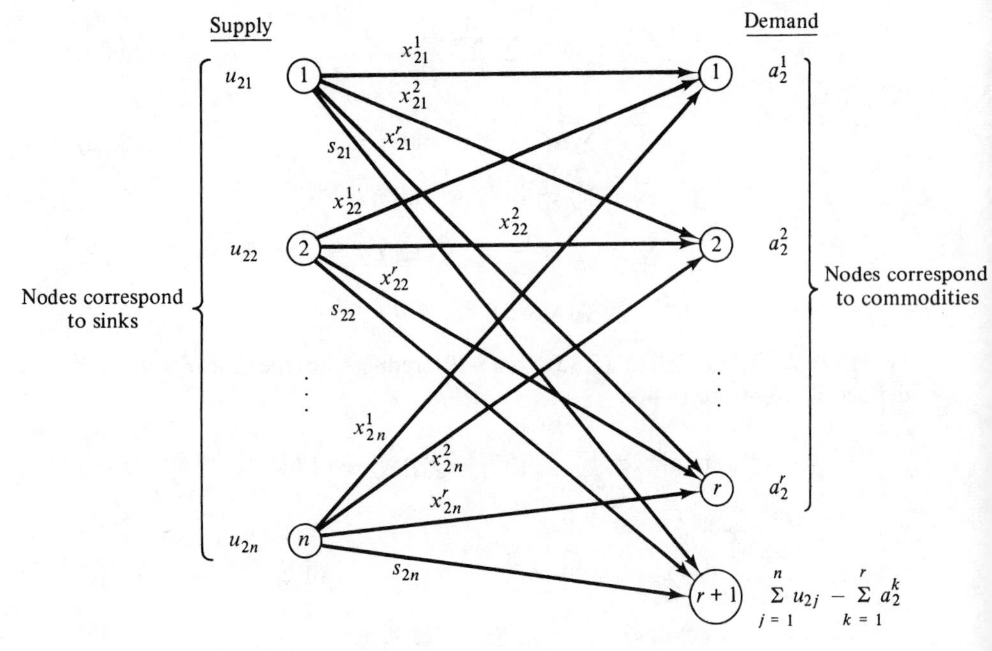

Figure 5.27. Single commodity formulation of 2-source MCTP.

This new formulation was obtained by performing row operations on the original constraints in a manner similar to the employment scheduling example in Section 2.1.7. It is a straightforward exercise to show that these two problems are equivalent. The interesting feature is that while the original linear program did not have the form of a network problem, we were able to transform the constraints into this form, thus making a rather difficult problem easy to solve.

5.19.1 Shipment of Automotive Transmissions

An automotive manufacturer operates two major transmission plants, one in Cincinnati and the other in St. Louis. Completed transmissions must be sent to assembly plants in Detroit, Dallas, and Atlanta. Three basic types of transmissions are produced: subcompact, compact, and midsize. Production capacities, demands, and transportation costs for each model are given in Table 5.10. Intercity freight capacities are also shown. The transformed network is given as an out-of-kilter problem in Figure 5.28.

Table 5.10 INPUT INFORMATION FOR SAMPLE PROBLEM

Unit Shipping Costs

		To			
From	*Model*	*Detroit*	*Dallas*	*Atlanta*	*Production (units/month)*
Cincinnati	SC	4.65	5.50	5.00	4000
	C	4.95	5.90	5.45	5000
	MS	5.30	6.30	6.00	2000
St. Louis	SC	5.15	4.85	5.25	2500
	C	5.70	4.95	5.90	3000
	MS	6.00	5.15	6.25	5000
Demand (units/month)	SC	3500	1500	1500	
	C	4000	1500	2500	
	MS	3000	2000	2000	

Monthly Transportation Capacities

	To		
From	*Detroit*	*Dallas*	*Atlanta*
Cincinnati	6000	3000	4000
St. Louis	6000	4000	3000

5.20 Approximate Solution of Multicommodity Transportation Problems through Aggregation

Often a decision maker is satisfied with good, though suboptimal, solutions to complex problems. This frequently occurs when software capable of performing true optimization is not available, or too costly when used on a repetitive basis, or when the timeliness of an answer precludes a major software development effort.

One method of simplifying a problem in order to make it easier to solve is by *aggregation*, that is, replacing sets of entities such as variables, nodes in a network, and so on, by single quantities. In this way, both the time necessary to solve the problem as well as computer storage requirements are reduced. However, the resulting solution is usually nonoptimal.

In this section we shall discuss the solution of multicommodity transportation problems by aggregating sources or destinations, drawing upon the results of Section 5.19. That is, we will consider cases in which the aggregated

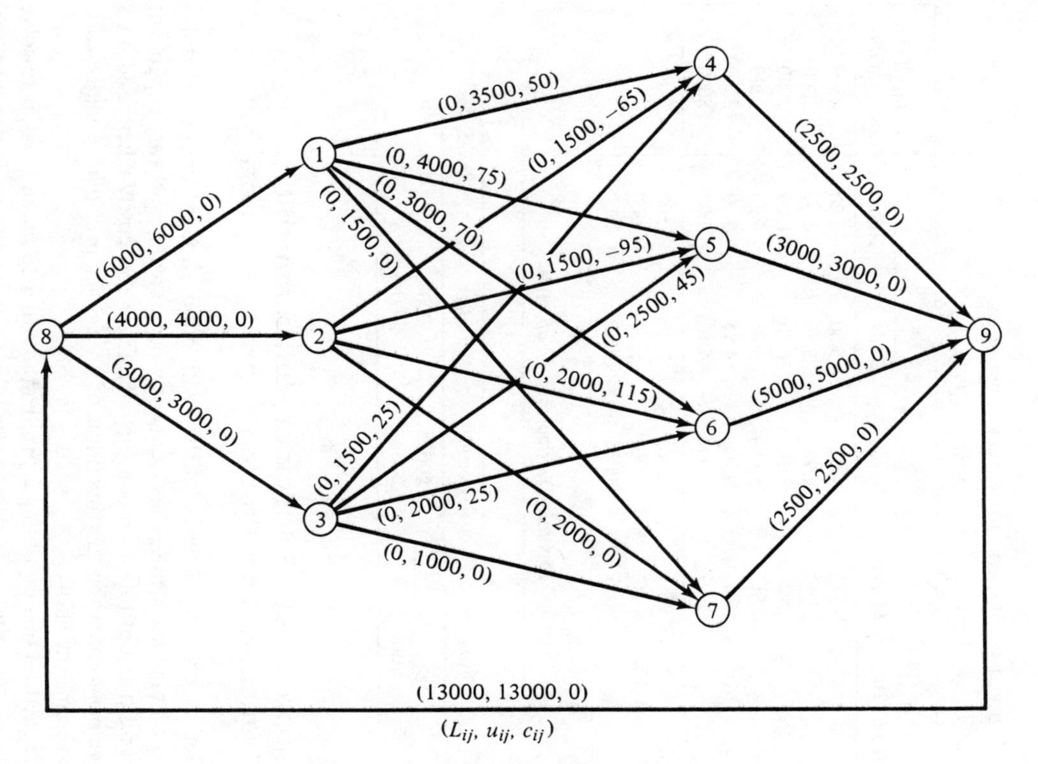

Figure 5.28. Out-of-kilter formulation (all costs × 100).

problem consists of exactly two origins and solve it by means of the single-commodity transformation.

Let us suppose that a multicommodity transportation problem consists of m origins, n destinations, and r commodities. Partition the origins into two sets S_1 and S_2, and replace the sets of nodes by two "pseudo nodes" as illustrated in Figure 5.29. This two-source network is the aggregated problem. We need to relate the parameters of this problem to those of the original problem in some meaningful fashion, and also be able to recover a feasible solution to the original problem after solving the aggregated problem.

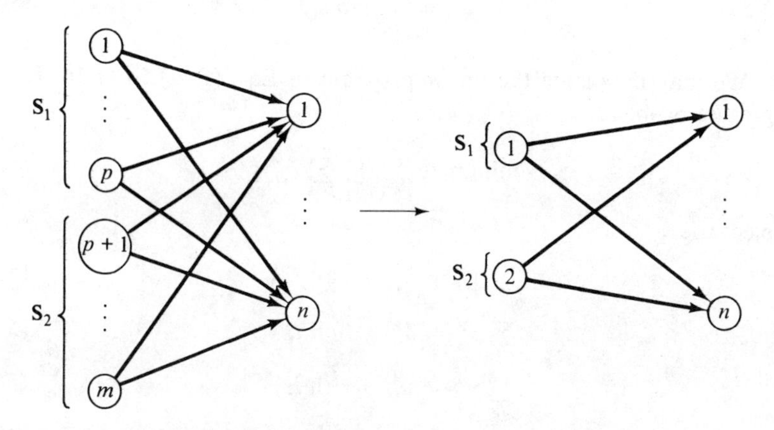

Figure 5.29. Aggregation of origins in the MCTP.

Since each of the two pseudo nodes represents a collection of original sources, it is natural to specify the supplies at these nodes as

$$\bar{a}_l^k = \sum_{i \in S_l} a_i^k \tag{5-47}$$

for $l = 1, 2$. Let y_{lj}^k represent the flow of commodity k from node l to node j in the aggregate problem. Since arc (l, j) is an aggregation of arcs from nodes $i \in S_l$ to destination j, we can write

$$y_{lj}^k = \sum_{i \in S_l} x_{ij}^k \tag{5-48}$$

There still remains the question of specifying costs and capacities on the aggregate arcs. There are several ways to do this; the method we shall use has certain advantages over others. To specify costs, we shall use a concept called *weighted aggregation*; that is, costs are weighted proportionate to the supplies of the origins represented by the pseudo node as follows:

$$\bar{c}_{lj}^k = \sum_{i \in S_l} c_{ij}^k \frac{a_i^k}{\bar{a}_l^k} \tag{5-49}$$

At first thought it would seem reasonable to define the capacity on an aggregated arc (l, j) as the sum of the capacities of the original arcs from $i \in \mathbf{S}_l$ to destination j. The problem with this is that it may be difficult to recover a feasible solution to the original problem. Therefore, the following method will be used. Define

$$\delta_i = \min_k \left[\frac{\bar{a}_l^k}{a_i^k}\right] \qquad \text{for } i \in \mathbf{S}_l \tag{5-50}$$

Now let

$$\bar{u}_{lj} = \min_{i \in \mathbf{S}_l} [\delta_i u_{ij}] \tag{5-51}$$

We may now state the linear program in Eqs. (5-52) to (5-56) for the aggregate problem:

$$\text{minimize} \sum_{k=1}^{r} \sum_{l=1}^{2} \sum_{j=1}^{n} \bar{c}_{lj}^k y_{lj}^k \tag{5-52}$$

subject to

$$\sum_l y_{lj}^k = b_j^k \qquad \text{all } j, k \tag{5-53}$$

$$\sum_j y_{lj}^k = \bar{a}_l^k \qquad \text{all } l, k \tag{5-54}$$

$$\sum_k y_{lj}^k \leq \bar{u}_{lj} \qquad \text{all } l, j \tag{5-55}$$

$$y_{lj}^k \geq 0 \qquad \text{all } l, j, k \tag{5-56}$$

We now need to be able to recover a solution to the original problem once we have an optimal solution to the aggregate problem defined in (5-52) to (5-56). Consider the following definition,

$$x_{ij}^k = y_{lj}^k \frac{a_i^k}{\bar{a}_l^k} \qquad \text{for } i \in \mathbf{S}_l \tag{5-57}$$

Note that Eq. (5-57) can be used to obtain

$$\begin{aligned}\sum_{i \in \mathbf{S}_l} x_{ij}^k &= y_{lj}^k \left(\sum_{i \in \mathbf{S}_l} \frac{a_i^k}{\bar{a}_l^k}\right) \\ &= y_{lj}^k \frac{\bar{a}_l^k}{\bar{a}_l^k} \\ &= y_{lj}^k\end{aligned}$$

which corresponds to our previous definition (5-48). The procedure defined in Eq. (5-57) is called *fixed-weight disaggregation.* With a little algebra it is not difficult to show that x_{ij}^k is feasible to the original problem (specifying

$\bar{u}_{lj}$ as we did was necessary in order to guarantee satisfying the original capacity constraints). Also,

$$\sum_k \sum_i \sum_j c_{ij}^k x_{ij}^k = \sum_k \sum_l \sum_j \bar{c}_{lj}^k y_{lj}^k \tag{5-58}$$

In other words, the fixed-weight disaggregation has the same objective value as the aggregate problem. The reader should be cautioned however, that the aggregate problem may not be feasible even though the original problem is. This is due to the nature of $\bar{u}_{lj}$. One may have to try different aggregations in this case.

5.20.1 A Fruit Distribution Problem

Let us solve the fruit distribution problem posed in the introduction of this part by aggregating sources 2 and 3. The supplies, costs, and capacities of the aggregate problem are given in Table 5.11.

Table 5.11. AGGREGATE PROBLEM PARAMETERS

Supplies

	Oranges	*Lemons*	*Limes*
Santa Barbara	800	700	700
Bakersfield/Sacramento	1500	1800	1000

Costs

	To		
From	*Denver*	*Seattle*	*Kansas City*
Santa Barbara			
Oranges	3.200	2.500	5.600
Lemons	2.400	1.900	4.300
Limes	2.300	1.700	4.000
Bakersfield/ Sacramento			
Oranges	3.033	2.067	5.567
Lemons	2.244	1.644	4.622
Limes	2.000	1.450	4.300

Capacities

	To		
From	*Denver*	*Seattle*	*Kansas City*
Santa Barbara	1200	900	1000
Bakersfield/ Sacramento	1500	1800	1000

The optimal solution to the original problem has a cost of $18,570 and is shown in Table 5.12. Table 5.13 gives the optimal solution to the aggregate problem having a cost of $18,799.10, an increase of 0.26% over the true optimal cost. It is left as an exercise to the reader to compute the disaggregate solution and verify that it has the same cost.

Table 5.12. SOLUTION TO FRUIT DISTRIBUTION PROBLEM

	To		
From	*Denver*	*Seattle*	*Kansas City*
Santa Barbara			
Oranges	800	0	0
Lemons	200	0	500
Limes	0	200	500
Bakersfield			
Oranges	100	200	700
Lemons	500	200	100
Limes	200	300	0
Sarcamento			
Oranges	0	500	0
Lemons	200	800	0
Limes	500	0	0

Table 5.13. SOLUTION TO AGGREGATED PROBLEM

	To		
From	*Denver*	*Seattle*	*Kansas City*
Santa Barbara			
Oranges	800	0	0
Lemons	200	0	500
Limes	0	400	300
Bakersfield/ Sacramento			
Oranges	100	700	700
Lemons	700	1000	100
Limes	700	100	200

5.21 Error Bounds for Aggregation[1]

By considering the dual to the aggregate problem formulated in (5-52) to (5-56), we shall derive a bound on the error resulting from the disaggregate solution. The bound is useful in determining the error resulting from aggrega-

[1]This section may be skipped without loss of continuity.

tion, although it should be realized that *actual* error is generally less than the bound.

The dual linear program to the aggregate problem is defined in Eqs. (5-59) to (5-61).

$$\text{maximize} \sum_k \sum_l \bar{a}_l^k \alpha_l^k + \sum_k \sum_j b_j^k \beta_j^k - \sum_j \sum_l \bar{u}_{lj} \gamma_{lj} \tag{5-59}$$

subject to

$$\alpha_l^k + \beta_j^k - \gamma_{lj} \leq \bar{c}_{lj} \qquad \text{all } l, j, k \tag{5-60}$$

$$\gamma_{lj} \geq 0 \qquad \text{all } l, j \tag{5-61}$$

Suppose that $\bar{\boldsymbol{\alpha}}$, $\bar{\boldsymbol{\beta}}$, and $\bar{\boldsymbol{\gamma}}$ is an optimal solution to the dual problem of Eqs. (5-59) to (5-61), with objective function value $\bar{z}$, and that $\bar{\mathbf{x}}$ is an optimal solution to the original MCTP of Eqs. (5-23) to (5-27), with objective function value z^*. Then

$$\begin{aligned} z^* \geq & \sum_k \sum_i \sum_j c_{ij}^k \bar{x}_{ij}^k + \sum_j \sum_k (b_j^k - \sum_i \bar{x}_{ij}^k) \bar{\beta}_j^k \\ & + \sum_l \sum_{i \in S_l} \sum_k (a_i^k - \sum_j \bar{x}_{ij}^k) \bar{\alpha}_l^k \\ & - \sum_j \sum_l \sum_{i \in S_l} (u_{ij} - \sum_k \bar{x}_{ij}^k) \bar{\gamma}_{lj} \end{aligned}$$

Adding and subtracting $\sum_l \sum_j \bar{u}_{lj} \bar{\gamma}_{lj}$ in the inequality above, rearranging terms, and using the definition of aggregate supplies, we have

$$\begin{aligned} z^* \geq & [\sum_k \sum_l \bar{a}_l^k \bar{\alpha}_l^k + \sum_k \sum_j b_j^k \bar{\beta}_j^k - \sum_l \sum_j \bar{u}_{lj} \bar{\gamma}_{lj}] \\ & - \sum_l \sum_{i \in S_l} \sum_j \sum_k \bar{x}_{ij}^k (\bar{\alpha}_l^k + \bar{\beta}_j^k - \bar{\gamma}_{lj} - c_{ij}^k) \\ & + \sum_l \sum_j (\bar{u}_{lj} - \sum_{i \in S_l} u_{ij}) \bar{\gamma}_{lj} \end{aligned}$$

We note that the term in the brackets is $\bar{z}$. Hence, we have

$$\begin{aligned} \bar{z} - z^* \leq & \sum_l \sum_{i \in S_l} \max_{j,k} [\bar{\alpha}_l^k + \bar{\beta}_j^k - \bar{\gamma}_{lj} - c_{ij}^k] \sum_j \sum_k \bar{x}_{ij}^k \\ & - \sum_l \sum_j (\bar{u}_{lj} - \sum_{i \in S_l} u_{ij}) \bar{\gamma}_{lj} \end{aligned}$$

Since $\sum_j \sum_k \bar{x}_{ij}^k = \sum_k a_i^k$, this reduces to

$$\begin{aligned} \bar{z} - z^* \leq & \sum_l \sum_{i \in S_l} \max_{j,k} [\bar{\alpha}_l^k + \bar{\beta}_j^k - \bar{\gamma}_{lj} - c_{ij}^k] \sum_k a_i^k \\ & - \sum_l \sum_j (\bar{u}_{lj} - \sum_{i \in S_l} u_{ij}) \bar{\gamma}_{lj} \triangleq \epsilon \end{aligned}$$

The term ϵ represents the maximum deviation of the aggregate solution value from the true optimum cost. Since $\bar{z}$ is an upper bound on the optimum cost, we have

$$\bar{z} - \epsilon \leq z^* \leq \bar{z}$$

5.22 Maximal Flows in Multicommodity Networks

For single-commodity problems it was shown that the value of the maximal flow is equal to that of the minimal cut set, and that there is an elegant and efficient algorithm for finding the maximal flow. At first, one might believe that perhaps these results can be generalized to multicommodity flow problems. But again, except for some special cases, this is not true. One of the earliest counterexamples is shown in Figure 5.30. Nodes s_k and t_k represent the source and sink for commodity k, respectively. Each arc has capacity of 1, and the problem is to maximize the *sum* of the flows of the three commodities from their sources to their respective sinks. For instance, if a flow of 1 unit is sent from s_1 to t_1 along the only path available, no other commodity flow can be sent through the network. This is not the maximal flow since we can send $\frac{1}{2}$ unit from s_k to t_k for each commodity, a noninteger flow.

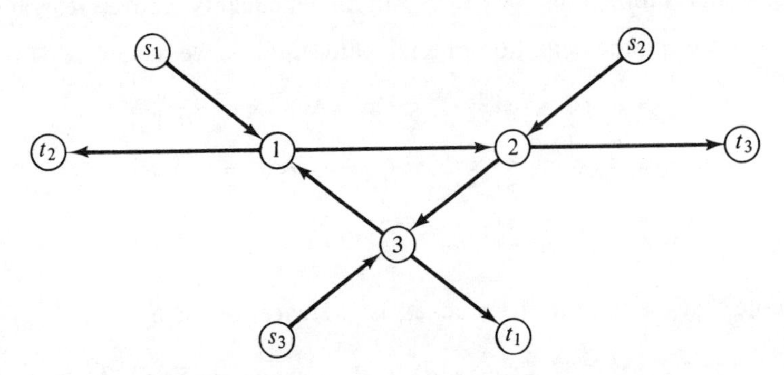

Figure 5.30. Multicommodity maximal flow problem.

In multicommodity problems, the analogous concept of a cut is called a *disconnecting set* and is defined as a set of arcs whose removal from the network destroys all paths from each source to its respective sink. In this example, the minimal disconnecting set consists of any two arcs between nodes 1, 2, and 3. The value of the minimal disconnecting set is 2; hence, we have that the maximal multicommodity flow can be *less than* the value of the minimal disconnecting set.

In general, the multicommodity maximal-flow problem, just as the minimal-cost problem, is difficult to solve. Most solution procedures are beyond the scope of this discussion. However, for special types of networks, it is true that maximum flow equals the value of the minimum disconnecting set and that all flow values are integer. This result can be shown to hold true for networks that are called *completely planar*. A network is *planar* if it can be drawn so that no two arcs intersect. For example, the network in Figure 5.31(a) is planar while the network in Figure 5.31(b) is not. A multicommodity network is completely planar if it can be drawn as in Figure 5.32, so that the resulting network is planar. An example with two commodities is given in Figure 5.33.

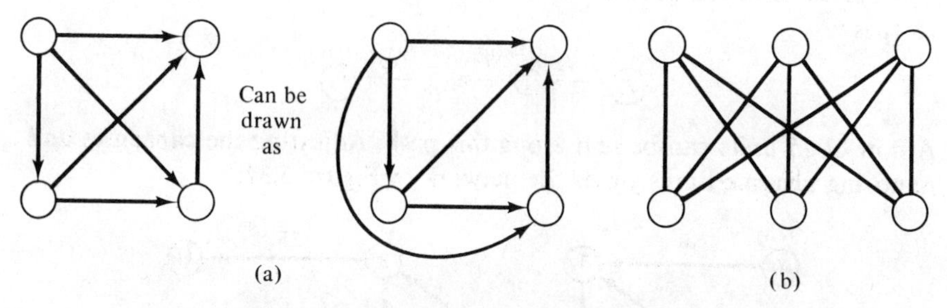

Figure 5.31. Planar and nonplanar networks.

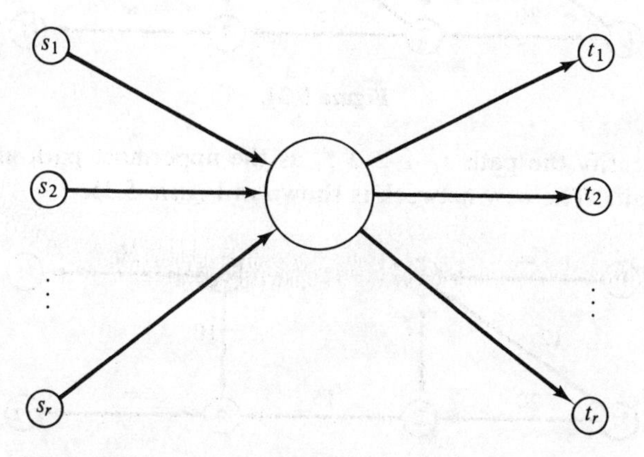

Figure 5.32. Structure of completely planar networks.

There is a very simple algorithm for solving problems of this type. Start with commodity 1. Identify the *uppermost* path from s_1 to t_1 and assign as much flow as possible. Adjust all capacities by subtracting the flow on the arcs. At least one arc will reach capacity; remove this and all other arcs whose total flow equals the capacity. When there are no flow-augmenting

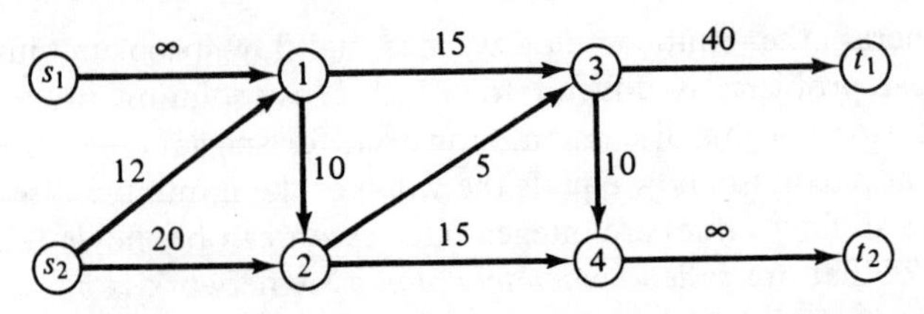

Figure 5.33. Two-commodity completely planar network.

paths remaining from s_1 to t_1, repeat this procedure for commodity 2, 3, and so on, until it is no longer possible to send any additional flow from a source to a sink. The resulting flow pattern is a maximal flow.

Let us solve the problem in Figure 5.33. The uppermost path from s_1 to t_1 is

$$(s_1) \xrightarrow{\infty} (1) \xrightarrow{15} (3) \xrightarrow{40} (t_1)$$

A flow of 15 units can be sent along this path. Adjusting the capacities and removing saturated arcs yields the network in Figure 5.34.

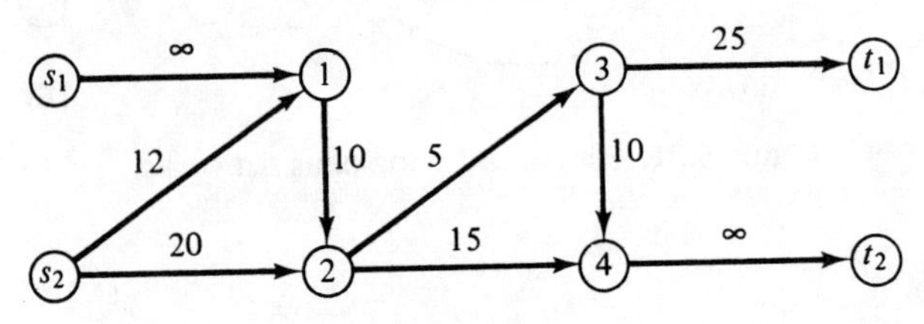

Figure 5.34.

We next identify the path s_1–1–2–3–t_1 as the uppermost path and assign a flow of 5 units. The new network is shown in Figure 5.35.

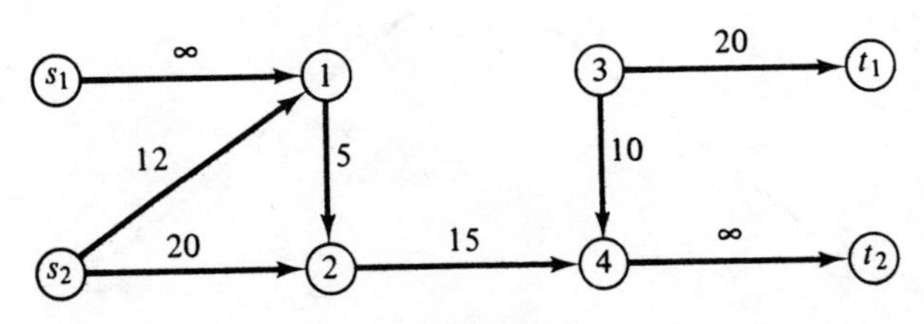

Figure 5.35.

No flow-augmenting paths exist from s_1 to t_1 in the modified network of Figure 5.35. Considering commodity 2, we start with the path s_2–1–2–4–t_2. Five units of flow can be sent along this path. The modified network is shown in Figure 5.36.

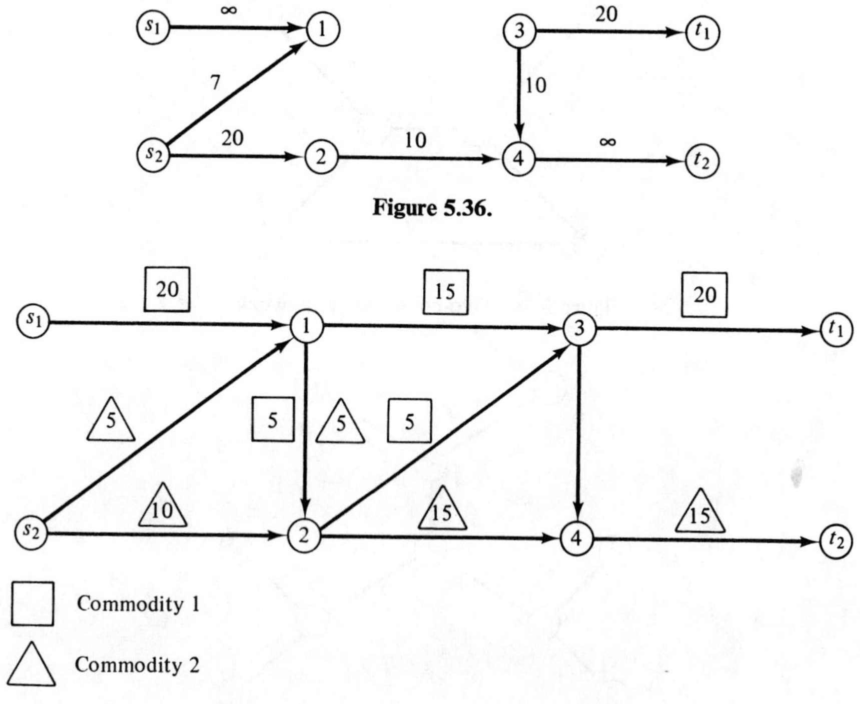

Figure 5.36.

Figure 5.37. Maximal two-commodity flow.

Finally, we may send 10 units along the path s_2–2–4–t_2. Combining these results, we have the maximal-flow pattern given in Figure 5.37. The maximal two-commodity flow is 35 and the minimal disconnecting set is given by {(1, 3), (2, 3), (2, 4)}.

5.23 Multicommodity Flows in Undirected Networks

In networks with undirected arcs, flow may traverse an arc in either direction. In the single-commodity case, one can replace an undirected arc by a pair of oppositely directed arcs. This can be done since flows in opposite directions cancel each other. However, for multicommodity problems, this cannot be done since flows of *different* commodities in opposite directions do not cancel out, but add to the total consumption of capacity of the arc. And as before, optimal flows need not be integer, nor does the maximum flow/minimum cut theorem always hold. Two examples to illustrate this are given in Figures 5.38 and 5.39. In Figure 5.38, we have a two-commodity flow problem with unit capacities. We leave it to the reader to find the maximal

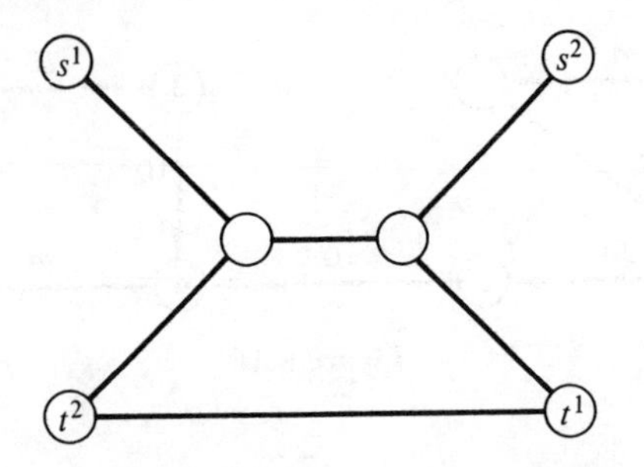

Figure 5.38. Two-commodity network.

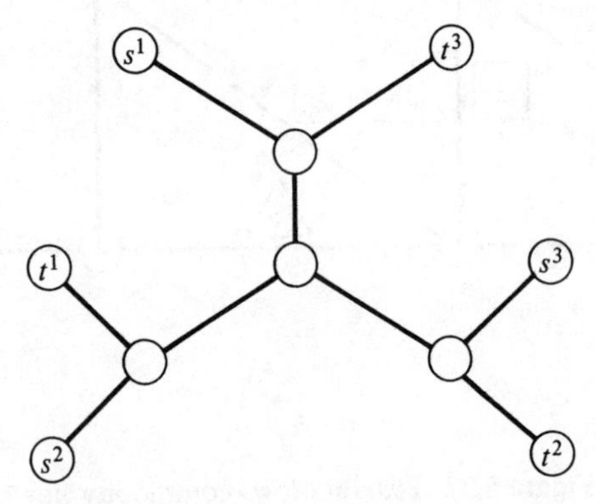

Figure 5.39. Three-commodity network (all capacities are equal to 1).

flow. (*Hint:* optimal flows are noninteger!) Figure 5.39 represents a three-commodity problem in which the maximal flow is strictly less than the minimal disconnecting set value. The reader should verify this.

Although the general problem is difficult to solve, we shall present an algorithm for the special case of two-commodity maximal flows. For this class of problems, it can be shown that optimal flows are always multiples of $\frac{1}{2}$, and that the maximum flow equals the value of the minimum disconnecting set.

To develop the algorithm, some new concepts and definitions need to be introduced. The flow of commodity k on arc (i, j) is denoted f_{ij}^k. Since flow can occur in either direction, a positive flow from i to j can be viewed as a negative flow from j to i; hence,

$$f_{ij}^k = -f_{ji}^k$$

Regardless of the direction of flow, the net amount of commodity k on arc (i, j) will be denoted $|f_{ij}^k|$. As before, s^k and t^k will be the source and sink, respectively, of commodity k, and u_{ij} is the capacity of arc (i, j), assumed to be integer.

Let us assume that we have a flow pattern in the network that satisfies conservation-of-flow restrictions. We will define two sets of nodes relative to the given flow pattern, **F** and **B**:

1. s^2 is defined to be in **F**. If $i \in \mathbf{F}$ and

$$f_{ji}^1 + f_{ij}^2 < u_{ij}$$

   then $j \in \mathbf{F}$.
2. s^2 is defined to be in **B**. If $i \in \mathbf{B}$ and

$$f_{ij}^1 + f_{ij}^2 < u_{ij}$$

   then $j \in \mathbf{B}$.

Suppose that we wish to send additional flow of commodity 1 from i to j. If $f_{ij}^1 > 0$, then the *residual capacity* is defined as in Eq. (5-62),

$$u_{ij}^1 = u_{ij} - f_{ij}^1 - |f_{ij}^2| \tag{5-62}$$

If we wish to send additional flow from j to i and $f_{ij}^1 > 0$, the residual capacity is given by

$$u_{ij}^1 = u_{ij} + f_{ij}^1 - |f_{ij}^2| \tag{5-63}$$

since the flow of the same commodity in opposite directions cancel. Similarly, for commodity 2, if $f_{ij}^2 > 0$, the residual capacity in the direction from i to j is

$$u_{ij}^2 = u_{ij} - |f_{ij}^1| - f_{ij}^2 \tag{5-64}$$

and in the direction from j to i,

$$u_{ij}^2 = u_{ij} - |f_{ij}^1| + f_{ij}^2 \tag{5-65}$$

Consider any path from s^2 to t^2 with arc (i, j) on the path. As we travel from s^2 to t^2 and node i is reached before node j, then arc (i, j) is said to be in the *positive direction* with respect to the path. In this case, a flow from i to j is said to be in the *forward direction*, and a flow from j to i in the *backward direction.*

Now, if $t^2 \in \mathbf{B}$, there is a path from s^2 to t^2 such that $u_{ij} + f_{ji}^1 - f_{ij}^2 > 0$ for all arcs on the path and where the positive direction of arc (i, j) is from i to j. We shall call this a *backward path* from s^2 to t^2. Similarly, if $t^2 \in \mathbf{F}$, then $u_{ij} + f_{ij}^1 - f_{ij}^2 > 0$ for all arcs on the path. This is called a *forward path.* A pair of forward and backward paths from s^2 to t^2 is called a *double path.* If the positive direction of all arcs in a path is from i to j, the *capacity of a backward path* is

$$\alpha_B = \min\,[u_{ij} + f^1_{ji} - f^2_{ij}] \tag{5-66}$$

and the *capacity of a forward path* is

$$\alpha_F = \min\,[u_{ij} + f^1_{ij} - f^2_{ij}] \tag{5-67}$$

These are simply the maximum amounts that can be shipped along these paths.

Let

$\Delta^1_F = \min\,[f^1_{ij} > 0]$, f^1_{ij} is on a forward path with f^1_{ij} in the forward direction.

$\Delta^1_B = \min\,[f^1_{ij} > 0]$, f^1_{ij} is on a backward path with f^1_{ij} in the backward direction.

Finally, for any double path, let

$$\Delta^1 = \min\,[\Delta^1_F, \Delta^1_B] \tag{5-68}$$

$$\alpha = \min\,[\alpha_B, \alpha_F] \tag{5-69}$$

We may now state an algorithm for finding maximal two-commodity flows in an undirected network.

Step 0: Find the maximal flow for commodity 1 using any single-commodity algorithm.

Step 1: Compute residual capacities u^2_{ij}, and find the maximal flow for commodity 2.

Step 2: Locate double paths from s^2 to t^2. If none exist, we have the maximal flow. Otherwise, let $\epsilon = \min\,[\Delta^1, \frac{1}{2}\alpha]$. Send ϵ units of commodity 1 from s^2 to t^2 along the backward path and from t^2 to s^2 along the forward path.

Step 3: Increase the total flow of commodity 2 by sending ϵ units of commodity 2 from s^2 to t^2 along both the forward and backward paths. Return to step 2.

What the algorithm essentially does is to redistribute the flow of commodity 1 along a double path so that the flow of commodity 2 can be increased along each of these paths.

5.23.1 A Two-Commodity Flow Problem

Let us solve the two-commodity problem for the network shown in Figure 5.38.

***Step 0*:** The maximal flow for commodity 1 is equal to 1(□→ denotes commodity 1, △→ denotes commodity 2):

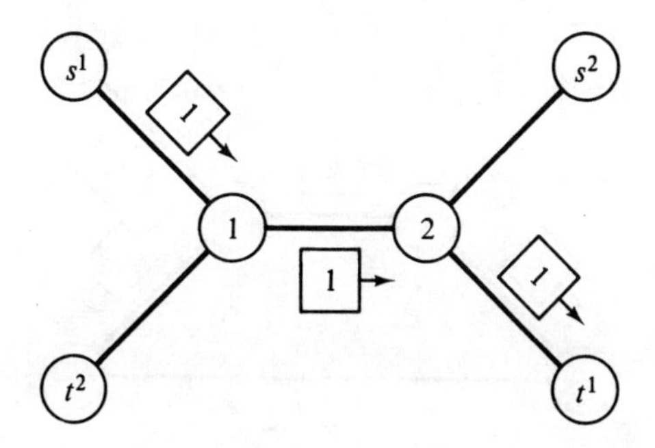

***Step 1*:** The residual capacities u_{ij}^2 are shown below. The maximal flow from s^2 to t^2 is zero.

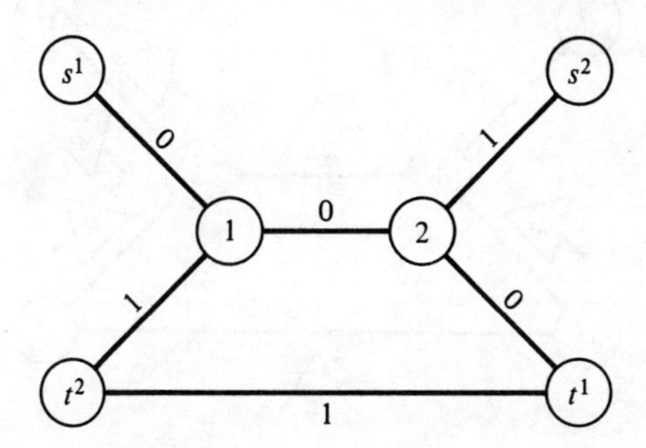

***Step 2*:** To locate double paths, we will determine the sets **F** and **B**,

$$\mathbf{F} = \{s^2, 2, t^1, t^2\}$$

$$\mathbf{B} = \{s^2, 2, 1, t^2\}$$

Since $t^2 \in \mathbf{F}$ and $t^2 \in \mathbf{B}$, we have a pair of double paths. The forward path is s^2–2–t^1–t^2 with capacity

$$\alpha_F = \min\,[1, 2, 1] = 1$$

The backward path is s^2–2–1–t^2 with capacity

$$\alpha_B = \min\,[1, 2, 1] = 1$$

Thus $\alpha = \min\,[\alpha_F, \alpha_B] = 1$. Also, $\Delta_F^1 = 1$, $\Delta_B^1 = 1$. Therefore, $\Delta^1 = \min\,[\Delta_F^1, \Delta_B^1] = 1$. We have $\epsilon = \frac{1}{2}$. To complete step 2, send $\frac{1}{2}$ unit of commodity 1 from s^2 to t^2 along the backward path and from t^2 to s^2 along the forward path as follows:

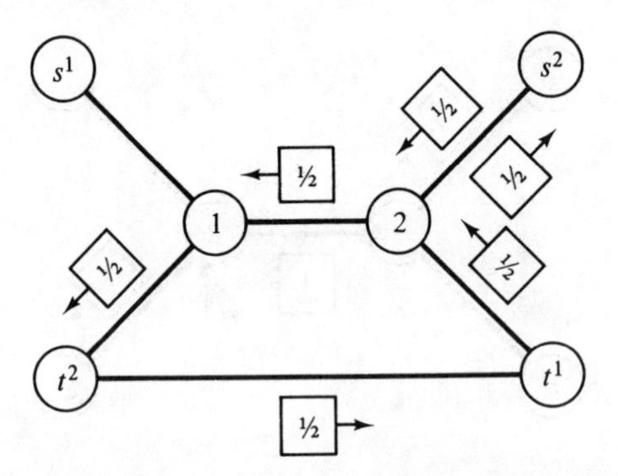

Step 3: Send $\frac{1}{2}$ unit of commodity 2 from s^2 to t^2 along both paths:

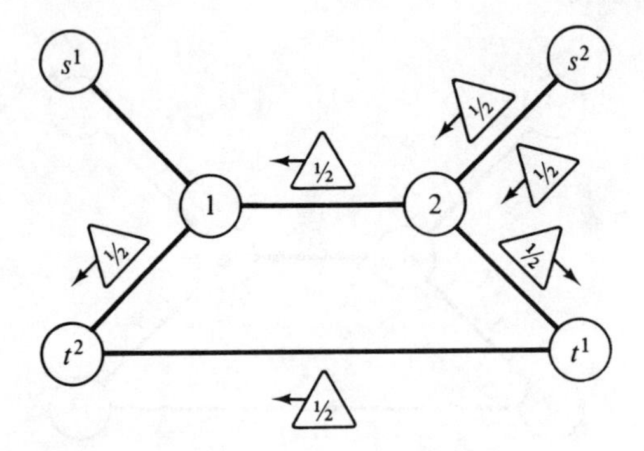

The current flow pattern is

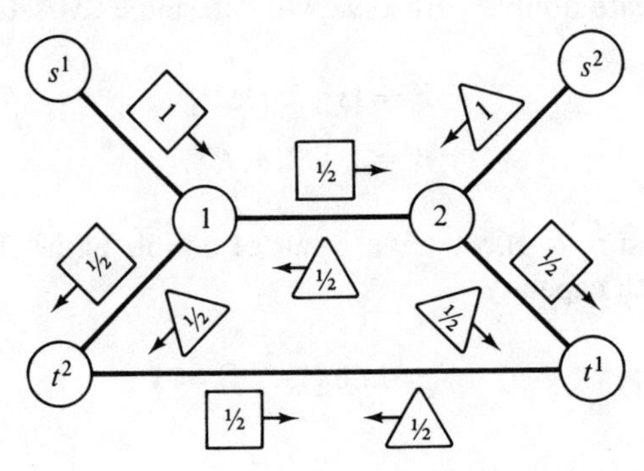

When we return to step 2, we find that no double paths exist; therefore, this is the optimal solution.

Notice that ϵ is either an integer or a multiple of $\frac{1}{2}$; thus, all flows will be integer or multiples of $\frac{1}{2}$. If all capacities are even integers, the optimal flow will be all-integer.

5.24 Maximal Flows and Funnel Nodes

A not uncommon logistics problem, particularly in military operations, can be described as follows. Supplies are located at one point for shipment to a destination. Vehicles that can transport the supplies are stationed at a different facility. Given a transportation network with limited capacity, one seeks an optimal routing of vehicles from their origin to the supply pickup point and then to the ultimate destination. A related problem that has a similar structure can be posed also: one desires to establish a central message center for an existing communication network through which all messages must pass. The maximum possible message flow must be maintained.

In both of these examples, a special node is singled out through which all flow from source to sink must pass. This node is called a *funnel node*. If we let f_{ij}^1 be the flow of the commodity from node i to j on its way *to* the funnel node, and f_{ij}^2 the flow from i to j after passing through the funnel node, we have the mathematical formulation in (5-70) to (5-75):

$$\text{maximize } v \tag{5-70}$$

subject to

$$\sum_j (f_{ij}^1 - f_{ji}^1) = \begin{cases} v^1 & \text{if } i = s \\ 0 & \text{if } i \neq s, a \\ -v^1 & \text{if } i = a \end{cases} \tag{5-71}$$

$$\sum_j (f_{ij}^2 - f_{ji}^2) = \begin{cases} v^2 & \text{if } i = a \\ 0 & \text{if } i \neq a, t \\ -v^2 & \text{if } i = t \end{cases} \tag{5-72}$$

$$|f_{ij}^1| + |f_{ij}^2| \leq u_{ij} \qquad \text{for all } i, j \tag{5-73}$$

$$v = v^1 = v^2 \tag{5-74}$$

$$f_{ij}^k \geq 0 \tag{5-75}$$

Here s, a, and t represent the source, funnel node, and sink, respectively. Notice that this is similar to the two-commodity flow problem on an undirected network. In this case, however, there are not really two commodities; they only represent different types of flow. In addition, the flows of "commodities" 1 and 2 are required to be equal. In the two-commodity problem we wish to maximize the sum of the commodity flows; in this case we maximize $\frac{1}{2}(v^1 + v^2)$.

We can show the following result: the maximal funnel node flow is equal to

$$\min [\bar{v}^1, \bar{v}^2, \tfrac{1}{2} \max [v^1 + v^2]]$$

where $\bar{v}^1$ and $\bar{v}^2$ are the maximal flows from s to a and from a to t, respectively. This result leads to a very simple algorithm, which depends only upon solving a sequence of ordinary maximum-flow problems.

***Step 1*:** Find $\bar{v}^1$ and $\bar{v}^2$ using a single-commodity flow algorithm.

***Step 2*:** Construct a new network G' by adding an additional node s' and arcs (s', s), (s', t) having infinite capacity. Let $\bar{v}$ be the maximal flow in this network from s' to a.

***Step 3*:** Determine

$$v^* = \min [\bar{v}^1, \bar{v}^2, \tfrac{1}{2}\bar{v}]$$

If $v^* = 0$, stop. No flow is possible.

***Step 4*:** Construct a new network G'' by adding a node s'' to the original network along with arcs (s'', s) and (s'', t), each with capacity v^*. Find the maximal flow in this network from s'' to a. Decompose the flow into a flow from s'' through s to a, and a flow from a through t to s''. Remove the additional nodes and arcs. The result is a maximal funnel node flow through a.

Notice that Step 2 determines $\bar{v} = \max [v^1 + v^2]$ by solving a single-commodity problem. It is easy to preserve the identities of "commodities" since any flow on a path from s^1 through s to a can be called "commodity 1" and from s^1 through t to a can be called "commodity 2." Step 4 simply finds a flow pattern having all of flow v^* passing through node a.

5.25 Applications of Multicommodity Networks

Multicommodity network models have obvious applications in communication systems, railway transportation, production and distribution planning, and military logistics to name just a few. In this section we shall examine some practical applications of multicommodity network models and introduce some new formulations of certain multicommodity problems.

5.25.1 Tanker Scheduling

One problem that often arises in the logistics area is optimal scheduling of a fleet of vehicles. Suppose that a company owns a fixed fleet of nonhomogeneous tankers which differ in speeds, capacities, and operating costs. We associate a utility with the delivery of a shipment by a particular type of tanker on a specific delivery date that reflects the desirability and cost of making that shipment. In addition, there is a cost (negative utility) of reassigning a tanker from a delivery port to a port of origin of a new shipment. The objective is to determine a schedule and routing of the fleet with maximal utility.

The problem may be viewed as a network where each node j corresponds to a receipt of a shipment at a delivery port on one of its allowable delivery dates; each node i corresponds to an origin port at a time equal to the delivery date minus the transit time. Transit times are assumed to be deterministic. Arcs (i, j) correspond to shipments and (j, i) to reassignments from delivery points to origins. Each type of tanker originates at a source node s^k; thus, arcs (s^k, i) correspond to initial deployment of the tankers. A pseudosink t is included and arcs (j, t) represent removal from service of all tankers. Finally, the total flow in all arcs corresponding to alternative shipping dates and tanker types for a given shipment must be restricted to the amount being shipped. Mathematically, we have

$$\text{maximize } z = \sum_k \sum_i \sum_j c_{ij}^k x_{ij}^k \tag{5-76}$$

subject to

$$\sum_k \sum_{\mathbf{A}_v} r_{ij}^k x_{ij}^k \leq b_v \tag{5-77}$$

$$\sum_i x_{ij}^k - \sum_j x_{ji}^k = \begin{cases} d^k & i = s^k \\ 0 & i \neq s^k, t \\ -d^k & i = t \end{cases} \tag{5-78}$$

$$0 \leq x_{ij}^k \leq u_{ij}^k \tag{5-79}$$

In this model, c_{ij}^k represents the utility associated with a particular tanker, shipment, and route. $\mathbf{A}_v$ represents the set of feasible routes for a given shipment; and r_{ij}^k represents the capacity of tanker type k over route (i, j). Thus, the total amount shipped cannot exceed the demand of the shipment. The remaining constraints represent conservation of flow on the number of tankers, and u_{ij}^k will either be 0 or ∞, depending on whether a particular tanker type is restricted from use or not. To illustrate, consider the problem in Table 5.14. Two tankers of each type are available. The network representation is given in Figure 5.40. Note that in this model class, nodes represent location and time combinations.

Table 5.14

Shipment	*Available at Time*	*Feasible Delivery Dates*
1	0	4, 5
2	2	6, 7, 8
3	3	7, 8

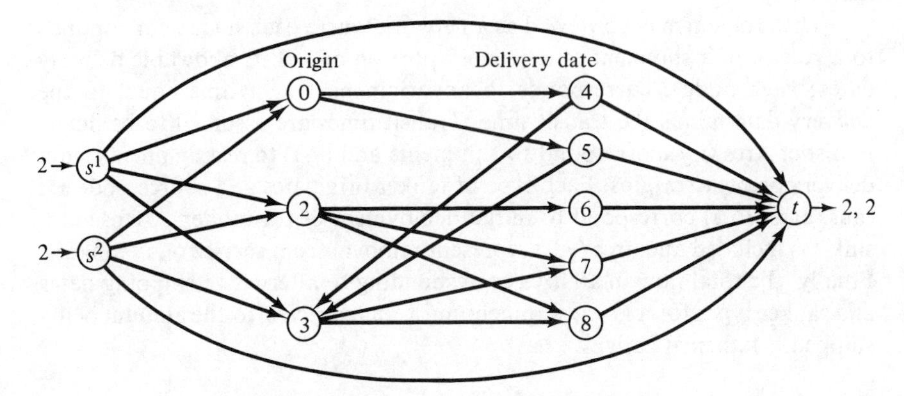

Figure 5.40. Tanker scheduling problem.

5.25.2 Urban Transportation Planning

Many multicommodity network models arise in planning urban transportation systems. Nodes represent zones or areas in a city while arcs correspond to streets or other types of roadways. Transportation demand is given by a trip matrix, **D**, in which d_{ij} denotes the number of vehicles that travel from zone i to zone j during a fixed time interval. Each arc has a fixed capacity u_{ij} which may be increased (by road improvements for example) at a cost of c_{ij}. Improvements are restricted by a total budget, B. "Commodities" are the flows from each origin to all destinations. Let x_{ij}^k be the number of vehicles on arc (i, j) originating at node k, y_{ij} the improvement in capacity, and let the travel time on arc (i, j) be some function of the flow:

$$f_{ij}(\sum_k x_{ij}^k)$$

The network design problem is to determine which arcs to improve, and to find the flows on each arc so that total travel time is minimized. Mathematically, we have

$$\text{minimize} \sum_{(i,j)} f_{ij}\left(\sum_k x_{ij}^k\right) \tag{5-80}$$

subject to

$$\sum_j x_{ji}^k - \sum_j x_{ij}^k = d_{ki} \tag{5-81}$$

$$\sum_k x_{ij}^k \leq u_{ij} + y_{ij} \tag{5-82}$$

$$\sum_{(i,j)} c_{ij} y_{ij} \leq B \tag{5-83}$$

$$x_{ij}^k, y_{ij} \geq 0 \tag{5-84}$$

The model formulated in (5-80) to (5-84) makes a critical assumption regarding user behavior. This is based on a classic principle of distribution of traffic presented by Wardrop [39]: users choose routes so that the sum of journey times over the system is minimal. Typically, the objective function (5-80) is nonlinear and convex.

5.25.3 Computer-Communication Models

Many applications in computer networks closely resemble the urban transportation model we have just discussed. In this case, arcs represent transmission lines and nodes represent terminals, systems, storage devices, and so on. Each arc has a fixed capacity u_{ij} which may be increased by y_{ij} units at a cost of c_{ij} up to a maximum b_{ij}. Commodities again represent flows (messages) from an origin to all destinations. The traffic rate matrix **D** represents the transmission rate d_{ij} from node i to node j. The unit cost of transmission over arc (i, j) is e_{ij}. The mathematical program may be formulated as follows:

$$\text{minimize} \sum_k \sum_{(i,j)} e_{ij} x_{ij}^k + \sum_{(i,j)} c_{ij} y_{ij} \tag{5-85}$$

subject to

$$\sum_j x_{ji}^k - \sum_j x_{ij}^k = d_{ij} \tag{5-86}$$

$$\sum_k x_{ij}^k \leq u_{ij} + y_{ij} \tag{5-87}$$

$$y_{ij} \leq b_{ij} \tag{5-88}$$

$$x_{ij}^k, y_{ij} \geq 0 \tag{5-89}$$

One use of this model is to determine the optimal capacity of a network that satisfies demand at least cost. In this case, e_{ij} is usually 0, $u_{ij} = 0$, and b_{ij} is very large. A second use is to consider a network with fixed capacities as already given, and to determine the optimal routing of messages to minimize cost. In this case, $b_{ij} = 0$.

5.26 Notes and Remarks

Surveys of multicommodity flow problems and various solution techniques have been written by Kennington [39] and Assad [40]. These also provide comprehensive bibliographies on the subject. Aspects of unimodularity in multicommodity flow problems and transformation to equivalent single-commodity problems were first proposed by Evans, Jarvis, and Duke [41], and extended by Evans and Jarvis [42] and Evans [43–45]. Aggregation results are generalizations of work performed by Geoffrion [46] and Zipkin [47] and were developed by Evans [48, 49]. Results for completely planar networks are due to Sakarovitch [50]. Hu [51] developed the algorithm for two-commodity flows in undirected networks. The funnel-node flow application is due to Jarvis and Miller [52]. Other applications in this section are discussed in Kennington's survey; specifically, the tanker scheduling problem is due to Bellmore, Bennington, and Lubore [53]. References to urban design and computer networks can be found in [39].

EXERCISES

1. A classical example of a multicommodity model is the route design problem with several trip-generation zones. Given a network $G = (\mathbf{N}, \mathbf{A})$ and certain original availabilities or requirements of each of L commodities at node i, designated as Q_{ik}, $k = 1, 2, \ldots, L$, find for each link (i, j) and each commodity k a flow x_{ijk} which would result in the minimization of the total route construction and travel cost. Assume that the cost of constructing the link (i, j) is p_{ij} and the variable cost per unit of flow over the arc (i, j) is c_{ij}.
2. For each of the following general types of problems, discuss possible interpretations of the nodes, the arcs, and the flows, in the context of a multicommodity network:
   (a) Freight distribution among cities by means of trucks, rail, air freight routes, and waterways.
   (b) Communication network synthesis with several message centers and several transmission lines.
   (c) Distribution of mail, including letters and parcels, among several post offices.
   (d) Production and inventory control with periodic supply and demand requirements.
   (e) Network design for collection, treatment, and discharge of solid waste disposal.
   (f) Regional transportation planning for passenger cars and hauling vehicles.

(g) Urban traffic planning.
(h) Mass transit design with several stations and several transit lines.

3. Two commodities, 1 and 2, can flow over the network shown. Assume that each commodity may flow along any arc in either direction. An unlimited supply of commodity 1 is available at nodes 1, 2, 3, and the final destination of this commodity is node 4. Additionally, an unlimited supply of commodity 2 is available at node 5, and the final destination of this commodity is node 1. The number on each arc represents the maximum total flow of both commodities allowed along the arc. Each unit of commodity 1 that reaches its final destination is worth 4, and each unit of commodity 2 that reaches its final destination is worth 6. The problem consists of identifying a feasible flow of commodities resulting in a maximum total value.

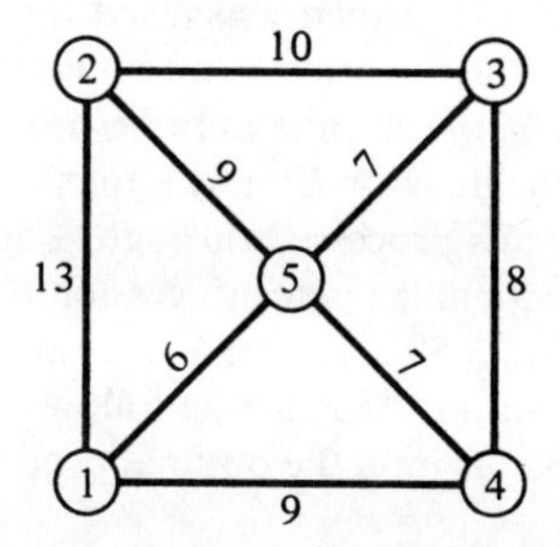

4. In meeting demand requirements, the standard transportation model does not distinguish among the supply centers. However, most multicommodity firms would determine a complete distribution system by taking account of the entire product line. Discuss how to use the regular single-commodity transportation model to solve a simple multicommodity model with a product mix required at each demand center. List all your assumptions.

5. Find the minimum-cost flow in the network shown with gains and losses. Each arc (i, j) is given four parameters, L_{ij}, U_{ij}, c_{ij}, and A_{ij}, representing the lower bound on flow, the upper bound on flow, the cost per unit of flow, and the gain or loss factor, respectively. *Hint*: Each arc with $L_{ij} > 0$ is equivalent to two arcs with parameters $(0, U_{ij} - L_{ij}, c_{ij})$ and $(0, L_{ij}, -\infty)$.

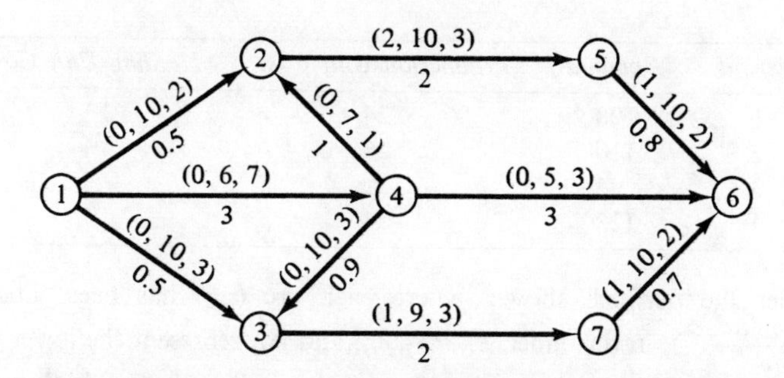

6. A financial secretary wants to establish an investment policy for a firm. The total amount to be invested is $1,000,000. There are six types of securities

with yields equal to 3.5%, 2.5%, 3.0%, 4.5%, 5.0%, and 4.0%, respectively. It is the firm's policy that at least 35% of the whole amount should be invested in either 1, or 2, or both, and that not more than 40% should be invested in either types 3 and 4 combined or types 5 and 6 combined. How should the money be invested?

7. In the manufacture of a certain product, three processes are involved. There are three machines, A, B, and C, which perform the operations of the first process. Similarly, there are two machines D and E for the second process, and three machines F, G, and H for the third process. All units processed through machine A must then be processed through machine D. All units processed through machine B are then processed through machine D or E. All units processed through machine C must then be processed through machine E. Units processed through D are then processed through either F or G. Units processed through E are then processed through either G or H. In any given week, a maximum of 50% of all units may be processed through A, B, or C; a maximum of 70% through D or E; and a maximum of 50% through F, G, or H. Only 95% of the units processed through A are acceptable and will continue to the next process; similar percentages for B, C, D, E, F, G, and H are 98%, 93%, 95%, 90%, 91%, 89%, and 94%, respectively. The costs for processing 1 unit through each machine are as follows: A, 10; B, 14; C, 8; D, 18; E, 14; F, 10; G, 9; and H, 11. Find the optimal production plan to manufacture 100 units per week.

8. A product can be manufactured in any of $T = 4$ time periods. Let D_t be the demand in period t, X_t the number of units produced in period t, c_t the unit cost in period t, h_t the holding cost from period t to period $t + 1$, and I_t the inventory level at the end of period t. Part I: (a) Construct the linear programming model which can be used to determine the minimum-cost production policy over the T periods. (b) Draw the equivalent network representation for this problem. Part II: Extend the basic model of part I to include backorders for stockouts in previous periods. Assume that the product can be backordered (delayed production) for only one period. Part III: Assume that 20% of the time a customer will cancel his order if items are backordered. Modify the network of part II to include this phenomenon. Solve the network problem of parts I, II, and III using the following data. Compare the results of parts II and III.

Period	*Demand*	*Production Unit Cost*	*Holding Unit Cost*
1	100	$24	$1
2	110	26	1
3	95	21	1
4	125	24	1

9. Consider the network shown, where each arc (i, j) has been labeled as $(i) \xrightarrow[A_{ij}]{(L_{ij}, U_{ij}, c_{ij})} (j)$; in this label, L_{ij}, U_{ij}, c_{ij}, and A_{ij} represent the lower bound, the upper bound, the cost per unit of flow, and the gain or loss multiplier, respectively. Find the best solution resulting in 5 units of flow at node 10.

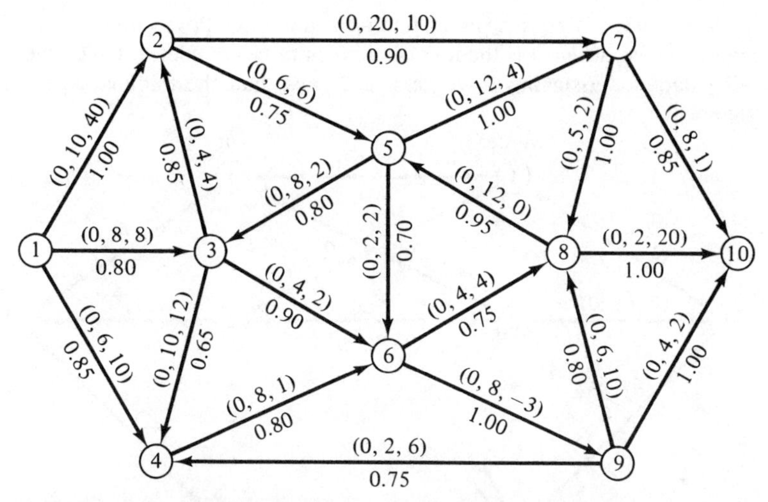

Initial network

10. We have $100,000 available for investment at the beginning of year 1, and $200,000 at the beginning of year 2. There is an investment opportunity at the beginning of each year 1, 2, and 3. Assume that it is possible to commit current cash to future investment by putting the money in a standard interim savings account with 5% interest. until the planned investment can be made. The amount of cash from any period that can be invested in each alternative should be at least 10,000 and at most 50,000. The return of the investment opportunities are 18%, 14%, and 10%, for years 1, 2, and 3, respectively. What is the investment policy that would maximize the total return?

11. The *generalized assignment* problem can be stated as follows. Assume that there are M workers who can be assigned to N jobs. A single worker is capable of performing all or a portion of any one job. In the latter case, the work on a single job must be partitioned among several (two or more) workers. Define the following notation.

(a) One hour of worker i is sufficient to complete a fraction K_{ij} of job j. Worker i cannot work more than U_i hours.

(b) Let X_{ij} = number of hours worker i works on job j
C_{ij} = cost per hour of worker i on job j

The problem is as follows:

$$\text{Minimize } \sum_i \sum_j C_{ij} X_{ij}$$

subject to:

$$\sum_j X_{ij} \leq U_i \qquad i = 1, 2, \ldots, M$$

$$\sum_i K_{ij} X_{ij} = 1 \qquad i = 1, 2, \ldots, N$$

$$X_{ij} \geq 0 \qquad \text{all } i, j$$

This formulation is equivalent to finding a minimum-flow value of J in the network shown, where J is the number of jobs to be completed. Given the following data for assigning 3 workers to 2 jobs, find the minimal-cost work assignment.

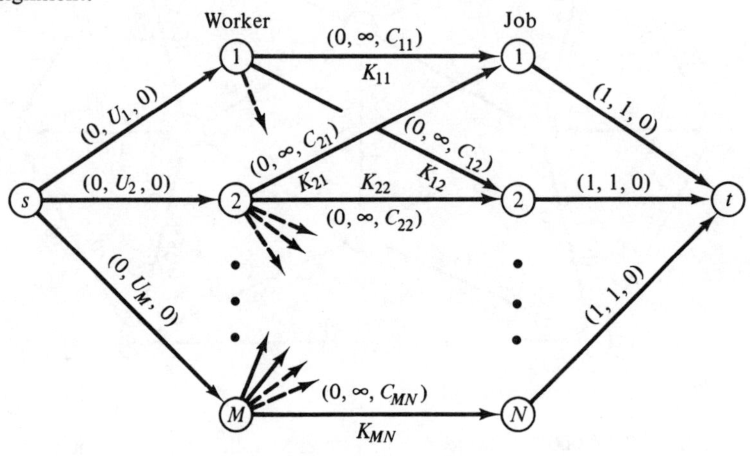

Worker	*Job*	*Cost*	*Gain Factor*
1	1	4	0.10
1	2	7	0.15
2	1	5	0.20
2	2	3	0.05
3	1	4	0.10
3	2	6	0.08

Worker	*Hours Available*
1	12
2	20
3	25

12. Solve the following generalized network problem using the loss/gain algorithm of Section 5.5.

Arc	*Lower*	*Upper*	*Cost*	*Arc Multiplier*
(1, 2)	0	3	1	4.0
(1, 3)	0	4	5	0.50
(2, 3)	0	3	4	1.00
(3, 2)	0	6	3	0.333
(2, 4)	0	5	1	0.25
(3, 4)	0	7	9	3.00

(a) Determine the input and costs necessary to supply 10 units of flow from node 4.
(b) Relate this problem to a conventional network flow model using the results of Section 5.4 and verify the solution to part (a).

13. Assume that a dart thrown at a board has a probability of hitting the target of 0.20. Develop the GERT network to find the expected number of throws to achieve two hits.

14. Donald-Do-Right has recently opened up a TV fix-it shop in College Station. Being an IE graduate from Texas A&M, he wishes to analyze his business. The procedure is as follows. Donald receives three basic types of jobs: (1) black-and-white portables, (2) black-and-white consoles, and (3) color sets. Recent experience indicates 42%, 31%, and 27% arrivals, respectively. The repair time for a B&W portable is exponential with mean 3 hours. A B&W console repair is exponential with mean 5 hours. He is not experienced in servicing color sets, and so he must send them to another shop to be repaired until he becomes more experienced. However, he does find that approximately one in three color sets have minor troubles, and so he repairs them immediately at a rate of 40 minutes 30% of the time, 55 minutes 30% of the time and 75 minutes 40% of the time. The outside contractor repairs sets according to a gamma probability law with a mean repair rate of 7 hours and a variance of 4 hours. Mr. Do-Right puts all his repairs on temporary test, and he has found that 10% of the outside contract jobs have to be returned for rework, and 5% of his jobs must be redone. The rework jobs take approximately $\frac{1}{2}$ of their original repair times and are also put on temporary test. Can you analyze this business?

15. Suppose that a poor merchant is cutting a diamond for King Richrocks and he requires a perfect cut for his crown. Any diamonds not perfectly cut will result in industrial tools. Suppose that the time to cut a diamond (in hours) is Poisson-distributed with mean equal to 18 hours. What is the expected time to cut and present to the King a diamond if the probability of success is 0.15 and delivery takes 1 hour?

16. In Exercise 15, suppose that the King is off fighting a war (as good kings do) and the diamond cutter is instructed to deliver the diamond personally. Suppose that the search time is exponential with mean equal to 9 hours. How long will total delivery take?

17. A steel ingot is being prepared for final processing at the Heavy Steel Company in Black Smoke, Texas. The raw steel comes from three different sources according to known probabilistic arrival percentages: $A = 50\%$, $B = 25\%$, and $C = 25\%$. The raw steel is melted in a cupola, and the processing time is normally distributed according to a normal density with $\mu = 25$ hours and $\sigma^2 = 9$ hours^2. A hot steel ingot is then sent to a cooling operation. The cooling operation takes 2.5 days for 30% of the ingots, 1.8 days for 15% of the ingots, and 2.7 days for 55% of the ingots. After cooling, the ingots are heat-treated. Heat-treatment times are exponentially distributed with $\mu = 46$ hours. After heat treating, an ingot is pulled and tested. 85% of the ingots pass testing and proceed to the cutting station. 15% are reprocessed, and the second heat-treatment time is exponentially distributed with mean of 14.6 hours. After the second heat treatment, the ingot is again reinspected. 95% of these ingots proceed to cutting, but 5% are rejected and sold as second-grade steel. All inspections take exactly 5 minutes by a computer scanner. After a cutting operation that requires an exponential time with $\mu = 6.2$ minutes, the new slabs (cut up ingots) are inspected: 85% are acceptable in present form; 10% are recut to enforce standards, and 5% are rejected as scrap.

(a) Draw the GERT network for this example.

(b) Construct the W-functions for each arc.

(c) What percentage of items starts as raw material A and B and wind up unacceptable slabs?

(d) What is the mean and variance of the time to produce a good slab?

18. Show that the MCTP defined in Eqs. (5-23) to (5-27) and the transportation problem given by Eqs. (5-40) to (5-46) are indeed equivalent; that is, any feasible solution to one is also a feasible solution to the other.

19. Compute the disaggregate solution to the fruit distribution problem in Table 5.11. Verify that it has the same cost as the aggregated solution.

20. Write down the constraints to the linear programming formulation of the multicommodity transshipment problem shown. What type of structure does the linear program have?

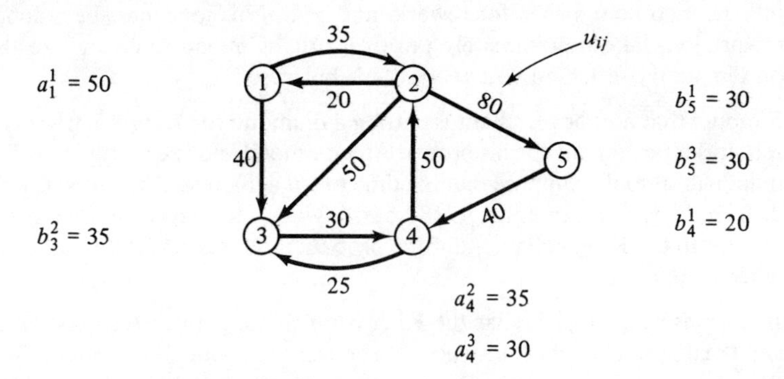

21. Convert the following problem to a single-commodity equivalent network problem and solve with the out-of-kilter algorithm. Use both formulations, that is, with and without capacitated arcs.

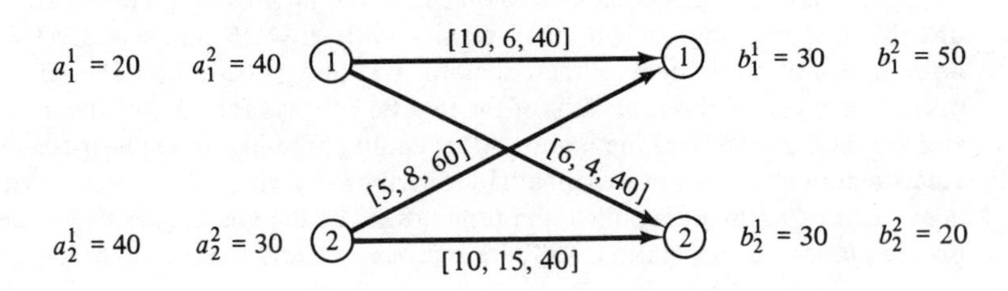

22. Resolve the fruit distribution problem by aggregating Santa Barbara and Bakersfield. Also try Santa Barbara and Sacramento. Which gives the best solution? Can you justify any reasons why?

23. Solve the multicommodity maximal-flow problem shown.

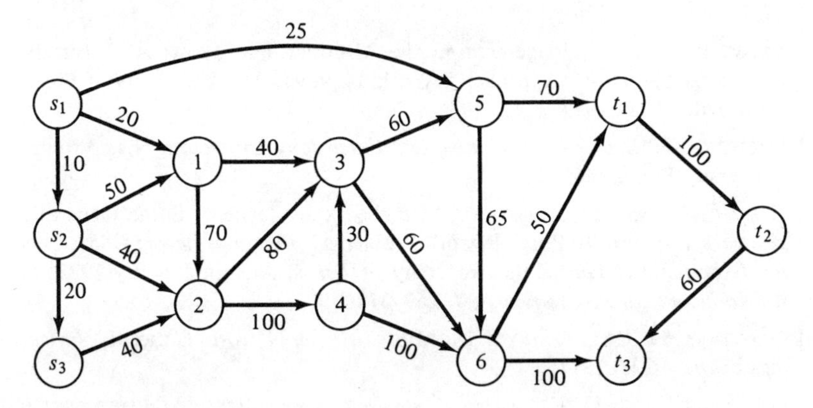

24. Find the maximal two-commodity flow.

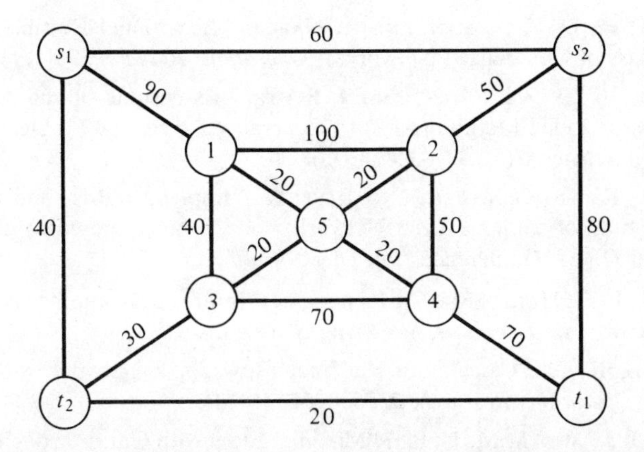

25. In the graph in Exercise 24, assume that s_1 and t_1 are the source and sink, respectively, for a single commodity and that node 5 is a funnel node. Find the maximal funnel-node flow.

REFERENCES

[1] BHAUMIK, G., "Optimum Operating Policies of a Water Distribution System with Losses." Unpublished Ph.D. dissertation, The University of Texas at Austin, 1973.

[2] CHARNES, A., N. KIRBY, AND W. RAIKE, "Chance-constrained Generalized Networks," *Operations Research*, *14*, 1113–1120 (1966).

[3] Charnes, A., and W. M. Raike, "One-Pass Algorithm for Some Generalized Network Problems," *Operations Research, 14*, 914–924 (1966).

[4] Charnes, A., and W. Cooper, *Management Models and Industrial Applications of Linear Programming*, Vols. 1 and 2. New York: John Wiley & Sons, Inc., 1961.

[5] Crum, R., "Cash Management in the Multinational Firm: A Constrained Generalized Network Approach," Working paper, The University of Florida, Gainesville, Fla., 1977.

[6] Dantzig, G., *Linear Programming and Extensions*. Princeton, N.J.: Princeton University Press, 1963.

[7] Glover, F., and D. Klingman, "On the Equivalence of Some Generalized Network Problems to Pure Network Problems." *Research Report CS81, Center for Cybernetic Studies*, The University of Texas, Austin, January 1972; also in *Mathematical Programming, 4*, 269–278 (1973).

[8] Glover, F., J. Hultz, and C. McMillan, "The Netform Concept," *Proceedings of the ACM*, 1977.

[9] Glover, F., and D. Klingman, "Network Applications in Government and Industry," *AIIE Transactions, 9*(4) (December 1977).

[10] Glover, F., D. Klingman, and A. Napier, "Basic Dual Feasible Solutions for a Class of Generalized Networks," *Operations Research, 20*(1) (1972).

[11] Glover, F., D. Klingman, and J. Stutz, "Extensions of the Augmented Predecessor Index Method (APS) to Generalized Network Problems," *Transportation Science, 7*(4), 377–384 (1973).

[12] Glover, F., D. Klingman, and J. Stutz, "Implementation and Computational Study of a Generalized Network Code," Paper presented at the 44th National ORSA Conference, San Diego, Calif., 1973.

[13] Glover, F., J. Hultz, and D. Klingman, "Improved Computer Based Planning Networks," *Interfaces, 8*(4) (August 1978).

[14] Grinold, R. C., "Calculating Maximal Flows in a Network with Positive Gains," *Operations Research, 21*, 528–541 (1973).

[15] Jarvis, J. J., and A. M. Jezior, "Maximal Flow with Gains through a Special Network," *Operations Research, 20*, 678–688 (1972).

[16] Jensen, P. A., and G. Bhaumik, "A Computationally Efficient Algorithm for the Network with Gains Problem," Paper presented at the 45th National Meeting of ORSA, Boston, April 1974.

[17] Jensen, P. A., and W. Barnes, *Network Flow Programming*. New York: John Wiley & Sons, Inc., 1980.

[18] Jewell, W. J., "Optimal Flows through Networks with Gains," *Operations Research, 10*, 476–499 (1962).

[19] Kim, Y., "An Optimal Computational Approach to the Analysis of a Generalized Network of Copper Refining Process," Joint ORSA/TIMS/AIIE Conference, Atlantic City, N.J., 1973.

[20] LAWLER, E. L., *Combinatorial Optimization: Networks and Matroids*. New York: Holt, Rinehart and Winston, 1976, pp. 134–138.

[21] MAURRAS, J. F., "Optimization of the Flow through Networks with Gains," *Mathematical Programming*, *3*, 135–144 (1972).

[22] MINIEKA, E. T., *Optimization Algorithms for Networks and Graphs*. New York: Marcel Dekker, Inc., 1978, pp. 151–174.

[23] MINIEKA, E. T., "Optimal Flow in a Network with Gains," *INFOR*, *10*, 171–178 (1972).

[24] OETTLI, W., AND W. PRAGER, "Flow Networks with Amplification and Coupling," *Unternehmensforschung*, *10*, 42–49 (1966).

[25] TRUEMPER, K., "On Max Flows with Gains and Pure Min. Cost Flows," *SIAM Journal of Applied Mathematics*, *32*, 450–456 (1973).

[26] TRUEMPER, K., "An Efficient Scaling Procedure for Gains Networks," *Networks*, *6*, 151–160 (1976).

[27] TRUEMPER, K., "Optimal Flows in Nonlinear Gains," *Networks*, *8*, 17–36 (1978).

[28] WAGNER, H., *Principles of Operations Research*. Englewood Cliffs, N.J.: Prentice-Hall, Inc., 1979.

[29] DUDLEY, N. A., "Work Time Distribution," *International Journal of Production Research*, *1*(2) (1963).

[30] KARGER, B. W., AND F. H. BAYTA, *Engineering Work Measurement*. New York: The Industrial Press, 1966.

[31] PHILLIPS, D. T. AND D. R. SMITH, "Determination of Probabilistic Time Standards for Tasks Performed Under Uncertainty," *Proceedings on the AIEE National Conference*, San Francisco, Ca., 1979.

[32] PHILLIPS, D. T., A. RAVINDRAN, AND J. J. SOLBERG, *Operations Research; Principles and Practice*. New York: John Wiley and Sons, Inc., 1976.

[33] PRITSKER AND ASSOCIATES, West Lafayette, Ind.—private consulting firm.

[34] PRITSKER, A. A. B., AND W. W. HAPP, "GERT: Graphical Evaluation and Review Technique, Part I, Fundamentals," *The Journal of Industrial Engineering* (May 1966).

[35] PRITSKER, A. A. B., AND G. E. WHITEHOUSE, "GERT: Graphical Evaluation and Review Technique, Part II," *Journal of Industrial Engineering* (June 1966).

[36] PRITSKER, A. A. B., *The Production Engineer*, pp. 499–506 (October 1968).

[37] WHITEHOUSE, GARY E., *System Analysis and Design Using Network Techniques*. Englewood Cliffs, N.J.: Prentice-Hall, Inc., 1973.

[38] WHITEHOUSE, G. E., AND A. A. B. PRITSKER, "GERT—Generating Functions, Conditional Distributions, Counters, Renewal Times, and Correlations," *AIIE Transactions* (March 1969).

[39] KENNINGTON, J. L., "A Survey of Linear Cost Multicommodity Network Flows," *Operations Research*, *26*, 209–236 (1978).

[40] ASSAD, A. A., "Multicommodity Network Flows—A Survey," *Networks 8*, 37–91 (1978).

[41] Evans, J. R., J. J. Jarvis, and R. A. Duke, "Graphic Matroids and the Multicommodity Transportation Problem," *Mathematical Programming*, *13*, 323–328 (1977).

[42] Evans, J. R., and J. J. Jarvis, "Network Topology and Integral Multicommodity Flow Problems," *Networks*, *8*, 107–119 (1978).

[43] Evans, J. R., "A Combinational Equivalence between a Class of Multicommodity Flow Problems and the Capacitated Transportation Problem," *Mathematical Programming*, *10*, 401–404 (1976).

[44] Evans, J. R., "A Single Commodity Transformation for Certain Multicommodity Networks," *Operations Research*, *26*, 673–680 (1978).

[45] Evans, J. R., "On Equivalent Formulations of Certain Multicommodity Networks as Single Commodity Flow Problems," *Mathematical Programming*, *15*, 92–99 (1978).

[46] Geoffrion, A., "Customer Aggregation in Distribution Modeling," Working Paper No. 259, Western Management Science Institute, University of California, Los Angeles, 1976.

[47] Zipkin, P. H., "Aggregation in Linear Programming," Ph.D. dissertation, Yale University, December 1977.

[48] Evans, J. R., "Solving Multicommodity Transportation Problems through Aggregation," ORSA/TIMS National Conference, Los Angeles, November 1978.

[49] Evans, J. R., "Model Simplification in Multicommodity Distribution Systems through Aggregation," *Proceedings, American Institute of Decision Sciences*, National Meeting, New Orleans, November 1979.

[50] Sakarovitch, M., "The Multicommodity Maximal Flow Problems," ORC 66–25, University of California, Berkeley, 1966.

[51] Hu, T. C., "Multicommodity Network Flows," *Operations Research*, *11*, 344–360 (1963).

[52] Jarvis, J. J., and D. D. Miller, "Maximal Funnel-Node Flow in an Undirected Network," *Operations Research*, *21*, 365–369 (1973).

[53] Bellmore, M., G. Bennington, and S. Lubore, "A Multivehicle Tanker Scheduling Problem," *Transporation Sciences*, *5*, 36–47 (1971).

APPENDIX

THE NETWORK OPTIMIZATION COMPUTER CODE

```
C***********************************************************************
C*                                                                     *
C*                     NETWORK OPTIMIZATION PACKAGE                    *
C*                                                                     *
C***********************************************************************
C*                                                                     *
C*       PURPOSE:                                                      *
C*                                                                     *
C*         TO OBTAIN OPTIMAL SOLUTIONS FOR VARIOUS DETERMINISTIC       *
C*         NETWORK PROBLEMS. THE FOLLOWING NETWORK PROBLEMS CAN BE     *
C*         SOLVED.                                                     *
C*                                                                     *
C*                 1) SHORTEST ROUTE PROBLEM                           *
C*                 2) MULTITERMINAL SHORTEST CHAINS PROBLEM            *
C*                 3) THE MULTITERMINAL MAXIMUM CAPACITY PROBLEM       *
C*                 4) THE MINIMAL SPANNING TREE PROBLEM                *
C*                 5) TRAVELLING SALESMAN PROBLEM                      *
C*                 6) MAX/FLOW PROBLEM                                 *
C*                 7) THE TRANSPORTATION PROBLEM                       *
C*                 8) THE ASSIGNMENT PROBLEM                           *
C*                 9) THE SHORTEST PATH TREE PROBLEM                   *
C*                10) THE TRANSHIPMENT PROBLEM                         *
C*                11) MIN/COST-MAX/FLOW PROBLEM                        *
C*                12) THE K SHORTEST ROUTE PROBLEM                     *
C*                13) GENERALIZED NETWORK MINIMIZATION                 *
C*                                                                     *
C*       USAGE:                                                        *
C*                                                                     *
C*         AT PRESENT THIS PACKAGE HANDLES PROBLEMS 1 THROUGH 6 AND 12 *
C*         WITH UP TO 50 NODES AND 50 ARCS. PROBLEMS 7 THROUGH 11 AND  *
C*         PROBLEM 13 ARE RESTRICTED TO NO MORE THAN 500 NODES AND     *
C*         500 ARCS. THIS NUMBER MAY BE INCREASED BY PROPER CHANGES    *
C*         IN DIMENSION STATEMENTS IN THE MAIN PROGRAM AND SUBROUTINES.*
C*                                                                     *
C*             NOTE: IF A SERIES OF PROBLEMS ARE AVAILABLE FOR THE     *
C*             COMPUTER, ALL OF THEM CAN BE HANDLED IN ONE RUN BY      *
C*             PLACING A "PEND" CARD (CARD SET 4) AFTER EACH PROBLEM,  *
C*             AND BY PLACING AN "EXIT" CARD (CARD SET 5) AS THE LAST  *
C*           CARD IN THE DATA SET. THE LIST OF ALGORITHMS WITH THEIR   *
C*           CALL NAMES AND SUBROUTINE NAMES IS GIVEN BELOW:           *
C*                                                                     *
C*           CALL NAME     SUBROUTINE NAME      ALGORITHM              *
C*                                                                     *
C*             DIJK          DIJKA                DIJKSTRA'S ALGORITHM *
C*                                                FOR SOLVING SHORTEST *
C*                                                ROUTE PROBLEM.       *
C*                                                                     *
C*             MTSC          MULTSH               ALGORITHM FOR SOLVING *
C*                                                MULTITERMINAL SHORTEST *
C*                                                CHAINS PROBLEM.      *
C*                                                                     *
C*             KSRT          KSHORT               ALGORITHM FOR FINDING *
C*                                                THE K SHORTEST PATH  *
C*                                                BETWEEN A GIVEN SOURCE *
C*                                                NODE AND A SINK      *
C*                                                                     *
C*             MSTR          MINSPA               ALGORITHM FOR SOLVING *
C*                                                MINIMAL SPANNING TREE *
C*                                                PROBLEM.             *
C*                                                                     *
```

```
C*                                                                     *
C*                                                                     *
C*                                                                     *
C*          MAXF        MAXFLO                  ALGORITHM FOR SOLVING  *
C*                                              MAX/FLOW PROBLEMS.     *
C*                                                                     *
C*          TSPR        TRASAL                  BRANCH AND BOUND       *
C*                                              ALGORITHM FOR SOLVING  *
C*                                              TRAVELLING SALESMAN    *
C*                                              PROBLEM.               *
C*                                                                     *
C*          OKAL        OKALG                   OUT-OF-KILTER          *
C*                                              ALGORITHM              *
C*                                                                     *
C*          MCRP        MAXCAP                  ALGORITHM FOR FINDING  *
C*                                              THE MAXIMUM CAPACITY   *
C*                                              ROUTE BETWEEN ALL      *
C*                                              PAIRS OF NODES         *
C*                                                                     *
C*          GNET        GENPAK                  ALGORITHM CODED BY     *
C*                                              JENSEN AND BHAUMIK FOR *
C*                                              SOLVING NETWORKS WITH  *
C*                                              GAINS AND LOSSES       *
C*                                                                     *
C*      DATA INPUT:                                                    *
C*                                                                     *
C*        THE INPUT DATA CONSISTS OF 5 SETS.                           *
C*                                                                     *
C*          SET 1: SINGLE CARD WITH THE CALL NAME OF THE ALGORITHM     *
C*                         - FORMAT(A4)                                *
C*          SET 2: THIS SET CONSISTS OF ONLY ONE CARD.                 *
C*          SINGLE CARD WHICH SPECIFIES THE NUMBER OF NODES AND        *
C*          THE NUMBER OF ARCS IN THE PROBLEM.                         *
C*          -FORMAT(2I10)                                              *
C*                                                                     *
C*          SET 3: THIS SET CONSISTS OF AS MANY CARDS AS THE NUMBER    *
C*          OF ARCS FOR THE PROBLEM                                    *
C*          *FOR PROBLEMS 1 THROUGH 5 THE FOLLOWING QUANTITIES ARE     *
C*          *READ FROM EACH CARD*                                      *
C*                                                                     *
C*                1) START NODE                                        *
C*                2) END NODE                                          *
C*                3) ARC VALUE                                         *
C*            -FORMAT(4X,I6,I10,F10.2)                                 *
C*                                                                     *
C*          *FOR PROBLEM 6 THE FOLLOWING QUANTITIES ARE READ*          *
C*          *FROM EACH CARD OF THIS SET.*                              *
C*                                                                     *
C*                1) START NODE                                        *
C*                2) END NODE                                          *
C*                3) CAPACITY OF THE ARC                               *
C*                4) INITIAL FLOW IN THE ARC (CONSERVATION OF FLOW     *
C*                   SHOULD BE PRESERVED) IF LEFT BLANK,ALL INITIAL    *
C*                   FLOWS WILL BE ZERO.                               *
C*              - FORMAT(4X,I6,I10,2F10.2)                             *
C*                                                                     *
C*          *FOR PROBLEMS 7 THROUGH 11 USING OUT-OF-KILTER             *
C*          *ALGORITHM THE FOLLOWING QUANTITIES ARE READ               *
C*          *FROM EACH CARD:                                           *
C*                                                                     *
C*                1) START NODE                                        *
```

```
C*               2) END NODE                                        *
C*               3) UPPER LIMIT ON ARC FLOW                         *
C*               4) LOWER LIMIT ON ARC FLOW                         *
C*               5) COST OF SHIPPING PER UNIT                       *
C*                  FORMAT(2I5,3F10.2)                              *
C*                                                                  *
C*        *FOR PROBLEM 12 THERE ARE THREE DISTINCT SECTIONS OF DATA *
C*        *INPUT IN THIS SET :                                      *
C*                                                                  *
C*         SECTION 1 :                                              *
C*            1) START NODE                                         *
C*            2) END NODE                                           *
C*            3) ARC LENGTH                                         *
C*             -FORMAT(3I10)                                        *
C*              A SINGLE CARD IS READ FOR EACH ARC IN THE NETWORK   *
C*              THE DATA IN THIS SECTION SHOULD BE PUNCHED IN       *
C*              INCREASING ORDER OF ENDING NODE.                    *
C*                                                                  *
C*        SECTION 2 :                                               *
C*           1) K,NS,IMAX                                           *
C*            -FORMAT(4X,3I5)                                       *
C*             K    =THE NUMBER OF DISTINCT PATH LENGTHS REQUIRED.  *
C*             IMAX =THE MAXIMUM NUMBER OF DOUBLE SWEEP ITERATIONS  *
C*                   ALLOWED.                                       *
C*             NS   =THE INITIAL NODE OF ALL K SHORTEST PATHS TO BE *
C*                   GENERATED.                                     *
C*                                                                  *
C*        SECTION 3 :                                               *
C*           1) NF,PMAX                                             *
C*            -FORMAT(4X,2I5)                                       *
C*             NF   =THE FINAL NODE OF ALL K SHORTEST PATHS TO BE   *
C*                   GENERATED.                                     *
C*             PMAX =THE MAXIMUM NUMBER OF PATHS TO BE GENERATED    *
C*                   BETWEEN NODES NS AND NF.                       *
C*                                                                  *
C*         NOTE: IF PROBLEM 12 IS TO BE SEQUENTIALLY REPEATED FOR   *
C*               DIFFERENT SOURCES AND/OR TERMINALS,THEN ONLY       *
C*               SECTIONS 2 AND 3 OF SET 4 SHOULD BE REPEATED.      *
C*                                                                  *
C*         FOR PROBLEM 13 THERE ARE TWO DISTINCT SECTIONS OF DATA   *
C*         *INPUT IN THIS SET:                                      *
C*                                                                  *
C*           SECTION 1 :                                            *
C*              1) START NODE                                       *
C*              2) END NODE                                         *
C*              3) LOWER BOUND ON FLOW                              *
C*              4) UPPER BOUND ON FLOW                              *
C*              5) SHIPPING COST/UNIT OF FLOW                       *
C*              6) GAIN FACTOR FOR THE ARC                          *
C*               -FORMAT(2I10,4F10.2)                               *
C*                A SINGLE CARD IS READ FOR EACH ARC IN THE NETWORK *
C*                                                                  *
C*           SECTION 2 :                                            *
C*                                                                  *
C*              1) SOURCE,SINK,IPRINT,OUTFLOW                       *
C*               -FORMAT(3I5,F10.2)                                 *
C*                IPRINT  =PRINT OPTION (0 FOR SHORT PRINT, 1 FOR   *
C*                         LONG PRINT, 2 FOR EXTRA LONG PRINT)      *
C*                OUTFLOW =DESIRED OUTPUT FLOW                      *
C*                                                                  *
C*        SET 4: FOR ALL PROBLEMS THIS SET CONSISTS OF ONLY ONE     *
```

```
C*          CARD AND INDICATES THE END OF A PROBLEM BY THE WORD            *
C*          'PEND'                                                         *
C*                FORMAT(A4)                                               *
C*                                                                         *
C*          SET 5: THIS SET CONSISTS OF ONLY ONE CARD AND INDICATES        *
C*          THE END OF NETWORK OPTIMIZATION (END OF ALL INPUT DATA)        *
C*          BY THE WORD 'EXIT'                                             *
C*                FORMAT(A4)                                               *
C*                                                                         *
C*        PACKAGE CONTENTS:                                                *
C*                                                                         *
C*          THIS PACKAGE CONSISTS OF 9 ALGORITHMS.                         *
C*                                                                         *
C*            1) DIJKSTRA'S ALGORITHM FOR FINDING THE SHORTEST             *
C*               CHAIN (ROUTE) FROM AN ORIGIN TO ALL OTHER POINTS          *
C*               IN A NETWORK                                              *
C*                                                                         *
C*            2) MULTITERMINAL SHORTEST CHAINS (ROUTES) ALGORITHM          *
C*               TO OBTAIN SHORTEST CHAINS BETWEEN EVERY PAIR OF           *
C*               NODES IN A NETWORK BY APPLYING TRIPLE OPERATION           *
C*                                                                         *
C*            3) MINIMAL SPANNING TREE ALGORITHM TO OBTAIN AN OPTIMAL      *
C*               SPANNING TREE FOR A NETWORK.                              *
C*                                                                         *
C*            4) MAX/FLOW ALGORITHM TO MAXIMIZE FLOW IN A CAPACITATED      *
C*               NETWORK BY LABELLING PROCEDURE.                           *
C*                                                                         *
C*            5) BRANCH AND BOUND ALGORITHM TO OBTAIN SOLUTIONS FOR        *
C*               TRAVELLING SALESMAN PROBLEMS                              *
C*                                                                         *
C*            6) OUT-OF-KILTER ALGORITHM TO OBTAIN OPTIMAL SOLUTIONS       *
C*               FOR TRANSPORTATION, TRANSHIPMENT, ASSIGNMENT ....ETC      *
C*               PROBLEMS.                                                 *
C*                                                                         *
C*            7) AN ALGORITHM DESIGNED TO USE A BRANCH AND TREE SEARCH     *
C*               FOR THE K SHORTEST PATHS BETWEEN ANY TWO NODES IN A       *
C*               NETWORK. CYCLES ARE PERMITTED.                            *
C*                                                                         *
C*            8) A MODIFIED TRIPLE OPERATION ALGORITHM SUGGESTED BY        *
C*               T.C. HU,WHICH FINDS THE MAXIMUM CAPACITY ROUTE            *
C*               BETWEEN ALL PAIRS OF NODES IN A GENERAL NETWORK.          *
C*               GIVEN A NETWORK WITH EACH ARC CAPACITATED (UPPER          *
C*               BOUNDED), A ROUTE BETWEEN NODE A AND B IS A MAXIMUM       *
C*               CAPACITY ROUTE IF THE MINIMUM ARC CAPACITY OF THIS        *
C*               ROUTE IS GREATER THAN THAT OF ANY OTHER ROUTE FROM A      *
C*               TO B.                                                     *
C*                                                                         *
C*            9) A MINIMUM COST FLOW LABELLING ALGORITHM TO SOLVE          *
C*               GENERALIZED NETWORKS WITH ARC MULTIPLIERS. (GAINS         *
C*               AND LOSSES). FLOW GENERATING CYCLES ARE PERMITTED.        *
C*               THE ALGORITHM IS ATTRIBUTED TO PAUL.A.JENSEN AND GORA     *
C*               BHAUMIK,AND IS FINITELY CONVERGENT.                       *
C*                                                                         *
C*        REFERENCES:                                                      *
C*                                                                         *
C*          E.W.DIJKSTRA, A NOTE ON TWO PROBLEMS IN CONNECTION WITH        *
C*          GRAPHS, NUM. MATH., 1,269-271,1959.                            *
C*                                                                         *
C*          T.C.HU, INTEGER PROGRAMMING AND NETWORK FLOWS,                 *
C*          ADDISON-WESLEY PUBLISHING CO., MASS.,1969.                     *
C*                                                                         *
```

```
C*          HILLIER,S.F.,LIEBERMAN,J.G.,OPERATIONS RESEARCH,2ND ED.,       *
C*          HOLDEN-DAY,INC.,CALIFORNIA,1974.                               *
C*                                                                         *
C*          HU,T.C., INTEGER PROGRAMMING AND NETWORK FLOWS,111-120,        *
C*          ADDISON-WESLEY PUBLISHING COMPANY,INC., MASSACHUSETTES,1969.*
C*                                                                         *
C*          LITTLE,J.D.C.,ET AL., AN ALGORITHM FOR THE TRAVELLING          *
C*          SALESMAN PROBLEM,OPERATIONS RESEARCH, 11(5),972-989,1963.      *
C*                                                                         *
C*          D.T.PHILLIPS AND P.A.JENSEN, NETWORK FLOW OPTIMIZATION         *
C*          WITH THE OUT-OF-KILTER ALGORITHM, RESEARCH MEMORANDUM          *
C*          NO.71-2, FEBRUARY 1971, PURDUE UNIVERSITY, W.LAFAYETTE         *
C*                                                                         *
C*          JENSEN,P.A. AND JORA BHAUMIK, "A FLOW AUGMENTATION APPROACH *
C*          TO THE NETWORK WITH GAINS MINIMUM COST FLOW PROBLEM",          *
C*          MANAGEMENT SCIENCE,VOL.23,NO.6,FEBRUARY,1979.                  *
C*                                                                         *
C*          DOUGLAS R. SHIER,'COMPUTATIONAL EXPERIENCE WITH AN             *
C*          ALGORITHM FOR FINDING THE K SHORTEST PATHS IN A NETWORK',      *
C*          J.OF RESEARCH OF THE NATIONAL BUREAU OF STANDARDS- VOL.78B, *
C*          NO.3,JULY-SEPT,1974.                                           *
C*                                                                         *
C*                                                                         *
C***************************************************************************
C***************************************************************************
C               MAIN PROGRAM
      DIMENSION NODE(50,50),A(50,50),LAB(50),LP(50),EPS(50),PL(50),
     1NICID(50,50),SCAN(50),TL(50)
      COMMON S(50,50),D(50,50),ROWUSD(50),COLUSD(50),TREE(50,2),
     1UP(50,50),NP(50),INFSTC(50,3),NODES,INF,LEVEL,M,N,MAX
      EQUIVALENCE (NODE,A,S),(LAB,LP,ROWUSD),(EPS,PL,COLUSD),(NICID,UP)
     1,(SCAN,TL,NP)
      INTEGER DIJK/'DIJK'/,MTSC/'MTSC'/,MSTR/'MSTR'/,MAXF/'MAXF'/,TSPR/
     1'TSPR'/,OKAL/'OKAL'/,EXIT/'EXIT'/,IPRO
      INTEGER PEND/'PEND'/,NPRO
      INTEGER KSRT/'KSRT'/,MCRP/'MCRP'/,GNET/'GNET'/
   60 READ(5,5)IPRO
    5 FORMAT (A4)
      IF(IPRO.EQ.EXIT) GO TO 90
      IF(IPRO.EQ.DIJK) GO TO 10
      IF(IPRO.EQ.MTSC) GO TO 15
      IF(IPRO.EQ.MSTR) GO TO 20
      IF(IPRO.EQ.MAXF) GO TO 25
      IF(IPRO.EQ.TSPR) GO TO 30
      IF(IPRO.EQ.OKAL) GO TO 40
      IF(IPRO.EQ.KSRT) GO TO 45
      IF(IPRO.EQ.MCRP) GO TO 50
      IF(IPRO.EQ.GNET) GO TO 55
      WRITE(6,2)
    2 FORMAT('UNIDENTIFIED ALGORITHM*****EXECUTION TERMINATED')
      GO TO 90
   10 CALL DIJKA
      GO TO 60
   15 CALL MULTSH
      GO TO 60
   20 CALL MINSPA
      GO TO 60
   25 CALL MAXFLO
      GO TO 60
   30 CALL TRASAL
      GO TO 60
```

```
   40 CALL OKALG
      READ(5,5) NPRO
      GO TO 60
   45 CALL KSHORT
      GO TO 60
   50 CALL MAXCAP
      GO TO 60
   55 CALL GENPAK
      READ(5,5) NPRO
      GO TO 60
   90 CONTINUE
      STOP
      END
C
      SUBROUTINE DIJKA
      DIMENSION NODE(50,50),A(50,50),LAB(50),LP(50),EPS(50),PL(50),
     1NICID(50,50),SCAN(50),TL(50)
      COMMON S(50,50),D(50,50),ROWUSD(50),COLUSD(50),FREE(50,2),
     1UP(50,50),NP(50),INFSTC(50,3),NODES,INF,LEVEL,M,N,MAX
      EQUIVALENCE (NODE,A,S),(LAB,LP,ROWUSD),(EPS,PL,COLUSD),(NICID,UP)
     1,(SCAN,TL,NP)
      INTEGER PEND/'PEND'/,NPRO
      LOGICAL LP
      READ(5,100)NODES
  100 FORMAT(2I10,F10.0)
C
C     INITIALIZES ARC DISTANCE MATRIX
C
      DO 1 I=1,NODES
      TL(I)=1.E10
      PL(I)=1.E10
      LP(I) = .FALSE.
      DO 1 J = 1,NODES
    1 D(I,J) = 1.E10
C
C     READ THE INPUT DATA
C
    2 READ(5,120) NPRO,I,J,VAL
  120 FORMAT(A4,I6,I10,F10.2)
      IF(NPRO.EQ.PEND) GO TO 10
    5 D(I,J) = VAL
      GO TO 2
   10 LP(1) = .TRUE.
      WRITE(6,500)
  500 FORMAT(1H1)
      WRITE(6,600)
  600 FORMAT(25X,'SHORTEST ROUTE PROBLEM'//)
      PL(1) =0.
      LASTLP = 1
C
C     PRINTS THE INPUT DATA
C
      WRITE(6,101)
  101 FORMAT(' ',7X,'DIJKSTRA S ALGORITHM TO FIND THE SHORTEST ROUTE'//'
     a         FROM AN ORIGIN TO ALL THE OTHER POINTS'//'         FROM NOD
     aE   TO NODE   IS A DISTANCE OF '//)
      DO 15 I = 1,NODES
      DO 15 J = 1,NODES
      IF(D(I,J).NE.1.E10) PRINT 200,I,J,D(I,J)
  200 FORMAT(5X,2I10,F10.2)
   15 CONTINUE
```

```
      DO 20 I= 2,NODES
C
C STEP A
C
      DO 16 J = 1,NODES
      IF(LP(J)) GO TO 16
      IF(PL(LASTLP) + D(LASTLP,J).LT.TL(J)) TL(J)=PL(LASTLP)+
     * D(LASTLP,J)
   16 CONTINUE
C
C STEP B
C
      MIN = 2
      XMIN = TL(2)
      DO 17 J = 3,NODES
      IF(TL(J).GT.XMIN) GO TO 17
      MIN = J
      XMIN = TL(J)
   17 CONTINUE
      PL(MIN) = XMIN
      TL(MIN) = 1.E10
      LP(MIN) = .TRUE.
   20 LASTLP = MIN
      DO 30 I = 2,NODES
   30 CALL TRACED(I)
   12 CONTINUE
      RETURN
      END
      SUBROUTINE TRACED(IT)
      DIMENSION NODE(50,50),A(50,50),LAB(50),LP(50),EPS(50),PL(50),
     1NICID(50,50),SCAN(50),TL(50)
      COMMON S(50,50),D(50,50),ROWUSD(50),COLUSD(50),TREE(50,2),
     1UP(50,50),NP(50),INFSTC(50,3),NODES,INF,LEVEL,M,N,MAX
      EQUIVALENCE (NODE,A,S),(LAB,LP,ROWUSD),(EPS,PL,COLUSD),(NICID,UP)
     1,(SCAN,TL,NP)
      INTEGER OPTPAT(50)
      EQUIVALENCE (OPTPAT,INFSTC(1,2))
      LOGICAL LP
      IX = IT
      OPTPAT(1) = IX
      K = 2
      PRINT 100, IX, PL(IX)
    1 DO 10 I = 1,NODES
      IF(IX.EQ.I) GO TO 10
      IF(PL(IX) - PL(I).EQ.D(I,IX)) GO TO 20
   10 CONTINUE
   20 OPTPAT(K) = I
      IF(I.EQ.1) GO TO 30
      IX = I
      K = K+1
      GO TO 1
   30 L = K-1
      DO 40 I = 1,L
      M = K-I
      N1 = OPTPAT(M+1)
      N2 = OPTPAT(M)
   40 PRINT 101, N1, N2, D(N1,N2)
      RETURN
  100 FORMAT(////,7X,'THE OPTIMAL ROUTE FROM NODE 1 TO NODE ',I5,
     @'HAS A SOLUTION OF ',F10.2,'AND'//,7X,'THIS IS ATTAINED BY THE
     *FOLLOWING ROUTE'//)
```

```
  101 FORMAT(' ',7X,'FROM NODE',I7,'TO NODE',I7,'IS A DISTANCE OF ',F10.
     92)
      END
C
      SUBROUTINE MULTSH
C
      DIMENSION NODE(50,50),A(50,50),LAB(50),LP(50),EPS(50),PL(50),
     1NICID(50,50),SCAN(50),TL(50)
      COMMON S(50,50),D(50,50),ROWUSD(50),COLUSD(50),TREE(50,2),
     1UP(50,50),NP(50),INFSTC(50,3),NODES,INF,LEVEL,M,N,MAX
      EQUIVALENCE (NODE,A,S),(LAB,LP,ROWUSD),(EPS,PL,COLUSD),(NICID,UP)
     1,(SCAN,TL,NP)
      INTEGER PEND/'PEND'/,NPRO
      IIN=5
      IOT=6
      N=0
      READ(IIN,400)NODES
  400 FORMAT(I10)
C
C     INITIALIZATION OF THE DISTANCE MATRIX.
C
      DO 1 L=1,NODES
      DO 1 M=1,NODES
      D(L,M)=9999999.99
      IF(L.EQ.M)  D(L,M)=0.0
    1 CONTINUE
C
C     DATA READ IN
C
    2 READ(IIN,3)  NPRO,I,J,B
      IF(NPRO.EQ.PEND)  GO TO 10
    3 FORMAT(A4,I6,I10,F10.2)
      D(I,J)=B
      IF(J.GT.N)  N=J
      IF(I.GT.N)  N=I
      GO TO 2
   10 CONTINUE
      WRITE(IOT,600)
  600 FORMAT('1',10X,' MULTITERMINAL SHORTEST CHAINS PROBLEM')
      WRITE(IOT,700)
  700 FORMAT('0',10X,'** ORIGINAL MATRIX**')
      CALL PRNTRX
C
C     INITIALIZATION OF THE NODES ON THE CHAINS MATRIX.
C
      DO 100 I=1,N
      DO 100 K=1,N
      NODE(I,K)=K
  100 CONTINUE
C
C     TRIPLE OPERATION TO UPDATE THE DISTANCE MATRIX.
C
      DO 20 J=1,N
      DO 20 I=1,N
      DO 20 K=1,N
      IF(I.EQ.J.OR.I.EQ.K.OR.J.EQ.K)  GO TO 20
      X=D(I,J)+D(J,K)
      Y=D(I,K)
      IF(Y.LE.X)  GO TO 20
      D(I,K)=X
C
```

```
C     UPDATING THE NODES ON THE CHAINS MATRIX.
C
      NODE(I,K)=NODE(I,J)
   20 CONTINUE
      WRITE(IOT,800)
  800 FORMAT('1',5X,'** SHORTEST DISTANCE MATRIX **')
      CALL PRNTRX
      WRITE(IOT,900)
  900 FORMAT('1',5X,'#MATRIX REPRESENTING THE NODES ON THE SHORTEST CHAI
     @NS')
      CALL PRNTRY
      RETURN
      END
C
      SUBROUTINE PRNTRX
      DIMENSION NODE(50,50),A(50,50),LAB(50),LP(50),EPS(50),PL(50),
     1NICID(50,50),SCAN(50),TL(50)
      COMMON S(50,50),D(50,50),ROWUSD(50),COLUSD(50),TREE(50,2),
     1UP(50,50),NP(50),INFSTC(50,3),NODES,INF,LEVEL,M,N,MAX
      EQUIVALENCE (NODE,A,S),(LAB,LP,ROWUSD),(EPS,PL,COLUSD),(NICID,UP)
     1,(SCAN,TL,NP)
      N1=1
      N2=5
   43 IF(N2-N) 45,45,44
   44 N2=N
   45 PRINT 911,(I,I=N1,N2)
  911 FORMAT(//,9X,5(I10,2X))
      WRITE(6,912)
  912 FORMAT(/)
      DO 48 L=1,N
   48 PRINT 913,L,(D(L,M),M=N1,N2)
      IF(N2-N) 52,55,55
   52 N1=N1+5
      N2=N2+5
      GO TO 43
  913 FORMAT(' ',5X,I2,2X,5(F10.2,2X))
   55 CONTINUE
      RETURN
      END
      SUBROUTINE PRNTRY
      DIMENSION NODE(50,50),A(50,50),LAB(50),LP(50),EPS(50),PL(50),
     1NICID(50,50),SCAN(50),TL(50)
      COMMON S(50,50),D(50,50),ROWUSD(50),COLUSD(50),TREE(50,2),
     1UP(50,50),NP(50),INFSTC(50,3),NODES,INF,LEVEL,M,N,MAX
      EQUIVALENCE (NODE,A,S),(LAB,LP,ROWUSD),(EPS,PL,COLUSD),(NICID,UP)
     1,(SCAN,TL,NP)
      N1=1
      N2=15
   43 IF(N2-N) 45,45,44
   44 N2=N
   45 PRINT 911,(I,I=N1,N2)
  911 FORMAT(//,10X,15(I3))
      WRITE(6,912)
  912 FORMAT(/)
      DO 48 L=1,N
   48 PRINT 913,L,(NODE(L,M),M=N1,N2)
      IF(N2-N) 52,55,55
   52 N1=N1+15
      N2=N2+15
      GO TO 43
  913 FORMAT(' ',4X,I2,3X,15(I3))
```

```
   55 CONTINUE
      RETURN
      END
      SUBROUTINE MINSPA
      DIMENSION NODE(50,50),A(50,50),LAB(50),LP(50),EPS(50),PL(50),
     1NICID(50,50),SCAN(50),TL(50)
      DIMENSION ICON(50)
      COMMON S(50,50),D(50,50),ROWUSD(50),COLUSD(50),TREE(50,2),
     1UP(50,50),NP(50),INFSTC(50,3),NODES,INF,LEVEL,M,N,MAX
      EQUIVALENCE (NODE,A,S),(LAB,LP,ROWUSD),(EPS,PL,COLUSD),(NICID,UP)
     1,(SCAN,TL,NP)
      EQUIVALENCE (ICON,UP)
      INTEGER PEND/'PEND'/,NPRO
      IIN=5
      READ(5,31) NODES
   31 FORMAT(I10)
C
C     INITIALIZATION AND REINITIALIZATION
C
      DO 1 I=1,NODES
      DO 1 J=1,NODES
      NICID(I,J)=0
    1 D(I,J)=0.0
      NN=0
C
C     DATA READ IN
C
  100 READ(IIN,2) NPRO,I,J,A1
    2 FORMAT(A4,I6,I10,F10.2)
      IF(NPRO.EQ.PEND) GO TO 1000
      NICID(I,J)=1
      D(I,J)=A1
      IF(J.GT.NN)NN=J
      GO TO 100
 1000 IOT=6
C
C     NODE ONE IS INCLUDED IN THE TREE TO START WITH
C
 1060 M=1
      ICON(1)=1
 1071 HOLD=999999.0
      IF(M.GT.NN) GO TO 1081
C
C     SEARCH FOR NEW CANDIDATES AND CONNECT THE BEST CANDIDATE
C
      DO 108 I=1,M
      K=ICON(I)
      DO 108 J=1,NN
      DO 1072 L=1,M
      IF(J.EQ.ICON(L)) GO TO 108
 1072 CONTINUE
      IF(K-J) 1073,108,1074
 1073 IF(NICID(K,J).EQ.0) GO TO 108
      IF(D(K,J).GE.HOLD) GO TO 108
      HOLD=D(K,J)
      GO TO 1075
 1074 IF(NICID(J,K).EQ.0) GO TO 108
      IF(D(J,K).GE.HOLD) GO TO 108
      HOLD=D(J,K)
 1075 IO=K
      JO=J
```

```
      GO TO 111
  110 EPS(I)=FLO(J,I)
  111 IF(LAB(NN).NE.0) GO TO 1145
  112 CONTINUE
      SCAN(J)=1
  114 CONTINUE
  113 CONTINUE
C
C     CHECK TO SEE IF ALL NODES ARE SCANNED
C
      DO 1131 I=1,NN
      IF(LAB(I).EQ.0) GO TO 1131
      IF(SCAN(I).NE.0) GO TO 1131
      GO TO 105
 1131 CONTINUE
      GO TO 990
C
C     SINK NODE LABELLED, INCREASE FLOW
C
 1145 ADD=EPS(NN)
  115 JK=LAB(NN)
      IF(LAB(NN).GT.0) GO TO 116
      FLO(JK,NN)=FLO(JK,NN)-ADD
      GO TO 117
  116 FLO(JK,NN)=FLO(JK,NN)+ADD
      NN=LAB(NN)
  117 IF(LAB(NN).EQ.9999) GO TO 1171
      GO TO 115
C
C     BACK AT THE BEGINNING, REINITIALIZE
C
 1171 NN=N
C
C     A FEASIBLE FLOW HAS BEEN FOUND PRINT SO.
C
      K=3
      GO TO 201
 2000 CONTINUE
C
C     REINITIALIZE THE REQUIRED
C
      DO 118 I=2,NN
      LAB(I)=0
  118 SCAN(I)=0
      SCAN(1)=0
      GO TO 105
  990 K=2
      NN=N
      GO TO 201
 3000 CONTINUE
      RETURN
      END
C
C
      SUBROUTINE OKALG
C
      DIMENSION I(500),J(500),HI(500),LO(500),FLOW(500),PI(500),COST
     @(500)
      COMMON S(50,50),D(50,50),ROWUSD(50),COLUSD(50),TREE(50,2),
     1UP(50,50),NP(50),INFSTC(50,3),NODES,INF,LEVEL,M,N,MAX
      EQUIVALENCE(I,S),(J,S(501)),(HI,S(1001)),(LO,S(1501)),
```

```
      CAP(I,J)=A1
      FLO(I,J)=A2
      IF(J.GT.NN)NN=J
      GO TO 102
  104 LAB(1)=9999.0
      EPS(1)=9999.0
      N=NN
      K=1
  201 WRITE(IOT,3)
      GO TO (10,20,30) ,K
   10 WRITE(IOT,4)
      GO TO 991
   20 WRITE(IOT,5)
      GO TO 991
   30 WRITE(IOT,8)
  991 FLOMAX=0.0
      WRITE(IOT,6)
      DO 992 I=1,NN
      DO 993 J=1,NN
      IF(NICID(I,J).EQ.0) GO TO 993
      WRITE(IOT,7) I,J,CAP(I,J),FLO(I,J)
      IF(J.EQ.NN) FLOMAX=FLOMAX+FLO(I,J)
  993 CONTINUE
  992 CONTINUE
      WRITE(IOT,1) FLOMAX
    1 FORMAT('0',5X,'MAXIMUM FLOW THROUGH THE ABOVE ARCS IS =',F14.2)
    3 FORMAT('1',30X,'MAX/FLOW PROBLEM')
    4 FORMAT('0',5X,'STATUS OF NETWORK AT START')
    5 FORMAT('0',5X,'STATUS OF NETWORK AT OPTIMAL SOLUTION')
    6 FORMAT('0',5X,'STARTING NODE',2X,'ENDING NODE',2X,'CAPACITY',
     14X,'FLOW IN ARC')
    7 FORMAT(1H0,3X,I10,3X,I10,F10.2,3X,F10.2)
    8 FORMAT('0',5X,'STATUS OF NETWORK AT A FEASIBLE SOLUTION')
      IF(K.EQ.2) GO TO 3000
      IF(K.EQ.3) GO TO 2000
  105 DO 113 J=1,NN
C
C     FIND A LABELED NODE.
C
  106 IF(LAB(J).EQ.0) GO TO 114
C
C     FOUND A LABELED NODE, IS IT SCANNED?, IF 'YES' CONTINUE,IF NO, SCAN IT.
C
      IF(SCAN(J).NE.0) GO TO 114
C     FOUND LABELLED,UNSCANNED NODE, SCAN IT .
  108 DO 112 I=1,NN
      IF(NICID(J,I).EQ.0) GO TO 112
  109 IF(LAB(I).NE.0) GO TO 112
      PS=CAP(J,I)-FLO(J,I)
      IF(PS.GT.0.0) GO TO 1091
      IF(NICID(I,J).EQ.0) GO TO 112
      IF(FLO(I,J).GE.0.0) GO TO 112
      LAB(I)=-J
      IF(FLO(J,I).LT.EPS(J)) GO TO 110
      EPS(I)=EPS(J)
      GO TO 111
 1091 LAB(I)=J
      IF(PS.LT.EPS(J)) GO TO 1092
      EPS(I)=EPS(J)
      GO TO 111
 1092 EPS(I)=PS
```

```
  108 CONTINUE
      M=M+1
      IF(IO.GT.JO) GO TO 1082
C
C     THE CONNECTED ARC IS INDEXED
C
      NICID(IO,JO)=2
 1082 NICID(JO,IO)=2
 1083 ICON(M)=JO
      GO TO 1071
 1081 CONTINUE
C
C     OUTPUT ROUTINE FOR MINIMAL SPANNING TREE
C
  300 WRITE(IOT,331)
  331 FORMAT('1',20X,'MINIMAL SPANNING TREE PROBLEM')
      WRITE(IOT,431)
  431 FORMAT('0',5X,' THE MINIMAL SPANNING TREE CONSISTS OF  THE FOLLOWI
     @NG ARCS')
      WRITE(IOT,34)
   34 FORMAT('0',5X,'STARTING NODE',2X,'ENDING NODE',2X,'DISTANCE')
      DO 302 I=1,NN
      DO 302 J=1,NN
      IF(I.GT.J) GO TO 302
      IF(NICID(I,J).EQ.2) WRITE(IOT,7) I,J,  D(I,J)
  302 CONTINUE
    7 FORMAT(1H0,3X,I10,3X,I10,F10.2,3X,F10.2)
      RETURN
      END
C
      SUBROUTINE MAXFLO
      DIMENSION FLO(50,50),CAP(50,50)
      DIMENSION NODE(50,50),A(50,50),LAB(50),LP(50),EPS(50),PL(50),
     1NICID(50,50),SCAN(50),TL(50)
      COMMON S(50,50),D(50,50),ROWUSD(50),COLUSD(50),TREE(50,2),
     1UP(50,50),NP(50),INFSTC(50,3),NODES,INF,LEVEL,M,N,MAX
      EQUIVALENCE (NODE,A,S),(LAB,LP,ROWUSD),(EPS,PL,COLUSD),(NICID,UP)
     1,(SCAN,TL,NP)
      EQUIVALENCE (FLO,S),(CAP,D)
      INTEGER SCAN,NICID
      INTEGER PEND/'PEND'/,NPRO
      IIN=5
      IOT=6
C
C     INITIALIZATION
C
      READ(5,100) N
  100 FORMAT(I10)
      DO 101 I=1,N
      LAB(I)=0
      EPS(I)=0
      SCAN(I)=0
      DO 101 J=1,N
      CAP(I,J)=0.0
      FLO (I,J)=0.0
  101 NICID(I,J)=0
      NN=0
  102 READ(IIN,9) NPRO,I,J,A1,A2
    9 FORMAT(A4,I6,I10,2F10.2)
      IF(NPRO.EQ.PEND) GO TO 104
  103 NICID(I,J)=1
```

```
     @(FLOW,S(2001)),(PI,D),(COST,D(501))
      INTEGER ARCS, FLOW, PI, COST, HI
      LOGICAL INFES
  100 FORMAT(2I10)
C
C     READS THE TOTAL NUMBER OF NODES AND ARCS FOR THE PROBLEM
C
      READ 100,NODES,ARCS
C
C     READS THE UPPER LIMIT, LOWER LIMIT AND COST FOR EACH
C     OF THE ARCS.
C
      DO 1 M=1,ARCS
      READ (5,250)I(M),J(M),T1,T2,T3
      HI(M)=T1+0.00001
      LO(M)=T2+0.00001
      COST(M)=T3+0.00001
    1 CONTINUE
  250 FORMAT(2I5,3F10.2)
C
C     INITIALIZES ARC FLOWS
C
       DO 5 M=1,ARCS
    5 FLOW(M)=0
C
C     INITIALIZES PI VALUES
C
      DO 10 M=1,NODES
   10 PI(M)=0
C
C     CALLS THE SUBROUTINE WHICH EVALUTES THE FEASIBLE NETFLOW
C          FOR EACH ARC
C
      CALL NETFLO(ARCS,INFES)
      IF(.NOT.INFES) WRITE(6,120)
C
C     PRINTS THE FINAL SUMMARY REPORT
C
      WRITE(6,801)
  801 FORMAT('1',12X,'****SOLUTION BY OUT-OF-KILTER ALGORITHM****')
      WRITE(6,112)
  112 FORMAT(//,20X,'FINAL SUMMARY REPORT ',///)
      WRITE(6,115) NODES,ARCS,(M,I(M),J(M),HI(M),LO(M),FLOW(M),COST(M),M=
     @1,ARCS)
  115 FORMAT(' ',8X,'NUMBER OF NODES =',I5,/,' ',8X,'NUMBER OF ARCS  =',
     1I5,///,12X,'M',5X,'I',5X,'J',10X,'HI',11X,'LO',10X,'FLOW',10X,'COS
     2T',///,(7X,3(2X,I4),4(3X,I10)))
      TCOST=0.0
      DO 140 M=1,ARCS
      TCOST=COST(M)*FLOW(M)+TCOST
  140 CONTINUE
      WRITE(6,150) TCOST
  150 FORMAT(//,'              TOTAL PROJECT COST =',F10.2)
      WRITE(6,125) (M,FLOW(M),M=1,ARCS)
      WRITE(6,130) (M,PI(M),M=1,NODES)
  120 FORMAT('  SOLUTION INFEASIBLE')
  125 FORMAT(///,5X,'  ARC             FLOW(ARC)',/,(5X,I4,6X,I10))
  130 FORMAT(///,5X,'  NODE       PI(NODE)',/,(4X,I4,2X,I10))
      WRITE(6,750)
  750 FORMAT(//,5X,'   SENSITIVITY ANALYSIS',///)
      WRITE(6,753)
```

```
753 FORMAT(12X,'M',5X,'I',5X,'J',10X,'HI',10X,'LO',10X,'FLOW',10X,'COS
   @T',//)
    DO 751 M=1,ARCS
    IZ=FLOW(M)
    IF(IZ.EQ.0)GO TO 751
    WRITE(6,752)M,I(M),J(M),HI(M),LO(M),FLOW(M),COST(M)
752 FORMAT(7X,3(2X,I4),4(3X,I10))
751 CONTINUE
    RETURN
    END
    SUBROUTINE NETFLO(ARCS,INFES)
    DIMENSION I(500),J(500),HI(500),LO(500),FLOW(500),PI(500),COST
   @(500),NA(500),NB(500)
    COMMON S(50,50),D(50,50),ROWUSD(50),COLUSD(50),TREE(50,2),
   1UP(50,50),NP(50),INFSTC(50,3),NODES,INF,LEVEL,M,N,MAX
    EQUIVALENCE(I,S),(J,S(501)),(HI,S(1001)),(LO,S(1501)),
   @(FLOW,S(2001)),(PI,D),(COST,D(501)),(NA,D(1001)),(NB,D(1501))
    LOGICAL INFES
    INTEGER A,AOK,C,COK,DEL,E,EPS,INF,LAB,N,NI,NJ,SRC,SNK,
   1FLOW,PI,NA, NODES,ARCS,I,J,COST,HI,LO,NB
C
C   CHECK FEASIBILITY OF FORMULATION
C
        INFES=.TRUE.
    DO 10 A=1,ARCS
    IF(LO(A).GT.HI(A)) GO TO 39
 10 CONTINUE
C
C   SET INF TO MAX AVAILABLE INTEGER
C
 16     INF=999999
        AOK=0
C
C   FIND OUT OF KILTER ARC
C
 20 DO 21 A=1,ARCS
    IA = I(A)
    JA = J(A)
      C=COST(A)+PI( IA )-PI( JA )
C
C   CHECKS THE CONDITIONS THE INDIVIDUAL ARCS ARE IN.
C
    IF((FLOW(A).LT.LO(A)).OR.(C.LT.0.AND.FLOW(A).LT.HI(A))) GO TO 22
    IF((FLOW(A).GT.HI(A)).OR.(C.GT.0.AND.FLOW(A).GT.LO(A))) GO TO 23
 21 CONTINUE
C
C   NO REMAINING OUT OF KILTER ARCS
C
    GO TO 38
 22     SRC=J(A)
        SNK=I(A)
        E=+1
    GO TO 24
 23     SRC=I(A)
        SNK=J(A)
        E=-1
    GO TO 24
 24 DO 99 N=1,NODES
        NA(N) = 0
        NB(N) =0
 99 CONTINUE
```

```
      IF((A.EQ.AOK).AND.(NA(SRC).NE.0)) GO TO 25
C
C     ATTEMPT TO BRING OUT OF KILTER ARCS INTO KILTER
C
         AOK=A
      DO 26 N=1,NODES
         NA(SRC)=IABS(SNK)*E
   26    NB(SRC)=IABS(AOK)*E
   25    COK=C
   27    LAB=0
      DO 30 A=1,ARCS
      IA = I(A)
      JA = J(A)
      IF((NA(IA).EQ.0.AND.NA(JA).EQ.0).OR.(NA(IA).NE.0..AND.NA(JA).NE.0)
     1)GO TO 30
      C= COST(A)+PI(IA) - PI(JA)
      IF(NA(IA).EQ.0) GO TO 28
      IF(FLOW(A).GE.HI(A).OR.(FLOW(A).GE.LO(A).AND.C.GT.0))GO TO 30
      NA(JA) = I(A)
      NB(JA) = A
      GO TO 29
   28 IF(FLOW(A).LE.LO(A).OR.(FLOW(A).LE.HI(A).AND.C.LT.0))GO TO 30
      IA = I(A)
      NA(IA) = -J(A)
      NB(IA) = -A
   29 LAB = 1
C
C NODE LABELED, TEST FOR BREAKTHRU
C
      IF(NA(SNK).NE.0) GO TO 33
   30 CONTINUE
C
C NO BREAKTHRU
C
      IF(LAB.NE.0) GO TO 27
C DETERMINE CHANGE TO PI VECTOR
      DEL = INF
      DO 31 A=1,ARCS
      IA = I(A)
      JA = J(A)
      IF((NA(IA).EQ.0.AND.NA(JA).EQ.0).OR.(NA(IA).NE.0.AND.NA(JA).NE.0))
     1GO TO 31
      C=COST(A)+PI(IA)-PI(JA)
      IF(NA(JA).EQ.0.AND.FLOW(A).LT.HI(A)) DEL= MIN0(DEL,C)
      IF(NA(JA).NE.0.AND.FLOW(A).GT.LO(A)) DEL= MIN0(DEL,-C)
   31 CONTINUE
      IF(DEL.EQ.INF.AND.(FLOW(AOK).EQ.HI(AOK).OR.FLOW(AOK).EQ.LO(AOK)))
     1DEL=IABS(COK)
      IF(DEL.EQ.INF) GO TO 39
C
C EXIT, NO FEASIBLE FLOW PATTERN
C CHANGE PI VECTOR BY COMPUTED DEL
C
      DO 32 N=1,NODES
   32 IF(NA(N).EQ.0) PI(N)=PI(N)+DEL
C
C FIND ANOTHER OUT-OF-KILTER ARC
C
      GO TO 20
C
C BREAKTHRU  COMPUTE INCREMENTAL FLOW
```

```
C
   33 EPS=INF
      NI=SRC
   34 NJ=IABS(NA(NI))
      A=IABS(NB(NI))
      C=COST(A)-ISIGN(IABS(PI(NI) -PI(NJ)),NB(NI))
      IF(NB(NI).LT.0) GO TO 35
      IF(C.GT.0.AND.FLOW(A).LT.LO(A)) EPS=MIN0(EPS,LO(A)-FLOW(A))
      IF(C.LE.0.AND.FLOW(A).LT.HI(A)) EPS=MIN0(EPS,HI(A)-FLOW(A))
      GO TO 36
   35 IF(C.LT.0.AND.FLOW(A).GT.HI(A)) EPS=MIN0(EPS,FLOW(A)-HI(A))
      IF(C.GE.0.AND.FLOW(A).GT.LO(A)) EPS=MIN0(EPS,FLOW(A)-LO(A))
   36 NI=NJ
      IF(NI.NE.SRC) GO TO 34
C
C CHANGE FLOW VECTOR BY COMPUTED EPS
C
   37 NJ=IABS(NA(NI))
      A=IABS(NB(NI))
      FLOW(A)=FLOW(A)+ISIGN(EPS,NB(NI))
      NI=NJ
      IF(NI.NE.SRC) GO TO 37
C
C FIND ANOTHER OUT OF KILTER ARC
C
      AOK=0
      GO TO 20
   39 INFES = .FALSE.
   38 CONTINUE
      RETURN
      END
      SUBROUTINE TRASAL
      COMMON S(50,50),D(50,50),ROWUSD(50),COLUSD(50),TREE(50,2),
     1UP(50,50),NP(50),INFSTC(50,3),NODES,INF,LEVEL,M,N,MAX
      INTEGER S,D,TREE,UP
      INTEGER PEND/'PEND'/,NPRO
      LOGICAL ROWUSD,COLUSD,FLAG
C
C     INITIALIZE
C
      INF=99999999
      MING=INF
      READ(5,32)NODES
   32 FORMAT(I10)
      DO 1 I=1,NODES
      INFSTC(I,1)=INF
      ROWUSD(I)=.FALSE.
      COLUSD(I)=.FALSE.
      TREE(I,1)=0
      TREE(I,2)=0
      DO 1 J=1,NODES
      UP(I,J)=0
      S(I,J)=INF
1     D(I,J)=INF
C
C          GET DATA
C
    2 READ(5,104) NPRO,I,J,VAL
      IF(NPRO.EQ.PEND) GO TO 11
    5 D(I,J)=VAL
      S(I,J)=VAL
```

```
      GO TO 2
   11 LEVEL=1
      WRITE(6,106)
  106 FORMAT('1',15X,'TRAVELLING SALESMAN PROBLEM****BY BRANCH AND BOUND
     1',////)
      WRITE(6,99)
C
      CALL PRTRXS
C
C
C           REDUCE THE MATRIX TO OBTAIN A LOWER BOUND ON THE SOLUTION
C
      MIN= MINCOL(I)+MINROW(I)
   13 M=0
      N=0
      MAX=0
C
C           THIS LOOP FINDS THE ARC OF VALUE 0  THAT ALLOWS
C           MAXIMUM SAVINGS
C
      DO 31 I=1,NODES
      IF(ROWUSD(I)) GO TO 31
      DO 20 J=1,NODES
      IF(COLUSD(J)) GO TO 20
      IF(D(I,J).NE.0) GO TO 20
      IF(M.NE.0) GO TO 17
      M=I
      N=J
   17 MT=INF
      DO 18 K=1,NODES
      IF(ROWUSD(K)) GO TO 18
      IF(D(K,J).GE.MT) GO TO 18
      IF(K.EQ.I) GO TO 18
      MT=D(K,J)
   18 CONTINUE
      MAXT=MT
      MT=INF
      DO 19 K=1,NODES
      IF(COLUSD(K)) GO TO 19
      IF(D(I,K).GE.MT) GO TO 19
      IF(K.EQ.J) GO TO 19
      MT=D(I,K)
   19 CONTINUE
      MAXT=MAXT+MT
      IF(MAXT.LE.MAX) GO TO 20
      MAX= MAXT
      M=I
      N=J
   20 CONTINUE
   31 CONTINUE
C
C           INDICATE THAT THE OPTIMUM ARC FROM M TO N IS IN THE SOLUTION
C
      ROWUSD(M)=.TRUE.
      COLUSD(N)=.TRUE.
      UP(M,N)= LEVEL
C
C           IF THE MATRIX HAS BEEN REDUCED TO A 2X2, NO MODIFICATION IS
C           NEEDED. HOWEVER, IF NOT, THE MATRIX MUST BE ALTERED SO THAT
C           CYCLES WILL NOT OCCUR, CAUSING A PREMATURE RETURN TO THE
C           ORIGINAL NODE
```

```
C
      IF(LEVEL.GE.(NODES-1)) GO TO 30
C
C           FIRST, THE CHAIN CURRENTLY BEING ADDED TO THE SOLUTION MUST BE
C           EXTENDED FORWARD AS FAR AS POSSIBLE  VIA THE ARCS  ALREADY IN
C           THE SOLUTION.
C
      L=N
   21 DO 22 I=1,NODES
      IF(UP(L,I).EQ.0) GO TO 22
      GO TO 23
   22 CONTINUE
      GO TO 25
   23 L=I
      GO TO 21
C
C           NEXT THE ARC CURRENTLY BEING ADDED  TO THE SOLUTION MUST BE
C           EXTENDED BACKWARD AS FAR AS POSSIBLE VIA THE ARCS ALREADY IN
C           THE SOLUTION.
C
   25 K=M
   26 DO 27 I=1,NODES
      IF(UP(I,K).EQ.0) GO TO 27
      GO TO 28
   27 CONTINUE
      GO TO 29
   28 K=I
      GO TO 26
C
C           AT THIS POINT , IF AN ARC FROM L TO K IS ADDED TO THE SOLUTION
C           A CYCLE WILL OCCUR, SO WE MUST BLOCK THAT ARC FROM ENTERING
C           THE SOLUTION
C
   29 D(L,K)=INF
C
C          WE MUST SET UP A TREE, THE LEFT HAND BRANCH (1) HAS THE VALUE
C          OF THE LOWER BOUND ON THE SOLUTION IF ARC (M,N) IS NOT USED
C          THE RIGHT HAND BRANCH (2) HAS THE VALUE OF THE LOWER BOUND
C          ON THE SOLUTION IF ARC (M,N) IS IN THE SOLUTION.
C
C
C           INDICATE THAT THE LEFT HAND BRANCH OF THE TREE HAS A VALUE
C           OF MIN+MAX, THAT IS, THE CURRENT LOWER BOUND PLUS THE
C           SAVINGS FROM USING ARC(M,N).
C
   30 TREE(LEVEL,1)=MIN+MAX
C
C          REDUCE THE MATRIX FURTHER(IF POSSIBLE) AND COMPUTE THE NEW
C          LOWER BOUND.
C
      MIN=MIN+MINCOL(I)+MINROW(I)
C
C           SET THE RIGHT HAND SIDE OF THE TREE TO THE NEW LOWER BOUND
C
      TREE(LEVEL,2)=MIN
      LEVEL=LEVEL+1
C
C           IF THE LOWER BOUND CURRENTLY IN EFFECT IS ALREADY GREATER
C           THAN THE GLOBAL MINIMUM ALREADY FOUND , THEN WE DO NOT
C           NEED TO PROCEED ANY FURTHER.
C
```

```
      IF(MIN.GE.MING) GO TO 60
C
C         IF THE MATRIX IS NOT FULLY ELIMINATED, GO BACK AND DO MORE
C         WORK
C
      IF(LEVEL.LE.NODES) GO TO 13
C
C        THE SOLUTION JUST FOUND IS BETTER THAN OUR PREVIOUS GLOBAL
C        MINIMUM. SO WE RESET MING AND SAVE THE ROUTE THAT WE TOOK.
C
C
      MING=MIN
      IS=1
      DO 50 I=1,NODES
      NP(I)=IS
      DO 45 J=1,NODES
      IF(UP(IS,J).EQ.0) GO TO 45
      GO TO 50
   45 CONTINUE
      PRINT 102,I,J,IS
   50 IS=J
      NP(NODES+1)=IS
      PRINT 105,MIN
      PRINT 103,(NP(I),I=1,NODES)
      GOTO 91
C
C         NOW WE MUST CHECK THE LEFT HAND BRANCHES OF THE TREE TO SEE
C         IF ANY OTHER ROUTE MIGHT FEASIBLY LEAD TO A BETTER SOLUTION.
C
   60 LEVEL1=LEVEL-1
      DO 65 I=1,LEVEL1
      IF(TREE(LEVEL-I,1).LT.MING) GO TO 70
   65 CONTINUE
C
C         SINCE ALL LEFT HAND BRANCHES OF THE TREE ARE GREATER THAN
C         THE GLOBAL MINIMUM, NO OTHER ROUTE IS FEASIBLE, AND WE ARE
C         FINISHED.
C
      GO TO 91
C
C        IT IS POSSIBLE THAT THIS BRANCH OF THE TREE WILL OFFER A
C        BETTER SOLUTION , SO WE MUST BLOCK THE ARC THAT WAS USED
C        AT THIS LEVEL FROM BEING USED IN THE NEW SOLUTION BY SETTING
C        THE DISTANCE TO INFINITY. RESET ALL INITIAL VALUES, AND GO
C        BACK TO THE START.
C
   70 LEVEL=LEVEL-I
      DO 75 I=1,NODES
      ROWUSD(I)=.FALSE.
      COLUSD(I)=.FALSE.
      TREE(I,1)=0
      TREE(I,2)=0
      DO 75 J=1,NODES
      D(I,J)=S(I,J)
      IF(UP(I,J).NE.LEVEL) GO TO 75
      M=I
      N=J
   75 UP(I,J)=0
C
C
C
```

```
C         IT IS IMPORTANT TO OBSERVE THAT IF ARC(I,J) HAS BEEN ELIMINATED
C         FROM THE SOLUTION, IT IS SET TO INFINITY. THEN AFTER PROCEEDING
C         DOWN, THE MIN WILL PERHAPS EXCEED MING AND IT IS TIME TO STOP
C         THAT TREE BUT AS YOU GO BACK AND CHECK THE LEFT HAND BRANCHES
C         PERHAPS AN ARC AFTER THE POINT IN THE TREE WHERE ARC (I,J) WAS
C         PREVIOUSLY SET TO INFINITY SHOULD NOW BE ELIMINATED. IN THIS
C         CASE ONE MUST BE CAREFUL TO LEAVE THE VALUE OF ARC(I,J) AT
C         INFINITY.
C
C
C
      PASS=0.
      DO 90  I=1,NODES
      IF(INFSTC(I,1).LE.LEVEL) GO TO 80
      IF(PASS.NE.0.) GO TO 85
      PASS=1.
      INFSTC(I,1)=LEVEL
      INFSTC(I,2)=M
      INFSTC(I,3)=N
   80 II=INFSTC(I,2)
      IJ=INFSTC(I,3)
      D(II,IJ)=INF
      GO TO 90
   85 INFSTC(I,1)=INF
   90 CONTINUE
      GO TO 11
   91 PRINT 100, MING
      DO 95 I=1,NODES
      NY=NP(I)
      NZ=NP(I+1)
   95 PRINT 101, NY, NZ, S(NY,NZ)
   99 FORMAT(' ',8X,'**ORIGINAL MATRIX**')
  100 FORMAT('1',8X,'WE HAVE AN OPTIMAL SOLUTION OF ',I14,//40X,'BY',/)
  101 FORMAT(' ',8X,'GO FROM NODE',I3,5X,'TO NODE',I3,5X,'AT A DISTANCE
     @OF',I10)
  102 FORMAT(' ','ROUTE UNDEFINED',3I10)
  103 FORMAT(' ',2X,10(3X,I4)/10(3X,I4)/10(3X,I4)/10(3X,I4)/10(3X,I4)/
     15(3X,I4)/)
  104 FORMAT(A4,I6,I10,F10.2)
  105 FORMAT(///,8X,'**WE HAVE A FEASIBLE SOLUTION **',I14,///)
      RETURN
      END
C
C           COLUMN REDUCTION
C
      FUNCTION MINCOL(KKK)
      COMMON S(50,50),D(50,50),ROWUSD(50),COLUSD(50),TREE(50,2),
     1UP(50,50),NP(50),INFSTC(50,3),NODES,INF,LEVEL,M,N,MAX
      INTEGER S,D,TREE,UP
      LOGICAL ROWUSD, COLUSD,FLAG
      MINCOL=0
      DO 50 I=1,NODES
      MIN=0
      IF(COLUSD(I)) GO TO 50
      MIN=INF
      DO 10 J=1,NODES
      IF(ROWUSD(J)) GO TO 10
      IF(D(J,I).GT.MIN) GO TO 10
      MIN=D(J,I)
   10 CONTINUE
      IF(MIN.EQ.0) GO TO 50
```

```
      DO 20 J=1,NODES
   20 D(J,I)=D(J,I)-MIN
   50 MINCOL=MINCOL+MIN
      RETURN
      END
C
C           ROW REDUCTION
C
      FUNCTION MINROW(KKK)
      COMMON S(50,50),D(50,50),ROWUSD(50),COLUSD(50),TREE(50,2),
     1UP(50,50),NP(50),INFSTC(50,3),NODES,INF,LEVEL,M,N,MAX
      INTEGER S,D,TREE,UP
      LOGICAL ROWUSD,COLUSD,FLAG
      MINROW=0
      DO 50 I=1,NODES
      MIN=0
      IF(ROWUSD(I)) GO TO 50
      MIN=INF
      DO 10 J=1,NODES
      IF(COLUSD(J)) GO TO 10
      IF(D(I,J).GT.MIN) GO TO 10
      MIN=D(I,J)
   10 CONTINUE
      IF(MIN.EQ.0) GO TO 50
      DO 20 J=1,NODES
   20 D(I,J)=D(I,J)-MIN
   50 MINROW=MINROW+MIN
      RETURN
      END
      SUBROUTINE PRTRXS
      COMMON S(50,50),D(50,50),ROWUSD(50),COLUSD(50),TREE(50,2),
     1UP(50,50),NP(50),INFSTC(50,3),NODES,INF,LEVEL,M,N,MAX
      INTEGER S,D,TREE,UP
      N=NODES
      N1=1
      N2=5
   43 IF(N2-N) 45,45,44
   44 N2=N
   45 PRINT 911,(I,I=N1,N2)
  911 FORMAT(//,9X,5(I10,2X))
      WRITE(6,912)
  912 FORMAT(/)
      DO 48 L=1,N
   48 PRINT 913,L,(D(L,M),M=N1,N2)
      IF(N2-N) 52,55,55
   52 N1=N1+5
      N2=N2+5
      GO TO 43
  913 FORMAT(' ',6X,I2,5(I10,2X))
   55 CONTINUE
      RETURN
      END
C
C
      SUBROUTINE MAXCAP
      DIMENSION NODE(50,50),A(50,50),LAB(50),LP(50),EPS(50),PL(50),
     1NICID(50,50),SCAN(50)
      COMMON S(50,50),D(50,50),ROWUSD(50),COLUSD(50),TREE(50,2),
     1UP(50,50),NP(50),INFSTC(50,3),NODES,INF,LEVEL,M,N,MAX
      EQUIVALENCE (NODE,A,S),(LAB,LP,ROWUSD),(EPS,PL,COLUSD),(NICID,UP)
     1,(SCAN,TL,NP)
```

```
      INTEGER PEND/'PEND'/,EXIT/'EXIT'/,IWRD
      TL=9999999.00
    4 N=0
      READ(5,400)NODES,NARCS
  400 FORMAT(2I10)
C
C     INITIALIZATION OF THE DISTANCE MATRIX
C
      DO 1 L=1,NODES
      DO 1 M=1,NODES
      D(L,M)=9999999.99
      IF(L.EQ.M)D(L,M)=0.0
    1 CONTINUE
C
C     DATA READ IN
C
    2 READ(5,3)IWRD,I,J,B
    3 FORMAT(A4,I6,I10,F10.2)
      IF(IWRD.EQ.PEND)GO TO 10
      D(I,J)=B
      IF(J.GT.N)N=J
      IF(I.GT.J)N=I
      GO TO 2
   10 WRITE(6,700)
  700 FORMAT('1',5X,' ** ORIGINAL MATRIX **')
      CALL PRNTRX
C
C     INITIALIZATION OF THE NODES ON THE CHAINS MATRIX
C
      DO 100 I=1,N
      DO 100 K=1,N
      NODE(I,K)=K
  100 CONTINUE
C
C     TRIPLE OPERATION TO UPDATE THE DISTANCE MATRIX
C
      DO 20 J=1,N
      DO 20 I=1,N
      DO 20 K=1,N
      IF(I.EQ.J.OR.I.EQ.K.OR.J.EQ.K)GO TO 20
      IF (D(I,J).EQ.TL.OR.D(J,K).EQ.TL)GO TO 20
      X=AMIN1(D(I,J),D(J,K))
      IF (D(I,K).EQ.TL)GO TO 19
      IF (X.LE.D(I,K))GO TO 20
   19 D(I,K)=X
C
C     UPDATING THE NODES ON THE CHAINS MATRIX
C
      NODE(I,K)=J
   20 CONTINUE
      WRITE(6,800)
  800 FORMAT('1',5X,'** ARC CAPACITY MATRIX **')
      CALL PRNTRX
      WRITE(6,900)
  900 FORMAT('1',5X,'*MATRIX REPRESENTING THE NODES ON THE MAX-CAP',
     1' PATH')
      CALL PRNTRY
  200 CONTINUE
      RETURN
      END
      SUBROUTINE KSHORT
```

```
      DIMENSION LLEN(40),LINC(50),LVAL(50)
      COMMON S(50,50),D(50,50),ROWUSD(50),COLUSD(50),TREE(50,2),
     1UP(50,50),NP(50),INFSTC(50,3),NODES,INF,LEVEL,M,N,MAX
      INTEGER ULEN(40),UINC(50),UVAL(50),START,VAL
      INTEGER PEND/'PEND'/,NPRO
      COMMON /BLK1/ MU,ML,LLEN,LINC,LVAL,ULEN,UINC,UVAL
      COMMON /BLK2/ K
      COMMON /BLK4/ START(41),INC(40),VAL(40)
C
C INF IS DEFINED.
C
      INF=99999999
C
C AS THE INPUT NETWORK IS READ IN, THE VARIABLES AND THE ARRAYS NEEDED
C BY DSWP AND TRACE ARE CREATED.
C
  110 FORMAT(2I10)
      J=0
      MU=0
      ML=0
      NPREV=0
      N=0
    1 READ(5,110) NODES,NARCS
      DO 30 I=1,NARCS
      READ (5,800) NB,NA,LEN
  800 FORMAT(3I10)
      IF(NA.GT.N) N=NA
      IF(NB.GT.N) N=NB
      IF(NA.EQ.NPREV)   GO TO 10
      IF(NA.EQ.NPREV+1) GO TO 3
      L1=NPREV+1
      L2=NA-1
      DO 2 L=L1,L2
      START(L)=0
      ULEN(L)=0
    2 LLEN(L)=0
    3 IF(J.EQ.0) GO TO 5
      ULEN(NPREV)=JU
      LLEN(NPREV)=JL
    5 START(NA)=J+1
      JU=0
      JL=0
      NPREV=NA
   10 J=J+1
      INC(J)=NB
      VAL(J)=LEN
      IF(NB.GT.NA) GO TO 20
      MU=MU+1
      UINC(MU)=NB
      UVAL(MU)=LEN
      JU=JU+1
      GO TO 30
   20 ML=ML+1
      LINC(ML)=NB
      LVAL(ML)=LEN
      JL=JL+1
   30 CONTINUE
      START(NPREV+1)=J+1
      ULEN(NPREV)=JU
      LLEN(NPREV)=JL
C
```

```
C THE (K,NS,IMAX) AND (NF,PMAX) DATA RECORDS ARE SUCCESSIVELY READ.
C
   40 READ (5,801) NPRO,I1,I2,I3
  801 FORMAT(A4,3I5)
      IF(NPRO.EQ.PEND) GO TO 100
      IF(I3.EQ.0) GO TO 50
      K=I1
      NS=I2
C
C THE K SHORTEST DISTINCT PATH LENGTHS FROM NODE NS TO ALL NODES OF
C THE NETWORK ARE CALCULATED.
C
      CALL DSWP(NS,I3)
      GO TO 40
C
C UP TO PMAX OF THE PATHS HAVING THE K SHORTEST PATH LENGTHS FROM NODE
C NS TO NODE NF ARE DETERMINED.
C
   50 CALL TRACE(NS,I1,I2)
      GO TO 40
  100 RETURN
      END
C
C
      SUBROUTINE DSWP(NS,IMAX)
      DIMENSION LLEN(40),LINC(50),LVAL(50)
      COMMON S(50,50),D(50,50),ROWUSD(50),COLUSD(50),TREE(50,2),
     1UP(50,50),NP(50),INPSTC(50,3),NODES,INF,LEVEL,M,N,MAX
      INTEGER ULEN(40),UINC(50),UVAL(50),X
      COMMON /BLK1/ MU,ML,LLEN,LINC,LVAL,ULEN,UINC,UVAL
      COMMON /BLK2/ K
      COMMON /BLK3/ X(40,5)
      N1=N-1
C
C THE INITIAL APPROXIMATION MATRIX X IS FORMED.
C
      DO 20 I=1,N
      DO 20 J=1,K
   20 X(I,J)=INF
      X(NS,1)=0
      ITNS=1
C
C THE CURRENT X IS MODIFIED THROUGH MATRIX MULTIPLICATION WITH THE
C LOWER TRIANGULAR PORTION OF THE ARC LENGTH MATRIX.
C
   30 IFIN=ML
      INDX=1
      DO 40 III=1,N1
      I=-III+N1+1
      IF(LLEN(I).EQ.0) GO TO 40
      IS=IFIN-LLEN(I)+1
      CALL XMULT(I,IS,IFIN,LINC,LVAL,INDX)
      IFIN=IS-1
   40 CONTINUE
      IF(ITNS.EQ.1) GO TO 50
C
C TEST FOR CONVERGENCE.
C
      IF(INDX.EQ.1) GO TO 100
C
C THE CURRENT X IS MODIFIED THROUGH MATRIX MULTIPLICATION WITH THE
```

```
C UPPER TRIANGULAR PORTION OF THE ARC LENGTH MATRIX.
C
   50 ITNS=ITNS+1
      IS=1
      INDX=1
      DO 60 I=2,N
      IF(ULEN(I).EQ.0) GO TO 60
      IFIN=IS+ULEN(I)-1
      CALL XMULT(I,IS,IFIN,UINC,UVAL,INDX)
      IS=IFIN+1
   60 CONTINUE
C
C TEST FOR CONVERGENCE.
C
      IF(INDX.EQ.1) GO TO 100
      ITNS=ITNS+1
C
C A TEST IS MADE TO SEE IF TOO MANY ITERATIONS HAVE BEEN PERFORMED.
C
      IF(ITNS.LT.IMAX) GO TO 30
      WRITE(6,900) IMAX
  900 FORMAT('NUMBER OF ITERATIONS EXCEEDS',I5)
      GO TO 200
C
C THE SOLUTION MATRIX X IS PRINTED OUT ON UNIT 6, TOGETHER WITH THE
C VALUES FOR K,NS AND ITNS.
C
  100 WRITE (6,901) K, NS
  901 FORMAT(1H1,10X,'K=',I3,1X,'SHORTEST PATH LENGTHS FROM NODE',I4//
     12X,'TO'/1X,'NODE'//)
      DO 130 I=1,N
  130 WRITE (6,902) I,(X(I,J),J=1,K)
  902 FORMAT(' ',I3,6X,(10I9))
      WRITE (6,903) ITNS
  903 FORMAT(//1H0,'NUMBER OF ITERATIONS REQUIRED FOR CONVERGENCE =',I5)
  200 RETURN
      END
C
C
      SUBROUTINE XMULT(I,IS,IFIN,INC,VAL,INDX)
      DIMENSION INC(50)
      COMMON S(50,50),D(50,50),ROWUSD(50),COLUSD(50),TREE(50,2),
     1UP(50,50),NP(50),INFSTC(50,3),NODES,INF,LEVEL,M,N,MAX
      INTEGER VAL(50),A(5),X
      COMMON /BLK2/ K
      COMMON /BLK3/ X(40,5)
C
C INITIALIZE  TO CURRENT K SHORTEST PATH LENGTHS FOR NODE I, IN
C STRICTLY INCREASING ORDER.
C
      DO 10 J=1,K
   10 A(J)=X(I,J)
      MAX=A(K)
C
C EACH NODE OF INC INCIDENT TO NODE I IS EXAMINED.
C
      DO 100  L=IS,IFIN
      II=INC(L)
      IV=VAL(L)
C
C TEST TO SEE WHETHER IXV IS TOO LARGE TO BE INSERTED INTO A.
```

```
C
      DO 90 M=1,K
      IX=X(II,M)
      IF(IX.GE.INF) GO TO 100
      IXV=IX+IV
      IF(IXV.GE.MAX) GO TO 100
C
C IDENTIFY THE POSITION INTO WHICH IXV CAN BE INSERTED.
C
      DO 30 JJJ=2,K
      J=-JJJ+K+2
      IF(IXV-A(J-1)) 30,90,50
   30 CONTINUE
      J=1
   50 JJ=K
   70 IF(JJ.LE.J) GO TO 80
      A(JJ)=A(JJ-1)
      JJ=JJ-1
      GO TO 70
   80 A(J)=IXV
C
C IF AN INSERTION HAS BEEN MADE IN A, SET INDX = 0.
C
      INDX=0
      MAX=A(K)
   90 CONTINUE
  100 CONTINUE
      IF(INDX.EQ.1) GO TO 120
C
C  UPDATE THE K SHORTEST PATH LENGTHS TO NODE I.
C
      DO 110 J=1,K
  110 X(I,J)=A(J)
  120 RETURN
      END
C
C
      SUBROUTINE TRACE(NS,NF,PMAX)
      COMMON S(50,50),D(50,50),ROWUSD(50),COLUSD(50),TREE(50,2),
     1UP(50,50),NP(50),INFSTC(50,3),NODES,INF,LEVEL,M,N,MAX
      INTEGER P(50),Q(50),PV(50),START,VAL,X,PMAX
      COMMON /BLK2/ K
      COMMON /BLK3/ X(40,5)
      COMMON /BLK4/ START(41),INC(40),VAL(40)
C
C INITIALIZATION PHASE.
C
      DO 10 I=1,50
      P(I)=0
      Q(I)=0
   10 PV(I)=0
      JJ=1
      IF(NS.EQ.NF)JJ=2
      NPH=0
      IF(X(NF,JJ).LT.INF) GO TO 15
      WRITE (6,909) NS,NF
  909 FORMAT(1H1,'THERE ARE NO PATHS FROM NODE',I4,'TO NODE',I4)
      GO TO 200
   15 WRITE(6,901) NS,NF
  901 FORMAT(1H1,'THE K SHORTEST PATHS FROM NODE',I3,1X, 'TO NODE',I4//
     11H0,'PATH    LENGTH    NODE SEQUENCE'//)
```

```
C
C THE JJ-TH DISTINCT PATH IS BEING EXPLORED.
C
   20 KK=1
      LAB=X(NF,JJ)
      IF(LAB.EQ.INF) GO TO 200
      LL=LAB
      P(1)=NF
   30 LAST=0
C
C NODES INCIDENT TO NODE P(KK) ARE SCANNED.
C
   40 NT=P(KK)
      IS=START(NT)
      DO 45 ND=NT,40
      IF(START(ND+1).NE.0) GO TO 48
   45 CONTINUE
   48 IF=START(ND+1)-1
      II=IS+LAST
   50 IF(II.GT.IF) GO TO 90
      NI=INC(II)
      NV=VAL(II)
      LT=LAB-NV
C
C A TEST IS MADE TO SEE IF THE CURRENT PATH CAN BE EXTENDED BACK TO
C NODE NI.
C
      DO 60 J=1,K
      IF(X(NI,J)-LT)60,80,70
   60 CONTINUE
   70 II=II+1
      GO TO 50
   80 KK=KK+1
      IF(KK.GT.50) GO TO 190
      P(KK)=NI
      Q(KK)=II-IS+1
      PV(KK)=NV
      LAB=LT
C
C TESTS ARE MADE TO SEE IF THE CURRENT PATH CAN BE EXTENDED FURTHER.
C
      IF(LAB.NE.0) GO TO 30
      IF(NI.NE.NS) GO TO 30
C
C A COMPLETE PATH FROM NS TO NF HAS BEEN GENERATED AND IS PRINTED
C OUT ON UNIT 6.
C
      NPH=NPH+1
      WRITE (6,902) NPH,LL,(P(J),J=1,KK)
  902 FORMAT(1X,I4,I8,5X,(20I5))
      IF(NPH.GE.PMAX) GO TO 200
   90 LAST=Q(KK)
      P(KK)=0
      LAB=LAB+PV(KK)
      KK=KK-1
      IF(KK.GT.0) GO TO 40
C
C THE EXPLORATION OF THE CURRENT JJ-TH DISTINCT PATH LENGTH IS ENDED.
C
      JJ=JJ+1
      IF(JJ.GT.K) GO TO 200
```

```
      GO TO 20
  190 WRITE(6,903)
  903 FORMAT(1H0,'NUMBER OF ARCS IN PATH EXCEEDS 50')
  200 RETURN
      END
C
      SUBROUTINE GENPAK
      DIMENSION IARC(500),JARC(500),COST(500),FLOW(500)
     1,AMP(500),LOWER(500),UPPER(500),TITLE(20)
      COMMON S(50,50),D(50,50),ROWUSD(50),COLUSD(50),TREE(50,2),
     1UP(50,50),NP(50),INFSTC(50,3),NODES,INF,LEVEL,M,N,MAX
      COMMON /G1/ IARC,JARC,NARC,IFS,IROOT,IPRINT
      COMMON /G2/ COST,FLOW,AMP
      COMMON /G3/ LOWER,UPPER,BIG
      COMMON /G5/ FLONET,OUTFLO,TOTCST,CSTNOW
      COMMON /G6/ TITLE,NDEG,NLOP,ITER,SINK
      COMMON /G7/ SOURCE,EPS,SIGH
      INTEGER  TITLE,SINK,SOURCE,FINISH
      INTEGER PEND/'PEND'/,NPRO
      REAL LOWER
      DATA FINISH/4H    /
COMPUTE THE MACHINE EPSILON.
      EPS = 1.0
    5 EPS = EPS/2.0
      TOL1 = 1.0   + EPS
      IF (TOL1 .GT. 1.0  ) GO TO 5
      EPS=SQRT(EPS)
COMPUTE THE MACHINE INFINITY.
      BIG=1.E+6
    8 BIG=BIG*BIG
      BIG1=1.+BIG
      IF(BIG1 .GT. BIG) GO TO 8
      BIG=BIG*BIG
C
      WRITE(6,195)
      READ(5,90) NODES,NARCS
      WRITE(6,130)
      WRITE(6,140)
      WRITE(6,150)
C   READ THE ARCS AND INITIALIZE THEIR FLOWS.
      DO 10 I=1,NARCS
      READ (5,160) IARC(I),JARC(I),LOWER(I),UPPER(I),COST(I),AMP(I)
      FLOW(I)=0.
      WRITE(6,170) I,IARC(I),JARC(I),LOWER(I),UPPER(I),COST(I),AMP(I)
   10 CONTINUE
      READ(5,110) SOURCE,SINK,IPRINT,OUTFLO
      WRITE(6,100) SOURCE,SINK
      WRITE(6,120) OUTFLO
   20 NDEG=0
      NLOP=0
      ITER=0
      FLONET=0.0
      TOTCST=0.0
      IFS=0
      NARC = NARCS + 1
CREATE A DUMMY ARC TO PROVIDE FEASIBLE OUTPUT FLOW IN
CASE THE DESIRED OUTPUT FLOW IS NOT FEASIBLE.
      IARC(NARC)=SOURCE
      JARC(NARC)=SINK
      LOWER(NARC)=0.
      UPPER(NARC)=BIG
```

```
      COST(NARC)=BIG
      AMP(NARC)=1.
      FLOW(NARC)=0.
C
C   DEFINE THE INVERSE NETWORK (OR THE MIRROR ARCS)
C
      DO 30 I=1,NARC
      NN = NARC + I
      IARC(NN) = JARC(I)
      JARC(NN) = IARC(I)
      AMP(NN)=1./AMP(I)
      LOWER(NN)=LOWER(I)*AMP(I)
      UPPER(NN)=UPPER(I)*AMP(I)
      COST(NN)=-COST(I)/AMP(I)
      FLOW(NN)=0.
 30   CONTINUE
      IF(IPRINT .GT. 0) WRITE(6,200)
 40   CALL SHORT (IENTER,ILEAV)
         IF (IENTER.EQ.0) GO TO 60
      CALL MAXFLW
      TOTCST=TOTCST+CSTNOW
      ITER=ITER+1
         IF (IPRINT.EQ.0) GO TO 50
      WRITE(6,180) ITER,FLONET,TOTCST
       WRITE(6,190) ILEAV,IENTER,SICH
 50   CONTINUE
         IF (ABS(FLONET-OUTFLO) .LE. EPS) GO TO 60
         GO TO 40
 60   CONTINUE
      CALL PROUT(IENTER)
 90   FORMAT(2I10)
 110  FORMAT(3I5,F10.2)
 100  FORMAT (' SOURCE NODE=',I5,5X,'SINK NODE =',I5,///)
 120  FORMAT (' OUTPUT FLOW REQUIREMENT',F10.2,///)
 130  FORMAT (21H *****INPUT DATA*****,//)
 140  FORMAT ('            ARC        START          END       LOWER       UPPER
     1 COST      AMPLIFICATION')
 150  FORMAT ('            NO.        NODE          NODE       BOUND       BOUND',//)
 160  FORMAT(2I10,4F10.2)
 170  FORMAT (3I10,6F11.2)
 180  FORMAT (' ITERATION ',I5,5X,' FLOW = ',F10.2,5X,' COST = ',F10.2)
  195 FORMAT(1H1,///,5X,'GENERALIZED NETWORK MINIMIZATION PROBLEM',//
     1//)
 190  FORMAT(15X,' REMOVE',I4,5X,'ENTER',I4,5X,'DELTA =',F10.3,//)
  200 FORMAT(1H1,'*****  INTERMEDIATE RESULTS   *****',///)
      RETURN
      END
C
      SUBROUTINE SHORT (IENTER,ILEAV)
C     SUBROUTINE TO FIND THE INITIAL AND SUBSEQUENT FLOW AUGMENTING TREE
      DIMENSION DISSET(300),BARC(300),FARC(300),RARC(300),ICHK(300)
     1 ,GAN(300),V(300),LOWER(500),UPPER(500)
     2  ,IARC(500),JARC(500),COST(500),FLOW(500),AMP(500)
      COMMON S(50,50),D(50,50),ROWUSD(50),COLUSD(50),TREE(50,2),
     1UP(50,50),NP(50),INFSTC(50,3),NODES,INF,LEVEL,M,N,MAX
      COMMON /G1/ IARC,JARC,NARC,IFS,IROOT,IPRINT
      COMMON /G2/ COST,FLOW,AMP
      COMMON /G3/ LOWER,UPPER,BIG
      COMMON /G5/ FLONET,OUTFLO,TOTCST,CSTNOW
      COMMON /G7/ SOURCE,EPS,SICH
      COMMON /G8/ BARC,FARC,RARC,GAN,V,ICHK
```

```
      INTEGER DISSET,BARC,FARC,RARC,SOURCE
      REAL LOWER
         IF (IFS.NE.0) GO TO 150
C     SET UP POINTERS TO FIND INITIAL TREE.
      DO 10 I=1,NODES
      BARC(I)=0
      FARC(I)=0
      RARC(I)=0
      DISSET(I)=1
      GAN(I)=1.
      V(I) =BIG
 10   CONTINUE
      V(SOURCE)=0.
      ICHANG=0
      IENTER=1
      ITF=0
      IFS=1
C     SET UP SHORTEST PATH TREE FOR FIRST ITERATION.
 20   CONTINUE
      DO 70 I=1,NARC
      JJ=JARC(I)
      II=IARC(I)
         IF( (LOWER(I)-FLOW(I)) .GT. EPS ) GO TO 40
      POT=(V(II)+COST(I))/AMP(I)
      GO TO 50
 40   POT=(V(II)-BIG)/AMP(I)
 50   CONTINUE
         IF ((V(JJ)-POT) .LT. EPS) GO TO 70
      V(JJ)=POT
      BARC(JJ)=I
         IF (ITF.EQ.0) GO TO 60
      CALL LOOP (I,JJ)
 60   CONTINUE
      ICHANG=1
 70   CONTINUE
         IF (ICHANG.EQ.0) GO TO 80
      ICHANG=0
      ITF=1
         GO TO 20
 80   CONTINUE
C     CALCULATE FORWARD POINTERS FOR FIRST ITERATION.
      DO 120 I=1,NODES
         IF (DISSET(I).EQ.0) GO TO 120
      KK=BARC(I)
         IF (KK.EQ.0) GO TO 120
      LL=IARC(KK)
         IF (FARC(LL).NE.0) GO TO 90
      FARC(LL)=KK
         GO TO 120
 90   CONTINUE
      MM=FARC(LL)
 100  CONTINUE
      MN=JARC(MM)
         IF (RARC(MN).NE.0) GO TO 110
      RARC(MN)=KK
         GO TO 120
 110  CONTINUE
      MM=RARC(MN)
         GO TO 100
 120  CONTINUE
 130  CONTINUE
```

```
      IF ( IPRINT .LT. 2 ) RETURN
      WRITE(6,320)  (BARC(I),I=1,NODES)
      WRITE(6,330)  (FARC(I),I=1,NODES)
      WRITE(6,340)  (RARC(I),I=1,NODES)
      WRITE(6,300)  (DISSET(I),I=1,NODES)
      WRITE(6,290)  (V(I),I=1,NODES)
      WRITE(6,310)  (GAN(I),I=1,NODES)
      RETURN
C     FIND NEW FLOW AUGMENTING TREE AFTER THE FIRST ITERATION.
  150 CONTINUE
      DO 160 I=1,NODES
      DISSET(I)=0
  160 CONTINUE
C     DELETE BRANCH FROM BASIS AFTER THE FIRST ITERATION.
      II=IROOT
      I=BARC(II)
      ILEAV=I
      IA=IARC(I)
      CALL DESUB (I,IA)
      RARC(II)=0
      BARC(II)=0
C     SET NEW NODE GAINS ON DISCONNECTED NODES.
  170 CONTINUE
      I=FARC(II)
         IF (I.EQ.0) GO TO 180
      JJ=JARC(I)
      GAN(JJ)=GAN(II)/AMP(I)
      II=JJ
         IF (II.EQ.IROOT) GO TO 180
         GO TO 170
  180 CONTINUE
      J=RARC(II)
      DISSET(II)=1
         IF (J.EQ.0) GO TO 190
      II=JARC(J)
      K=BARC(II)
      JJ=IARC(K)
      GAN(II)=GAN(JJ)/AMP(K)
         GO TO 170
  190 CONTINUE
      K=BARC(II)
         IF (II.EQ.IROOT) GO TO 200
      II=IARC(K)
         GO TO 180
  200 CONTINUE
C     DETERMINE THE NEW BRANCH TO ENTER THE BASIS.
      IENTER=0
      SICH=BIG
      DO 250 K=1,NARC
         IF ((UPPER(K)-FLOW(K)).GT.EPS) GO TO 210
C     LOOKING AT A BACKWARD BRANCH
         IF((FLOW(K)-LOWER(K)) .LT. EPS) GO TO 250
      JJ=IARC(K)
         IF (DISSET(JJ).EQ.0) GO TO 250
      II=JARC(K)
      I=K+NARC
         IF (DISSET(II).EQ.0) GO TO 240
C     NEW BRANCH FORMS A LOOP
         GO TO 220
  210 CONTINUE
      I=K
```

```
      JJ=JARC(I)
         IF (DISSET(JJ).EQ.0) GO TO 250
      II=IARC(I)
         IF (DISSET(II).EQ.0) GO TO 240
C     NEW BRANCH FORMS A LOOP.
  220 CONTINUE
      GALPIV=GAN(II)/(GAN(JJ)*AMP(I))
      IF ( GALPIV .GE. (1.-EPS)) GO TO 250
      POTCH=(((V(II)+COSTF(I))/AMP(I))-V(JJ))/((1.-GALPIV)*GAN(JJ))
  230 CONTINUE
         IF (POTCH.GE.SICH) GO TO 250
      SICH=POTCH
      IENTER=I
         GO TO 250
C     NEW BRANCH DOES NOT FORM A LOOP.
  240 CONTINUE
      POTCH=(((V(II)+COSTF(I))/AMP(I))-V(JJ))/GAN(JJ)
         GO TO 230
  250 CONTINUE
         IF (IENTER.EQ.0) GO TO 270
C     CHANGE NODE LABELS AND POINTERS TO REFLECT ENTERING BRANCH.
C     CHANGE POINTERS.
      CALL TRECHG (IENTER,ILEAV)
C     CHANGE NODE POTENTIALS.
      DO 260 II=1,NODES
         IF (DISSET(II).EQ.0) GO TO 260
      V(II)=SICH*GAN(II)+V(II)
  260 CONTINUE
         GO TO 130
  270 CONTINUE
      IF (IPRINT .NE. 0) WRITE(6,280) FLONET
                        II=IARC(ILEAV)
      CALL ADSUB (ILEAV,II)
                       JJ=JARC(ILEAV)
      BARC(JJ)=ILEAV
         GO TO 130
  280 FORMAT (' MAXIMUM FLOW FOUND ',F20.10)
  290 FORMAT ( ' V         ',(21(F5.2,1X)))
  300 FORMAT ( ' DISSET ',(21I5))
  310 FORMAT ( ' GAIN      ',(21(F5.2,1X)))
  320 FORMAT ( ' BARC      ',(21I5))
  330 FORMAT ( ' FARC      ',(21I5))
  340 FORMAT ( ' RARC      ',(21I5))
       END
C
      SUBROUTINE TRECHG (IENTER,ILEAV)
C     SUBROUTINE TO FORM THE FLOW AUGMENTING TREE BY DELETING ONE ARC
C     FROM THE TREE AND INSERTING A NEW ARC INTO THE TREE.
      DIMENSION IARC(500),JARC(500),BARC(300),FARC(300),RARC(300)
     1,GAN(300) , V(300),ICHK(300)
      COMMON /G1/ IARC,JARC,NARC,IFS,IROOT,IPRINT
      COMMON /G8/ BARC,FARC,RARC,GAN,V,ICHK
      INTEGER   BARC,  RARC
      IROOT=JARC(ILEAV)
      NLIS=0
      JJ=JARC(IENTER)
C     DELETE PATH FROM JARC(IENTER) TO IROOT
   10 CONTINUE
      JB=BARC(JJ)
         IF (JJ.EQ.IROOT) GO TO 20
      NLIS=NLIS+1
```

```
      II = IARC(JB)
      CALL DESUB (JB,II)
      IF(JB.LE.NARC) I=JB+NARC
      IF(JB.GT.NARC) I=JB-NARC
      ICHK(NLIS)=I
      BARC(JJ)=0
      RARC(JJ)=0
      JJ=II
         GO TO 10
 20   CONTINUE
C     ADD IN THE REVERSE OF THE PATH JUST DELETED
         IF (NLIS.EQ.0) GO TO 30
      I=ICHK(NLIS)
      NLIS=NLIS-1
      II=IARC(I)
      JJ=JARC(I)
      CALL ADSUB (I,II)
      BARC(JJ)=I
         GO TO 20
 30   CONTINUE
      II = IARC(IENTER)
      CALL ADSUB (IENTER,II)
      JJ=JARC(IENTER)
      BARC(JJ)=IENTER
      RARC(JJ)=0
      RETURN
      END
C
      SUBROUTINE FLCHG (II,FLONOW)
C     SUBROUTINE TO INCREASE THE FLOW IN AN ARC BY A GIVEN AMOUNT.
      DIMENSION IARC(500),JARC(500),COST(500),FLOW(500),AMP(500)
      COMMON /G1/ IARC,JARC,NARC,IFS,IROOT,IPRINT
      COMMON /G2/ COST,FLOW,AMP
      COMMON /G5/ FLONET,OUTFLO,TOTCST,CSTNOW
      IF(II.GT.NARC) GO TO 10
C     CHANGE FLOW IN A FORWARD ARC.
      FLOW(II)=FLOW(II)+FLONOW
      CSTNOW=CSTNOW+FLONOW*COST(II)
         GO TO 20
C     CHANGE FLOW IN A MIRROR ARC.
 10   CONTINUE
      KK=II-NARC
      FLOW(KK)=FLOW(KK)-FLONOW/AMP(KK)
      CSTNOW=CSTNOW-FLONOW*COST(KK)/AMP(KK)
 20   CONTINUE
      RETURN
      END
C
      SUBROUTINE FLOP (JJ,FLMAX,GN)
C     SUBROUTINE TO DETERMINE THE GAIN, MAXIMUM FLOW CHANGE, AND ROOT OF
C     A FLOW GENERATING CYCLE IN THE FLOW AUGMENTING TREE.
      DIMENSION IARC(500),JARC(500),BARC(300),FARC(300),RARC(300)
     1,GAN(300) , V(300),ICHK(300) ,COST(500),FLOW(500),AMP(500)
     2,LOWER(500),UPPER(500)
      COMMON /G1/ IARC,JARC,NARC,IFS,IROOT,IPRINT
      COMMON /G2/ COST,FLOW,AMP
      COMMON /G3/ LOWER,UPPER,BIG
      COMMON /G8/ BARC,FARC,RARC,GAN,V,ICHK
      INTEGER   BARC
      FLMAX=BIG
      GN=1.
```

```
      IJ=JJ
 10   CONTINUE
      IJK=BARC(IJ)
      GN=GN*AMP(IJK)
      FLMXT=FLMXC(IJK)*GN
         IF (FLMXT.GT.FLMAX) GO TO 20
      FLMAX=FLMXT
      IROOT=JARC(IJK)
 20   CONTINUE
       IF (IARC(IJK) .EQ. JJ ) GO TO 30
      IJ=IARC(IJK)
      GAN(IJ)=GAN(JJ)*GN
         GO TO 10
 30   CONTINUE
      FLMAX=FLMAX*(1.-1./GN)
      RETURN
      END
C
      SUBROUTINE ADSUB (I,II)
C     SUBROUTINE TO ADD AN ARC TO THE TRIPLE LABEL REPRESENTATION OF THE
C     FLOW AUGMENTING TREE.
      DIMENSION IARC(500),JARC(500),BARC(300),FARC(300),RARC(300)
     1,GAN(300) , V(300),ICHK(300)
      COMMON /G1/ IARC,JARC,NARC,IFS,IROOT,IPRINT
      COMMON /G8/ BARC,FARC,RARC,GAN,V,ICHK
      INTEGER     FARC, RARC
C     ADDS ARC I TO THE LIST OF SUBSEQUENT ARCS TO NODE II.
         IF (FARC(II).NE.0) GO TO 10
      FARC(II)=I
         GO TO 40
 10   CONTINUE
      MM=FARC(II)
 20   CONTINUE
      MN=JARC(MM)
         IF (RARC(MN).NE.0) GO TO 30
      RARC(MN)=I
         GO TO 40
 30   CONTINUE
      MM=RARC(MN)
      GO TO 20
 40   CONTINUE
      RETURN
      END
C
      SUBROUTINE DESUB (I,II)
C     SUBROUTINE TO DELETE AN ARC FROM THE TRIPLE LABEL REPRESENTATION
C     OF THE FLOW AUGMENTING TREE.
      DIMENSION IARC(500),JARC(500),BARC(300),FARC(300),RARC(300)
     1,GAN(300) , V(300),ICHK(300)
      COMMON /G1/ IARC,JARC,NARC,IFS,IROOT,IPRINT
      COMMON /G8/ BARC,FARC,RARC,GAN,V,ICHK
      INTEGER     FARC, RARC
C     DELETES ARC I FROM THE LIST OF SUBSEQUENT ARCS TO NODE II.
      JJ = JARC(I)
         IF (FARC(II).NE.I) GO TO 10
      FARC(II)=RARC(JJ)
      RETURN
 10   CONTINUE
      MM=FARC(II)
 20   CONTINUE
      MN=JARC(MM)
```

```
      IF (RARC(MN).NE.I) GO TO 30
      RARC(MN)=RARC(JJ)
      RETURN
30    CONTINUE
      MN=RARC(MN)
      GO TO 20
      END
      FUNCTION FLMXC (I)
C     FUNCTION TO DETERMINE MAXIMUM FLOW CHANGE IN AN ARC
      DIMENSION IARC(500),JARC(500),BARC(300),FARC(300),RARC(300)
     1 ,V(300),ICHK(300),COST(500),FLOW(500),AMP(500),LOWER(500)
     2 ,GAN(300), UPPER(500)
      COMMON /G1/ IARC,JARC,NARC,IFS,IROOT,IPRINT
      COMMON /G2/ COST,FLOW,AMP
      COMMON /G3/ LOWER,UPPER,BIG
      COMMON /G7/ SOURCE,EPS,SICH
      COMMON /G8/ BARC,FARC,RARC,GAN,V,ICHK
      REAL      LOWER
      II=IARC(I)
      JJ=JARC(I)
      IF(I.GT.NARC) GO TO 80
      IF (FLOW(I).GE.UPPER(I)) GO TO 110
      IF (FLOW(I).LT.LOWER(I)) GO TO 70
      FLMXC=UPPER(I)-FLOW(I)
      RETURN
70    CONTINUE
      FLMXC=LOWER(I)-FLOW(I)
      RETURN
80    CONTINUE
      K=I-NARC
      IF (FLOW(K).GT.UPPER(K)) GO TO 100
      IF (FLOW(K).LE.LOWER(K)) GO TO 110
      FLMXC=(FLOW(K)-LOWER(K))*AMP(K)
      RETURN
100   CONTINUE
      FLMXC=(FLOW(K)-UPPER(K))*AMP(K)
      RETURN
110   CONTINUE
      FLMXC=0.0
      RETURN
      END
      FUNCTION COSTF(I)
C   FUNCTION TO CALCULATE THE COSTS ON AN ARC CONSIDERING
C   UPPER AND LOWER BOUNDS.
      DIMENSION IARC(500),JARC(500),COST(500),FLOW(500)
     1,AMP(500),LOWER(500),UPPER(500)
      COMMON /G1/ IARC,JARC,NARC,IFS,IROOT,IPRINT
      COMMON /G2/ COST,FLOW,AMP
      COMMON /G3/ LOWER,UPPER,BIG
      REAL LOWER
      IF(I .GT. NARC) GO TO 50
      IF(FLOW(I)-LOWER(I)) 10,30,30
   10 COSTF=-BIG
      RETURN
   20 COSTF=COST(I)
      RETURN
   30 IF(UPPER(I)-FLOW(I)) 40,40,20
   40 COSTF=BIG
      RETURN
   50 K=I-NARC
      IF(FLOW(K)-LOWER(K)) 80,80,70
```

```
   60 COSTF=-COST(K)/AMP(K)
      RETURN
   70 IF(UPPER(K)-FLOW(K)) 90,60,60
   80 COSTF=BIG/AMP(K)
      RETURN
   90 COSTF=-BIG/AMP(K)
      RETURN
      END
C
C
      SUBROUTINE LOOP (I,JJ)
C      SUBROUTINE TO DETERMINE IF THE FLOW AUGMENTING TREE INCLUDES A
C     FLOW GENERATING CYCLE.  IF SO NODE POTENTIALS ARE ADJUSTED
C     ACCORDINGLY. USED ONLY IN THE FIRST ITERATION.
      DIMENSION IARC(500),JARC(500),BARC(300),FARC(300),RARC(300)
     1,GAN(300)  , V(300),ICHK(300)  ,COST(500),FLOW(500),AMP(500)
      COMMON S(50,50),D(50,50),ROWUSD(50),COLUSD(50),TREE(50,2),
     1UP(50,50),NP(50),INFSTC(50,3),NODES,INF,LEVEL,M,N,MAX
      COMMON /G1/ IARC,JARC,NARC,IFS,IROOT,IPRINT
      COMMON /G2/ COST,FLOW,AMP
      COMMON /G8/ BARC,FARC,RARC,GAN,V,ICHK
      INTEGER    BARC,  FARC, RARC
C     DETERMINE IF POINTERS INDICATE A LOOP.
      DO 10  K=1,NODES
      ICHK(K)=0
 10   CONTINUE
      ICHK(JJ)=1
      IJ=JJ
 20   CONTINUE
      IJK=BARC(IJ)
      IF  (IJK.EQ.0)  RETURN
      IA=IARC(IJK)
         IF  (IA.EQ.JJ)  GO TO 30
      IF  (ICHK(IA).EQ. 1)  RETURN
      ICHK(IA)=1
      IJ=IA
         GO TO  20
 30   CONTINUE
      TEMP=0.
      GN=1.
      IJ=JJ
C     CALCULATE THE COST TO OBTAIN ONE UNIT OF FLOW INTO JJ
 40   CONTINUE
      IJK=BARC(IJ)
      GN=GN*AMP(IJK)
      TEMP = TEMP + COSTF(IJK)  /  GN
      IA=IARC(IJK)
         IF  (IA.EQ.JJ)  GO TO 50
      IJ=IA
         GO TO 40
 50   CONTINUE
C     CALCULATE COST TO OBTAIN ONE UNIT OF FLOW OUT OF LOOP AT JJ
      V(JJ)=TEMP/(1.-1./GN)
 60   CONTINUE
      IJ=IA
      IJK=BARC(IJ)
      IA=IARC(IJK)
      IF  ( IA. EQ.  JJ )  GO TO 70
      V(IA)  =  V(IJ)  *  AMP(IJK)  -  COSTF(IJK)
         GO TO 60
   70 CONTINUE
```

```
      RETURN
      END
C
      SUBROUTINE MAXFLW
C     SUBROUTINE TO CALCULATE THE MAXIMUM FLOW INCREASE INTO THE SINK.
C     ARC TO LEAVE THE TREE IS ALSO DETERMINED.  THE FLOW IS CHANGED IN
C     THE AUGMENTING PATH.
      DIMENSION IARC(500),JARC(500),BARC(300),FARC(300),RARC(300)
     1,GAN(300)  , V(300),ICHK(300)  ,TITLE(20)
     2,COST(500),FLOW(500),AMP(500),LOWER(500),UPPER(500)
      COMMON S(50,50),D(50,50),ROWUSD(50),COLUSD(50),TREE(50,2),
     1UP(50,50),NP(50),INFSTC(50,3),NODES,INF,LEVEL,M,N,MAX
      COMMON /G1/ IARC,JARC,NARC,IFS,IROOT,IPRINT
      COMMON /G2/ COST,FLOW,AMP
      COMMON /G3/ LOWER,UPPER,BIG
      COMMON /G5/ FLONET,OUTFLO,TOTCST,CSTNOW
      COMMON /G6/ TITLE,NDEG,NLOP,ITER,SINK
      COMMON /G7/ SOURCE,EPS,GICH
      COMMON /G8/ BARC,FARC,RARC,GAN,V,ICHK
      INTEGER   SOURCE, SINK, BARC, TITLE
C     FIND OUT IF THERE IS A LOOP. IF SO JJ IS THE JUNCTION OF THE LOOP.
      DO 10 I=1,NODES
      ICHK(I)=0
 10   CONTINUE
      JJ=SOURCE
      I=SINK
 20   CONTINUE
      ICHK(I)=1
         IF (I.EQ.SOURCE) GO TO 40
      II=BARC(I)
         IF (II.EQ.0) GO TO 130
      I=IARC(II)
         IF (ICHK(I).EQ.1) GO TO 30
         GO TO 20
 30   CONTINUE
      JJ=I
C   INCREASE THE NUMBER OF LOOP ITERATIONS.
      NLOP=NLOP+1
C     FIND MAXIMUM FLOW CHANGE POSSIBLE.
 40   CONTINUE
      FLMX=BIG
      GN=1.
      I=SINK
      GAN(I)=1.
 50   CONTINUE
         IF (I.EQ.JJ) GO TO 60
      KK=BARC(I)
      GN=GN*AMP(KK)
      FLMXT=FLMXC(KK)*GN
      I=IARC(KK)
      GAN(I)=GN
         IF (FLMXT.GT.FLMX) GO TO 50
      FLMX=FLMXT
      IROOT=JARC(KK)
         GO TO 50
 60   CONTINUE
         IF (JJ.EQ.SOURCE) GO TO 70
      CALL FLOP (JJ,FLMAX,GLOOP)
      FLMXT=FLMAX*GN
         IF (FLMXT.GT.FLMX) GO TO 70
      FLMX=FLMXT
```

```
 70     CONTINUE
C       INCREASE TOTAL FLOW BY THE MAXIMUM FLOW CHANGE.
        FLO=OUTFLO-FLONET
        IF (FLO.GT.FLMX) FLO=FLMX
        CSTNOW=0.
        IF (FLO .GT. EPS) GO TO 75
C    INCREASE THE NUMBER OF DEGENERATE ITERATIONS.
        NDEG=NDEG+1
        GO TO 120
   75 FLONET=FLONET+FLO
C       CALCULATE FLOW CHANGE ON EACH ARC.
        I=SINK
 80     CONTINUE
           IF (I.EQ.JJ) GO TO 90
        II=BARC(I)
           IF (II.EQ.0) GO TO 130
        I=IARC(II)
        FLONOW=FLO/GAN(I)
        CALL FLCHG (II,FLONOW)
           GO TO 80
 90     CONTINUE
           IF (JJ.EQ.SOURCE) GO TO 120
        FLOOP=FLO/(GAN(I)*(1-(1/GLOOP)))
        I=JJ
        FLGA=FLOOP*GAN(JJ)
 100    CONTINUE
        II=BARC(I)
        I=IARC(II)
        FLONOW=FLGA/GAN(I)
           IF (I.NE.JJ) GO TO 110
        J=JARC(II)
        FLONOW=FLGA/(GAN(J)*AMP(II))
 110    CONTINUE
        CALL FLCHG (II,FLONOW)
           IF (I.EQ.JJ) GO TO 120
           GO TO 100
 120    CONTINUE
        RETURN
 130    CONTINUE
        WRITE(6,140) FLONET,TOTCST
        CALL EXIT
        RETURN
 140    FORMAT (///,' PROBLEM IS INFEASIBLE. THE MAXIMUM FLOW IS  ',F10.2,
       1'   AT A TOTAL COST OF  ',F10.2)
        END
C
        SUBROUTINE PROUT(IENTER)
C SUBROUTINE TO PRINT OUT THE OPTIMAL SOLUTION.
        DIMENSION IARC(500),JARC(500),BARC(300),FARC(300),RARC(300)
       1  ,ICHK(300),COST(500),FLOW(500),AMP(500),LOWER(500)
       2 ,GAN(300),V(300),UPPER(500),TITLE(20)
        COMMON /G1/ IARC,JARC,NARC,IFS,IROOT,IPRINT
        COMMON /G2/ COST,FLOW,AMP
        COMMON /G3/ LOWER,UPPER,BIG
        COMMON /G5/ FLONET,OUTFLO,TOTCST,CSTNOW
        COMMON /G6/ TITLE,NDEG,NLOP,ITER,SINK
        COMMON /G8/ BARC,FARC,RARC,GAN,V,ICHK
        INTEGER TITLE
        REAL LOWER
        WRITE(6,800)
        IF(FLOW(NARC)) 30,30,10
```

```
10 WRITE(6,110)
   WRITE(6,120)
   NARC=NARC-1
   DO 20 I=1,NARC
   IF(FLOW(I) .LT. UPPER(I)) GO TO 20
   WRITE(6,130) IARC(I),JARC(I)
20 CONTINUE
   WRITE(6,170)
   GO TO 60
30 DO 35 I=1,NARC
   IF(FLOW(I) .GE. LOWER(I)) GO TO 35
   IENTER=1
   GO TO 38
35 CONTINUE
   IF(IENTER .NE. 0 .AND. FLONET .EQ. OUTFLO) GO TO 50
38 WRITE(6,110)
   WRITE(6,140)
   NARC=NARC-1
   DO 40 I=1,NARC
   IF(FLOW(I) .GE. LOWER(I)) GO TO 40
   WRITE(6,130) IARC(I),JARC(I)
40 CONTINUE
   WRITE(6,170)
   GO TO 60
50 NARC=NARC-1
   WRITE(6,160)
60 WRITE(6,400)
   DO 70 I=1,NARC
   ACOST = COST(I)*FLOW(I)
   WRITE(6,500)I,IARC(I),JARC(I),LOWER(I),UPPER(I),COST(I),AMP(I)
  1     , FLOW(I), ACOST
70 CONTINUE
   WRITE(6,600) FLONET,TOTCST
   WRITE(6,700) ITER,NDEG,NLOP
   WRITE(6,800)
   RETURN
110 FORMAT(///,'********** THE OUTPUT FLOW REQUIREMENT IS NOT FEASIBLE'
   1 , '*********',//)
120 FORMAT(/,5X,' THE FOLLOWING ARCS ARE SATURATED :')
130 FORMAT(10X,I3,2X,I3)
140 FORMAT(/,5X,'THE LOWER BOUNDS CONSTRAINTS ARE VIOLATED FOR THE'
   1 ,'FOLLOWING ARCS :')
160 FORMAT(///, '******** OPTIMAL FLOW PATTERN ***********',//)
170 FORMAT(///,'********** FLOW PATTERN OBTAINED AT TERMINATION **'
   1 ,'*********',//)
400 FORMAT(/,'   ARC   START END    LOWER     UPPER       COST
   1,'GAIN       FLOW          ARC COST',/)
500 FORMAT (I5,2X,2I5,5F10.2,F15.2)
600 FORMAT(///,' TOTAL OUTPUT FLOW = ',F20.4,/,' TOTAL COST       =
   1        F20.4,/)
700 FORMAT (' NUMBER OF ITERATIONS ',I10,/,
   1        ' NUMBER OF DEGENERATE ITERATIONS ',I10,/,
   2        ' NUMBER OF LOOP ITERATIONS ',I10,/)
800 FORMAT(1H1)
   END
```

INDEX

A

Activity, 273
- earliest finish, 280
- earliest start, 279
- latest finish, 280
- latest start, 280

Activity crashing, 279
Activity float, 280, 281, 283, 285, 286
- free float, 285, 291
- independent float, 285, 291
- safety float, 286, 291
- total float, 283, 290

Activity-on-arc model, 26
Activity-on-node model, 26, 287-91
Activity slack, 273
Acyclic network, 8, 9
Addition (generalized), 73
Adjacency matrix, 11
Aggregation, 393
Allocation of highway funds, 95
Analysis and complexity of shortest-path algorithms, 88-90
Applications of network-flow models, 24-37
Arborescence, 9
Arc (DEFN), 5, 206
- arc failure, 179
- arc multiplier, 344

Assad, A.A., 414
Assignment problem, 30, 41, 132, 141, 234-35.
Automotive transmission, 392

B

Backtracking, 105
Balotski, G.R., 96
Barnes, W., 4, 350, 354
Bazzara, M., 4
Bellmore, M., 414
Bennington, G.E., 24, 414
Benson, L.A., 315
Bhaumik, G., 344, 350, 351, 354

Bidirected arcs, 5
Bipartite network, 232
Bottleneck assignment problem, 247-50
Bounding, 101
Bradley, H., 4
Branch and bound, 97
Branching, 101
Busacker, R.G., 4
Buying a new car, 49

C

Capacitated arc, 207
Capacitated network, 27, 207
Caterer problem, 30
Chain, 7
Chain change, 346
Chain gain, 346
Chain loss, 346
Charnes, A., 4, 345
Chebyshev's inequality, 384
Circuit, 8
Circuitless network, 8
Circulation network, 207
Closed subpaths, 141
Complementary slackness condition, 212, 302
Computational complexity of Dijksta's method, 89
Computational complexity of double-sweep method, 90
Computational complexity of Floyd's algorithm, 89
Computer-communication models, 413
Condensed network, 159, 160
Connected network, 8
Conservation of flow, 14
Constrained resourses, 322
Cooper, W.W., 4, 345
Cost crashing, 296-300
Cost matrix, 11
CPM/PERT, 5, 268-330
CPM/PERT computer programs, 327-30
CPM/PERT network construction, 273-76
Critical activity, 272
Critical path, 272, 279
Critical path method (CPM), 26, 271, 276-91
Crude oil distribution, 255-59
Crum, R., 345
Cut (DEFN), 15, 149
Cut capacity, 16, 149
Cut tree, 162
Cut value, 16, 149
Cycle, 8
Cycle change, 346
Cycle (flow-generating), 347

D

Dantzig, G.B., 4
Davis, E.W., 295, 327
Dei-Rossi, J.A., 96
Dijkstra, E.W., 46, 89
Dijkstra's shortest path algorithm, 46, 89
Directed arcs, 5
Disconnected network, 179
Disconnecting sets, 400
Disjoint subpaths, 141
Distance matrix, 98
Distance networks, 27
DNC/NC, 107
Donor route, 124
Double-sweep method, 73-77, 90
Dryfus, S., 72
Duke, R.A., 414

E

Earliest finish time, 280
Earliest start time, 279
Elmaghraby, S., 4
Employment scheduling, 31
Equipment replacement, 24
Error bounds, 398
Euhler, Leonard, 3
Evans, J.R., 386, 414

Event, 274
 earliest possible time, 276
 latest possible time, 277
 slack, 278

F

Fixed charge problem, 68
Fixed weight disaggregation, 396
Fleet scheduling problem, 32
Float, 280, 281, 283, 285, 286
 free, 285, 291
 independent, 285, 291
 safety, 286, 291
 total, 283, 290
Flow (DEFN), 14
Flow-augmentation process, 351, 353-55
Flow-augmenting path, 146
Flow conservation constraints, 15, 40, 208, 209
Flow conservation property, 15
Flow-generating cycle, 347
Flowgraphs, 370, 371
Floyd, R.W., 53, 89
Floyd's algorithm, 53-56, 89
Fondahl, J.W., 315
Forbidden links, 103
Ford, L.R., 4, 8, 16, 144
Forward arc, 145, 207, 215
Frank, H., 4, 180
Free float, 285, 291
Frisch, I.T., 4, 180
Fruit distribution problem, 397
Fulkerson, D.R., 4, 8, 16, 144, 304, 314
Funnel nodes, 409

G

Gains (network with), 343
Garfinkel, R.S., 40
Generalized addition, 73
Generalized minimization, 73
Generalized network, 343
Generalized network problem, 345-63
Geoffrion, A., 414
GERT, 363-86
 applications, 376-86
 mean and variance calculation, 376
 network, 364
 W-functions, 366
Glover, F.J., 350
Golden, B.L., 4
Gomory, R.E., 97, 159, 160
Gormory–Hu algorithm, 159-65
Grain shipment/storage, 156
Graphical evaluation and review technique (GERT), 363-86
Graph theory, 3
Grinold, R.C., 350

H

Heiser, R.S., 96
Held, M., 97
Heuristic approaches, 322
Highway fund allocation, 95
Hitchcock, 4, 29
Hollander, G., 315
Hu, T.C., 4, 53, 97, 159, 160, 166, 414
Hungarian algorithm, 133

I

Incidence matrix, 12, 44
Independent float, 285, 291
In-kilter arc, 213
International travel (scheduling), 109
Isomorphic, 11

J

Jarvis, J.J., 4, 350, 414
Jensen, P., 4, 239, 350, 351, 354

Jewell, W.J., 350, 351
Jezior, A.M., 350
Johnson, L., 35
Johnson, T.J.R., 327

K

K-shortest route problem, 72
Karp, R., 97
Kelley, J.E., 314, 322
Kennington, J., 414
Kim, Y., 344
King, N.S., 96
Kirchhoff, G.R., 4
Kleitman, D.H., 180, 181
Klingman, D., 350
Konigsberg bridge problem, 3
Koopmans, 4

L

Labeling procedure, 146, 214, 307
  for forward arcs, 216, 307
  for reverse arcs, 217, 307
Lagrangian multipliers, 68
Latest finish time, 280
Latest start time, 280
Lawler, E.L., 142, 143
Levy, F.K., 320
Linear programming and networks, 39-45, 345-46
Link, 98
Little, J.D.C., 97
Loop (*see also* Self-loop), 8, 371
Losses (networks with), 343
Lower bound, 98
Lubore, S., 414

M

Magnanti, T.L., 4
Mahler, P., 327
Mail distribution system design, 61
Marginal network, 350, 352-53
Mason, S.J., 370
Mason's rule, 373-76
Matrix representation, 11
Maximal-flow/minimum-cut theorem, 16, 149, 150
Maximal spanning tree, 160, 161, 162
Maximum-flow problem, 28, 41, 144, 235, 409
Maxwell, J.C., 3
Miller, D.D., 414
Minieka, E., 4, 350
Minimal spanning tree, 9, 91, 96
Minimal spanning tree algorithm, 92
Minimal spanning tree problem, 91
Minimization (generalized), 73
Minimum-cost flow-augmenting chain, 351
Minimum-cost flow problems, 28, 40, 208, 209, 345
Minimum cut, 16
Mirror arc, 351
Moder, J.J., 315, 320, 328
Multicommodity networks—maximal flows, 400
Multicommodity transportation problem, 388
Multiterminal maximal flows, 158
Multiterminal maximum-capacity problem, 165-67
Multiterminal shortest chain (route), 53, 57

N

NC/DNC, 107
Nemhauser, G.L., 40
Network algorithms, 5
Network (DEFN), 5, 206
Network interpretation of transportation model, 126
Network models, 3, 4, 5
Node (DEFN), 5, 206
Node-arc incidence matrix, 12, 44
Node disjoint paths, 180

Node failures, 179
Node potential, 351
Non-breakthrough, 217
Noninteger flows, 404
Nonlinear costs, 238
Northwest corner method, 117-19
NP-complete, 106

O

Oettli, W., 350
Oil transport technology, 52
One-commodity flow, 14
Optimal approaches, 324
Optimal movement of bulk freight, 167
Out-of-kilter arc, 213
Out-of-kilter method, 28, 204-59
    algorithmic steps, 223-24
    applications, 247-59
    graphical interpretation, 219-23
    modeling concepts, 230-47
    theoretical concepts, 205-29

P

Partial disconnection, 179
Path, 8
Penalties (turn), 68
PERT computer programs, 327-30
Phillips, C.R., 315, 320, 328
Phillips, D.T., 52, 239
Pivotal column, 55
Pivotal row, 55
Planar networks, 401
Plant allocation model, 139
Prager, W., 350
Precedence relationships, 274
Pritsker, A.A.B., 4, 377
Pritsker and Associates, Inc., 387
Production distribution model, 128, 239-46
Production planning model, 35
Product model, 139
Program evaluation and review technique (PERT), 26, 270, 291-95
Project management with CPM/PERT, 269-95
Project planning, 6, 25
Pseudo-nodes, 395

R

Rail cars (routing), 65
Recipient route, 124
Residual capacity, 146
Resource allocation in project networks, 295-327
    resource leveling, 318
    resource limits, 320
    resource loading, 316
Return arc, 207
Reverse arc, 145, 207, 215
Root (tree), 9
Route matrix, 55
Routing (rail cars), 65
Ruthberg, Z. G., 96

S

Saaty, T., 4
Safety float, 286, 291
Sakrovitch, M., 414
Saltman, R.G., 96
Scanned node, 216
Scheduling international travel, 109
Scheduling tanker voyages, 26
Scheduling workers to tasks, 250-52
Secondary penalties, 102
Self-loop, 8, 368
Sewall, R.F., 315
Shortest-path matrix, 55
Shortest-path tree problem, 236
Shortest-route problem, 24, 41, 46, 49, 56, 235-36
Siemens, N., 315
Single commodity flow, 14

Sink node, 7
Slack, 273, 278
Smith, L.A., 327
Smythe, W.R., 35
Source node, 7
Spanning tree, 8, 9, 91, 96
Stepping-stone path, 124
Stochastic networks, 363
Storage and marketing problem, 252-53
Subgraph, 8
Subtour, 103
Super source (sink) node, 7
Switching costs, 65

T

Tanker scheduling, 411
Thomas, W. H., 269, 281
Time/cost trade-offs in CPM, 300-14
Total float, 283-90
Total unimodular matrix, 42
Total unimodularity conditions, 45
Tour, 98
Tracing procedure, 49, 55, 76, 167
Transportation model, 29, 40, 111-13, 231-35
Transportation simplex algorithm, 115
Transporting the space shuttle, 173
Transshipment problem, 130, 237-38
Traveling salesman problem, 97, 141
Tree, 8
Tree (arborescent), 9
Tree (spanning), 9
Triple operation, 54, 166
Truemper, K., 347
Turn penalties, 68
Two-commodity flow problem, 407

U

Undirected arcs, 5
Unimodularity conditions, 45
Unimodular matrix, 42
Unscanned node, 216
Upper triangular matrix, 141
Urban transportation planning, 412

V

Vogel approximation method (VAM), 119-22

W

Walker, M.R., 314, 322
Waste treatment, 151
Weight (tree), 9
Weighted aggregation, 395
Whitehouse, G., 4
Wiest, J.D., 320, 326
Woolpert, B., 328

Z

Zipken, P.H., 414